Edited by
San Ping Jiang and Yushan Yan

Materials for High-Temperature Fuel Cells

Related Titles

Stolten, D., Emonts, B. (eds.)

Fuel Cell Science and Engineering

Materials, Processes, Systems and Technology

2012

ISBN: 978-3-527-33012-6

Stolten, D. (ed.)

Hydrogen and Fuel Cells

Fundamentals, Technologies and Applications

2010

ISBN: 978-3-527-32711-9

Wieckowski, A., Norskov, J. (eds.)

Fuel Cell Science

Theory, Fundamentals, and Biocatalysis

ISBN: 978-0-470-41029-5

Vielstich, W., Gasteiger, H.A., Yokokawa, H. (eds.)

Handbook of Fuel Cells

Advances in Electrocatalysis, Materials, Diagnostics and Durability, Volumes 5 & 6

2 Volumes

ISBN: 978-0-470-72311-1

Bagotsky, V. S.

Fuel Cells

Problems and Solutions

ISBN: 978-0-470-23289-7

Sundmacher, K., Kienle, A., Pesch, H.J., Berndt, J.F., Huppmann, G. (eds.)

Molten Carbonate Fuel Cells

Modeling, Analysis, Simulation, and Control

2007

ISBN: 978-3-527-31474-4

Edited by San Ping Jiang and Yushan Yan

Materials for High-Temperature Fuel Cells

The Editors

Prof. San Ping Jiang
Curtin University
Fuels and Energy Technology Institute &
Department of Chemical Engineering
1, Turner Avenue
Perth, WA 6845
Australia

Prof. Yushan Yan
Dept. Chem. Engineering
University of Delaware
150 Academy Street
Newark, DE 19716
USA

Library of Congress Card No.: applied for

British Library Cataloguing-in-Publication Data
A catalogue record for this book is available from the British Library.

Bibliographic information published by the Deutsche Nationalbibliothek
The Deutsche Nationalbibliothek lists this publication in the Deutsche Nationalbibliografie; detailed bibliographic data are available on the Internet at <http://dnb.d-nb.de>.

Typesetting Laserwords Private Limited, Chennai, India

Printing and Binding Markono Print Media Pte Ltd, Singapore
Cover Design Formgeber, Eppelheim, Germany

Print ISBN: 978-3-527-33041-6
ePDF ISBN: 978-3-527-64428-5
ePub ISBN: 978-3-527-64427-8
mobi ISBN: 978-3-527-64429-2
oBook ISBN: 978-3-527-64426-1

Editorial Board

Contents

Series Editor Preface

The Wiley Series on New Materials for Sustainable Energy and Development

Sustainable energy and development are attracting increasing attention from scientific research communities and industries alike, with an international race to develop technologies for clean fossil energy, hydrogen, and renewable energy, as well as water reuse and recycling. According to the REN21 (Renewables Global Status Report 2012, p. 17), total investment in renewable energy reached $257 billion in 2011, up from $211 billion in 2010. The top countries for investment in 2011 were China, Germany, the United States, Italy, and Brazil. In addressing the challenging issues of energy security, oil price rise, and climate change, innovative materials are essential enablers.

In this context, there is a need for an authoritative source of information, presented in a systematic manner, on the latest scientific breakthroughs and knowledge advancement in materials science and engineering, as they pertain to energy and the environment. The aim of the *Wiley Series on New Materials for Sustainable Energy and Development* is to serve the community in this respect. This has been an ambitious publication project on materials science for energy applications. Each volume of the series will include high-quality contributions from top international researchers and is expected to become the standard reference for many years to come.

This book series covers advances in materials science and innovation for renewable energy, clean use of fossil energy, and greenhouse gas mitigation and associated environmental technologies. Current volumes in the series are

Supercapacitors. Materials, Systems, and Applications
Functional Nanostructured Materials and Membranes for Water Treatment
Materials for High-Temperature Fuel Cells
Materials for Low-Temperature Fuel Cells
Advanced Thermoelectric Materials. Fundamentals and Applications
Advanced Lithium-Ion Batteries. Recent Trends and Perspectives
Photocatalysis and Water Purification. From Fundamentals to Recent Applications

In presenting this volume on Materials for High-Temperature Fuel Cells, I would like to thank the authors and editors of this important book, for their tremendous effort and hard work in completing their manuscripts in a timely manner. The quality of the chapters reflects well the caliber of the contributing authors to this book, and will no doubt be recognized and valued by readers.

Finally, I would like to thank the editorial board members. I am grateful to them for their excellent advice and help in terms of examining coverage of topics and suggesting authors, and evaluating book proposals.

I would also like to thank the editors from the publisher Wiley-VCH with whom I have worked since 2008, Dr. Esther Levy, Dr. Gudrun Walter, and Dr. Bente Flier for their professional assistance and strong support during this project.

I hope you will find this book interesting, informative, and valuable as a reference in your work. We will endeavor to bring to you further volumes in this series or update you on the future book plans in this growing field.

Brisbane, Australia
31 July 2012

G.Q. Max Lu

Preface

Electricity is the most convenient form of energy today. For the past 100 years, electricity is primarily produced by combustion of fossil fuels, which have an intrinsically low conversion efficiency and emit carbon dioxide and other air pollutants. Carbon dioxide contributes to climate changes. With increasing energy demand, depleting fossil fuel reserves, and growing concern about the climate and environment, there is an urgent need to increase electricity generation efficiency and to develop renewable energy sources. Fuel cell is an energy conversion device to electrochemically convert directly the chemical energy of fuels such as hydrogen, methanol, ethanol, natural gas, and hydrocarbons to electricity, and hence, fuel cells inherently have a significantly higher efficiency than conventional energy conversion technologies such as internal combustion engine (ICE). Among the various types of fuel cells, high-temperature solid oxide fuel cells (SOFCs) and molten carbonate fuel cells (MCFCs) are considered to be the most efficient, as they allow internal reforming, promote rapid kinetics with nonprecious materials, and offer high flexibilities in fuel choice.

During the past 20 years, SOFCs and MCFCs have received enormous attention worldwide as alternative electrical energy conversion systems. The book starts with Chapters 1–3 on the materials, ionic transport process, conductivity, electrocatalytic characteristics, and synthesis of key SOFC components of anode, cathode, and oxide ion electrolyte. Chapter 4 presents an overview of proton-conducting electrolytes, a separate class of ion-conducting material. The use of proton-conducting electrolyte materials can have some advantages such as water generation at the cathode side, prevention of fuel dilution at the anode, and the formation of NO_x or SO_x can be avoided when ammonia or H_2S is used as the fuel. This is followed by Chapters 5 and 6 on the materials, processing, and thermal and electrical properties of metallic interconnect and sealants, the most important stack material for an SOFC. Chapter 7 is dedicated to degradation and durability of SOFC electrodes, one of the most challenging problems associated with an SOFC system over a 5 year lifetime. Chapter 8 presents the materials, processing, and status of metal-supported SOFCs or MS-SOFCs, an alternative cell configuration developed to address the critical issues of SOFC systems on cost, durability, and thermal cyclability. Last but not the least, Chapter 9 covers the brief history, operating principles, and status of the state-of-the-art materials and components of MCFCs, one of the most mature and technologically advanced fuel cell technologies.

All chapters were written by leading international experts. It was the intension of the editors and authors that the book be designed to help those involved in the research and development of high-temperature fuel cells and, at the same time, to serve as a reference book for students, materials engineer, and researchers interested in fuel cells technology in general.

Perth, Australia — *San Ping Jiang*
Newark, USA — *Yushan Yan*

About the Series Editor

Professor Max Lu
Editor, New Materials for Sustainable Energy and Development Series

Professor Lu's research expertise is in the areas of materials chemistry and nanotechnology. He is known for his work on nanoparticles and nanoporous materials for clean energy and environmental technologies. With over 500 journal publications in high-impact journals, including Nature, *Journal of the American Chemical Society, Angewandte Chemie, and Advanced Materials,* he is also coinventor of 20 international patents. Professor Lu is an Institute for Scientific Information (ISI) Highly Cited Author in Materials Science with over 17 500 citations (h-index of 63). He has received numerous prestigious awards nationally and internationally, including the Chinese Academy of Sciences International Cooperation Award (2011), the Orica Award, the RK Murphy Medal, the Le Fevre Prize, the ExxonMobil Award, the Chemeca Medal, the Top 100 Most Influential Engineers in Australia (2004, 2010, and 2012), and the Top 50 Most Influential Chinese in the World (2006). He won the Australian Research Council Federation Fellowship twice (2003 and 2008). He is an elected Fellow of the Australian Academy of Technological Sciences and Engineering (ATSE) and Fellow of Institution of Chemical Engineers (IChemE). He is editor and editorial board member of 12 major international journals including *Journal of Colloid and Interface Science and Carbon.*

Max Lu has been Deputy Vice-Chancellor and Vice-President (Research) since 2009. He previously held positions of acting Senior Deputy Vice-Chancellor (2012), acting Deputy Vice-Chancellor (Research), and Pro-Vice-Chancellor (Research

Linkages) from October 2008 to June 2009. He was also the Foundation Director of the ARC Centre of Excellence for Functional Nanomaterials from 2003 to 2009.

Professor Lu had formerly served on many government committees and advisory groups including the Prime Minister's Science, Engineering and Innovation Council (2004, 2005, and 2009) and the ARC College of Experts (2002–2004). He is the past Chairman of the IChemE Australia Board and former Director of the Board of ATSE. His other previous board memberships include Uniseed Pty Ltd., ARC Nanotechnology Network, and Queensland China Council. He is currently Board member of the Australian Synchrotron, National eResearch Collaboration Tools and Resources, and Research Data Storage Infrastructure. He also holds a ministerial appointment as member of the National Emerging Technologies Forum.

About the Volume Editors

Dr. San Ping Jiang is a professor at the Department of Chemical Engineering, the Deputy Director of Fuels and Energy Technology Institute, Curtin University, Australia, and an Adjunct Professor of the University of the Sunshine Coast, Australia. He also holds Visiting/Guest Professorships at the Harbin Institute of Technology, Guangzhou University, Huazhong University of Science and Technology, Wuhan University of Technology, University of Science and Technology of China, Sichuan University, and Shandong University. After receiving his BEng from the South China University of Technology and Ph. D. from the City University, London, Dr. Jiang worked at the CSIRO Manufacturing Science and Technology Division, Ceramic Fuel Cells Ltd (CFCL) in Australia and the Nanyang Technological University in Singapore. His research interests encompass solid oxide fuel cells, proton exchange and direct methanol fuel cells, direct alcohol fuel cells, and electrolysis. With an h-index of 44, he has published over 240 journal papers, which have accrued ~6500 citations.

Yushan Yan has been a Distinguished Engineering Professor at the University of Delaware since 2011. He received his B. S. from the University of Science and Technology of China and Ph. D. from the California Institute of Technology. He worked for AlliedSignal Inc. as Senior Staff Engineer before joining the faculty at the University of California at Riverside (Assistant Professor 1998, Associate Professor 2002, Professor 2005, University Scholar 2006, Department Chair 2008, Presidential Chair 2010). He is a Fellow of the American Association for the Advancement of Science. He was recognized with the Donald Breck Award by the International Zeolite Association. He was one of the 37 awardees in ARPA-E OPEN

2009 for his fuel cell technology and one of the 66 awardees in ARPA-E OPEN 2012 for his redox flow battery concept by the US Department of Energy. His patents were licensed to form startup companies, for example, NanoH2O, Full Cycle Energy, Zeolite Solution Materials, and OH-Energy. He has published more than 140 journal articles (h-index = 46, total number of citations = more than 6700).

List of Contributors

Hua Bin
Huazhong University of
Science & Technology
School of Materials Science and
Engineering
1037 Luo Yu Road
Wuhan, Hubei 430074
China

Kongfa Chen
Curtin University
Fuels and Energy Technology
Institute & Department of
Chemical Engineering
1, Turner Avenue
Perth, WA 6845
Australia

Ping-Hsun Hsieh
Illinois Institute of Technology
Department of Chemical and
Biological Engineering
10 W. 33rd Street
Chicago, IL 60616
USA

Rob Hui
National Research Council
Institute for Fuel Cell Innovation
4250 Wesbrook Mall
Vancouver, BC V6T 1W5
Canada

Tatsumi Ishihara
Kyushu University
Department of Applied Chemistry
Faculty of Engineering
International Institute for Carbon
Neutral Energy (I^2CNER)
Motooka 744
Fukuoka, 819-0395
Japan

Li Jian
Huazhong University of
Science & Technology
School of Materials Science and
Engineering
1037 Luo Yu Road
Wuhan, Hubei 430074
China

San Ping Jiang
Curtin University
Fuels and Energy Technology
Institute & Department of
Chemical Engineering
1, Turner Avenue
Perth, WA 6845
Australia

Chan Kwak
Samsung Advanced Institute of Technology (SAIT)
14-1 Nongseo-dong
Yongin-si
Gyunggi-do 446-712
Korea

Rong Lan
University of Strathclyde
Department of Chemical and Process Engineering
75 Montrose Street
Glasgow G1 1XJ
UK

Steven McIntosh
Lehigh University
Department of Chemical Engineering
111 Research Drive
Bethlehem, PA 18013
USA

Stephen J. McPhail
ENEA– Italian National Agency for New Technologies, Energy, and Sustainable Economic Development
Unit Renewable Sources - Hydrogen and Fuel Cells
Via Anguillarese 301
00123 Rome
Italy

Hee Jung Park
Samsung Advanced Institute of Technology (SAIT)
14-1 Nongseo-dong
Yongin-si
Gyunggi-do 446-712
Korea

Lian Peng
Chinese Academy of Sciences
Institute of Process Engineering
State Key Laboratory of Multiphase Complex Systems
Zhong Guan Cun, Bei Er Tiao 1
Beijing 100190
China

Jan Robert Selman
Illinois Institute of Technology
Department of Chemical and Biological Engineering
10 W. 33rd Street
Chicago, IL 60616
USA

Zongping Shao
Nanjing University of Technology
College of Chemistry & Chemical Engineering
State Key Laboratories of Materials-Oriented Chemical Engineering
No. 5, Xin Mofan Road
Nanjing 210009
China

Shanwen Tao
University of Strathclyde
Department of Chemical and Process Engineering
75 Montrose Street
Glasgow G1 1XJ
UK

Zhang Wenying
Huazhong University of Science & Technology
School of Materials Science and Engineering
1037 Luo Yu Road
Wuhan, Hubei 430074
China

Tao Zhang
Chinese Academy of Sciences
Institute of Process Engineering
State Key Laboratory of
Multiphase Complex Systems
Zhong Guan Cun, Bei Er Tiao 1
Beijing 100190
China

Wei Zhou
Nanjing University of Technology
College of Chemistry & Chemical
Engineering
State Key Laboratories of
Materials-Oriented Chemical
Engineering
No. 5, Xin Mofan Road
Nanjing 210009
China

Qingshan Zhu
Chinese Academy of Sciences
Institute of Process Engineering
State Key Laboratory of
Multiphase Complex Systems
Zhong Guan Cun, Bei Er Tiao 1
Beijing 100190
China

1
Advanced Anodes for Solid Oxide Fuel Cells

Steven McIntosh

1.1
Introduction

The solid oxide fuel cell (SOFC) anode must perform four basic functions: (i) transport oxygen anions from the 2D electrolyte–electrode interface out into the higher surface area 3D electrode structure, (ii) transport gas-phase fuel to the reaction site and products from the reaction site, (iii) catalyze the electrochemical oxidation of the fuel, and (iv) transport the product electrons from the reaction site to the current collector at the electrode surface. These requirements are in addition to considerations of material stability, manufacturability, redox tolerance, and resistance to possible poisons in the fuel.

The most developed and widely utilized SOFC anode is a porous ceramic–metallic (cermet) composite of Ni with the oxide electrolyte, most commonly 8 mol% yttria-stabilized zirconia (YSZ). Ni provides electronic conductivity and electrocatalytic activity, while YSZ provides oxygen anion conductivity. The reaction in the Ni–yttria-stabilized zirconia (Ni–YSZ) anode can thus only occur at the triple phase boundary (TPB) involving the Ni, YSZ, and gas phases – these are the regions of the anode where all the electrode requirements are met. When the relative fractions of these materials and the cermet porosity are optimized, the Ni–YSZ cermet electrode can provide performance sufficient for commercialization [1]. The required performance metric for the anode is a function of cost and the relative performance of the other cell components. A rough rule of thumb performance target for an SOFC anode is a polarization resistance of less than 0.15 Ω cm [2]. In addition, Ni and YSZ can be cosintered at the sintering temperatures required to create a dense YSZ electrolyte, 1400–1550 °C, thus simplifying cell fabrication [3]. This combination of performance and manufacturability makes Ni–YSZ cermets the anode of choice for current commercial SOFC technologies.

Ni-based cermet electrodes are not without disadvantages. Perhaps the most significant of them is that the use of Ni places significant restrictions on the fuel choice for SOFCs. Unlike other fuel cells, the SOFC operating principle is based on the transport of the oxidant to the fuel; as such, SOFCs can theoretically operate on any oxidizable fuel [4]. This makes SOFCs an attractive option for efficient power

Materials for High-Temperature Fuel Cells, First Edition. Edited by San Ping Jiang and Yushan Yan.

generation from current fossil and future renewable hydrocarbon fuels. However, Ni catalyzes the formation of graphitic carbon in dry hydrocarbon atmospheres [5–7], limiting the fuel choice to H_2 and CO. This problem can be partially surmounted by steam reforming of the hydrocarbon fuel to form H_2 and CO either before feeding to the cell (external steam reforming) or on the Ni-based anode itself (internal steam reforming). The first solution increases cost by requiring an additional reactor in the system. The second places large thermal stresses on the cells and stack by combining the highly endothermic reforming reaction with the exothermic fuel oxidation reaction. Another potential disadvantage of Ni anodes is their potential instability toward reduction and oxidation cycling. Accidental oxidation of Ni to form NiO is accompanied by a large lattice expansion, potentially leading to mechanical failure of the cell [8]. There are also concerns regarding the tolerance of Ni toward impurities in the feed, including sulfur and heavy metals, although many of these impurities can be removed by fuel pretreatment. These disadvantages and the potential advantages of direct hydrocarbon utilization have led to significant research efforts to develop new anode compositions. The majority of these are mixed metal oxides with the perovskite or a related structure. However, the development of these oxide materials that can provide the cell performance levels required for economic implementation is a significant challenge.

SOFC anode understanding and optimization are the topics of a large number of studies around the world: a citation search for the concept "SOFC anode" yields more than 900 results for 2011 alone. It is not possible to cover all this material in a single chapter, requiring focus. In this chapter, we first provide a background overview of Ni–YSZ anodes before discussing the most recent advances in our understanding of their function, from techniques that enable study of real microstructures to recent insights into the anode reaction mechanism. We then focus on the challenge of developing novel anode materials to facilitate SOFCs operating with hydrocarbon fuels. The reader is directed to the large number of excellent review articles available in the literature for further discussion of these and related topics [3, 4, 9–19].

1.2 Ni–YSZ Anode Overview

On the basis of the overall reaction mechanism, there are some basic requirements of the Ni–YSZ microstructure (Figure 1.1) [20]. First, there must be sufficient porosity for gas transport. Second, there must be continuous connectivity within the YSZ phases and Ni phases to facilitate ion and electron transport, respectively. Third, the microstructure should be optimized to achieve a high density of TPB regions to facilitate reaction. The anode must also provide sufficient mechanical strength, as most SOFC designs utilize a relatively thick anode cermet as the physical support structure for a thin electrolyte film.

The Ni–YSZ cermet is typically fabricated by tape casting a physical mixture of NiO and YSZ powders. A thinner YSZ-based tape is then joined with the

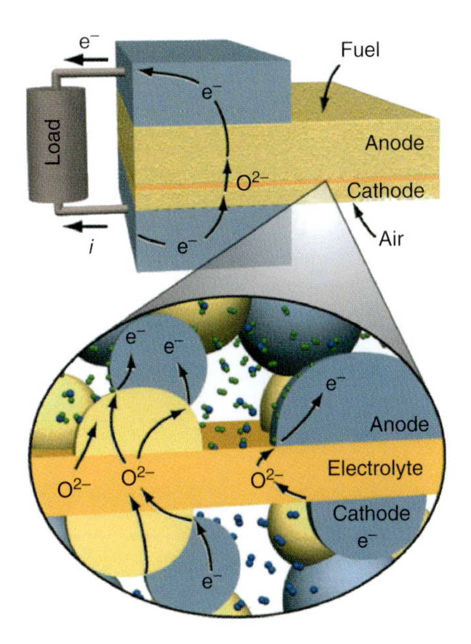

Figure 1.1 Illustration of the roles of the Ni (gray) and YSZ (yellow) phases in the SOFC anode [20].

green anode tape either by casting the two layers one on top of the other or by laminating them together. Both layers are then cosintered between 1400 and 1550 °C [3] to yield a thin, dense YSZ electrolyte (typically $<50\,\mu m$ thick) supported on a NiO–YSZ anode substrate (typically $>100\,\mu m$ thick). Alternative approaches are often utilized to further minimize the electrolyte thickness. They focus on deposition of the electrolyte layer onto a preformed anode structure by chemical or physical vapor deposition, pulsed laser deposition, DC magnetron sputtering, or spray pyrolysis [21]. The cathode is then added to the free electrolyte surface and the cell assembled for operation. Reduction of NiO to Ni by H_2 during cell start-up forms the active Ni–YSZ anode. The volume change associated with the reduction of NiO creates pores in the cermet to facilitate gas transport (Figure 1.2) [8, 22]. While tape casting is an industrially scalable approach, it limits microstructure optimization to variation of this tape composition in terms of particle sizes, size distributions, and pore formers. All of these parameters must be optimized to create a high-performance anode.

Early work by Dees *et al.* [2] determined the breakthrough point for a continuous Ni phase to be around 30 vol% Ni, setting a minimum Ni content for an active electrode. This was determined by measuring the electrode conductivity as a function of Ni loading, with electrodes fabricated using the typical tape casting process. This 30 vol% Ni loading is also the point at which the coefficient of thermal expansion (CTE) for the oxidized cermet closely matches that of YSZ. CTE matching between components is essential during fabrication, when the electrolyte and anode structure are co-fired.

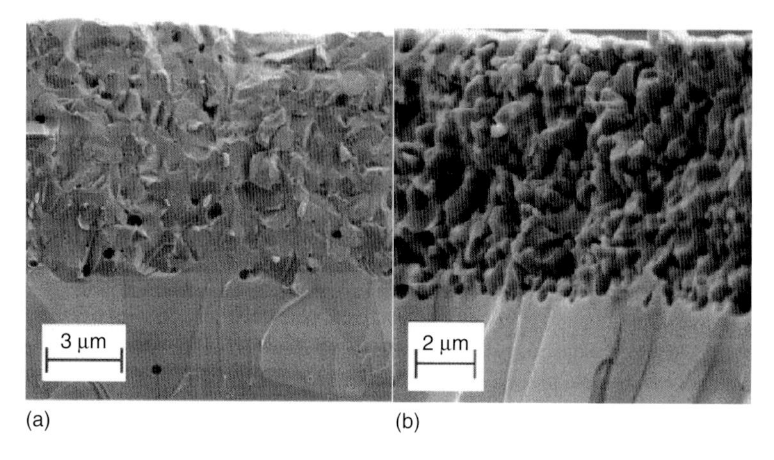

Figure 1.2 Example of (a) NiO–YSZ composite after fabrication and (b) the same cermet following reduction in H_2 [22].

The breakthrough point for a continuous Ni phase and the CTE match depend strongly on the porosity, relative Ni–YSZ particle sizes, and Ni distribution, all of which can be influenced by the fabrication method. For example, Clemmer and Corbin [23] compared sintering, CTE, and conductivity between composites fabricated with a physical mixture of Ni and graphite and with Ni-coated graphite, where graphite is used as the pore former during cermet fabrication. They found that Ni-coated graphite led to high electrical conductivity and only 10 vol% Ni (including pores) compared to more than 15 vol% Ni obtained using the physical mixture [24]. This concept of an optimal distribution of phases was also considered by Lee *et al.* [25], who demonstrated that the most important factors in determining the anode performance are pore interconnectivity and uniform distribution of the conducting phase and pores. The existence of closed pores leads to poor connectivity of both conducting and pore phases, creating barriers for both electron and gas transport. Poor distribution of the Ni and YSZ phases leads to a low density of active TPB regions.

Zhao and Virkar conducted a thorough experimental study that included the influence of anode porosity and anode thickness on the area-specific resistance (ASR) and maximum power density (MPD) of anode-supported SOFCs (Figure 1.3). They found that the MPD increases with increasing anode porosity, but that it goes through a maximum in the porosity range 57–76%. This suggests that there is an optimum microstructure that facilitates gas transport while still providing a high enough density of active reaction sites. A high density of large pores will decrease the surface area of the electrode. While the ASR was lower for 76% porosity, these cells showed lower than theoretical open circuit potential (OCP), attributed to pinholes in the electrolyte caused by the lack of anode support. Zhao and Virkar modeled their results, including gas transport and Tafel-based electrochemical reaction kinetics, for these cells. There was good agreement in the gas-phase transport parameters extracted from numerous experiments, and also between

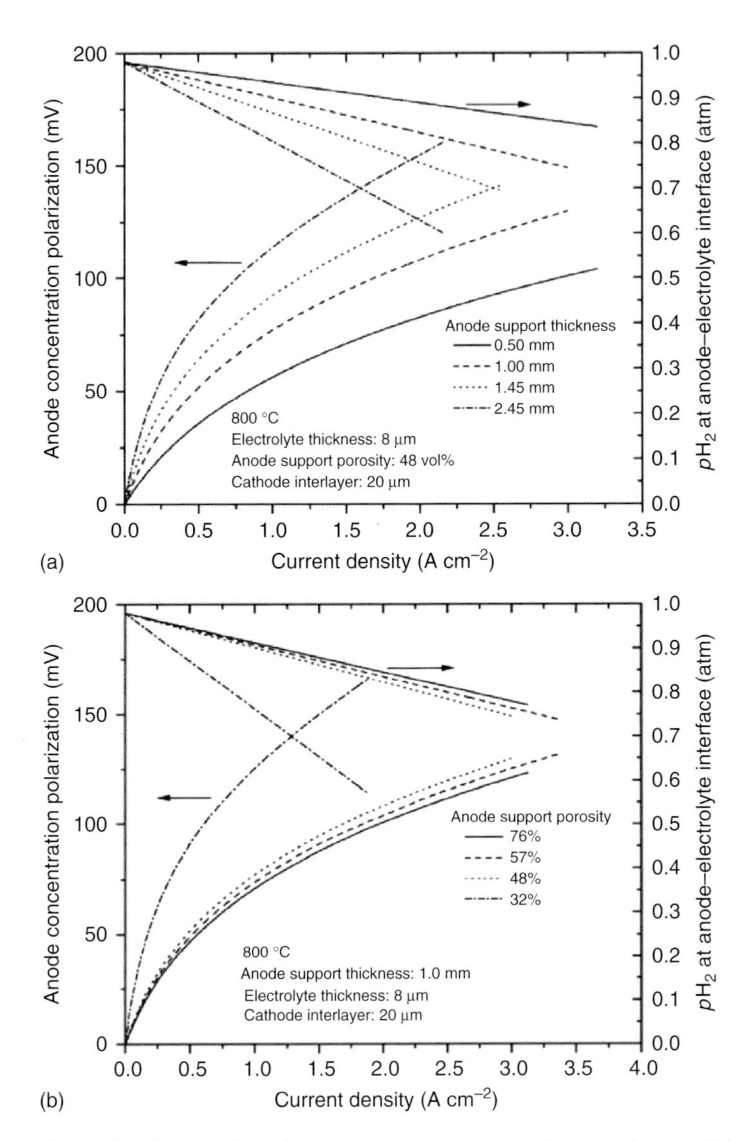

Figure 1.3 Calculated anode concentration polarization losses and H_2 partial pressure at the anode–electrolyte interface as a function of (a) anode thickness and (b) anode porosity [26].

exchange current densities fit for different anode thicknesses when keeping the anode microstructure constant. From this model, they were able to attribute performance differences in these structures to differences in hydrogen partial pressure at the anode–electrolyte interface, demonstrating a strong influence of mass transport on cell performance [26].

Jensen *et al.* have performed a systematic study of some of the SOFC operating parameters in an attempt to separate and confirm the individual impedance

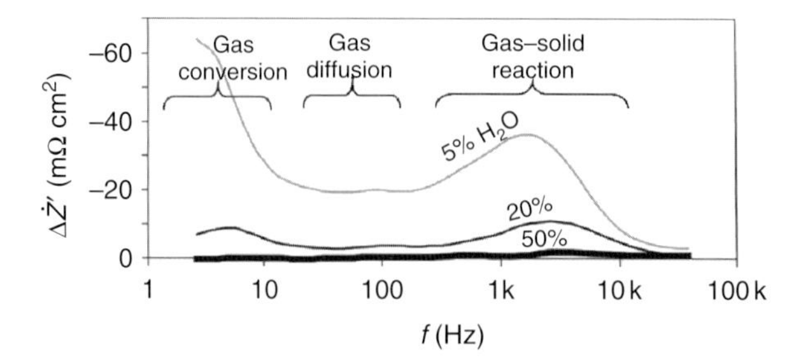

Figure 1.4 $\Delta \dot{Z}'$ as a function of frequency on shifting the H_2O concentration in the H_2 feed to the anode. The separate distinct contributions to the total impedance are labeled [27].

contributions of a whole cell. In order to clarify the individual contributions to the spectra, they utilized a parameter, $\Delta Z'$, reflecting the change in the impedance on changing the operating conditions. In this way, they confirmed the presence of gas conversion, gas diffusion, and gas–solid reaction resistances in the anode (Figure 1.4) [27].

While the SOFC anode may be several hundred micrometers thick, the active region is suggested to be confined to several tens of micrometers from the anode–electrolyte interface [28, 29]; that is, the anions migrating through the electrolyte are typically close to the anode–electrolyte interface. This has led to the development of functionally graded anodes, where the Ni/YSZ ratio, particle size, and pore size are varied as we move from the anode–electrolyte interface into the bulk anode. The goal is to maximize the active TPB area in the electrochemically functional region while still facilitating gas and electron transport in the mechanically supported anode bulk. This is typically achieved by fabricating the anode as two (or more) distinct layers – an anode functional layer (AFL) close to the anode–electrolyte interface combined with a thicker, more porous support layer [22, 30, 31]. The AFL consists of smaller particles, with smaller pores to maximize the TPB density and hence electrochemical activity. Outside this layer, the particle sizes and pores increase to facilitate electron transport and gas transport (Figure 1.5). One potential issue with the AFL is reduction in redox stability. The small particles and small pores in the AFL can generate large stresses if the Ni phase is oxidized [32]. Fine grain structures lead to higher expansion and cracking on oxidation, where coarse-grained materials are more tolerant. Cracking can be avoided by lowering the Ni content, but this will influence performance [32].

The general trends in performance noted for Ni–YSZ electrodes hold when other electrolyte materials are utilized. Promising alternatives to YSZ are $Ce_{1-x}Gd_xO_{3-\delta}$ (CGO) and $La_{1-x}Sr_xGa_{1-y}Mg_yO_{3-\delta}$ (LSGM). For example, Ni and CGO composite anodes show increased activity as the grain size is decreased, which is likely due to an increase in the TPB density [33, 34]. There are reports of performance enhancement by the use of mixed conductors (including reduced CGO) [35, 36]. This may be due to extension of the TPB onto the mixed conductor surface (especially, if the mixed conductor shows catalytic activity), but this requires further

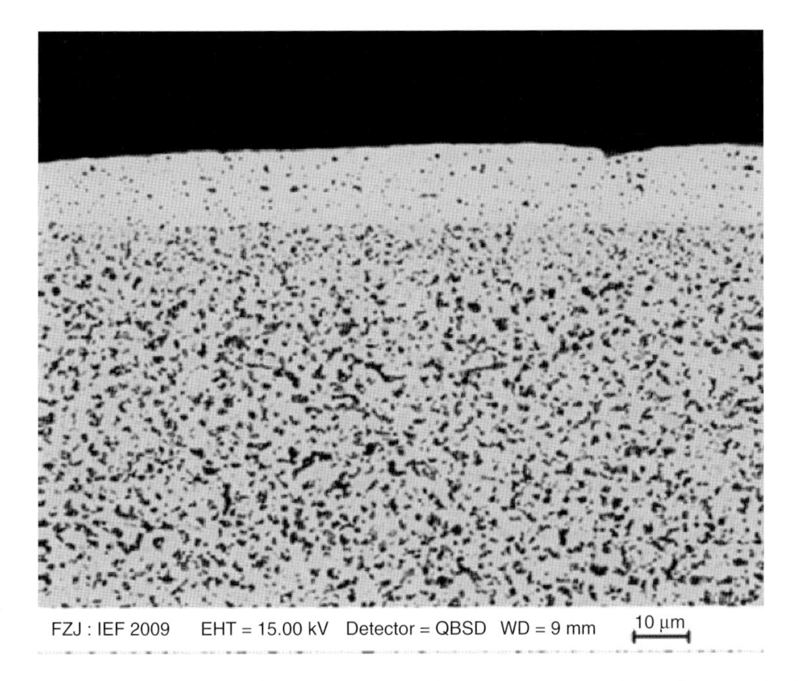

Figure 1.5 SEM image showing a ~10 μm thick anode functional layer between the bulk anode and electrolyte layer [31].

study to verify. Ni–CGO cermets have been demonstrated to show higher activity toward methane steam reforming than Ni–YSZ cermets, enabling these electrodes to operate via internal steam reforming of CH_4 [37–39], or possibly by direct CH_4 oxidation [40]. Again, this needs further research into the electrocatalytic mechanism to verify. The mechanisms of hydrogen oxidation and methane re-formation in the comparatively well-researched Ni–YSZ system have only recently been determined.

1.3 Insights from Real Ni–YSZ Microstructures

Many studies support the concept that the anode microstructure must be optimized to achieve two competing requirements: high TPB density and high porosity. Further understanding of the influence of microstructure on performance requires detailed knowledge of the three-dimensional structure of the anode. Two approaches have been utilized to obtain this information as an input for modeling: focused ion beam-scanning electron microscopy (FIB-SEM) and X-ray computed tomography (XCT). Grew *et al.* [41] have provided a thorough discussion on the analysis methods and information that can be extracted from these techniques.

In FIB-SEM [42], FIB is utilized to mill a layer of material from the electrode cross section before imaging with SEM. A sequence of 2D images through the electrode

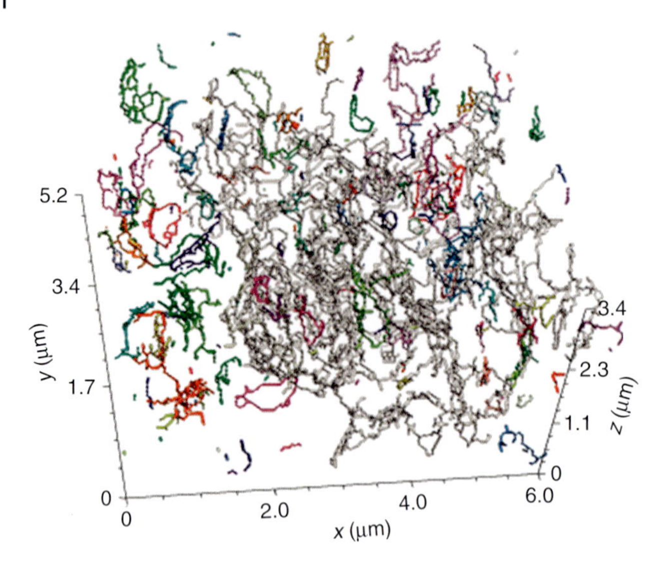

Figure 1.6 Representation of TPB regions in a Ni–YSZ anode from FIB-SEM measurements. White/gray (63%) are connected TPBs, the remainder with other colors are disconnected [42].

can then be combined to create a 3D reconstruction. This can be utilized to quantify the distribution of each phase, connectivity between phases, and the tortuosity of transport pathways. Figure 1.6 shows the TPBs identified within a Ni–YSZ anode by Wilson *et al.* [42]. They found that the majority (63%) of the TPB segments were well connected to the rest of the anode; that is, there were continuous electronic and ionic conducting pathways to/from the TPB region, indicating that it could be electrochemically active. However, a significant fraction (19%) of the TPBs was isolated segments that would not be active for electrochemical reactions. This information can only be extracted from these 3D reconstructions, indicating that standard 2D imaging may not give sufficient information to fully interpret the role of microstructure in determining anode performance. The connectivity of the remaining TPBs could not be determined, as they intersected with the measured sample boundaries. In a subsequent study, Wilson and Barnett [43] found that the density of TPBs within the active region was not directly correlated with low polarization resistance. While they confirmed that the lowest polarization resistance and highest TPB density occur for 50 wt% NiO in the initial cermet (corresponding to ~34 vol% Ni in the reduced state), polarization resistances measured for various Ni loadings did not correlate with TPB density. The measured polarization resistances varied much more than predicted from a simple model. They thus suggested that in addition to TPB density, other factors, including phase continuity and tortuosity, were a significant influence on anode performance.

The importance of facile ionic and electronic transport pathways and TPB distribution was further emphasized by Shikazono *et al.* [44]. Building on the previous work in their group [45], they combined FIB-SEM reconstruction with a detailed three-dimensional electrochemical model to show that the electrochemical potential and associated current density distributions in the anode are highly nonuniform due to nonuniformity in these transport pathways and TPB distribution.

An earlier quantitative microscopy (although not 3D reconstruction) analysis by Lee *et al.* [46] indicates that optimizing this phase contiguity may be a challenge. They found that while the self-contiguity of phases was directly related to the quantity of that phase present, the contact between phases was related to the amount of Ni present. The Ni cermet conductivity shows a breakthrough between 40 and 45 vol% Ni, corresponding to a Ni–Ni contiguity between 0.16 and 0.22. Thus, it may be impossible to fully optimize these competing factors of contiguity, porosity, tortuosity, and TPB density, all of which clearly influence the anode performance and are influenced by particle sizes and pore former size and shape [25].

Izzo *et al.* have explored the alternative, nondestructive XCT technique to generate 3D reconstructions of a Ni–YSZ anode with 42.7 nm spatial resolution. This structural information was then utilized as an input to a multicomponent lattice Boltzmann method (LBM) to generate a detailed picture of local gas concentrations within the anode. A thorough analysis was presented by the same group [41], demonstrating an ability to generate detailed models of the anode, capturing details of phase distribution and transport pathways, and identifying where resistive losses occur within the anode microstructure.

These detailed microstructure studies emphasize the role of the 3D aspects of the anode, such as tortuosity and contiguity, in influencing performance. A comprehensive study is required to fully map these parameters as a function of anode processing and definitively link them to cell performance. Furthermore, these 3D microstructures should become the standard input into modeling efforts, although this may increase the computational intensity of such work.

1.4 Mechanistic Understanding of Fuel Oxidation in Ni-Based Anodes

1.4.1 Hydrogen Oxidation

The overall steps of the fuel oxidation reaction in the SOFC anode can be derived from the anode-operating material properties. The reaction in the Ni–YSZ anode occurs at the TPB between the Ni, YSZ, and gas phases. Oxygen anions are transported to the reaction site through the YSZ, where they react with hydrogen absorbed on the Ni or YSZ surface to form H_2O. The liberated electrons flow through the Ni phase to the external circuit and the product H_2O diffuses out of the anode. While this overall process is generally agreed upon, there is significant debate in the literature regarding which of these processes is rate determining and the specific details of the electrocatalytic reaction mechanism, for example, the nature of the active reaction site and the rate-limiting step. The detailed steps of this reaction include gas diffusion of reactants and products, adsorption/desorption of species onto the catalytically active surface, surface diffusion of adsorbed species and intermediates, charge transfer between solid phases, and catalytic and electrocatalytic reaction steps. Gas diffusion limitations can typically be identified and remedied through modifications of the electrode microstructure. Identification of other mechanistic steps as rate determining is otherwise more complex. Similarly, identifying on which surface these steps occur is nontrivial. As discussed in the review by Horita *et al.* [6], the reported activation energies and reaction orders with respect to H_2 and H_2O vary greatly between reports.

Some possible anode reaction mechanisms are shown in Figure 1.7 [47]: (a) hydrogen spillover from the Ni surface to either an oxygen ion or a hydroxyl ion on the YSZ surface, (b) oxygen spillover from the oxide surface to the Ni surface, (c) hydroxyl spillover from the YSZ surface to the Ni surface, (d) interstitial proton transport from the Ni to the YSZ through the bulk, and (e) all chemical reactions occurring on the oxide surface. We restrict the following discussion to studies that utilize model experimental electrodes and the analysis of these results through simulation. This combination of experimental and theoretical modeling provides the clearest insights into the reaction mechanism.

While porous 3D anode microstructures yield high performance by maximizing the TPB area, the interpreting mechanism from these studies is complicated by the need to also interpret the influence of microstructure. This has led to a number of model electrode studies where the electrode structure is simplified, including point electrodes where a Ni wire is pressed against a YSZ pellet, pattern electrodes where the Ni is lithographically patterned onto the YSZ surface, and pure Ni electrodes [48, 49]. However, despite a large number of studies, there is no comprehensive agreement on the reaction mechanism or rate-determining step. Most recently, there have been a number of efforts to combine experimental results with simulation. These combined efforts can provide substantial additional insight by offering alternative explanations, or by verifying conclusions reached from experiment.

One experimental study that clearly demonstrates that the reaction occurs at the TPB is the work of Mizusaki *et al.* [50], who used patterned Ni strip electrodes on a dense YSZ substrate. Their results demonstrate an almost linear relationship between *interface conductivity* (defined as the inverse of the electrode area-specific resistance) and TPB length (Figure 1.8); that is, more TPB implies lower resistance. Furthermore, based on no observed difference in interface conductivity with open YSZ surface area around the TPB, they went on to conclude that the rate-determining reaction occurs on the Ni surface, and not on the YSZ. In their second publication, the same group further analyzed their data and suggested that the rate-determining step in the reaction mechanism was the direct reaction of $H_2(g)$ with an O atom adsorbed on the Ni surface [51].

Goodwin *et al.* [52] recently utilized computational modeling to revisit this interpretation, comparing the Mizusaki data with three possible mechanisms.

1) *Oxygen spillover from the YSZ surface to the Ni.* First, one electron is transferred from an O^{2-} ion to the Ni, leaving O^- on the YSZ surface, this singly ionized

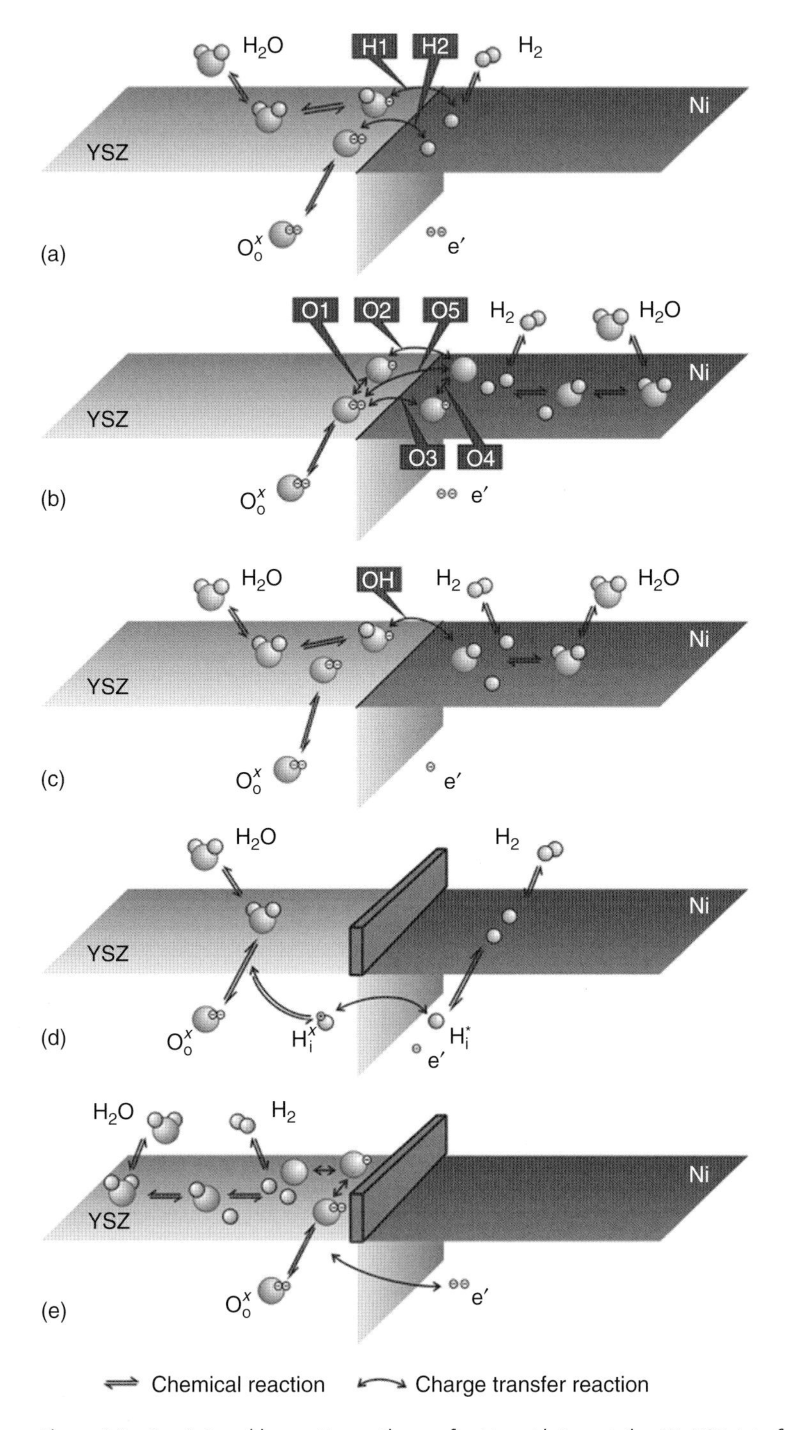

Figure 1.7 (a–e) Possible reaction pathways for H_2 oxidation at the Ni–YSZ interface [47].

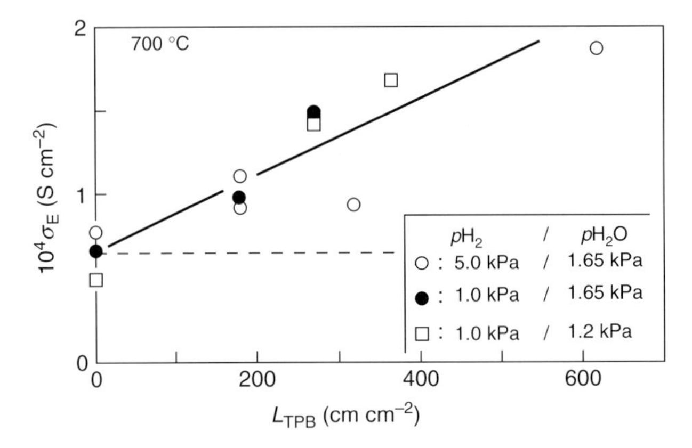

Figure 1.8 Dependence of Ni–YSZ–gas interfacial conductivity with TPB length for Ni-patterned anodes on YSZ substrates [50].

species is then transferred across the TPB to the Ni surface as the rate-limiting step. This step includes the final charge transfer to the Ni phase and leaves an oxygen vacancy on the YSZ surface.

2) *Hydrogen spillover through a single channel.* A proton is transferred from the Ni surface to an oxygen anion on the YSZ surface, generating a surface hydroxyl, OH^-, ion, and transferring a single electron to the Ni. The rate-limiting step is the transfer of a second proton from the Ni surface to form H_2O on the YSZ surface and then transferring the second electron to the Ni.
3) *Hydrogen spillover through a dual channel.* This follows the same mechanistic steps as the single-channel process; however, the two proton-transfer steps are assigned differing rate constants. This is physically feasible if we consider, for example, two separate surface sites for the absorbed species.

In contrast to the conclusions of Mizusaki and coworkers, Goodwin proposes that a hydrogen spillover mechanism is able to adequately describe the experimental data of Mizusaki, under both anodic and cathodic polarization. The data correlates with this mechanism over a range of applied potentials and the resulting ionic flux (Figure 1.9).

Another set of patterned electrode studies performed by Bieberle *et al.* [53] confirmed that the anode kinetics are limited by the available TPB length. This study confirmed the result obtained by Mizusaki *et al.* that the electrode conductivity scales with TPB length. Bieberle *et al.* also observed an increase in anode kinetics when there is an increase in the gas-phase water partial pressure above the electrode. Their analysis led to a suggestion that water leads to formation of hydroxyl groups on the YSZ surface that are then deprotonated by oxygen anions emerging from the YSZ surface, enabling proton migration (hydrogen spillover) from the Ni surface to the YSZ surface. The catalytic enhancement with water is thus suggested to be due to enhanced hydrogen spillover. Another study supports a proton transport mechanism in adsorbed water on the YSZ surface at lower temperatures but

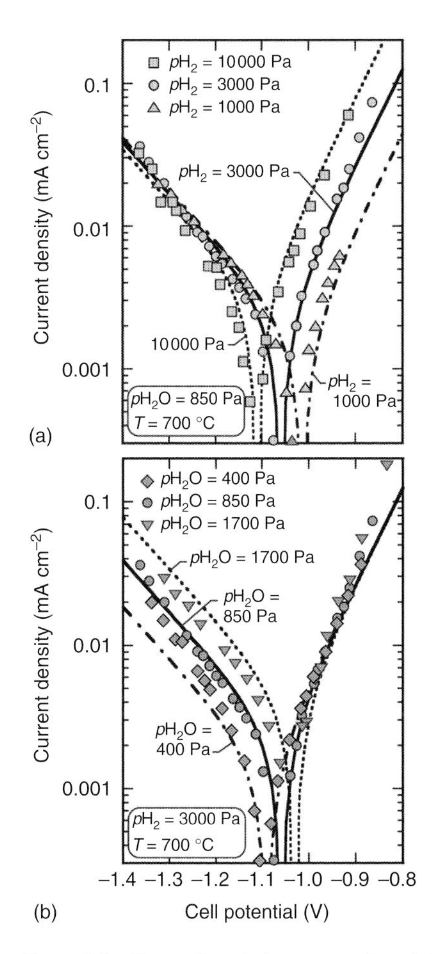

Figure 1.9 Comparison between experimental (symbols) [51] and calculated (lines) [52] (a) anodic and (b) cathodic overpotentials for patterned Ni–YSZ electrodes following a single-channel hydrogen spillover charge transfer model.

indicates that surface hydroxyl coverage will be low at SOFC temperatures [54]; however, this coverage will depend on the water partial pressure in the gas phase.

Bessler *et al.* developed a model that extends from an elementary kinetic description of the electrochemical and nonelectrochemical reactions through physical descriptions of the electrical double layers and multiscale mass transport. This is formulated as a fully transient description to enable calculation of dynamic performance and cell impedance [55]. The experimental work of Bieberle *et al.* was simulated [47] using the model of Bessler *et al.* [55]. When compared with a number of possible anode charge transfer mechanisms, qualitative agreement between experiment and simulation is only found for models that include a hydrogen spillover mechanism. Unfortunately, the role of surface diffusion could not be probed due to limited experimental data.

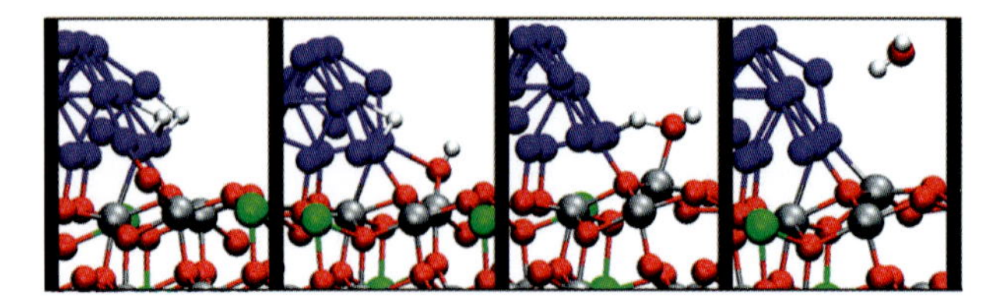

Figure 1.10 Proposed H_2 oxidation mechanism at the Ni–YSZ interface [59].

A recent combined experimental and theoretical study of lithographically patterned Ni–YSZ anodes [56] interpreted an observed decrease in electrode resistance with increasing pH_2, indicating that the dominant mechanism is hydrogen spillover from the Ni to the YSZ. This latest study also confirmed the trend of decreasing resistance with increasing pH_2O, finding that this is due to the change in both local electrode potential and surface species concentration, in-line with the interpretation proposed previously by Bessler *et al.* [57]. Finally, the authors of this study also point to limitations in previously reported experimental data, where critical parameters such as structural changes, impurities, or activation/relaxation effects are not included. These factors could lead to discrepancies in the experimental results and thus confusion regarding the reaction pathway.

Shishkin and Ziegler [58] examined reactions at the Ni–YSZ interface using the density functional theory (DFT), suggesting a necessary modification to these mechanisms. Their results confirm the feasibility of both the oxygen and hydrogen spillover reactions. Additionally, they find that oxygen spillover from the YSZ to Ni is only favorable from a YSZ surface that is enriched with oxygen from the bulk. A subsequent study by the same authors [59] suggested that neither mechanism is correct and that the most facile mechanism is water formation on an oxygen atom adsorbed to both the Ni and YSZ phases (Figure 1.10). However, it must be noted that this mechanism would suggest no extension of the TPB from a 2D line and is not correlated with experimental studies.

In summary, the most recent comprehensive studies that attempt to directly link experimental and simulation work all point to a hydrogen spillover mechanism from the Ni surface to the YSZ surface as the dominant mechanism of hydrogen oxidation in Ni–YSZ cermets.

1.4.2 Hydrocarbon Fuels in Ni-Based Anodes

Owing to the operating mechanism of oxygen transport from the cathode to the anode, SOFCs can theoretically utilize a wide range of fuels, including hydrocarbons. The primary hydrocarbon fuel targeted is methane due to its abundance, availability in many metropolitan areas, and cost. While there is still some debate over the mechanism of H_2 oxidation in these anodes, the case of hydrocarbons is even more complex. Unlike H_2 that can only be oxidized to form H_2O, hydrocarbons can undergo a number of electrochemical and nonelectrochemical reactions in the anode, including partial oxidation, total oxidation, steam reforming, dry reforming, cracking, and polymerization to form tars. The gas composition at the anode, and thus electrochemical reactions occurring at the TPB, is dictated by the complex interaction between all these reaction steps [16]. It should be noted that only the electrochemical reactions directly contribute to the current generated by the cell; however, the nonelectrochemical reactions may control the composition of the gas at the TPB. The desired reaction is the complete oxidation of hydrocarbons to CO_2 and H_2O, as this represents maximum fuel utilization. Some possible reactions of methane are

$$CH_4 + 4O^{2-} \rightarrow CO_2 + 2H_2O + 8e^- \quad \text{Total electrochemical oxidation} \tag{1.1}$$

$$CH_4 + O^{2-} \rightarrow CO + 2H_2 + 2e^- \quad \text{Partial electrochemical oxidation} \tag{1.2}$$

$$CH_4 + 3O^{2-} \rightarrow CO + 2H_2O + 6e^- \quad \text{Partial electrochemical oxidation} \tag{1.3}$$

$$CH_4 + H_2O \rightarrow CO + 3H_2 \quad \text{Steam reforming} \tag{1.4}$$

$$CH_4 + CO_2 \rightarrow 2CO + 2H_2 \quad \text{Dry reforming} \tag{1.5}$$

$$CH_4 \rightarrow C + 2H_2 \quad \text{Cracking} \tag{1.6}$$

The primary issue with utilizing Ni-based anodes with hydrocarbon fuels is the propensity of Ni to catalyze the formation of graphite from hydrocarbons [5–7]. This limits the operating conditions of the cell, requiring steam reforming of the hydrocarbon before feeding (external reforming), the addition of steam to the feed (internal steam reforming [60]), the addition of oxygen to the feed [61], or the maintenance of a relatively high oxygen ionic flux to produce sufficient oxidation products within the anode itself [62]. In all cases, the goal is to supply enough oxygen to the anode compartment to operate outside the carbon-forming regime of the C–H–O ternary phase diagram. However, even this may not be sufficient to suppress all carbon formation due to the rapid kinetics of graphitic carbon filaments on Ni surfaces [16, 61]. Offer and coworkers [63] have written a comprehensive review of sulfur and carbon poisoning, which makes the explicit point that thermodynamic models are not always sufficient to capture the influence of these poisons. Kinetic models are required as deposition kinetics appear to control the extent of poisoning.

More recent studies have focused on *in situ* observation of the poisoning kinetics and characterization of the deposited material. Pomfret *et al.* [64] utilized *in situ* Raman spectroscopy to study graphite formation. They observed the formation of ordered graphitic carbon after 2 min in 14% CH_4–Ar mixture at 715 °C under OCV (open circuit voltage) conditions, with a monotonic increase in deposited carbon as a function of time. This carbon formation was suppressed by application of a DC bias, utilizing the supplied oxygen anions to oxidize the deposited carbon. Butane shows more rapid carbon formation, with carbon detected only after a small pulse of butane is applied, with both ordered graphite and disordered carbon detected. As with the carbon deposited from CH, both peaks were reduced on application of a DC bias. Similar trends were reported for ethylene and propylene.

This indicates that long-term operation can be sustained with a sufficiently high ionic flux, but this approach has the danger of catastrophic failure of the cell if there is a change in operating conditions. It must be feasible to "idle" the fuel cell stack under open circuit (zero current) conditions and to operate at a wide range of current densities. Hence, while experimentally demonstrated, this approach is not commercially viable.

The primary potential advantage of internal reforming is energy integration – the thermal energy from the exothermic fuel oxidation reactions can be utilized as the energy source for the endothermic steam reforming reactions. However, the rates of these two reactions differ significantly in the hydrocarbon-rich inlet section of the fuel cell system leading to large temperature excursions close to the fuel inlet, resulting in large temperature gradients across the cell stack [65]. One option to suppress this endothermic reaction is to reduce the amount of steam in the feed and utilize the steam generated through reaction in the cell stack to complete the steam reforming of the fuel; however, this increases the risk of carbon formation in the stack [66]. Alternatively, it may be possible to suppress the rate of steam reforming in this inlet section while maintaining activity for fuel oxidation. This would spread the heat load further into the stack, but again increases the risk of carbon deposition. Another alternative is to include an additional layer in the anode itself, which is optimized for steam reforming of the incoming hydrocarbon fuel [61], and such separation of the reforming and electrochemical reaction regions may enable their separate optimization.

External steam reforming is the most commonly adopted approach for industrial applications because it overcomes these complications by separating the reactions, enabling optimization of the steam reforming reactor system independent of the fuel cell system. This can then yield a full or partially reformed feed mixture of H_2 and CO for the SOFC stack. This also enables the removal of higher hydrocarbons that may be present in a natural gas feed. These higher hydrocarbons contribute significantly toward carbon fouling in the anode. A recent study by Powell *et al.* [67] demonstrated a methane-fueled SOFC system with external reformer that achieves a system efficiency of ~57% based on the lower heating value of the fuel. This high efficiency was achieved by recirculating the anode effluent gas to provide heat and water for the endothermic steam reforming reactions. This demonstrates that significant energy integration can be achieved, even with external steam reforming reactors. Figure 1.11 shows a schematic of the system.

Unless performed externally and taken to 100% hydrocarbon conversion, steam reforming will yield fuel mixtures containing hydrocarbons, H_2, H_2O, CO, and CO_2. It is necessary to understand the kinetics of the dominant reactions in these systems in order to design the most active anodes for these fuel supplies. A number of studies, both experimental work and simulation, have sought to understand the kinetics and mechanism of hydrocarbon and CO oxidation on Ni–YSZ anodes.

If we first consider direct reaction with CH_4, Abudula *et al.* [68] demonstrated that a smaller fraction of the anode is active for electrochemical reactions with hydrocarbon feeds when compared with pure H_2 fuel. They observed that cell performance showed a dependence on Ni–YSZ anode thickness up to 120 μm for

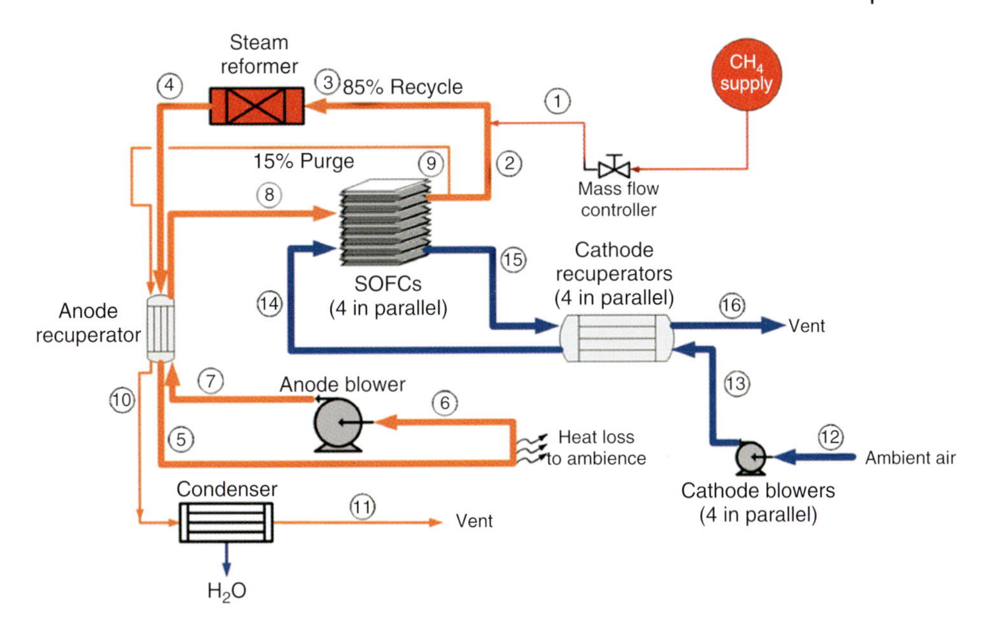

Figure 1.11 Schematic of SOFC power system operating with external reforming of CH_4 [67].

H_2 fuel, but only up until 76 μm for slightly humidified CH_4. This was interpreted by proposing that the active TPB region does not extend as far for CH_4. It should be noted that the current density for CH_4 was significantly lower than that for H_2. However, this is evidence that the activation of CH_4 is likely the rate-determining step for full-thickness anodes. It is perhaps not surprising that H_2 is more readily oxidized than CH_4. For hydrocarbon fuels, we must also consider the reaction products. Abudula *et al.* [68] reported CO_2 formation at all current densities and were able to directly correlate the rate of oxygen supplied through the electrolyte to the rate of CO_2 production. Horita *et al.* [69] demonstrated a correlation between reaction mechanism and the cell overpotential (and hence the cell current density) for CH_4 at H_2O/C ratio between 0 and 1. At low overpotential, by the direct oxidation of carbon, it is deposited through nonelectrochemical cracking of CH_4, but at higher overpotential, the mechanism switches over to the oxidation of CO and H_2. Hotlappels *et al.* [70] demonstrated that the activity of Ni–YSZ cermets toward CO oxidation is more than one order of magnitude lower than for H_2. Matsuzaki and Yasuda [71] quantified a factor of 2 difference between H_2 and CO oxidation rates and suggested that this was due to slower diffusion of CO on the electrode surface. This correlates with the work of Weber *et al.* [72] who showed a decrease in cell performance because of high CO content in CO–H_2 mixtures.

Zhan *et al.* [73] characterized the observed product distributions from CH_4-fueled Ni–YSZ anodes in terms of O^{2-}/CH_4 ratio; they observed methane partial oxidation to CO at $O^{2-}/CH_4 \approx 1$, but there was an increase in the total oxidation product, CO_2, content as this ratio increased. More recent work from Buccheri and Hill [74] reported a similar shift, showing that at low current density the primary reaction of CH_4 on Ni-based anodes is the electrochemical oxidation of C to CO. H_2O and CO_2 were observed at higher current densities, formed through further oxidation of the H_2 and CO.

Kleis *et al.* [75] performed a computational study to examine two possible mechanistic pathways on Ni – either a CH+O or a C+O oxidation pathway. They found that the preferred pathway depends on the surface orientation of the Ni phase. The CH+O pathway is preferred on the stepped Ni(211) surface, but the C+O route is preferred on the terraced (111) surface. The lowest barriers for the entire pathway are found on the undercoordinated surfaces. This may suggest a computational verification of the experimental results of Horita *et al.* [69] and Buccheri and Hill [74], which also indicated this pathway. Of course, experimental studies on real anodes will measure the results from multiple orientations, and it should be noted that these calculations are performed at zero overpotential. Work by Pillai *et al.* [76] suggests an alternative explanation: the product distribution is dictated entirely by reforming reaction equilibria within the anode, which shifts the product distribution to equilibrium values. This implies that the observed product distribution may not be indicative of the electrochemical reactions occurring at the anode TPB.

The proposed changes in mechanism may be due to changes in the gas composition within the anode as the oxygen flux to the anode is increased. At high current densities, the concentrations of the products CO_2 and H_2O in the anode channels may reach a sufficient level to form significant amounts of H_2 and/or CO via steam and/or dry reforming of the hydrocarbon. Compared to hydrocarbons, these fuels are more readily oxidized in the TPB, leading to an observed shift in mechanism. At low current densities, the low activity toward CH_4 oxidation leads to a mechanism based on the oxidation of carbon deposits. The observation of these mechanistic shifts may be a result of operating these cells with low humidity in the feed. This concept of changing mechanism with changing gas atmosphere is supported by Hao and Goodwin, who studied the reactions occurring in an SOFC anode upon co-feeding of CH_4 and O_2. Their results indicate that the anode can be considered to consist of three zones: a near-surface region where the oxygen is consumed through methane and hydrogen oxidation, followed by a mid-region where reforming reactions dominate, and finally, a region deeper in the anode that is dominated by the water gas shift reaction [77]. The only change through the anode is the gas composition. Fully unraveling this mechanism thus requires a combination of careful experimentation to measure the kinetics of each species present in the anode gas and whole-anode simulations to account for the product and reactant distributions within the anode channels.

Numerous studies have sought to understand the hydrocarbon steam reforming and product oxidation reactions. One important consideration is whether the reactions occur as homogenous gas-phase reactions or as heterogeneous reactions on the solid Ni–YSZ cermet surface. The homogenous reactions for methane do not contribute significantly to the anode chemistry below 900 °C; however, higher hydrocarbons will undergo homogenous cracking and polymerization at lower temperatures [63, 64, 78–80]. This is often witnessed directly as tar deposits in SOFC systems operating with these fuels [78, 81]. It is also necessary to distinguish between the catalytic formation of graphitic carbon filaments from a Ni surface and the formation of tarlike deposits through gas-phase chemistry. Graphitic carbon formation on Ni can lead to catastrophic failure of the cell due to the large volume expansion and can lead to "dry corrosion" of the anode cermet [82]. The formation of tars occurs in the gas phase, with the tar depositing in and physically clogging the anode pores once it reaches a high enough molecular weight [81]. Tar formation or dry corrosion will lead to slow degradation of the anode.

Experimental studies of dry (CO_2) reforming kinetics are less prevalent. A study by Kim *et al.* [83] indicates that as with the low H_2O/CH_4 ratio steam reforming studies, carbon is deposited on the anode under open circuit conditions. Under operating conditions, they suggest an operating mechanism based on oxidation of carbon deposits at the TPB. Experimental details regarding competitive kinetics are also scarce; however, Sukeshini *et al.* [84] did report inhibition of CO oxidation by H_2O on Ni–YSZ pattern electrodes.

The most detailed kinetic mechanism for methane reforming and the associated reaction kinetics over Ni–YSZ cermets has been experimentally determined and modeled by Hecht *et al.* [85]. Their resulting set of kinetic parameters is perhaps the most comprehensive available and forms the basis for further computational modeling of complete anodes with internal steam reforming. These comprehensive models from the Kee group [86] can accurately capture experimental results; key aspects of the anode, including an electrochemically active region that extends $\sim$10 μm from the anode–electrolyte interface [87]; and the electrochemical impedance behavior of the cell [88]. These impedance studies note that due to the strong dependence of the reforming chemistry on the local species concentration, the electrochemical cell response and relative rates of the reforming reaction will vary significantly with the current density. These agree with the previously discussed results that indicate a variation in product distribution with current density [69, 73, 74, 76] but argue that it is an indirect consequence of the increased oxygen flux through the cell.

1.5 Poisoning of Ni-Based Anodes

Sulfur poisoning time is independent of the sulfur concentration in the fuel, with anodes poisoning at approximately the same rate for all concentrations. The concentration required to poison the cell is a function of temperature. For example, impedance increases for a sulfur content above 0.5 ppm at 1173 K but increases at 0.05 ppm at 1023 K. This is sulfur as H_2S in H_2/H_2O (79/21 mol%). This poisoning is reversed on removal of sulfur from the fuel [89]. Other reports have indicated that sulfur poisoning up to 100 ppm H_2S is reversible [90]. This is in agreement with the work of Kromp *et al.* [91] who demonstrated that sulfur poisoning primarily occurs through chemisorption of sulfur on the catalytically active Ni sites, blocking both H_2

oxidation and reforming reactions. Cheng and Liu [92] further demonstrated this point using both *in situ* and *ex situ* analysis, and comparison with thermodynamic data, indicating that bulk Ni–S compounds were not responsible for the long-term degradation of Ni-based anodes in H_2S-containing atmospheres. The researchers suggest that surface species lead to the initial rapid performance deterioration of S-containing fuels but indicate that the longer term degradation mechanism is not well resolved at this time. These poisoning studies can also provide insights into reaction mechanisms for the various fuels. For example, the influence of sulfur poisoning on methane and hydrogen oxidation rates differ, indicating that the reaction sites differ [93].

Other poisons can be present and can create significant poisoning, especially if the fuel is generated from coal gasification. For example, nickel phosphide is formed as a bulk phase when it is exposed to phosphorus in the gas phase [94]. Arsenic also interacts strongly with Ni, forming Ni–As solid solution, Ni_5As_2, and $Ni_{11}As_8$. Loss of electrical connectivity is caused by Ni–As phases. Eventual cell failure was due to these Ni–As phases reaching the anode–electrolyte interface. Less than 10 ppb As is required to limit degradation [95].

Silicon impurities in the starting materials can also create problems under long-term operation, as these impurities can accumulate at the electrode–electrolyte interface, altering the microstructure and forming a secondary sodium silicate glass phase [96–98]. In the presence of water vapor, this damage can extend to grain corrosion in the electrolyte and lead to Ni transport from the electrode to the electrolyte phase [96]. Figure 1.12 shows the interface between a Si-containing Ni wire and a YSZ surface (Ni removed) following exposure to 3% humidified H_2 for approximately 1 week [99].

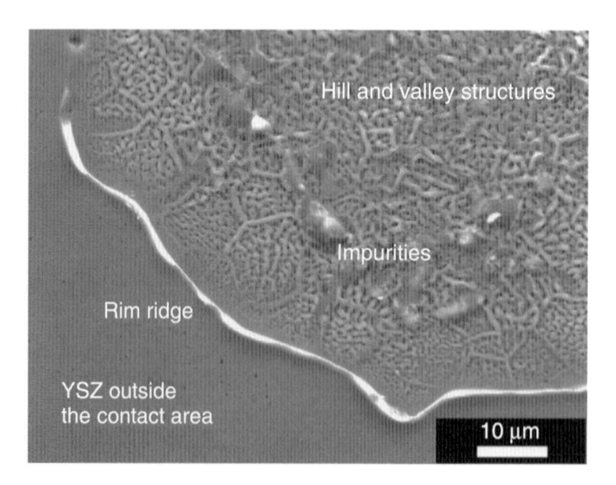

Figure 1.12 SEM image showing impurity ridges formed beneath a Ni wire contact point [99].

1.6
Alternative Anode Materials for Direct Hydrocarbon Utilization

The primary concern with the addition of steam to the system is the reduction in cell potential that accompanies the addition of oxygen to the anode chamber. The cell open circuit potential, OCV, should be the Nernst potential determined by the equilibrium oxygen chemical potential at either side of the cell. For hydrogen, the oxygen chemical potential in the anode is set by the equilibrium of H_2, H_2O, and O_2. As discussed earlier, the gas composition in the anode for hydrocarbon feeds is dictated by a complex interacting set of reactions, making it difficult to determine what the cell OCV "should" be. However, increasing the steam supply to the anode to suppress carbon formation will lower the OCV by increasing the oxygen chemical potential in the anode chamber. The typical industrial steam reforming steam/carbon ratio is $\sim$3. It is often stated that the ideal situation is to find or control the anode catalyst to suppress all but the direct electrochemical oxidation of dry hydrocarbons. This leads to the highest thermodynamically predicted open circuit potential, and hence, potentially, the highest efficiency of the cell. This will require the design of Ni-free alternative anode materials that are capable of direct operation with hydrocarbon fuels.

The complexity of realizing this selective catalysis is considerable. Even for the simplest hydrocarbon fuel, CH_4, the number of possible heterogeneous reaction steps involving reactants, products, and intermediates is very large and is compounded by the possibility of a large number of gas-phase homogenous reactions [85]. This large number of reaction equilibria and the resulting mixture of species set the anode oxygen fugacity and thus the OCP. We also note that the total oxidation of CH_4, reaction 1.1, involves four oxygen anions and eight electrons. The possibility of electrochemically catalyzing all these reaction steps while suppressing all other pathways seems remote. This is especially true if we consider that the gas-phase composition in the fuel cell stack will vary from the reactant-rich entrance region to the product-rich exit. Also, the reaction system complexity increases rapidly as the carbon number of the fuel increases. Finally, this complexity complicates the analysis of cell performance, and the literature reports discuss general trends rather than details of the reaction mechanism. This is an area ripe for future research.

The material requirements for direct hydrocarbon anodes are the same as for Ni-based electrodes: high electronic conductivity, high ionic conductivity, and high electrocatalytic activity toward fuel oxidation. Again, this is coupled to requirements of manufacturability and durability. A number of potential materials have been proposed, but as of the time of writing, none of these have proven to meet all the electrode requirements at a commercially feasible cost. Most of these materials are implemented as direct replacements for Ni in the Ni–YSZ (or other electrolyte material) system, with the new oxide taking the role of electrocatalyst and electron conductor. Intriguingly, mixed metal oxides can show mixed electronic and ionic conductivity, with such materials commonly implemented in SOFC cathodes [100]. This leads to the possibility that a single material could potentially be developed that would meet all the SOFC anode requirements. This could render the entire electrode surface active (all the necessary functional requirements would be met at all points), creating a high-performance electrode not limited by the concept of a TPB. However, no such material has been developed at this time; again, this is an area for future research and innovation.

One of the most promising approaches to removing Ni was proposed by Gorte, Vohs, and coworkers [4, 101–103], who developed a Cu–CeO_2–YSZ anode system. In this case, the Cu provides high electronic conductivity, while CeO_2 provides catalytic activity. Cu does not catalyze graphitic carbon formation from dry hydrocarbons, thus enabling operation of the resulting SOFCs on fuels ranging from CH_4 to *n*-decane. One of the most researched oxide systems is the perovskite structured material $La_{0.75}Sr_{0.25}Cr_{0.5}Mn_{0.5}O_3$ (LSCM), first proposed by Tao and Irvine [104]. In this case, the LSCM phase replaces the Ni to provide both electronic conductivity and catalytic activity. The LSCM–YSZ composite anodes have shown sufficient conductivity and corresponding SOFC performance in CH_4 fuel at 900 °C. Other suggested material systems include the double perovskites, especially those based on $Sr_2MgMnO_{6-\delta}$ (SMMO) [105], doped strontium titanates [106, 107], pyrochlores such as the series $Gd_2Mo_xTi_{2-x}O_7$ [108], tungsten bronzes with the general formula $A_{0.6}BO_3$ [109, 110], and vanadates [111–113]. Table 1.1 is a summary of a number of proposed materials. It should be noted that these characterizations regarding properties can often be circumvented. For example, low catalytic activity can be overcome by incorporating an additional catalyst, or chemical compatibility can be overcome by preparing electrodes using infiltration. Furthermore, all these properties are functions of temperature, pressure, and oxygen chemical and electrochemical potential.

These anode compositions vary considerably during the course of research on each material system as researchers strive to improve the performance of the base materials or utilize additional materials in the anode to optimize electrode performance. In the following, we discuss the most promising alternative anode materials developed at the time of writing, grouping the discussion into sections based on the anode requirements. A full discussion of the crystal structure, defect chemistry, and ionic and electronic transport mechanisms in this class of materials is outside the scope of this chapter and the reader is referred to a number of excellent reviews in this area [114–117].

1.6.1 Electronic Conductivity of Alternative Materials

Most measurements of electronic conductivity are performed on pure materials, which are then integrated into the final anode composite. While these measurements are essential when comparing alternative anode materials and studying the origin and magnitude of their conductivity, care must be taken when extrapolating to composite electrode performance. The electronic conductivity of the composite may be significantly lower than that of pure materials due to the blocking influence of the pores and the electrolyte material. A typical rule of thumb

Table 1.1 Comparison of some proposed SOFC anode materials.

Structure	Typical materials	Stability in reducing atmosphere	Ionic conductivity	Electronic conductivity	Chemical compatibility with YSZ	Thermal compatibility with YSZ	Performance using H_2 fuel	Performance using CH_4 fuel	Redox stability
Mixture	Ni–YSZ	✓	✓	✓	✓	✓	✓	×	×
Mixture	Cu–YSZ	?	✓	✓	✓	✓	✓	✓	✓
Fluorite	YZTScYZT CGO	✓	✓	×	✓	✓	Ok	Ok	✓
Cr–perovskite	$La_{1-x}Sr_xCr_{1-y}TM_yO_3$	✓	?	✓	✓	✓	✓	✓	✓
Ti–perovskite	$La_{1-x}Sr_xTi_{1-y}TMO_3$	✓	×	✓	✓	✓	×	×	✓
Double perovskite	Sr_2MgMoO_6	✓	?	Ok	×	✓	✓	×	✓
Pyrochlore	Gd_2TiMoO_7	×	Ok	✓	✓	?	✓	?	×
Tungsten bronze	$Sr_{0.6}Ti_{0.2}N_bO_3$	✓	×	✓	✓	?	×	?	✓
Monoclinic space group *C*2/*m*	Nb_2TiO_7	✓	×	✓	✓	×	?	?	×

Question mark indicates unknown.
Adapted from the review by Tao and Irvine [18].

is that the composite should have an electronic conductivity of >100 S cm^{-1} [118]. The electronic conductivity of the composite is a function of material percolation and constriction, depending significantly on morphology, relative loadings, relative particle sizes, and pore size. Thus, it should be common practice to also include the conductivity of the resulting electrode when analyzing cell performance. For example, Cu–YSZ composites fabricated by infiltration of Cu precursors into a preformed YSZ scaffold can show conductivities above 1000 S cm^{-1} at Cu loadings of only 15 vol%. This is much lower than the >40 vol% Ni typically required for percolation in Ni–YSZ anodes fabricated by random mixing of NiO and YSZ powders. This high conductivity at low loading was attributed to selectively coating the YSZ pore surfaces to create a contiguous surface layer [119].

The total composite electronic conductivity required is dependent on the anode thickness. Ideally, the ohmic resistance of the electrode will not contribute significantly to the allowable total electrode resistance. This can be achieved either by increasing electronic conductivity or decreasing the length of the electronic pathways by decreasing the electrode thickness (the electronic path length is the distance from the active electrode region next to the electrolyte to the current collector placed on the top electrode surface). Thus, it is possible to utilize lower conductivity materials as the primary electronic conductor in either a thin electrode or within a thin functional layer in a thicker electrode [120–122]. Low conductivity can thus be overcome by utilizing thin anodes, but this requires one of the other components, either cathode or electrolyte, to be thicker and provide the mechanical support for the cell.

The electronic conductivity of oxide materials of SOFC anodes may be either n- or p-type, with p-type materials being more common, although this does lead to a decrease in electronic conductivity within the reducing atmosphere of the anode. The conductivity of these materials is typically tailored by aliovalent doping on one site. This doping is charge compensated either by formation of oxygen vacancies (most of the oxide materials of interest are oxygen substoichiometric under SOFC anode conditions) or by the generation of electronic charge carriers through changes in transition-metal cation valence state [114, 123, 124].

As an example, we examine efforts to improve the electronic conductivity of LSCM. The nominal starting composition, $La_{0.75}Sr_{0.25}Cr_{0.50}Mn_{0.50}O_{3-\delta}$, is ~40 S cm^{-1} in air and 1.5 S cm^{-1} in 5% H_2 at 900 °C [125, 126]. This reduction in total conductivity with a reduction in pO_2 is typical of p-type conductors. Taking $LaCr_{0.5}Mn_{0.5}O_3$ as a base material, substitution of Sr^{2+} with La^{3+} is charge compensated either through changing the oxidation state of Mn or via oxygen vacancy formation. High pO_2 atmospheres are unfavorable for oxygen vacancy formation, and under these conditions, generation of electron holes on Mn dominates, leading to high conductivity [127]. As pO_2 decreases, the conductivity decreases as charge compensation shifts to oxygen vacancy formation. The exact values that classify pO_2 as "high" and "low" depend on the reducibility of the transition-metal cations and the temperature of interest.

$$4SrO + 4La_{La}^{\times} + 4Mn_{Mn}^{\times} + O_2(g) \rightleftharpoons 4Sr'_{La} + 2La_2O_3 + 4Mn_{Mn}^{\bullet} \quad \text{High } pO_2 \tag{1.7}$$

$$4Mn_{Mn}^{\bullet} + 2O_O^{\times} \rightleftharpoons 4Mn_{Mn}^{\times} + 2V_O^{\bullet\bullet} + O_2\,(g) \qquad \text{Low } pO_2 \tag{1.8}$$

The most obvious route to increase conductivity is thus to increase substitution of La^{3+} with more Sr^{2+} [128]. However, it was found that more than 25 mol% Sr does not significantly enhance electronic conductivity but increasing the Sr content above 30 mol% Sr resulted in higher solid-state reactivity to form secondary phases with YSZ electrolytes [128]. An alternative approach is to increase the ratio of Mn/Cr; however, this leads to decreased stability in SOFC anode atmospheres [129]. These last two points illustrate the holistic approach that must be taken when developing SOFC anode materials; it is not sufficient to consider one required property in isolation.

The strontium titanate oxide ($SrTiO_3$, STO) provides another well-researched example of the use of aliovalent doping to modify the electronic conductivity of oxides. Discussion of this system is useful to illustrate the complexities associated with doping, in particular, concerns regarding the stability of the resulting materials. In contrast to the LSCM example discussed earlier, where we doped a lower valence cation onto the A site of the perovskite, STO doping focuses on doping higher valence cations onto the A or B sites. The resulting charge compensation can occur by the formation of either electronic carriers or cation vacancies in the structure. At low pO_2 and high temperature, the primary compensation route is reduction of Ti^{4+} in the base material to Ti^{3+}. Thus, doping of a trivalent cation such as Y^{3+} or La^{3+} on the Sr site [130–132] leads to the formation of n-type electronic carriers via

$$\frac{1}{2}X_2O_3 + Sr_{Sr}^{\times} + Ti_{Ti}^{\times} \rightarrow X_{Sr}^{'} + Ti_{Ti}^{'} + SrO + \frac{1}{4}O_2 \tag{1.9}$$

where X represents a trivalent cation dopant. From Eq. (1.9), we may anticipate that the concentration of carriers is directly related to the concentration of dopants, while the general trend holds that the association of carriers with dopant cations can lead to deviations from this simple assumption [133]. Nevertheless, this approach can lead to materials with conductivities in the desirable 100 S cm^{-1} range [107], but these high conductivities may require harsh reducing conditions because of the difficulty in reducing Ti^{4+} [106, 134, 135]. As such, under normal operating conditions, this may require high-temperature pretreatment of the cell before use, thus adding to fabrication costs for the cells or placing an additional materials burden on the stack if the cells are treated *in situ*. This is also a concern if slow reoxidation of the material, and subsequent loss of conductivity, occur during long-term operation, particularly toward the exhaust of the cell stack where the pO_2 is higher.

High pO_2 is unfavorable for Ti^{4+} reduction, requiring charge compensation via a different mechanism. The formation of cation vacancies on the Sr sublattice [133, 136] leads to charge neutrality but at the expense of electronic carriers:

$$Sr_{Sr}^{\times} + 2Ti_{Ti}^{'} + \frac{1}{2}O_2 \rightleftharpoons SrO + V_{Sr}^{''} + 2Ti_{Ti}^{\times} \tag{1.10}$$

This leads to the formation of a secondary SrO- or Sr-rich phase [137, 138]. While the anode should not encounter such high pO_2 under operating conditions, it is possible for this to occur during operational transients or over a long period. The full thermodynamic diagram under SOFC conditions has not been mapped out for these materials. An option to avoid secondary-phase formation is to deliberately introduce Sr vacancies at a level to compensate for the charge introduced through donor doping [78, 110, 139, 140]. This does yield high-conductivity materials, but may also lead to the development of Ti-rich secondary phases. While these materials show good electronic conductivity, recent work has suggested that A-site deficiency leads to precipitation of secondary Ti-rich phases [106, 141].

Figure 1.13 illustrates the differences in conductivity described above [107]. Figure 1.13a shows the conductivity as a function of pO_2 for samples previously treated in air. In general, the conductivity increases with doping concentration (this is not obvious due to the logarithm scale), and also as pO_2 decreases due to a shift

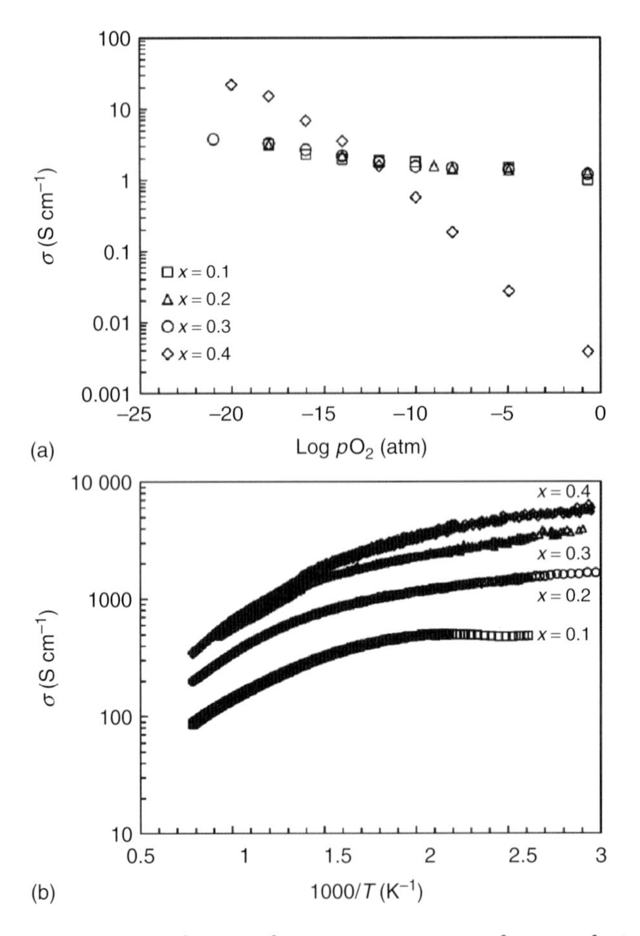

Figure 1.13 Conductivity of $La_{1-x}Sr_xTiO_{3-\delta}$ (a) as a function of pO_2 at 1000 °C for samples previously sintered in air at 1650 °C and (b) as a function of temperature on heating for samples previously reduced in 4% H_2/Ar with a CO_2 buffer to a pO_2 of 10^{-18} atm at 1000 °C.

from ionic to electronic conductivity. The increase in conductivity with increasing dopant concentration is clearer, and the measured conductivities are significantly higher, for the samples previously reduced in hydrogen (Figure 1.13b).

An alternative approach to replacing divalent Sr on the A site with a trivalent cation is to replace the tetravalent Ti on the B site with pentavalent Nb [142–144]. As with trivalent A-site dopants, this approach can lead to considerable electronic conductivity at high temperatures and low pO_2 via charge carriers introduced by reduction of Ti^{4+}:

$$\frac{1}{2}Nb_2O_5 + 2Ti_{Ti}^{\times} \rightarrow Nb_{Ti}^{\bullet} + Ti'_{Ti} + TiO_2 + \frac{1}{4}O_2 \quad (1.11)$$

However, again analogous to the trivalent doping, achieving this conductivity requires harsh reduction conditions, although there is evidence that subsequent reoxidation is very slow [145]. Slow reoxidation of both the Nb-doped and other materials is likely due to low oxygen ionic conductivity.

Both LSCM and STO are perovskite structure oxides. Another approach to enhancing electronic conductivity is to explore other crystal structures. Huang *et al.* and others proposed the double-perovskite SMMO as a potential anode material, showing conductivities of up to 10 S cm^{-1} at 800 °C in H_2 [105, 146, 147]. In this case, n-type charge carriers are generated by charge compensation for reduction of the Mo(VI) to Mo(V) cations. This conductivity may not be sufficient for full-thickness anodes, but reasonable fuel cell performance has been reported for thin electrode cells [105]. Doping of this material to generate higher carrier concentrations will likely lead to disruption of the double-perovskite ordering; however, it is not clear if this double-perovskite structure is necessary to achieve high conductivity.

A number of other materials have been proposed, but all lack the level of conductivity required to fabricate the thick electrodes necessary to form an anode-supported SOFC. The fluorite structured oxide SMMO has been demonstrated to provide sufficient conductivity and good performance as a redox-stable anode, providing an MPD of 850 mW cm^{-2} for a symmetric cell with LSGM electrolyte in H_2 [148]. Again, this is when utilizing thin electrodes. Pyrochlore-structured $Gd_2Mo_xTi_{2-x}O_7$ can show sufficient electronic conductivity and yield acceptable SOFC performance, but these materials have low stability in the anode environment [108, 149].

The discussion above is not meant to be exhaustive but rather illustrative of the challenge of developing alternative oxide-based electronic conductors to replace Ni. There are numerous other examples of possible candidate materials [18]. One approach to overcome the limited electronic conductivity of these oxides is to incorporate a metal, in particular, Cu, as the alternative electronic conducting phase. The use of Cu follows the pioneering work of Gorte, Vohs, and coworkers who utilized Cu as the primary electronic conductor in Cu–CeO_2–YSZ anodes. As discussed, the conductivity of these Cu-based anodes is sufficient to enable their use as thick electrodes (>200 μm) in anode-supported cells [4]. This approach has been utilized with other oxides as catalysts [150], including Cu–LSCM–YSZ electrodes [151]. An alternative approach, also first utilized by Gorte, Vohs, and coworkers, is the use of Cu–Ni, Cu–Co, and Cu–Cr mixed metal electrodes [120, 152, 153]. These bimetallics can be stable in hydrocarbons, with the Cu suppressing carbon formation on the other metal. The primary concerns with utilizing Cu-based anodes is the necessary complexity in fabrication, as Cu cannot be co-fired with YSZ due to the relatively low melting point of CuO (1336 °C). As such, Cu must be added through infiltration or other means after forming the YSZ structure [102]. The thermal stability of Cu is also a point of concern, as the cells operate relatively close to the melting point of Cu (1085 °C), with the thermal stability of the Cu phase being a function of the infiltration technique. The bimetallics may increase the melting point and aid stability. The addition of another solid phase also complicates the concept of an anode TPB. In this new case, we have four phases, namely, electrolyte, electronic conductor, electrocatalyst, and gas phase, all of which must be in close proximity to create a functioning anode.

In conclusion, while the search for high conductivity oxides is ongoing, the currently known materials can provide sufficient conductivity for use in thin electrodes. Their use will require a shift from the traditional anode-supported cell architecture to cathode-supported cells. Electrolyte-supported cells will suffer from low performance due to the ohmic resistance from a thick electrolyte. For anode-supported cells, it is currently necessary to add a second metal phase, typically Cu.

1.6.2 Electrocatalytic Activity of Alternative Anode Materials

The development of alternative anode materials has not reached the same level of detail concerning the electrode reaction mechanism as discussed for Ni–YSZ. Instead, we can draw broad conclusions regarding the activity of the currently developed materials and point to directions for further development.

The primary conclusion that can be drawn from almost all studies regarding direct hydrocarbon utilization is the necessity of a highly active electrocatalyst for hydrocarbon oxidation. Apart from a few studies that include highly active precious metal catalysts, all other studies show significantly lower cell performance for hydrocarbon fuels when compared with H_2 fuel under the same operating conditions. Since all the cells are the same and the only change is the anode environment, we can conclude that the performance difference is due to differences in anode electrocatalytic activity toward H_2 and hydrocarbons. H_2 is much more easily oxidized than hydrocarbons, in particular, CH_4, which is relatively stable at typical SOFC operating temperatures. An alternative explanation is that the gas-phase transport of hydrocarbons may be slower than that of H_2; however, this is unlikely to be true across all the anode microstructures utilized in these studies, and the number of electrons consumed by each mole of hydrocarbon (the lowest is 8 mol of electrons per mole of CH_4) would likely offset any mass transfer problems.

There is a large amount of evidence that a highly active electrocatalyst is required in the anode. Perhaps, the most studied system is the Cu–CeO_2–YSZ-based system, where CeO_2 is the primary electrocatalyst [4, 154]. Cu–YSZ anodes show very low performance, particularly in hydrocarbon fuels [79]. Replacing CeO_2 with other lanthanide catalysts further demonstrates that the cell performance is directly linked to the catalytic activity of the lanthanide component [79]. The cell performance of Cu–CeO_2–YSZ anode SOFCs is lower for hydrocarbon fuels than that for H_2 in the data available for these cells in the literature (see, for example, studies by Park *et al.* [101, 102] and Kim *et al.* [103]). An exception is a study that utilized precious metal dopants in the anode [155] to yield the same OCP, polarization resistance, and hence, power density for CH_4 and H_2 fuels. This suggests that the addition of the precious metal removed the electrocatalytic barrier to hydrocarbon oxidation (precious metals supported on CeO_2 are very active catalysts for hydrocarbon oxidation) and that the cell performance was then limited by other factors.

LSCM-based anodes show the same trend of lower performance (higher polarization resistances) with hydrocarbon fuels when compared to hydrogen [104, 156]. Figure 1.14 shows the fuel cell performance and impedance spectra measured for LSCM-based anodes in wet H_2 and CH_4 fuels [104]. Note that the cell OCP and MPD are lower, and the polarization resistance is greater, in CH_4 when compared with H_2 fuels. Catalytic studies of LSCM conducted under continuous flow conditions indicate that it is active toward CH_4 oxidation, has some activity toward dry reforming of CH_4 with CO_2, and has only limited CH_4 steam reforming activity [157, 158]. Mn has been shown to be the catalytically active element in LSCM. X-ray absorption near-edge structure (XANES) measurements demonstrated that the Cr K-edge energies remain constant, while the Mn energy levels change (indicating a difference in oxygen coordination) on reduction of the sample [127]. This conclusion is in agreement with previous reports indicating that the Mn-free materials $La_{1-x}Cr_xO_{3-\delta}$ are catalytically inactive [158, 159]. This is also in agreement with the proposed Mars-van Krevelen reaction typically observed for perovskites. This mechanism is based on the involvement of lattice oxygen in the reaction. The lattice

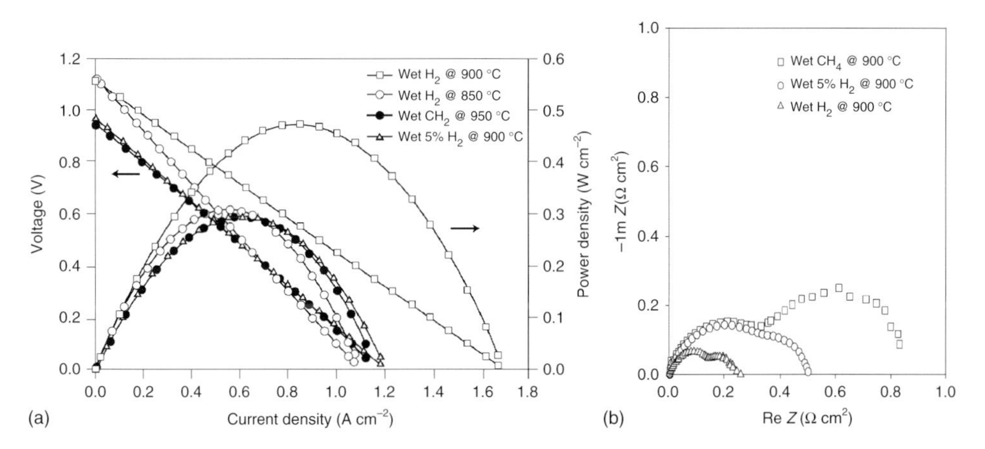

Figure 1.14 (a,b) SOFC performance curves for LSCM-based anodes in H_2 and CH_4 fuel [104].

oxygen is consumed to oxidize the reactant in a reduction step, before reoxidation in a subsequent step to complete the catalytic cycle. A number of reports link the activity of these perovskites to the redox ability of the B-site transition-metal cations [117, 158, 160–162], and based on this concept, a number of studies have sought to increase catalytic activity by further substitution on the B site [163–171]. While further doping does influence catalytic activity, it can often lead to exsolution of the dopant from the lattice [162, 164, 167, 168, 170].

As an example, van den Bossche and McIntosh [162] replaced 10 mol% of Mn with Co, Fe, or Ni and found that CH_4 oxidation rates were greatly enhanced in LSCMNi and LSCMCo up to an order of magnitude compared to those of LSCM. However, this occurred only after the exsolution of the Co and Ni metals. The addition of catalytically active metal particles to the LSCM surface, either by addition of a separate phase or through exsolution from a doped substrate, is the most successful route to enhance the activity of these oxides. Relatively small amounts (1–5 wt%) of active transition or precious metals are required to significantly increase the performance of these SOFCs with hydrocarbon fuels, often more than doubling the power output [151, 172–175]. This is analogous to the addition of precious metals to the CeO_2–YSZ-based anodes discussed previously.

STO-based SOFCs suffer from similarly poor electrocatalytic performance, even for H_2 [176], again requiring the addition of secondary catalysts to the anode to achieve reasonable SOFC performance [176–179]. As with LSCM, doping more reducible transition metals onto the B site can lead to enhanced catalytic activity [179]. Results for other oxide-based anode materials are more limited but demonstrate the same trends: lower performance for CH_4 when compared with H_2 and enhanced performance in both fuels when an additional catalyst is added [148, 180–186].

SMMO is the one material that shows conflicting reports of performance in the literature. Initial reports demonstrated significant performance in H_2 and CH_4 fuels, indicating that the material had significant activity toward CH_4 oxidation [105]. However, subsequent direct measurements of the CH_4 oxidation activity [187] indicated that the material is inactive. One note made in the initial report relates to the use of a Pt (a highly active oxidation catalyst) current collector on the anode. A recent study on the use of different current collector materials [188] demonstrated negligible power output for CH_4 fuel when catalytically inert Au, Ag, or $La_{0.3}Sr_{0.7}TiO_{3-\delta}$ current collectors were used. Only a Pt current collector enabled operation with CH_4 fuel, clearly indicating that this is the source of the catalytic activity. This enhanced performance is in agreement with the numerous studies discussed earlier, which demonstrated the deliberate use of additional catalytically active materials in the anode, and reports regarding the activity of Pt/SMMO catalysts [187]. Care must be taken when selecting current collector materials and evaluating performance with CH_4. Operation of SMMO-based anodes with H_2 fuel suggests that it may be a promising material for use with this fuel [189, 190]. As with the single-perovskite materials, improved electrocatalytic activity can be achieved through doping with transition metals, but this must, as always, be balanced with material stability [190–193].

Etsell and coworkers [194, 195] have examined the use of WC as an alternative anode material and proposed a different reaction mechanism for these materials. While WC provides sufficient electronic conductivity, its electrocatalytic performance toward methane oxidation is quite poor. Addition of Ni as an additional electrocatalyst (5 wt% in the anode phase) enhances activity toward methane oxidation and, if the WC and Ni phases are not required to provide structural integrity, can form a redox-stable anode. The resulting anode was shown to provide stable performance for 24 h at 850 °C. The researchers suggest that fuel oxidation occurs via a WC oxidation and recarburization redox couple and that Ni may act to enhance the recarburization rate to stabilize the WC phase.

The mechanism of electrochemical hydrocarbon oxidation in these materials is poorly understood. It is quite well established that reactions proceed by the Mars-van Krevelen mechanism and that transition metals in the oxide are the active redox centers. However, when compared to the Ni–YSZ system, we have very little knowledge of the detailed reaction mechanism or rates of the individual steps. In part, this is due to the wide variety of materials proposed for these direct hydrocarbon anodes and the absence of materials that demonstrate high activity toward hydrocarbons without the addition of cocatalysts.

An interesting trend reported in the literature is the variability in anode electrocatalytic activity and selectivity with varying cell overpotential and corresponding oxygen anion flux to the anode. Studies of oxide-based anode materials suggest that the observed shift in performance for these anodes may be due to changes in the oxidation state of the electrocatalyst. This concept comes from measurements of the cell product distribution as a function of the current density [79, 156]. It is observed that at low oxygen anion flux (lower current density and low overpotential), partial oxidation of the fuel dominates. In the case of CH_4, CO is observed in the outlet. As the ionic flux increases, the cell polarization resistance decreases and the product distribution shifts to the total oxidation product CO_2, indicating a shift in the reaction mechanism. A similar observation has been made for Ni–YSZ anodes, which may be explained by a change in the nonelectrochemical reforming reaction mechanism caused by changing gas composition above the Ni surface. This explanation may also hold for the change in product distribution reported for LSCM; however, measurements of the oxidation activity and reaction selectivity as a function of oxygen stoichiometry ($3-\delta$) in these materials [158, 162, 196, 197] indicate that the former decreases and the latter shifts to partial oxidation as the oxygen stoichiometry decreases. This suggests that the change in product distribution and decrease in polarization resistance may be linked to local dynamic changes in the material oxygen stoichiometry. At low oxygen flux, the LSCM is highly reduced due to the reducing gas atmosphere. This leads to low oxygen stoichiometry, low oxidation rates, high polarization resistance, and selectivity toward partial oxidation. As the ionic flux increases, the local oxygen electrochemical potential increases, leading to increased oxygen stoichiometry, increased reaction rate, decreased polarization resistance, and selectivity toward total oxidation. Further experimental work and modeling studies are required to fully understand this phenomenon. Control of the product distribution by controlling

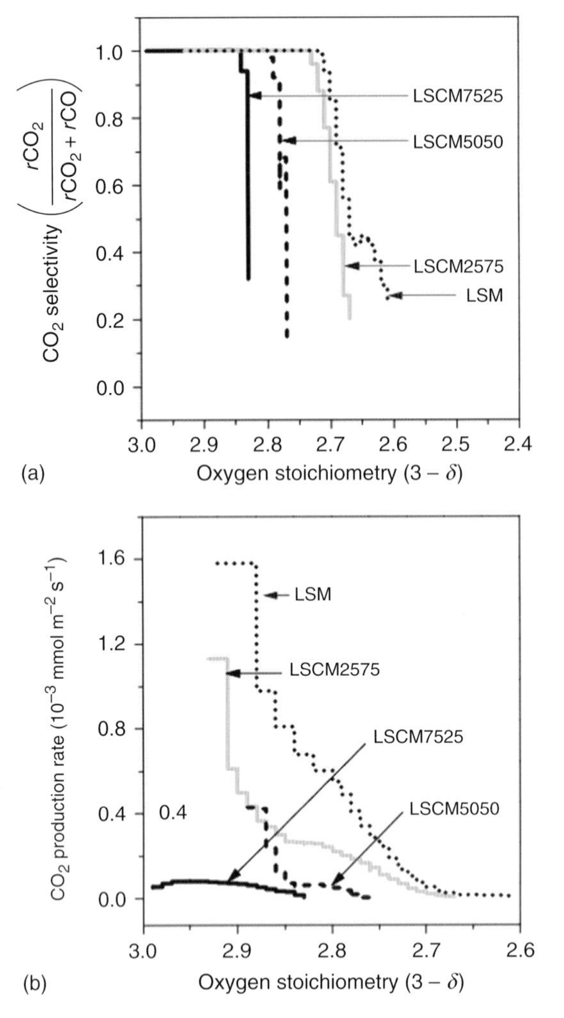

Figure 1.15 CO_2 versus CO selectivity (a) and CO_2 formation rate (b) as a function of oxygen stoichiometry measured for CH_4 oxidation at 700 °C over $La_{0.75}Sr_{0.25}MnO_{3-\delta}$ (LSM), $La_{0.75}Sr_{0.25}Cr_{0.25}Mn_{0.75}O_{3-\delta}$ (LSCM2575), $La_{0.75}Sr_{0.25}Cr_{0.5}Mn_{0.5}O_{3-\delta}$ (LSCM5050), and $La_{0.75}Sr_{0.25}Cr_{0.75}Mn_{0.25}O_{3-\delta}$ (LSCM7525) [158].

the ionic flux opens up the possibility of utilizing the SOFC technology in selective oxidation electrochemical reactors.

Figure 1.15 shows experimental evidence for this shift in reaction mechanism collected utilizing a pulse reactor system that enables the reaction selectivity (a) and rate (b) to be determined as a function of oxygen stoichiometry [158]. It can be observed that for all compositions, the CH_4 total oxidation (CO_2) selectivity and reaction rate decrease as the oxygen stoichiometry decreases. Increasing the Mn content leads to both higher rate and a larger range of oxygen stoichiometry where total oxidation of the fuel is favored. This is in agreement with the results of

Plint *et al.* [127], as discussed previously. These trends were consistent from 700 to 900 °C. This data suggests that the shift in reaction mechanism is a fundamental change in the mechanism occurring on the oxide as it moves from a more oxidized to a reduced state. One would expect a larger amount of CO to be present if the shift in product composition were due to a secondary fuel dry reforming reaction. The steam reforming rates measured on these materials are very low [158].

1.6.3
Poisoning of Alternative Anode Materials

A significant complexity in the development of direct hydrocarbon SOFC anodes is the tarlike carbon deposits formed through the gas-phase free radical polymerization of hydrocarbon fuels. Gas-phase cracking of CH_4 is not typically a problem below ~800 °C. However, as the carbon number of the hydrocarbon increases and the C–C bond strength decreases, the onset temperature for these reactions decreases. The primary products are long-chain and aromatic hydrocarbons that will nonselectively deposit on the electrode surface, even at elevated SOFC operating temperatures [81]. While these deposits may ultimately block the entire electrode, limited deposition has been shown to enhance the electronic conductivity in the electrode [78]. It was suggested that the limited conductivity of these tarlike hydrocarbons can link previously isolated metal particles, leading to enhanced electronic connectivity.

One potential advantage of alternative anode materials is increased resistance to sulfur poisoning. Sulfur poisoning at SOFC temperatures is dictated by the thermodynamics. This has been clearly established in the Cu–CeO_2–YSZ systems. Kim *et al.* [198] and He *et al.* [199] demonstrated that stable operation could be achieved if the sulfur level in the fuel is kept below the thermodynamic limit for Ce_2O_2S formation. Other oxides, including STO and $La_{1-x}Sr_xVO_3$, are quite tolerant to sulfur poisoning [112, 113, 200, 201].

1.7
Infiltration as an Alternative Fabrication Method

Gaining finer control over the microstructure and introducing alternative anode materials may require a shift from the traditional mixed powder processing. The most researched alternative approach is to prefabricate a porous electrolyte backbone before infiltration of salt solutions of the active components. These salts are then decomposed through calcination in air before reduction during cell start-up. This approach was initially utilized by Gorte and Vohs for the fabrication of Cu-based anodes [102]. Traditional mixed powder approaches are not applicable for Cu due to the low melting point of CuO compared with the required sintering temperatures of YSZ. These anodes were prepared by tape casting a slurry containing YSZ and graphite pore formers on top of a thinner, pore-former-free YSZ layer. On sintering, the graphite burns out, leaving a dense

YSZ electrolyte film on a porous YSZ substrate. These pores are then infiltrated with, in this case, Ce and Cu nitrate solutions to form a Cu–CeO_2–YSZ anode. The Cu–CeO_2–YSZ-infiltrated anodes fabricated by Gorte, Vohs, and coworkers [4, 154] are perhaps the widest studied set of infiltrated anodes, although the same approach has been performed with other anode materials and a range of cathode materials [202]. Infiltration has also been utilized extensively as a means to add small amounts of secondary catalytic materials, in particular, precious metals, to SOFC anodes [151, 155, 174, 175, 203, 204].

Gorte and coworkers have also investigated the use of infiltration to fabricate full SOFC anodes and SOFC AFLs from a range of alternative anode materials [11, 120–122, 172, 176, 205–208]. One of their most intriguing findings is that the morphology of these infiltrated materials can change from being an almost continuous coating to an interconnected array of nanometer-scale particles (Figure 1.16) [208]. This was shown to be a function of the Mn content of Mn-containing perovskites and is attributed to an interaction between the infiltrated anode and the underlying oxide framework. Furthermore, the degree of this interaction and associated "wetting" of the underlying oxide is a function of the extent of reduction of the perovskite. This could be an intriguing technique to tailor the anode structure to create very high TPB densities.

Another area of interest in case of infiltrated anodes is in the metal-supported SOFC designs [209]. The use of a metal alloy as the primary structural support leads to reduced material costs, increased strength, and tolerance to rapid thermal cycling. However, these metal supports required the SOFC components to be co-fired under reducing atmospheres to avoid oxidation of the metal, but exposing Ni to high temperatures leads to Ni coarsening and reduction of cell performance. Tucker *et al.* [210] utilized infiltration of a YSZ scaffold to form Ni–YSZ anodes in metal-supported cells that showed close to commercially viable performance. Unfortunately, this performance decreased rapidly due to rapid coarsening of the fine Ni structure induced by infiltration [211]. Blennow *et al.* [212] sought to overcome this limitation by infiltrating the active components, $Ce_{0.8}Gd_{0.2}O_{2-\delta} + 10\,wt\%$ Ni, into a preformed porous ceramic (YSZ) – metallic (Fe–Cr) cermet. The motivation was to avoid the need for a contiguous Ni phase by providing sufficient electronic conductivity in the backbone or the porous structure. This led to significant durability improvement over 1000 h of testing.

As an alternative to infiltration of all the active materials, a number of researchers have sought to enhance anode performance through infiltration of small (typically $<1\,\mu m$) particles of oxides or known catalysts. Jiang *et al.* [213] demonstrated a reduction in Ni–YSZ anode polarization of $\sim$3.5 times due to the addition of YSZ particles and up to 7 times due to the addition of $Gd_{0.2}Ce_{0.8}(NO_3)_x$ solution (to form $Gd_{0.2}Ce_{0.8}O_{1.9}$ particles). Similarly, the same group [214] enhanced the performance of LSCM-based anodes for direct hydrocarbon utilization through infiltration of $Gd_{0.2}Ce_{0.8}O_{1.9}$. Jiang *et al.* [215] also showed that infiltration of a significant reduction in Ni coarsening during sintering and NiO reduction steps led to a high TPB density and associated high performance.

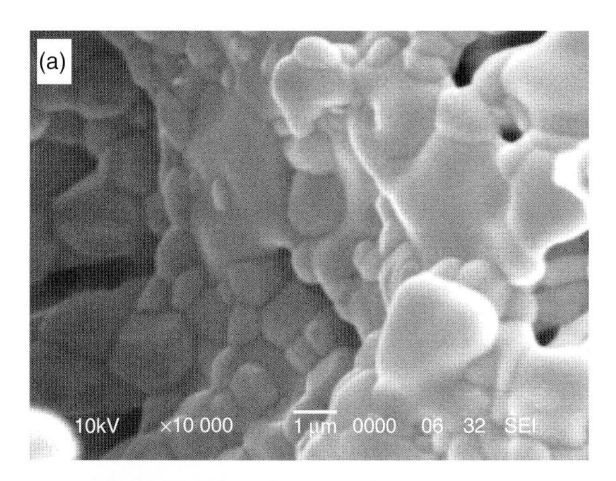

Figure 1.16 SEM images of YSZ infiltrated with $La_{0.75}Sr_{0.25}Cr_{0.5}Mn_{0.5}O_{3-\delta}$ following reduction in 3% H_2O/H_2 for 5 h at (a) 700 °C, (b) 800 °C, and (c) 900 °C [208].

The primary problem with these infiltration approaches is the labor-intensive processing involved. This is especially true when multiple infiltration steps are required to build up the total amount of active material in the anode to the levels required for a contiguous phase. However, it should be noted that the amount required is a function of the morphology of the infiltrated material. This morphology is influenced by the infiltration method, precursor selection, and postinfiltration treatment [119].

1.8 Summary and Outlook

The development of SOFC anodes to their current status has occurred over several decades. For Ni–YSZ cermets, long-term stability is the primary remaining technological challenge with regard to their operation with clean hydrogen or reformed hydrocarbon fuel. These cermets can provide sufficient performance in the laboratory environment, but these levels must be sustained for thousands of hours of commercial operation. Slow degradation of anode performance occurs due to the slow formation of secondary impurity phases and coarsening of the Ni phase. While it appears that the mechanism of H_2 oxidation in the Ni–YSZ cermet is reasonably well resolved, we currently do not have the same level of understanding regarding CO oxidation. Internal reforming of CH_4 can be accurately modeled at the system level, but the nature of the reaction sites for each step on the Ni–YSZ surfaces is still unknown.

Significant breakthroughs in both fundamental science and engineering are required if we are to realize SOFC anodes that can be commercially operated with hydrocarbon fuels. While a large number of materials have been investigated, direct hydrocarbon SOFCs remain a laboratory curiosity. We have yet to discover a single material that can adequately meet all the anode requirements; indeed, it may be argued that we do not have a single material that can match the electronic conductivity and electrocatalytic activity of Ni in H_2 fuel. It seems most likely that multicomponent electrodes will be required, with each component meeting only one or perhaps two functional requirements, for example, an ion-conducting YSZ matrix partially filled with a redox-stable chemically and thermally compatible material with high electronic conductivity. Catalytic activity can then be added by adding a secondary transition metal on the surface. This would suggest that material development should focus on optimizing specific functionality, rather than on attempting to develop a single multifunctional material.

The primary disadvantage of this approach is that we now move from a TPB to a more complex reaction mechanism that involves more than two solid components to meet all the material requirements. Microstructure optimization to optimize the relative loadings, particle sizes, layer thicknesses, and distribution of these components would be a critical component in developing highly active multicomponent anodes. This is feasible if we couple the recent advances in microstructure analysis with the advances made in microstructure control via infiltration.

We need greater insight into the reaction mechanism of hydrocarbon fuels in these anodes. At the time of writing, there are no comprehensive studies of these anodes that include real electrodes, simplified model electrodes, and computational modeling. As such, we know relatively little about the reaction mechanism, making optimization of materials a significant challenge.

References

1. Weber, A. and Ivers-Tiffée, E. (2004) Materials and concepts for solid oxide fuel cells (SOFCs) in stationary and mobile applications. *J. Power Sources*, **127** (1–2), 273–283.
2. Dees, D.W., Claar, T.D., Easler, T.E., Fee, D.C., and Mrazek, F.C. (1987) Conductivity of porous Ni/ZrO_2-Y_2O_3 cermets. *J. Electrochem. Soc.*, **134** (9), 2141–2146.
3. Minh, N.Q. (1993) Ceramic fuel-cells. *J. Am. Ceram. Soc.*, **76** (3), 563–588.
4. McIntosh, S. and Gorte, R.J. (2004) Direct hydrocarbon solid oxide fuel cells. *Chem. Rev.*, **104** (10), 4845–4865.
5. Finnerty, C.M., Coe, N.J., Cunningham, R.H., and Ormerod, R.M. (1998) Carbon formation on and deactivation of nickel-based/zirconia anodes in solid oxide fuel cells running on methane. *Catal. Today*, **46** (2–3), 137–145.
6. Horita, T., Kishimoto, H., Yamaji, K., Xiong, Y., Sakai, N., Brito, M.E., and Yokokawa, H. (2006) Materials and reaction mechanisms at anode/electrolyte interfaces for SOFCs. *Solid State Ionics*, **177** (19–25), 1941–1948.
7. Triantafyllopoulos, N.C. and Neophytides, S.G. (2003) The nature and binding strength of carbon adspecies formed during the equilibrium dissociative adsorption of CH_4 on Ni–YSZ cermet catalysts. *J. Catal.*, **217** (2), 324–333.
8. Sarantaridis, D. and Atkinson, A. (2007) Redox cycling of Ni-based solid oxide fuel cell anodes: a review. *Fuel Cells*, **7** (3), 246–258.
9. Brandon, N.P., Skinner, S., and Steele, B.C.H. (2003) Recent advances in materials for fuel cells. *Annu. Rev. Mater. Res.*, **33** (1), 183–213.
10. Jiang, S.P. and Chan, S.H. (2004) A review of anode materials development in solid oxide fuel cells. *J. Mater. Sci.*, **39** (14), 4405–4439.
11. Gorte, R.J. and Vohs, J.M. (2009) Nanostructured anodes for solid oxide fuel cells. *Curr. Opin. Colloid Interface Sci.*, **14** (4), 236–244.
12. Jacobson, A.J. (2009) Materials for solid oxide fuel cells. *Chem. Mater.*, **22** (3), 660.
13. Orera, A. and Slater, P.R. (2009) New chemical systems for solid oxide fuel cells. *Chem. Mater.*, **22** (3), 675.
14. Sun, C.W. and Stimming, U. (2007) Recent anode advances in solid oxide fuel cells. *J. Power Sources*, **171** (2), 247–260.
15. Fergus, J.W. (2006) Oxide anode materials for solid oxide fuel cells. *Solid State Ionics*, **177** (17–18), 1529–1541.
16. Atkinson, A., Barnett, S., Gorte, R.J., Irvine, J.T.S., McEvoy, A.J., Mogensen, M., Singhal, S.C., and Vohs, J. (2004) Advanced anodes for high-temperature fuel cells. *Nat. Mater.*, **3** (1), 17–27.
17. Mogensen, M. and Kammer, K. (2003) Conversion of hydrocarbons in solid oxide fuel cells. *Annu. Rev. Mater. Res.*, **33**, 321–331.
18. Tao, S.W. and Irvine, J.T.S. (2004) Discovery and characterization of novel oxide anodes for solid oxide fuel cells. *Chem. Rec.*, **4** (2), 83–95.
19. van den Bossche, M. and McIntosh, S. (2011) in *Direct Hydrocarbon Solid Oxide Fuel Cells – Fuel Cells for Today's Fuels* (ed. K.D. Kreuer), Encyclopedia of Sustainable Science and Technology, Springer.
20. Kee, R.J., Zhu, H., and Goodwin, D.G. (2005) Solid-oxide fuel cells with hydrocarbon fuels. *Proc. Combust. Inst.*, **30** (2), 2379.

21. Will, J., Mitterdorfer, A., Kleinlogel, C., Perednis, D., and Gauckler, L.J. (2000) Fabrication of thin electrolytes for second-generation solid oxide fuel cells. *Solid State Ionics*, **131** (1–2), 79.
22. Müller, A.C., Herbstritt, D., and Ivers-Tiffée, E. (2002) Development of a multilayer anode for solid oxide fuel cells. *Solid State Ionics*, **152–153**, 537–542.
23. Clemmer, R.M.C. and Corbin, S.F. (2009) The influence of pore and Ni morphology on the electrical conductivity of porous Ni/YSZ composite anodes for use in solid oxide fuel cell applications. *Solid State Ionics*, **180** (9–10), 721.
24. Clemmer, R.M.C. and Corbin, S.F. (2004) Influence of porous composite microstructure on the processing and properties of solid oxide fuel cell anodes. *Solid State Ionics*, **166** (3–4), 251–259.
25. Lee, J., Heo, J., Lee, D., Kim, J., Kim, G., Lee, H., Song, H.S., and Moon, J. (2003) The impact of anode microstructure on the power generating characteristics of SOFC. *Solid State Ionics*, **158** (3–4), 225–232.
26. Zhao, F. and Virkar, A.V. (2005) Dependence of polarization in anode-supported solid oxide fuel cells on various cell parameters. *J. Power Sources*, **141** (1), 79–95.
27. Jensen, S.H., Hauch, A., Hendriksen, P.V., Mogensen, M., Bonanos, N., and Jacobsen, T. (2007) A method to separate process contributions in impedance spectra by variation of test conditions. *J. Electrochem. Soc.*, **154** (12), B1325–B1330.
28. Kong, J., Sun, K., Zhou, D., Zhang, N., Mu, J., and Qiao, J. (2007) Ni–YSZ gradient anodes for anode-supported SOFCs. *J. Power Sources*, **166** (2), 337–342.
29. Brown, M., Primdahl, S., Mogensen, M., and Sammes, N.M. (1998) Minimum active layer thickness for a Ni/YSZ cermet SOFC anode. *J. Australas. Ceram. Soc.*, **34** (1), 248–253.
30. Meulenberg, W.A., Menzler, N.H., Buchkremer, H.P., and Stöver, D. (2002) Manufacturing routes and state-of-the-art of the planar Jülich anode-supported concept for solid oxide fuel cells. *Ceram. Trans.*, **127**, 99–108.
31. Büchler, O., Bram, M., Mücke, R., and Buchkremer, H.P. (2009) Preparation of thin functional layers for anode supported SOFC by roll coating process. *ECS Trans.*, **25** (2), 655–663.
32. Waldbillig, D., Wood, A., and Ivey, D.G. (2005) Thermal analysis of the cyclic reduction and oxidation behaviour of SOFC anodes. *Solid State Ionics*, **176** (9–10), 847–859.
33. Muecke, U.P., Akiba, K., Infortuna, A., Salkus, T., Stus, N.V., and Gauckler, L.J. (2008) Electrochemical performance of nanocrystalline nickel/gadolinia-doped ceria thin film anodes for solid oxide fuel cells. *Solid State Ionics*, **178** (33–34), 1762–1768.
34. Zha, S., Rauch, W., and Liu, M. (2004) Ni-$Ce_{0.9}Gd_{0.1}O_{1.95}$ anode for GDC electrolyte-based low-temperature SOFCs. *Solid State Ionics*, **166** (3–4), 241–250.
35. Nakamura, T., Kobayashi, T., Yashiro, K., Kaimai, A., Otake, T., Sato, K., Mizusaki, J., and Kawada, T. (2008) Electrochemical behaviors of mixed conducting oxide anodes for solid oxide fuel cell. *J. Electrochem. Soc.*, **155** (6), B563–B569.
36. Holtappels, P., Bradley, J., Irvine, J.T.S., Kaiser, A., and Mogensen, M. (2001) Electrochemical characterization of ceramic SOFC anodes. *J. Electrochem. Soc.*, **148** (8), A923–A929.
37. Timmermann, H., Fouquet, D., Weber, A., Ivers-Tiffee, E., Hennings, U., and Reimert, R. (2006) Internal reforming of methane at Ni/YSZ and Ni/CGO SOFC cermet anodes. *Fuel Cells (Weinheim)*, **6** (3–4), 307–313.
38. Marina, O.A. and Mogensen, M. (1999) High-temperature conversion of methane on a composite gadolinia-doped ceria-gold electrode. *Appl. Catal., A: Gen.*, **189** (1), 117–126.
39. Marina, O.A., Bagger, C., Primdahl, S., and Mogensen, M. (1999) A solid oxide fuel cell with a gadolinia-doped ceria anode: preparation and performance. *Solid State Ionics*, **123** (1–4), 199.

40. Murray, E.P., Tsai, T., and Barnett, S.A. (1999) A direct-methane fuel cell with a ceria-based anode. *Nature*, **400** (6745), 649.
41. Grew, K.N., Peracchio, A.A., Joshi, A.S., Izzo, J.R. Jr., and Chiu, W.K.S. (2010) Characterization and analysis methods for the examination of the heterogeneous solid oxide fuel cell electrode microstructure. Part 1: volumetric measurements of the heterogeneous structure. *J. Power Sources*, **195** (24), 7930–7942.
42. Wilson, J.R., Kobsiriphat, W., Mendoza, R., Chen, H., Hiller, J.M., Miller, D.J., Thornton, K., Voorhees, P.W., Adler, S.B., and Barnett, S.A. (2006) Three-dimensional reconstruction of a solid-oxide fuel-cell anode. *Nat. Mater.*, **5** (7), 541–544.
43. Wilson, J.R. and Barnett, S.A. (2008) Solid oxide fuel cell Ni–YSZ anodes: effect of composition on microstructure and performance. *Electrochem. Solid-State Lett.*, **11** (10), B181–B185.
44. Shikazono, N., Kanno, D., Matsuzaki, K., Teshima, H., Sumino, S., and Kasagi, N. (2010) Numerical assessment of SOFC anode polarization based on three-dimensional model microstructure reconstructed from FIB-SEM images. *J. Electrochem. Soc.*, **157** (5), B665–B672.
45. Iwai, H., Shikazono, N., Matsui, T., Teshima, H., Kishimoto, M., Kishida, R., Hayashi, D., Matsuzaki, K., Kanno, D., Saito, M., Muroyama, H., Eguchi, K., Kasagi, N., and Yoshida, H. (2010) Quantification of SOFC anode microstructure based on dual beam FIB-SEM technique. *J. Power Sources*, **195** (4), 955–961.
46. Lee, J., Moon, H., Lee, H., Kim, J., Kim, J., and Yoon, K. (2002) Quantitative analysis of microstructure and its related electrical property of SOFC anode, Ni–YSZ cermet. *Solid State Ionics*, **148** (1–2), 15–26.
47. Vogler, M., Bieberle-Hutter, A., Gauckler, L., Warnatz, J., and Bessler, W.G. (2009) Modelling study of surface reactions, diffusion, and spillover at a Ni/YSZ patterned anode. *J. Electrochem. Soc.*, **156** (5), B663–B672.
48. Nakagawa, N., Sakurai, H., Kondo, K., Morimoto, T., Hatanaka, K., and Kato, K. (1995) Evaluation of the effective reaction zone at Ni(NiO)/zirconia anode by using an electrode with a novel structure. *J. Electrochem. Soc.*, **142** (10), 3474–3479.
49. Jiang, S.P. and Badwal, S.P.S. (1997) Hydrogen oxidation at the nickel and platinum electrodes on yttria-tetragonal zirconia electrolyte. *J. Electrochem. Soc.*, **144** (11), 3777–3784.
50. Mizusaki, J., Tagawa, H., Saito, T., Kamitani, K., Yamamura, T., Hirano, K., Ehara, S., Takagi, T., Hikita, T., Ippommatsu, M., Nakagawa, S., and Hashimoto, K. (1994) Preparation of nickel pattern electrodes on YSZ and their electrochemical properties in H_2-H_2O atmospheres. *J. Electrochem. Soc.*, **141** (8), 2129–2134.
51. Mizusaki, J., Tagawa, H., Saito, T., Yamamura, T., Kamitani, K., Hirano, K., Ehara, S., Takagi, T., Hikita, T., Ippommatsu, M., Nakagawa, S., and Hashimoto, K. (1994) Kinetic-studies of the reaction at the nickel pattern electrode on YSZ in H_2-H_2O atmospheres. *Solid State Ionics*, **70**, 52–58.
52. Goodwin, D.G., Zhu, H., Colclasure, A.M., and Kee, R.J. (2009) Modeling electrochemical oxidation of hydrogen on Ni–YSZ pattern anodes. *J. Electrochem. Soc.*, **156** (9), B1004–B1021.
53. Bieberle, A., Meier, L.P., and Gauckler, L.J. (2001) The electrochemistry of Ni pattern anodes used as solid oxide fuel cell model electrodes. *J. Electrochem. Soc.*, **148** (6), A646–A656.
54. Raz, S., Sasaki, K., Maier, J., and Riess, I. (2001) Characterization of adsorbed water layers on Y_2O_3-doped ZrO_2. *Solid State Ionics*, **143** (2), 181–204.
55. Bessler, W.G., Gewies, S., and Vogler, M. (2007) A new framework for physically based modeling of solid oxide fuel cells. *Electrochim. Acta*, **53** (4), 1782–1800.
56. Bessler, W.G., Vogler, M., Stormer, H., Gerthsen, D., Utz, A., Weber, A., and Ivers-Tiffee, E. (2010) Model anodes and anode models for understanding the mechanism of hydrogen oxidation

in solid oxide fuel cells. *Phys. Chem. Chem. Phys.*, **12** (42), 13888–13903.

57. Bessler, W.G., Warnatz, J., and Goodwin, D.G. (2007) The influence of equilibrium potential on the hydrogen oxidation kinetics of SOFC anodes. *Solid State Ionics*, **177** (39–40), 3371–3383.
58. Shishkin, M. and Ziegler, T. (2009) Oxidation of H_2, CH_4, and CO molecules at the interface between nickel and yttria-stabilized zirconia: a theoretical study based on DFT. *J. Phys. Chem. C*, **113** (52), 21667–21678.
59. Shishkin, M. and Ziegler, T. (2010) Hydrogen oxidation at the Ni/yttria-stabilized zirconia interface: a study based on density functional theory. *J. Phys. Chem. C*, **114** (25), 11209–11214.
60. Clarke, S.H., Dicks, A.L., Pointon, K., Smith, T.A., and Swann, A. (1997) Catalytic aspects of the steam reforming of hydrocarbons in internal reforming fuel cells. *Catal. Today*, **38** (4), 411–423.
61. Zhan, Z. and Barnett, S.A. (2005) An octane-fueled solid oxide fuel cell. *Science*, **308** (5723), 844–847.
62. Liu, J.A. and Barnett, S.A. (2003) Operation of anode-supported solid oxide fuel cells on methane and natural gas. *Solid State Ionics*, **158** (1–2), 11–16.
63. Offer, G.J., Mermelstein, J., Brightman, E., and Brandon, N.P. (2009) Thermodynamics and kinetics of the interaction of carbon and sulfur with solid oxide fuel cell anodes. *J. Am. Ceram. Soc.*, **92** (4), 763–780.
64. Pomfret, M.B., Marda, J., Jackson, G.S., Eichhorn, B.W., Dean, A.M., and Walker, R.A. (2008) Hydrocarbon fuels in solid oxide fuel cells: in situ Raman studies of graphite formation and oxidation. *J. Phys. Chem. C*, **112** (13), 5232–5240.
65. Aguiar, P., Adjiman, C.S., and Brandon, N.P. (2005) Anode-supported intermediate-temperature direct internal reforming solid oxide fuel cell: II. Model-based dynamic performance and control. *J. Power Sources*, **147** (1–2), 136–147.
66. Klein, J., Bultel, Y., Georges, S., and Pons, M. (2007) Modeling of a SOFC fuelled by methane: from direct internal reforming to gradual internal reforming. *Chem. Eng. Sci.*, **62** (6), 1636–1649.
67. Powell, M., Meinhardt, K., Sprenkle, V., Chick, L., and McVay, G. (2012) Demonstration of a highly efficient solid oxide fuel cell power system using adiabatic steam reforming and anode gas recirculation. *J. Power Sources*, **205**, 377–384.
68. Abudula, A., Ihara, M., Komiyama, H., and Yamada, K. (1996) Oxidation mechanism and effective anode thickness of SOFC for dry methane fuel. *Solid State Ionics*, **86–88** Part 2, 1203–1209.
69. Horita, T., Sakai, N., Kawada, T., Yokokawa, H., and Dokiya, M. (1996) Oxidation and steam reforming of CH_4 on Ni and Fe anodes under low humidity conditions in solid oxide fuel cells. *J. Electrochem. Soc.*, **143** (4), 1161–1168.
70. Holtappels, P., De Haart, L.G.J., and Stimming, U. (1999) Reaction of CO/CO_2 gas mixtures on Ni-YSZ cermet electrodes. *J. Appl. Electrochem.*, **29** (5), 561–568.
71. Matsuzaki, Y. and Yasuda, I. (2000) Electrochemical oxidation of H_2 and CO in a H_2-H_2O-CO-CO_2 system at the interface of a Ni-YSZ cermet electrode and YSZ electrolyte. *J. Electrochem. Soc.*, **147** (5), 1630–1635.
72. Weber, A., Sauer, B., Müller, A.C., Herbstritt, D., and Ivers-Tiffée, E. (2002) Oxidation of H_2, CO and methane in SOFCs with Ni/YSZ-cermet anodes. *Solid State Ionics*, **152–153**, 543–550.
73. Zhan, Z., Lin, Y., Pillai, M., Kim, I., and Barnett, S.A. (2006) High-rate electrochemical partial oxidation of methane in solid oxide fuel cells. *J. Power Sources*, **161** (1), 460–465.
74. Buccheri, M.A. and Hill, J.M. (2012) Methane electrochemical oxidation pathway over a Ni/YSZ and $La_{0.3}Sr_{0.7}TiO_3$ bi-layer SOFC anode. *J. Electrochem. Soc.*, **159** (4), B361–B367.
75. Kleis, J., Jones, G., Abild-Pedersen, F., Tripkovic, V., Bligaard, T., and Rossmeisl, J. (2009) Trends for

methane oxidation at solid oxide fuel cell conditions. *J. Electrochem. Soc.*, **156** (12), B1447–B1456.

76. Pillai, M.R., Jiang, Y., Mansourian, N., Kim, I., Bierschenk, D.M., Zhu, H.Y., Kee, R.J., and Barnett, S.A. (2008) Solid oxide fuel cell with oxide anode-side support. *Electrochem. Solid State Lett.*, **11** (10), B174–B177.
77. Hao, Y. and Goodwin, D.G. (2008) Numerical study of heterogeneous reactions in an SOFC anode with oxygen addition. *J. Electrochem. Soc.*, **155** (7), B666–B674.
78. McIntosh, S., Vohs, J.M., and Gorte, R.J. (2003) Role of hydrocarbon deposits in the enhanced performance of direct-oxidation SOFCs. *J. Electrochem. Soc.*, **150** (4), A470–A476.
79. McIntosh, S., Vohs, J.M., and Gorte, R.J. (2002) An examination of lanthanide additives on the performance of Cu-YSZ cermet anodes. *Electrochim. Acta*, **47** (22–23), 3815–3821.
80. Pomfret, M.B., Owrutsky, J.C., and Walker, R.A. (2010) In situ optical studies of solid-oxide fuel cells. *Annu. Rev. Anal. Chem.*, **3** (1), 151–174.
81. McIntosh, S., He, H.P., Lee, S.I., Costa-Nunes, O., Krishnan, V.V., Vohs, J.M., and Gorte, R.J. (2004) An examination of carbonaceous deposits in direct-utilization SOFC anodes. *J. Electrochem. Soc.*, **151** (4), A604–A608.
82. Toh, C.H., Munroe, P.R., Young, D.J., and Foger, K. (2003) High temperature carbon corrosion in solid oxide fuel cells. *Mater. High Temp.*, **20** (2), 129–136.
83. Kim, T., Moon, S., and Hong, S. (2002) Internal carbon dioxide reforming by methane over Ni-YSZ-CeO_2 catalyst electrode in electrochemical cell. *Appl. Catal., A: Gen.*, **224** (1–2), 111–120.
84. Sukeshini, A.M., Habibzadeh, B., Becker, B.P., Stoltz, C.A., Eichhorn, B.W., and Jackson, G.S. (2006) Electrochemical oxidation of H_2, CO, and CO/H_2 mixtures on patterned Ni anodes on YSZ electrolytes. *J. Electrochem. Soc.*, **153** (4), A705–A715.
85. Hecht, E.S., Gupta, G.K., Zhu, H., Dean, A.M., Kee, R.J., Maier, L., and Deutschmann, O. (2005) Methane reforming kinetics within a Ni–YSZ SOFC anode support. *Appl. Catal., A: Gen.*, **295** (1), 40–51.
86. Zhu, H., Kee, R.J., Janardhanan, V.M., Deutschmann, O., and Goodwin, D.G. (2005) Modeling elementary heterogeneous chemistry and electrochemistry in solid-oxide fuel cells. *J. Electrochem. Soc.*, **152** (12), A2427–A2440.
87. Zhu, H. and Kee, R.J. (2008) Modeling distributed charge-transfer processes in SOFC membrane electrode assemblies. *J. Electrochem. Soc.*, **155** (7), B715–B729.
88. Zhu, H. and Kee, R.J. (2006) Modeling electrochemical impedance spectra in SOFC button cells with internal methane reforming. *J. Electrochem. Soc.*, **153** (9), A1765–A1772.
89. Matsuzaki, Y. and Yasuda, I. (2000) The poisoning effect of sulfur-containing impurity gas on a SOFC anode: part I. Dependence on temperature, time, and impurity concentration. *Solid State Ionics*, **132** (3–4), 261–269.
90. Rasmussen, J.F.B. and Hagen, A. (2009) The effect of H_2S on the performance of Ni–YSZ anodes in solid oxide fuel cells. *J. Power Sources*, **191** (2), 534–541.
91. Kromp, A., Dierickx, S., Leonide, A., Weber, A., and Ivers-Tiffee, E. (2012) Electrochemical analysis of sulfur-poisoning in anode supported SOFCs fuelled with a model reformate. *J. Electrochem. Soc.*, **159** (5), B597–B601.
92. Cheng, Z. and Liu, M. (2007) Characterization of sulfur poisoning of Ni–YSZ anodes for solid oxide fuel cells using in situ Raman microspectroscopy. *Solid State Ionics*, **178** (13–14), 925–935.
93. Rostrup-Nielsen, J.R., Hansen, J.B., Helveg, S., Christiansen, N., and Jannasch, A.K. (2006) Sites for catalysis and electrochemistry in solid oxide fuel cell (SOFC) anode. *Appl. Phys. A: Mater. Sci. Process.*, **V85** (4), 427.
94. Marina, O.A., Coyle, C.A., Thomsen, E.C., Edwards, D.J., Coffey, G.W., and Pederson, L.R. (2010) Degradation mechanisms of SOFC anodes in coal gas containing phosphorus. *Solid State Ionics*, **181** (8–10), 430–440.

95. Coyle, C.A., Marina, O.A., Thomsen, E.C., Edwards, D.J., Cramer, C.D., Coffey, G.W., and Pederson, L.R. (2009) Interactions of nickel/zirconia solid oxide fuel cell anodes with coal gas containing arsenic. *J. Power Sources*, **193** (2), 730–738.

96. Liu, Y.L., Primdahl, S., and Mogensen, M. (2003) Effects of impurities on microstructure in Ni/YSZ–YSZ half-cells for SOFC. *Solid State Ionics*, **161** (1–2), 1–10.

97. Liu, Y.L. and Jiao, C. (2005) Microstructure degradation of an anode/electrolyte interface in SOFC studied by transmission electron microscopy. *Solid State Ionics*, **176** (5–6), 435–442.

98. Utz, A., Hansen, K.V., Norrman, K., Ivers-Tiffée, E., and Mogensen, M. (2011) Impurity features in Ni-YSZ-H_2-H_2O electrodes. *Solid State Ionics*, **183** (1), 60–70.

99. Mogensen, M., Jensen, K.V., Jørgensen, M.J., and Primdahl, S. (2002) Progress in understanding SOFC electrodes. *Solid State Ionics*, **150** (1–2), 123–129.

100. Adler, S.B. (2004) Factors governing oxygen reduction in solid oxide fuel cell cathodes. *Chem. Rev.*, **104** (10), 4791.

101. Park, S., Vohs, J.M., and Gorte, R.J. (2000) Direct oxidation of hydrocarbons in a solid-oxide fuel cell. *Nature*, **404** (6775), 265.

102. Park, S., Gorte, R.J., and Vohs, J.M. (2001) Tape cast solid-oxide fuel cells for the direct oxidation of hydrocarbons. *J. Electrochem. Soc.*, **148** (5), A443–A447.

103. Kim, H., Park, S., Vohs, J.M., and Gorte, R.J. (2001) Direct oxidation of liquid fuels in a solid oxide fuel cell. *J. Electrochem. Soc.*, **148** (7), A693–A695.

104. Tao, S.W. and Irvine, J.T.S. (2003) A redox-stable efficient anode for solid-oxide fuel cells. *Nat. Mater.*, **2** (5), 320–323.

105. Huang, Y.H., Dass, R.I., Xing, Z.L., and Goodenough, J.B. (2006) Double perovskites as anode materials for solid-oxide fuel cells. *Science*, **312** (5771), 254–257.

106. Kolodiazhnyi, T. and Petric, A. (2005) The applicability of Sr-deficient n-type $SrTiO_3$ for SOFC anodes. *J. Electroceram.*, **15** (1), 5–11.

107. Marina, O.A., Canfield, N.L., and Stevenson, J.W. (2002) Thermal, electrical, and electrocatalytical properties of lanthanum-doped strontium titanate. *Solid State Ionics*, **149** (1–2), 21–28.

108. Porat, O., Heremans, C., and Tuller, H.L. (1997) Stability and mixed ionic electronic conduction in $Gd_2(Ti_{1-x}Mox)_2O_7$ under anodic conditions. *Solid State Ionics*, **94** (1–4), 75.

109. Kaiser, A., Bradley, J.L., Slater, P.R., and Irvine, J.T.S. (2000) Tetragonal tungsten bronze type phases $(Sr_{1-x}Bax)_{0.6}Ti_{0.2}Nb_{0.8}O_{3-\delta}$: material characterisation and performance as SOFC anodes. *Solid State Ionics*, **135** (1–4), 519–524.

110. Slater, P.R. and Irvine, J.T.S. (1999) Niobium based tetragonal tungsten bronzes as potential anodes for solid oxide fuel cells: synthesis and electrical characterisation. *Solid State Ionics*, **120** (1–4), 125–134.

111. Cheng, Z., Zha, S.W., Aguilar, L., and Liu, M.L. (2005) Chemical, electrical, and thermal properties of strontium doped lanthanum vanadate. *Solid State Ionics*, **176** (23–24), 1921–1928.

112. Cheng, Z., Zha, S.W., Aguilar, L., Wang, D., Winnick, J., and Liu, M.L. (2006) A solid oxide fuel cell running on H_2S/CH_4 fuel mixtures. *Electrochem. Solid State Lett.*, **9** (1), A31–A33.

113. Aguilar, L., Zha, S., Cheng, Z., Winnick, J., and Liu, M. (2004) A solid oxide fuel cell operating on hydrogen sulfide (H_2S) and sulfur-containing fuels. *J. Power Sources*, **135** (1–2), 17–24.

114. Mizusaki, J. (1992) Nonstoichiometry, diffusion, and electrical properties of perovskite-type oxide electrode materials. *Solid State Ionics*, **52** (1–3), 79–91.

115. Jacobson, A.J. (2010) Materials for solid oxide fuel cells†. *Chem. Mater.*, **22** (3), 660–674.

116. Riess, I. (1997) in *CRC Handbook of Solid State Electrochemistry* (eds P.J. Gellings and H.J.M. Bouwmeester), CRC Press, Boca Raton, FL, pp. 223–294.
117. Gellings, P.J. and Bouwmeester, H.J.M. (1992) Ion and mixed conducting oxides as catalysts. *Catal. Today*, **12** (1), 1.
118. Steele, B.C.H., Middleton, P.H., and Rudkin, R.A. (1990) Material science aspects of SOFC technology with special reference to anode development. *Solid State Ionics*, **40–41**(Part 1), 388.
119. Jung, S., Lu, C., He, H., Ahn, K., Gorte, R.J., and Vohs, J.M. (2006) Influence of composition and Cu impregnation method on the performance of Cu/CeO_2/YSZ SOFC anodes. *J. Power Sources*, **154** (1), 42–50.
120. Gross, M.D., Vohs, J.M., and Gorte, R.J. (2007) Recent progress in SOFC anodes for direct utilization of hydrocarbons. *J. Mater. Chem.*, **17** (30), 3071–3077.
121. Kim, G., Gross, M.D., Wang, W., Vohs, J.M., and Gorte, R.J. (2008) SOFC anodes based on LST-YSZ composites and on $Y_{0.04}Ce_{0.48}Zr_{0.48}O_2$. *J. Electrochem. Soc.*, **155** (4), B360–B366.
122. Gross, M.D., Vohs, J.M., and Gorte, R.J. (2007) An examination of SOFC anode functional layers based on ceria in YSZ. *J. Electrochem. Soc.*, **154** (7), B694–B699.
123. Anderson, H.U. (1992) Review of p-type doped perovskite materials for SOFC and other applications. *Solid State Ionics*, **52** (1–3), 33.
124. Pena, M.A. and Fierro, J.L.G. (2001) Chemical structures and performance of perovskite oxides. *Chem. Rev.*, **101** (7), 1981–2017.
125. Tao, S.W. and Irvine, J.T.S. (2008) Structural and electrochemical properties of the perovskite oxide $Pr_{0.7}Sr_{0.3}Cr_{0.9}Ni_{0.1}O_{3-\delta}$. *Solid State Ionics*, **179** (19–20), 725–731.
126. Kharton, V.V., Tsipis, E.V., Marozau, I.P., Viskup, A.P., Frade, J.R., and Irvine, J.T.S. (2007) Mixed conductivity and electrochemical behavior of $(La_{0.75}Sr_{0.25})(0.95)Cr_{0.5}Mn_{0.5}O_{3-\delta}$. *Solid State Ionics*, **178** (1–2), 101–113.
127. Plint, S.M., Connor, P.A., Tao, S., and Irvine, J.T.S. (2006) Electronic transport in the novel SOFC anode material $La_{1-x}Sr_xCr_{0.5}Mn_{0.5}O_{3+/-d}$. *Solid State Ionics*, **177** (19–25), 2005–2008.
128. Fonseca, F.C., Muccillo, E.N.S., Muccillo, R., and de Florio, D.Z. (2008) Synthesis and electrical characterization of the ceramic anode $La_{1-x}Sr_xMn_{0.5}Cr_{0.5}O_3$. *J. Electrochem. Soc.*, **155** (5), B483–B487.
129. Zha, S.W., Tsang, P., Cheng, Z., and Liu, M.L. (2005) Electrical properties and sulfur tolerance of $La_{0.75}Sr_{0.25}Cr_{1-x}Mn_xO_3$ under anodic conditions. *J. Solid State Chem.*, **178** (6), 1844–1850.
130. Moos, R., Bischoff, T., Menesklou, W., and Hardtl, K. (1997) Solubility of lanthanum in strontium titanate in oxygen-rich atmospheres. *J. Mater. Sci.*, **32** (16), 4247–4252.
131. Moos, R. and Hardtl, K.H. (1997) Defect chemistry of donor-doped and undoped strontium titanate ceramics between 1000° and 1400°C. *J. Am. Ceram. Soc.*, **80** (10), 2549–2562.
132. Huang, X.L., Zhao, H.L., Shen, W., Qiu, W.H., and Wu, W.J. (2006) Effect of fabrication parameters on the electrical conductivity of $Y_xSr_{1-x}TiO_3$ for anode materials. *J. Phys. Chem. Solids*, **67** (12), 2609–2613.
133. Balachandran, U. and Eror, N.G. (1982) Electrical conductivity in lanthanum-doped strontium titanate. *J. Electrochem. Soc.*, **129** (5), 1021–1026.
134. Hui, S.Q. and Petric, A. (2002) Evaluation of yttrium-doped $SrTiO_3$ as an anode for solid oxide fuel cells. *J. Eur. Ceram. Soc.*, **22** (9–10), 1673–1681.
135. Fu, Q.X., Mi, S.B., Wessel, E., and Tietz, F. (2008) Influence of sintering conditions on microstructure and electrical conductivity of yttrium-substituted $SrTiO_3$. *J. Eur. Ceram. Soc.*, **28** (4), 811–820.
136. Flandermeyer, B.F., Agarwal, A.K., Anderson, H.U., and Nasrallah, M.M. (1984) Oxidation-reduction behaviour of La-doped $SrTiO_3$. *J. Mater. Sci.*, **19** (8), 2593–2598.
137. Battle, P.D., Bennett, J.E., Sloan, J., Tilley, R.J.D., and Vente, J.F. (2000)

A-site cation-vacancy ordering in $Sr_{1-3x/2}La_xTiO_3$: a study by HRTEM. *J. Solid State Chem.*, **149** (2), 360–369.
138. Meyer, R., Waser, R., Helmbold, J., and Borchardt, G. (2002) Cationic surface segregation in donor-doped $SrTiO_3$ under oxidizing conditions. *J. Electroceram.*, **9** (2), 101–110.
139. Slater, P.R., Fagg, D.P., and Irvine, J.T.S. (1997) Synthesis and electrical characterisation of doped perovskite titanates as potential anode materials for solid oxide fuel cells. *J. Mater. Chem.*, **7** (12), 2495–2498.
140. Mitchell, B.J., Rogan, R.C., Richardson, J.W., Ma, B., and Balachandran, U. (2002) Stability of the cubic perovskite $SrFe_{0.8}Co_{0.2}O_{3-\delta}$. *Solid State Ionics*, **146** (3–4), 313–321.
141. Blennow, P., Hansen, K.K., Reine Wallenberg, L., and Mogensen, M. (2006) Effects of Sr/Ti-ratio in $SrTiO_3$-based SOFC anodes investigated by the use of cone-shaped electrodes. *Electrochim. Acta*, **52** (4), 1651–1661.
142. Blennow, P., Hagen, A., Hansen, K.K., Wallenberg, L.R., and Mogensen, M. (2008) Defect and electrical transport properties of Nb-doped $SrTiO_3$. *Solid State Ionics*, **179** (35–36), 2047–2058.
143. Blennow, P., Hansen, K.K., Wallenberg, L.R., and Mogensen, M. (2009) Electrochemical characterization and redox behavior of Nb-doped $SrTiO_3$. *Solid State Ionics*, **180** (1), 63.
144. Blennow, P., Hansen, K.K., Wallenberg, L.R., and Mogensen, M. (2007) Synthesis of Nb-doped $SrTiO_3$ by a modified glycine-nitrate process. *J. Eur. Ceram. Soc.*, **27** (13–15), 3609–3612.
145. Hashimoto, S., Poulsen, F.W., and Mogensen, M. (2007) Conductivity of $SrTiO_3$ based oxides in the reducing atmosphere at high temperature. *J. Alloys Compd.*, **439** (1–2), 232–236.
146. Huang, Y.H., Dass, R.I., Denyszyn, J.C., and Goodenough, J.B. (2006) Synthesis and characterization of Sr_2MgMoO_{6-d} – An anode material for the solid oxide fuel cell. *J. Electrochem. Soc.*, **153** (7), A1266–A1272.
147. Marrero-López, D., Peña-Martínez, J., Ruiz-Morales, J.C., Gabás, M., Núñez, P., Aranda, M.A.G., and Ramos-Barrado, J.R. (2009) Redox behaviour, chemical compatibility and electrochemical performance of Sr_2MgMoO_{6-d} as SOFC anode. *Solid State Ionics*, **180** (40), 1672.
148. Liu, Q., Dong, X., Xiao, G., Zhao, F., and Chen, F. (2010) A novel electrode material for symmetrical SOFCs. *Adv. Mater.*, **22** (48), 5478–5482.
149. Sprague, J.J. and Tuller, H.L. (1999) Mixed ionic and electronic conduction in Mn/Mo doped gadolinium titanate. *J. Eur. Ceram. Soc.*, **19** (6–7), 803–806.
150. Ruiz-Morales, J.C., Canales-Vazquez, J., Marrero-Lopez, D., Irvine, J.T.S., and Nunez, P. (2007) Improvement of the electrochemical properties of novel solid oxide fuel cell anodes, $La_{0.75}Sr_{0.25}Cr_{0.5}Mn_{0.5}O_{3-\delta}$ and $La_4Sr_8Ti_{11}Mn_{0.5}Ga_{0.5}O_{37.5-\delta}$, using Cu-YSZ-based cermets. *Electrochim. Acta*, **52** (25), 7217–7225.
151. Lu, X.C. and Zhu, J.H. (2007) Cu(Pd)-impregnated $La_{0.75}Sr_{0.25}Cr_{0.5}Mn_{0.5}O_{3-\delta}$ anodes for direct utilization of methane in SOFC. *Solid State Ionics*, **178** (25–26), 1467–1475.
152. Lee, S., Vohs, J.M., and Gorte, R.J. (2004) A study of SOFC anodes based on Cu-Ni and Cu-Co bimetallics in CeO_2-YSZ. *J. Electrochem. Soc.*, **151** (9), A1319–A1323.
153. Kim, H., Lu, C., Worrell, W.L., Vohs, J.M., and Gorte, R.J. (2002) Cu-Ni cermet anodes for direct oxidation of methane in solid-oxide fuel cells. *J. Electrochem. Soc.*, **149** (3), A247–A250.
154. Gorte, R.J., Vohs, J.M., and McIntosh, S. (2004) Recent developments on anodes for direct fuel utilization in SOFC. *Solid State Ionics*, **175** (1–4), 1–6.
155. McIntosh, S., Vohs, J.M., and Gorte, R.J. (2003) Effect of precious-metal dopants on SOFC anodes for direct utilization of hydrocarbons. *Electrochem. Solid State Lett.*, **6** (11), A240–A243.
156. Bruce, M.K., van den Bossche, M., and McIntosh, S. (2008) The influence of current density on the electrocatalytic

activity of oxide-based direct hydrocarbon SOFC anodes. *J. Electrochem. Soc.*, **155** (11), B1202–B1209.

157. Tao, S.W., Irvine, J.T.S., and Plint, S.M. (2006) Methane oxidation at redox stable fuel cell electrode $La_{0.75}Sr_{0.25}Cr_{0.5}Mn_{0.5}O_{3-\delta}$. *J. Phys. Chem. B*, **110** (43), 21771–21776.
158. van den Bossche, M. and McIntosh, S. (2008) Rate and selectivity of methane oxidation over $La_{0.75}Sr_{0.25}Cr_xMn_{1-x}O_{3-\delta}$ as a function of lattice oxygen stoichiometry under solid oxide fuel cell anode conditions. *J. Catal.*, **255** (2), 313–323.
159. Doshi, R., Alcock, C.B., Gunasekaran, N., and Carberry, J.J. (1993) Carbon-monoxide and methane oxidation properties of oxide solid-solution catalysts. *J. Catal.*, **140** (2), 557–563.
160. Yamazoe, N. and Teraoka, Y. (1990) Oxidation catalysis of perovskites — relationships to bulk structure and composition (valency, defect, etc.). *Catal. Today*, **8** (2), 175.
161. Gellings, P.J. and Bouwmeester, H.J.M. (2000) Solid state aspects of oxidation catalysis. *Catal. Today*, **58** (1), 1–53.
162. van den Bossche, M. and McIntosh, S. (2010) Pulse reactor studies to assess the potential of $La_{0.75}Sr_{0.25}Cr_{0.5}Mn_{0.4}X_{0.1}O_{3-\delta}$ (X = Co, Fe, Mn, Ni, V) as direct hydrocarbon solid oxide fuel cell anodes. *Chem. Mater.*, **22**, 5856–5865.
163. Sfeir, J., Buffat, P.A., Mockli, P., Xanthopoulos, N., Vasquez, R., Mathieu, H.J., Van herle, J., and Thampi, K.R. (2001) Lanthanum chromite based catalysts for oxidation of methane directly on SOFC anodes. *J. Catal.*, **202** (2), 229–244.
164. Danilovic, N., Vincent, A., Luo, J., Chuang, K.T., Hui, R., and Sanger, A.R. (2009) Correlation of fuel cell anode electrocatalytic and ex situ catalytic activity of perovskites $La_{0.75}Sr_{0.25}Cr_{0.5}X_{0.5}O_{3-\delta}$ (X = Ti, Mn, Fe, Co). *Chem. Mater.*, **22** (3), 957–965.
165. Primdahl, S., Hansen, J.R., Grahl-Madsen, L., and Larsen, P.H. (2001) Sr-doped $LaCrO_3$ anode for solid oxide fuel cells. *J. Electrochem. Soc.*, **148** (1), A74–A81.
166. Vernoux, P., Djurado, E., and Guillodo, M. (2001) Catalytic and electrochemical properties of doped lanthanum chromites as new anode materials for solid oxide fuel cells. *J. Am. Ceram. Soc.*, **84** (10), 2289–2295.
167. Kobsiriphat, W., Madsen, B.D., Wang, Y., Marks, L.D., and Barnett, S.A. (2009) $La_{0.8}Sr_{0.2}Cr_{1-x}Ru_xO_{3-\delta}Gd_{0.1}Ce_{0.9}O_{1.95}$ solid oxide fuel cell anodes: Ru precipitation and electrochemical performance. *Solid State Ionics*, **180** (2–3), 257.
168. Madsen, B.D., Kobsiriphat, W., Wang, Y., Marks, L.D., and Barnett, S.A. (2007) Nucleation of nanometer-scale electrocatalyst particles in solid oxide fuel cell anodes. *J. Power Sources*, **166** (1), 64–67.
169. Tao, S.W. and Irvine, J.T.S. (2004) Catalytic properties of the perovskite oxide $La_{0.75}Sr_{0.25}Cr_{0.5}Fe_{0.5}O_{3-\delta}$ in relation to its potential as a solid oxide fuel cell anode material. *Chem. Mater.*, **16** (21), 4116–4121.
170. Jardiel, T., Caldes, M.T., Moser, F., Hamon, J., Gauthier, G., and Joubert, O. (2010) New SOFC electrode materials: the Ni-substituted LSCM-based compounds $(La_{0.75}Sr_{0.25})(Cr_{0.5}Mn_{0.5-x}Ni_x)O_{3-\delta}$ and $(La_{0.75}Sr_{0.25})(Cr_{0.5-x}Ni_xMn_{0.5})O_{3-\delta}$. *Solid State Ionics*, **181** (19–20), 894.
171. Pudmich, G., Boukamp, B.A., Gonzalez-Cuenca, M., Jungen, W., Zipprich, W., and Tietz, F. (2000) Chromite/titanate based perovskites for application as anodes in solid oxide fuel cells. *Solid State Ionics*, **135** (1–4), 433.
172. Kim, G., Lee, S., Shin, J.Y., Corre, G., Irvine, J.T.S., Vohs, J.M., and Gorte, R.J. (2009) Investigation of the structural and catalytic requirements for high-performance SOFC anodes formed by infiltration of LSCM. *Electrochem. Solid State Lett.*, **12** (3), B48–B52.
173. Liu, J., Madsen, B.D., Ji, Z.Q., and Barnett, S.A. (2002) A fuel-flexible ceramic-based anode for solid oxide fuel cells. *Electrochem. Solid State Lett.*, **5** (6), A122–A124.

174. Jiang, S.P., Ye, Y.M., He, T.M., and Ho, S.B. (2008) Nanostructured palladium-$La_{0.75}Sr_{0.25}Cr_{0.5}Mn_{0.5}O_3/Y_2O_3$-$ZrO_2$ composite anodes for direct methane and ethanol solid oxide fuel cells. *J. Power Sources*, **185** (1), 179–182.

175. Ye, Y.M., He, T.M., Li, Y., Tang, E.H., Reitz, T.L., and Jiang, S.P. (2008) Pd-promoted $La_{0.75}Sr_{0.25}Cr_{0.5}Mn_{0.5}O_3$/YSZ composite anodes for direct utilization of methane in SOFCs. *J. Electrochem. Soc.*, **155** (8), B811–B818.

176. Lee, S., Kim, G., Vohs, J.M., and Gorte, R.J. (2008) SOFC anodes based on infiltration of $La_{0.3}Sr_{0.7}TiO_3$. *J. Electrochem. Soc.*, **155** (11), B1179–B1183.

177. Stevenson, J.W., Armstrong, T.R., Carneim, R.D., Pederson, L.R., and Weber, W.J. (1996) Electrochemical properties of mixed conducting perovskites $La_{1-x}M_xCo_{1-y}Fe_yO_{3-\delta}$ (M = Sr, Ba, Ca). *J. Electrochem. Soc.*, **143** (9), 2722.

178. Fu, Q.X., Tietz, F., Sebold, D., Tao, S.W., and Irvine, J.T.S. (2007) An efficient ceramic-based anode for solid oxide fuel cells. *J. Power Sources*, **171** (2), 663–669.

179. Yang, L.M., DeJonghe, L.C., Jacobsen, C.P., and Visco, S.J. (2007) B-site doping and catalytic activity of $Sr(Y)TiO_3$. *J. Electrochem. Soc.*, **154** (9), B949–B955.

180. Vincent, A., Luo, J., Chuang, K.T., and Sanger, A.R. (2010) Effect of Ba doping on performance of LST as anode in solid oxide fuel cells. *J. Power Sources*, **195** (3), 769.

181. Sun, X., Wang, S., Wang, Z., Ye, X., Wen, T., and Huang, F. (2008) Anode performance of LST–$xCeO_2$ for solid oxide fuel cells. *J. Power Sources*, **183** (1), 114–117.

182. Sun, X., Wang, S., Wang, Z., Qian, J., Wen, T., and Huang, F. (2009) Evaluation of $Sr_{0.88}Y_{0.08}TiO_3$–CeO_2 as composite anode for solid oxide fuel cells running on CH_4 fuel. *J. Power Sources*, **187** (1), 85–89.

183. Fu, Q.X., Tietz, F., and Stover, D. (2006) $La_{0.4}Sr_{0.6}Ti_{1-x}Mn_xO_{3-\delta}$ perovskites as anode materials for solid oxide fuel cells. *J. Electrochem. Soc.*, **153** (4), D74–D83.

184. Fu, Q.X., Tietz, F., Lersch, P., and Stover, D. (2006) Evaluation of Sr- and Mn-substituted $LaAlO_3$ as potential SOFC anode materials. *Solid State Ionics*, **177** (11–12), 1059–1069.

185. Ruiz-Morales, J., Canales-Vázquez, J., Savaniu, C., Marrero-López, D., Zhou, W., and Irvine, J.T.S. (2006) Disruption of extended defects in solid oxide fuel cell anodes for methane oxidation. *Nature*, **439** (7076), 568–571.

186. He, B., Zhao, L., Song, S., Liu, T., Chen, F., and Xia, C. (2012) $Sr_2Fe_{1.5}Mo_{0.5}O_{6-d}$ – $Sm_{0.2}Ce_{0.8}O_{1.9}$ composite anodes for intermediate-temperature solid oxide fuel cells. *J. Electrochem. Soc.*, **159** (5), B619–B626.

187. Bossche, M.v.d. and McIntosh, S. (2011) On the methane oxidation activity of $Sr_2(MgMo)_2O_{6-d}$: a potential anode material for direct hydrocarbon solid oxide fuel cells. *J. Mater. Chem.*, **21** (20), 7443–7451.

188. Bi, Z.H. and Zhu, J.H. (2011) Effect of current collecting materials on the performance of the double-perovskite Sr_2MgMoO_{6-d} anode. *J. Electrochem. Soc.*, **158** (6), B605–B613.

189. Marrero-Lopez, D., Pena-Martinez, J., Ruiz-Morales, J.C., Gabas, M., Nunez, P., Aranda, M.A.G., and Ramos-Barrado, J.R. (2010) Redox behaviour, chemical compatibility and electrochemical performance of $Sr_2MgMoO_{6-\delta}$ as SOFC anode. *Solid State Ionics*, **180** (40), 1672–1682.

190. Xie, Z., Zhao, H., Du, Z., Chen, T., Chen, N., Liu, X., and Skinner, S.J. (2012) Effects of Co doping on the electrochemical performance of double perovskite oxide $Sr_2MgMoO_{6-\delta}$ as an anode material for solid oxide fuel cells. *J. Phys. Chem. C*, **116** (17), 9734–9743.

191. Huang, Y., Liang, G., Croft, M., Lehtimaki, M., Karppinen, M., and Goodenough, J.B. (2009) Double-perovskite anode materials Sr_2MMoO_6 (M = Co, Ni) for solid oxide fuel cells. *Chem. Mater.*, **21** (11), 2319–2326.

192. Vasala, S., Lehtimäki, M., Huang, Y.H., Yamauchi, H., Goodenough, J.B., and

Karppinen, M. (2010) Degree of order and redox balance in B-site ordered double-perovskite oxides, $Sr_2MMoO_{6-\delta}$ (M = Mg, Mn, Fe, Co, Ni, Zn). *J. Solid State Chem.*, **183** (5), 1007.

193. Marrero-Lopez, D., Pena-Martinez, J., Ruiz-Morales, J.C., Martin-Sedeno, M.C., and Nunez, P. (2009) High temperature phase transition in SOFC anodes based on $Sr_2MgMoO_{6-\delta}$. *J. Solid State Chem.*, **182** (5), 1027–1034.

194. Torabi, A. and Etsell, T.H. (2012) Ni modified WC-based anode materials for direct methane solid oxide fuel cells. *J. Electrochem. Soc.*, **159** (6), B714–B722.

195. Torabi, A., Etsell, T.H., Semagina, N., and Sarkar, P. (2012) Electrochemical behaviour of tungsten carbide-based materials as candidate anodes for solid oxide fuel cells. *Electrochim. Acta*, **67**, 172–180.

196. McIntosh, S. and van den Bossche, M. (2011) Influence of lattice oxygen stoichiometry on the mechanism of methane oxidation in SOFC anodes. *Solid State Ionics*, **192** (1), 453–457.

197. van den Bossche, M., Matthews, R., Lichtenberger, A., and McIntosh, S. (2010) Insights into the fuel oxidation mechanism of $La_{0.75}Sr_{0.25}Cr_{0.5}Mn_{0.5}O_{3-\delta}$ SOFC anodes. *J. Electrochem. Soc.*, **157** (3), B392–B399.

198. Kim, H., Vohs, J.M., and Gorte, R.J. (2001) Direct oxidation of sulfur-containing fuels in a solid oxide fuel cell. *Chem. Commun.*, **22**, 2334–2335.

199. He, H., Gorte, R.J., and Vohs, J.M. (2005) Highly sulfur tolerant Cu-ceria anodes for SOFCs. *Electrochem. Solid-State Lett.*, **8** (6), A279–A280.

200. Cheng, Z., Zha, S.W., and Liu, M.L. (2006) Stability of materials as candidates for sulfur-resistant anodes of solid oxide fuel cells. *J. Electrochem. Soc.*, **153** (7), A1302–A1309.

201. Mukundan, R., Brosha, E.L., and Garzon, F.H. (2004) Sulfur tolerant anodes for SOFCs. *Electrochem. Solid State Lett.*, **7** (1), A5–A7.

202. Jiang, S.P. (2006) A review of wet impregnation – An alternative method for the fabrication of high performance and nano-structured electrodes of solid oxide fuel cells. *Mater. Sci. Eng., A*, **418** (1–2), 199–210.

203. Kurokawa, H., Yang, L., Jacobson, C.P., De Jonghe, L.C., and Visco, S.J. (2007) Y-doped $SrTiO_3$ based sulfur tolerant anode for solid oxide fuel cells. *J. Power Sources*, **164** (2), 510–518.

204. Lu, X.C., Zhu, J.H., Yang, Z., Xia, G., and Stevenson, J.W. (2009) Pd-impregnated SYT/LDC composite as sulfur-tolerant anode for solid oxide fuel cells. *J. Power Sources*, **192** (2), 381–384.

205. Kim, G., Corre, G., Irvine, J.T.S., Vohs, J.M., and Gorte, R.J. (2008) Engineering composite oxide SOFC anodes for efficient oxidation of methane. *Electrochem. Solid State Lett.*, **11** (2), B16–B19.

206. He, H.P., Huang, Y.Y., Vohs, J.M., and Gorte, R.J. (2004) Characterization of YSZ-YST composites for SOFC anodes. *Solid State Ionics*, **175** (1–4), 171.

207. He, H.P., Huang, Y.Y., Regal, J., Boaro, M., Vohs, J.M., and Gorte, R.J. (2004) Low-temperature fabrication of oxide composites for solid-oxide fuel cells. *J. Am. Ceram. Soc.*, **87** (3), 331.

208. Corre, G., Kim, G., Cassidy, M., Vohs, J.M., Gorte, R.J., and Irvine, J.T.S. (2009) Activation and ripening of impregnated manganese containing perovskite SOFC electrodes under redox cycling. *Chem. Mater.*, **21** (6), 1077–1084.

209. Tucker, M., Sholklapper, T., Lau, G., DeJonghe, L., and Visco, S. (2009) Progress in metal-supported SOFCs. *ECS Trans.*, **25** (2), 673–680.

210. Tucker, M.C., Lau, G.Y., Jacobson, C.P., DeJonghe, L.C., and Visco, S.J. (2007) Performance of metal-supported SOFCs with infiltrated electrodes. *J. Power Sources*, **171** (2), 477–482.

211. Tucker, M.C., Lau, G.Y., Jacobson, C.P., DeJonghe, L.C., and Visco, S.J. (2008) Stability and robustness of metal-supported SOFCs. *J. Power Sources*, **175** (1), 447–451.

212. Blennow, P., Hjelm, J., Klemensø, T., Persson, A., Brodersen, K., Srivastava, A., Frandsen, H., Lundberg, M., Ramousse, S., and Mogensen, M. (2009) Development of planar metal supported SOFC with novel cermet anode. *ECS Trans.*, **25** (2), 701–710.

213. Jiang, S.P., Zhang, S., Zhen, Y.D., and Wang, W. (2005) Fabrication and performance of impregnated Ni anodes of solid oxide fuel cells. *J. Am. Ceram. Soc.*, **88** (7), 1779–1785.

214. Jiang, S.P., Chen, X.J., Chan, S.H., and Kwok, J.T. (2006) GDC-impregnated, $(La_{0.75}Sr_{0.25})(Cr_{0.5}Mn_{0.5})O_3$ anodes for direct utilization of methane in solid oxide fuel cells. *J. Electrochem. Soc.*, **153** (5), A850–A856.

215. Jiang, S.P., Duan, Y.Y., and Love, J.G. (2002) Fabrication of high-performance $Ni/Y_2O_3ZrO_2$ cermet anodes of solid oxide fuel cells by Ion impregnation. *J. Electrochem. Soc.*, **149** (9), A1175–A1183.

2
Advanced Cathodes for Solid Oxide Fuel Cells

Wei Zhou, Zongping Shao, Chan Kwak, and Hee Jung Park

2.1
Introduction

Solid oxide fuel cells (SOFCs) are a type of high-temperature electrochemical devices that directly convert chemical energy to electric power with high efficiency and low emissions. A single cell of SOFC is composed of a porous cathode, a porous anode, and a dense electrolyte sandwiching them. Cathode is the place where electrochemical reduction of oxygen to oxide ion happens. The desired cathode should possess not only high electronic conductivity for efficient current collection and high catalytic activity toward oxygen reduction but also an appropriate microstructure to avoid mass transport limitations. Furthermore, the cathode should be thermomechanically and chemically compatible with the electrolyte under the operation and fabrication conditions; that is, it should exhibit minimum thermochemical expansion mismatch with the electrolyte and negligible chemical reaction with the electrolyte during the high-temperature fabrication and operation conditions.

The current commercialization-oriented SOFCs are typically composed of yttrium-stabilized zirconia (YSZ) electrolyte, Ni–YSZ cermet anode, and $La_{0.8}Sr_{0.2}MnO_3$ (LSM) oxide cathode, and they operate at elevated temperatures of 850–1000 °C [1]. For such electrochemical systems, high efficiencies can be achieved by integrating with gas turbines for large-scale stationary applications. However, the high operating temperature also introduces several severe problems or drawbacks, such as a high possibility of interfacial reaction between the cell components, easy densification of the electrode layer, possible crack formation due to mismatch in thermal expansion of cell components, and the requirement for high-cost $LaCrO_3$ ceramic as the interconnect material [1]. High-temperature operation also means that the stack components need to be mainly ceramic and a tubular or box section design is commonly used, which results in low volumetric power density [2].

For smaller scale applications, such as microcombined heat and power, auxiliary power units, and small electric generators for domestic use, there is a general trend to lower the operating temperatures to the range 450–750 °C [3] with the

Materials for High-Temperature Fuel Cells, First Edition. Edited by San Ping Jiang and Yushan Yan.

development of low- to intermediate-temperature solid oxide fuel cells (LIT-SOFCs), since such reduction can introduce several important beneficial effects such as [4]

1) offering the choice of low-cost cell materials, such as stainless steels for the interconnectors and construction materials, with better mechanical properties and thermal conductivity;
2) allowing more rapid start-up and shutdown of the fuel cells;
3) increasing flexibility of the cell design and material selections;
4) reducing the interfacial reaction between cell components and prolonging the lifetime of fuel cells;
5) simplifying the thermal management process.

Typically, both the ionic conductivity of electrolyte and the electrochemical activity of cathode for oxygen reduction drop quickly with decreasing operating temperature because of the large activation energies associated with the oxygen-ion transport and the activation of molecular oxygen. It is a big challenge to keep the ohmic resistance of electrolyte and the polarization resistance of electrode at a low level at reduced temperatures in order to achieve a high cell power output. The state-of-the-art SOFCs with thick YSZ electrolyte and LSM cathode show high ohmic resistance and cathodic polarization resistance at reduced operating temperatures, making them impractical for use in low-temperature operation. Reducing the thickness of YSZ electrolyte and/or developing alternative electrolyte materials with higher ionic conductivity is the general way to realize low ohmic resistance at reduced temperatures [5]. Among the various electrolyte materials of SOFCs, doped ceria shows much higher conductivity than YSZ at reduced temperatures. It was proposed that thin film electrolytes of doped ceria with thickness around 10–20 μm can be operated at temperatures down to 500 °C without causing a large ohmic drop [6, 7]. With the advances in fabrication technique, it is now facile to prepare anode-supported SOFCs with electrolyte thickness down to the micrometer level and to achieve mass production capability. On the other hand, porous Ni–cermet anode substrates with well-developed microstructure can bring about negligible polarization loss in many cases. How to decrease polarization loss from cathode becomes the key point to realize the LIT-SOFC technology.

Two strategies can be applied to improve the cathodic performance of SOFCs at reduced temperature: developing novel cathode materials with high low-temperature performance and optimizing the microstructure of the cathodes to maximize the oxygen reduction sites and minimize the gas diffusion polarization resistance. This chapter reviews the most recent advance in the development of cathode materials for LIT-SOFCs based on different electrolytes, that is, doped ceria, stabilized zirconia, and also the most recently developed proton-conducting oxides. The drawbacks of the various available cathodes are discussed. The advanced fabrication techniques of the cathodes with a specific microstructure are also presented in a separate section. The future research and development directions are proposed.

2.2
Cathodes on Oxygen-Ion-Conducting Electrolytes

The mechanism of oxygen reduction over an SOFC cathode is highly dependent on the conducting mechanism of the electrolyte that is in contact with the cathode. By applying an oxygen-ion-conducting electrolyte, the oxygen should be finally reduced to oxygen ions over the cathode surface, which then migrate through the electrolyte to the anode side to react with the fuel. The oxygen electrochemical reductions over an SOFC cathode on an oxygen-ion-conducting electrolyte may comprise a number of elemental steps, such as diffusion, adsorption, dissociation, ionization, and incorporation of oxygen ions into the crystal lattice of the electrolyte. The sum reaction of the oxygen reduction over the cathode that is in contact with an oxygen-ion-conducting electrolyte can be expressed as

$$O_2 + 4e' + 2V_O^{\bullet\bullet} \rightarrow 2O_O^{\times} \quad (2.1)$$

If the electrolyte is a proton conductor, the cathode reactions are even more complicated, which may also involve the reactions between hydrogen and oxygen both in atomic and ionic forms.

For simplicity, the oxygen electrochemical reduction processes can be separated into two general paths: the surface and the bulk paths. If the electrode material is a pure electronic conductor with negligible oxygen ionic conductivity (Pt, Pd, Rh, etc.), the surface path is the only possible mechanism [8]. In this case, oxygen molecules from the gas phase adsorb onto the surface and diffuse to the electrolyte–electrode–air triple-phase boundary (TPB) region, where the formed O^{2-} species are finally incorporated into the vacancies of the electrolyte (assuming oxygen-ion-conducting mechanism) and diffuse to the anode side driven by the oxygen potential differential across the fuel cells (Figure 2.1a). In another case, if the electrode material by itself is also an oxygen-ion conductor (i.e., a mixed conductor), an alternative reaction path for the diffusion of oxygen ions through the electrode interior also becomes possible [9]. With the presence of bulk ionic transport pathway in the electrode, molecular oxygen can be reduced to O^{2-} over a significant portion of the electrode surface, thereby greatly extending the size of active region and improving the oxygen reduction kinetics at reduced temperatures (Figure 2.1b).

Oxygen-ion conduction can be performed in a cathode by adding an ion-conducting second phase to a pure electron-conducting electrode with the formation of a composite electrode, such as YSZ–LSM. It can also be created by introducing lattice defect (oxygen vacancy) by tailoring the elemental composition of a single-phase composite oxide with a defined lattice structure. Among the many crystallized compounds, the families with an ABO_3 perovskite-type phase structure, a layered perovskite structure, and K_2NiF_4 structure have received the most attention as cathode materials for LIT-SOFCs because of their versatile composition and good potential to possess both high electronic conductivity and favorable oxygen ionic conductivity. Among the various mixed conducting oxides, cobalt-containing ones usually show superior ionic conductivity and catalytic activity for oxygen reduction and therefore have received particular attention recently.

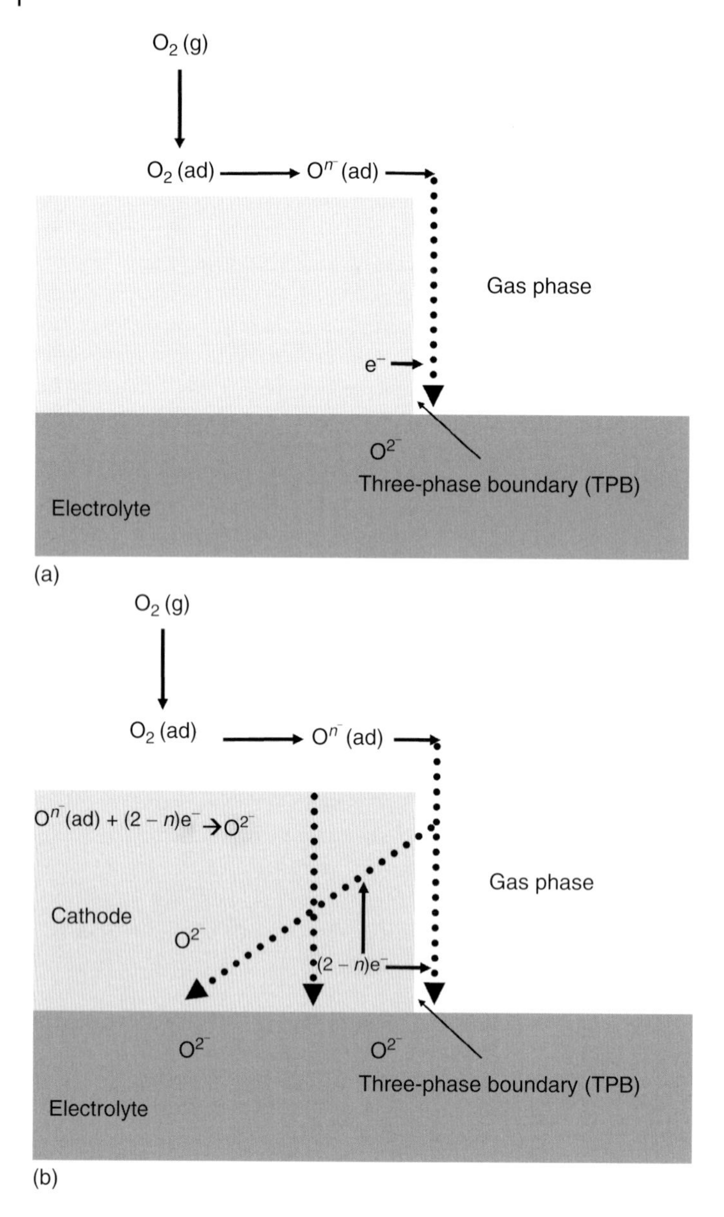

Figure 2.1 Schematic comparison of the oxygen reduction reaction on Pt (a) and MIEC (b) cathodes.

2.2.1
Cathodes on Doped Ceria Electrolytes

Currently, the most popular materials used as electrolytes of SOFCs include fluorite-type stabilized zirconia (ScSZ, YSZ), fluorite-type gadolinium-doped ceria

(GDC) or samarium-doped ceria (SDC), perovskite-type strontium, magnesium-doped lanthanum gallate (LSGM), and the more recently developed perovskite-type proton-conducting oxides, such as $BaCe_{1-x}Y_xO_3$. Pure ceria has a fluorite-type phase structure. The doping of CeO_2 with a cation having a lower valence state cation than Ce^{4+}, such as Sm^{3+}, could significantly increase the ionic conductivity of the oxide by creating oxygen vacancies within the oxide lattice. The highest oxygen ionic conductivity is reached if the dopants have the minimum size difference with Ce^{4+}, that is, Sm^{3+} and Gd^{3+} [10–12]. Both GDC and SDC show much higher ionic conductivity than stabilized zirconia; however, they show partial electronic conductivity at elevated temperatures due to the thermochemical reduction of Ce^{4+} to Ce^{3+} [13]. Such partial electronic conductivity is significantly reduced at lower temperatures, making them applicable as electrolytes at temperatures lower than 600 °C.

In order to achieve low polarization resistance of the electrode for LIT-SOFCs with doped ceria electrolytes at such low operating temperatures, many mixed conducting oxides have been exploited as potential cathode materials. Doped ceria electrolytes were found to have good chemical compatibility with a wide variety of electrodes, including many well-investigated cobalt-containing materials such as $SrCoO_{3-\delta}$-based perovskites, A-site-cation-ordered $LnBaCo_2O_{5+\delta}$-type double perovskites, and K_2NiF_4-type composite oxides.

2.2.1.1 **Perovskite**

Lanthanum cobaltite-based perovskite-type oxides are typical mixed conductors, which have been widely investigated as potential cathodes of SOFCs with doped ceria electrolytes [14–17]. $La_{1-x}Sr_xCoO_{3-\delta}$ (LSC) perovskites possess favorable oxygen ionic conductivity and ultrahigh electronic conductivity of over 1000 S cm^{-1} [18], but poor phase stability [19]. One strategy to solve the stability problem is partial substitution of cobalt ion with iron, leading to the development of $La_{1-x}Sr_xCo_{1-y}Fe_yO_{3-\delta}$ (LSCF) perovskites [20–26]. After the substitution, phase stability is greatly improved; however, electronic conductivity is also reduced somewhat, but its maximum conductivity can still exceed 300 S cm^{-1} for certain compositions (e.g., $La_{0.6}Sr_{0.4}Fe_{0.8}Co_{0.2}O_{3-\delta}$) at 750 °C [27], which is sufficient for application as SOFC cathodes. The ionic conductivity of LSCF is also favorably high, reaching $\sim 1 \times 10^{-3}$ S cm^{-1} at 750 °C for $La_{0.6}Sr_{0.4}Fe_{0.8}Co_{0.2}O_{3-\delta}$, measured with the help of an electronic blocking electrode technique [28]. The mixed conductivity of LSCF was further supported by the low electrode resistance of dense, 0.5 μm thick LSCF cathodes [29].

An area-specific resistance (ASR) of 0.7 Ω cm^2 was reported for a porous LSCF cathode at 600 °C, based on the symmetric cell test [30]. By applying a thin (~1 μm) dense LSCF interlayer between the porous LSCF cathode and the GDC electrolyte and by etching the electrolyte surface to eliminate the influence of SiO_2 impurity, the electrode performance was improved effectively and the resistivity values of the corresponding electrode were decreased by a factor of 2–3 as compared to the typical porous LSCF electrode [31, 32]. The formation of composite electrodes with doped ceria is another effective way to improve the electrode performance of LSCF at reduced temperatures. Perry Murray *et al.* [33] studied the electrochemical performance of $La_{0.6}Sr_{0.4}Fe_{0.8}Co_{0.2}O_{3-\delta}$–GDC composites. By addition of 50 vol% GDC to $La_{0.6}Sr_{0.4}Fe_{0.8}Co_{0.2}O_{3-\delta}$, the polarization resistance reached as low as 0.01 Ω cm^2 at 750 °C and 0.33 Ω cm^2 at 600 °C, a 10-fold reduction as compared to that of pure $La_{0.6}Sr_{0.4}Fe_{0.8}Co_{0.2}O_{3-\delta}$. Dusastre and Kilner [34] measured the polarization resistance of LSCF and various $La_{0.6}Sr_{0.4}Fe_{0.8}Co_{0.2}O_{3-\delta}$–GDC composite cathodes on a GDC electrolyte over a temperature range 500–700 °C in air. It was found that polarization resistance decreased by four times after introducing 36 vol% GDC into $La_{0.6}Sr_{0.4}Fe_{0.8}Co_{0.2}O_{3-\delta}$ electrode; this composition is very close to that predicted by the effective medium percolation theory to give the best performance [34]. From the impedance spectroscopy data, it was found that the formation of LSCF–GDC composite electrode improved the rates of oxygen diffusion and oxygen-ion charge transfer at the electrode–electrolyte interface. Esquirol *et al.* [35] study the oxygen tracer diffusion coefficient, D^*, and the surface exchange coefficient k of the $La_{0.6}Sr_{0.4}Fe_{0.8}Co_{0.2}O_{3-\delta}$–GDC composite. The measured D^* values of the composite material are higher than those of pure $La_{0.6}Sr_{0.4}Fe_{0.8}Co_{0.2}O_{3-\delta}$, while the k values are similar at all temperatures. Good low-temperature performance of various LSCF–doped ceria composite cathodes on doped ceria electrolyte was also reported by a number of other researchers [36–38]. Another significant advantage in forming a composite electrode with doped ceria is the improved thermomechanical compatibility between the electrode and the doped ceria electrolyte. Both LSC and LSCF oxides have high thermal expansion coefficient (TEC) values of $>20 \times 10^{-6}$ K^{-1} [39–41] because of their large chemical expansion associated with the valence state and spin state change in the cobalt ions in the perovskite lattice, whereas the doped ceria electrolytes have much smaller TEC values of around 12×10^{-6} K^{-1} [42]. The introduction of doped ceria as a second phase of the electrode effectively reduces the apparent TEC value of the composite electrode.

As mentioned previously, the oxygen reduction over an SOFC cathode is really complicated, involving many substeps such as surface adsorption, dissociation, surface diffusion, and charge transfer. It is well known that precious metals such as Pd and Pt have high catalytic activity for oxygen activation [43], which may show superior performance to oxide electrodes for some of those substep reactions during the oxygen reduction process. As a result, the surface modification of LSCF electrodes with some precious metals may also be an effective way to improve electrode performance. Christie *et al.* [44] demonstrated improved performance in $La_{0.6}Sr_{0.4}Fe_{0.8}Co_{0.2}O_{3-\delta}$ cathodes by introducing microcrystalline Pt particles. Sahibzada *et al.* [45] investigated Pd as a promoter of $La_{0.6}Sr_{0.4}Fe_{0.8}Co_{0.2}O_{3-\delta}$ cathodes, and a decrease in overall cell resistance by 15% at 650 °C and 40% at 550 °C was demonstrated. Hwang *et al.* [46] found that the introduction of a small amount (0.5 vol%) of Pt effectively reduced the polarization resistance of the $La_{0.6}Sr_{0.4}Fe_{0.8}Co_{0.2}O_{3-\delta}$ electrode over the whole range of operating temperatures (500–800 °C). Two concerns about the precious-metal-promoted electrodes are the high cost and the easy sintering of precious metals.

Although the catalytic activity of perovskite oxides is directly related to the B-site cations, the modification of the A-site cation could lead to a change in the valence state and redox properties of the B-site cations [47, 48], thus altering the catalytic activity of the oxides. Nowadays, a lot of efforts are under way to modify phase stability and electrochemical activity of LSC by replacing La with other rare earth metals [49–52]. Takeda *et al.* [49] demonstrated the substitution of La with Gd with the formation of $Gd_{1-x}Sr_xCoO_{3-\delta}$ oxides that actually worsened the electrode performance. However, the replacement of La with Sm with the development of $Sm_{1-x}Sr_xCoO_{3-\delta}$ (SSC) perovskites was beneficial [50]. These materials show high conductivity with the maximum conductivity up to 1000 S cm^{-1} at $x = 0.5$ [51]. It was further demonstrated that the overpotentials of SSC cathodes are relatively low [52]. Dense thin film $Sm_{0.5}Sr_{0.5}CoO_{3-\delta}$ electrode showed ~50% lower overpotential than $La_{0.6}Sr_{0.4}CoO_{3-\delta}$ under the similar operating conditions [53]. The rate-determining step of oxygen reduction over the dense $Sm_{0.5}Sr_{0.5}CoO_{3-\delta}$ thin film cathode is the adsorption-desorption of oxygen at the electrode surface, the same as in $La_{0.6}Sr_{0.4}CoO_{3-\delta}$ electrode, but the adsorption and desorption rate constants of $Sm_{0.5}Sr_{0.5}CoO_{3-\delta}$ were approximately one order of magnitude larger than those of $La_{0.6}Sr_{0.4}CoO_{3-\delta}$ [53].

Despite the high surface exchange kinetics, the low-temperature electrode performance of SSC is still not sufficiently high. An ASR of ~20 Ω cm^2 was observed for a single-phase SSC electrode at 500 °C measured by the symmetric cell test [54]. High-temperature permeation study demonstrated that dense SSC membrane had only modest oxygen flux [55]. It suggested that the oxygen ionic conductivity of SSC is relatively low, especially at reduced temperatures, and as a result, the TPB length is greatly reduced at low operating temperatures.

The cathodic performance of porous $Sm_{0.5}Sr_{0.5}CoO_{3-\delta}$ was improved by introducing doped ceria as the second phase [56]. Besides the creation of additional ionic conduction path within the electrode, the addition of SDC phase introduces additional benefit by suppressing the grain growth of $Sm_{0.5}Sr_{0.5}CoO_{3-\delta}$, thereby effectively increasing the TPB length. Xia *et al.* [56] systematically investigated the SSC–SDC composite electrodes and demonstrated that the $Sm_{0.5}Sr_{0.5}CoO_{3-\delta}$–SDC composite containing 30% SDC in weight exhibited the lowest interfacial resistance or the highest catalytic activity for oxygen reduction among the others, which has an ASR of less than 0.18 Ω cm^2 at 600 °C under open-circuit conditions, as a comparison of ~2.0 Ω cm^2 for a single-phase SSC electrode.

Mixed conducting $Ba_{0.5}Sr_{0.5}Co_{0.8}Fe_{0.2}O_{3-\delta}$ (BSCF) was initially developed by Shao *et al.* [57–61] as a material of ceramic oxygen-separation membrane or membrane reactor. The oxygen permeation flux of a mixed conducting membrane is closely related to the electronic and oxygen ionic conductivity of the membrane material. BSCF turned out to have much higher oxygen permeability than SSC [57]. Thus, BSCF may be a promising cathode material of LIT-SOFCs. Shao and Haile [62] first investigated the performance of BSCF as a cathode of IT-SOFCs (intermediate-temperature solid oxide fuel cells) and demonstrated an outstanding low-temperature electrode performance. In a single cell based on a 20 μm thick SDC electrolyte film, with humidified (3% water vapor) hydrogen as the fuel and air as the oxidant, maximum power densities of 1010 mW cm^{-2} were achieved at 600 °C, as shown in Figure 2.2. Liu *et al.* [63] reported that by further reducing

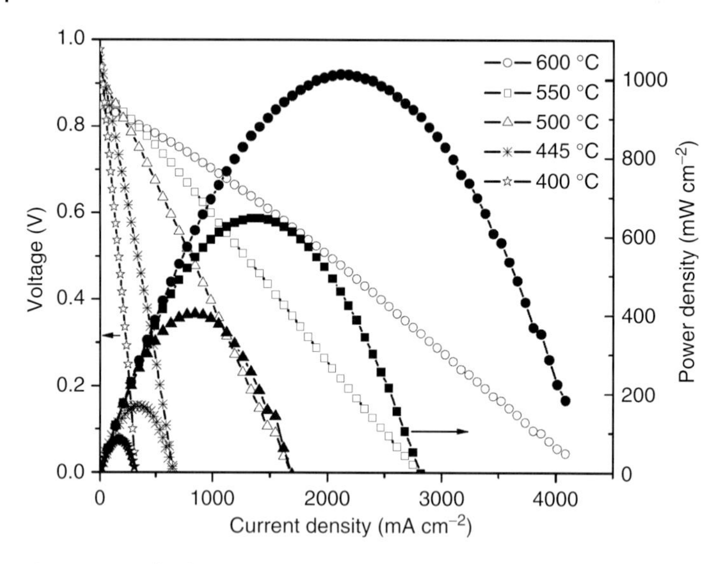

Figure 2.2 Cell voltage and power density as functions of current density performance obtained from a BSCF|SDC|Ni–SDC fuel cell. (Source: Reproduced with permission from Ref. [63].)

the thickness of GDC electrolyte to around 10 μm, an anode-supported single cell with the BSCF cathode and a Ni–GDC cermet anode achieved the peak power densities of 1329, 863, 454, 208, and 83 mW cm^{-2} at 600, 550, 500, 450, and 400 °C, respectively. A combined *in situ* neutron diffraction and thermogravimetric study yielded unusually high values for the oxygen nonstoichiometry in BSCF of $\delta = 0.7{-}0.8$ between 600 and 900 °C in the pO_2 range 0.001–1 atm [64]. Exceptionally high concentrations of mobile oxygen vacancies pronouncedly improved the oxygen reduction reaction (ORR) process on BSCF cathode at low temperatures. Baumann *et al.* [65, 66] reported an extremely low absolute value of the electrochemical surface exchange resistance of ~0.09 Ω cm^2 at 750 °C in air, which is more than a factor of 50 lower than the corresponding value measured for $La_{0.6}Sr_{0.4}Co_{0.8}Fe_{0.2}O_{3-\delta}$ microelectrodes with the same geometry. The progress in understanding and development of BSCF-based cathodes for IT-SOFCs was reviewed by Zhou *et al.* [67].

It is well known that surface oxygen vacancies are beneficial to oxygen adsorption, dissociation, and diffusion. However, high bulk oxygen vacancy concentration would create a suppressing effect on the electron polaron hopping in the perovskite lattice, and as a result, a decrease in electronic conductivity with increasing oxygen nonstoichiometry was often observed. Insufficient electrical conductivity may introduce a large charge-transfer resistance. For example, BSCF has a relatively low electrical conductivity of only ~40 S cm^{-1} [68]. Recently, it was demonstrated that some donor-doped $SrCoO_{3-\delta}$ perovskites possess both high oxygen vacancy concentration and electrical conductivity. Aguadero *et al.* [69] stabilized a tetragonal perovskite phase successfully by doping the $SrCoO_{3-\delta}$ system with 10 mol% of Sb^{5+} in the cobalt site. The thermal evolution of the electrical conductivity of

$SrCo_{0.9}Sb_{0.1}O_{3-\delta}$ exhibits a maximum value of 300 S cm^{-1} at 400 °C. Lin *et al.* [70] reported that on SDC electrolyte, the $SrCo_{0.9}Sb_{0.1}O_{3-\delta}$ electrode had a low polarization resistance of 0.09 Ω cm^2 in air at 700 °C under open-circuit conditions based on symmetric cell test. A detailed study of $SrCo_{1-y}Sb_yO_{3-\delta}$ perovskites showed that the composition with $y = 0.05$ has the lowest ASR values ranging from 0.009 to 0.23 Ω cm^2 in the 900–600 °C temperature interval with an activation energy of only 0.82 eV [71]. It is likely that the introduction of Sb^{5+} into $SrCoO_{3-\delta}$ drives an electron-doping effect, enhancing the mixed valence over the cobalt ion and thus promoting the electrical conductivity. Zhou *et al.* [72, 73] reported on the Nb-doped perovskite with the compositions of $SrNb_{0.1}Co_{0.9}O_{3-\delta}$. The material showed both high conductivity and oxygen vacancy concentration at 400–600 °C. A cell with $SrNb_{0.1}Co_{0.9}O_{3-\delta}$ as a cathode delivered a power density as high as 561 mW cm^{-2} at 500 °C in an SDC-based fuel cell, as shown in Figure 2.3.

As mentioned earlier, several perovskite compounds with cobalt as the predominant B-site cation have high electrochemical performance. However, most cobalt-rich compositions have high TEC, and there is concern with respect to delamination of the cathode from the electrolyte on thermal cycling. This disadvantage is much serious in BSCF perovskites. The TEC of BSCF is much larger than most of the electrolytes [74]. Owing to this mismatch, in some cases, the high TEC even resulted in the BSCF cathode being peeled off from the GDC–ScSZ electrolyte [75]. The chemical and thermal expansion of BSCF between 600 and 900 °C and at oxygen partial pressures of 1×10^{-3} to 1 atm was determined using *in situ* neutron diffraction by McIntosh *et al.* [64]. In the range covered by the experiments, the thermal and chemical expansion coefficients are $19.0(5)–20.8(6) \times 10^{-6}$ K^{-1} and 0.016(2)–0.026(4), respectively. The thermal expansion was attributed to crystal expansion from harmonic atomic vibrations, which depend on the electrostatic

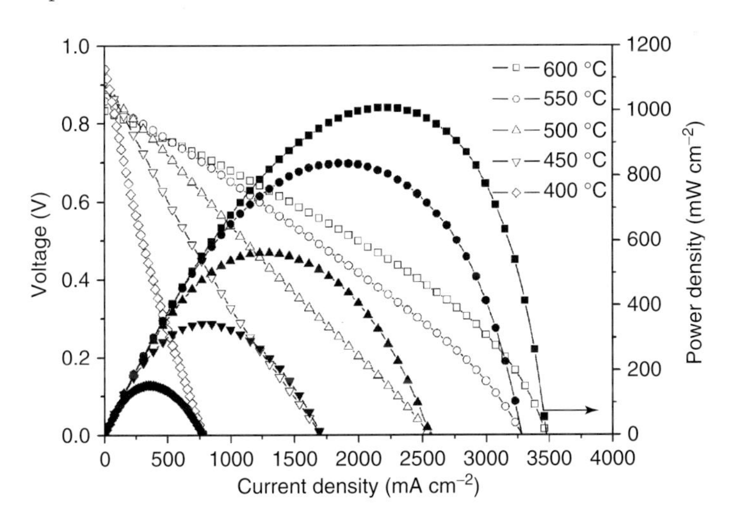

Figure 2.3 Cell voltage and power density as functions of current density. Performance obtained from an $SrNb_{0.1}Co_{0.9}O_{3-\delta}$|SDC|Ni–SDC fuel cell. (Source: Reproduced with permission from Ref. [72].)

attraction forces within the lattice [76], while the chemical expansion was induced by both the cobalt ion spin transition and the thermochemical reduction of cobalt ions to lower oxidation states [77].

A decrease in cobalt content could lead to a decrease in TEC of the cobalt-containing materials, partially due to the depressed oxidation of cobalt ion from Co^{3+} to Co^{4+} [40, 78]. However, oversubstitution of the cobalt ion may lead to a sharp increase in cathodic polarization resistance for ORR. It is a big challenge to maintain a high electrocatalytic activity at low operating temperatures as well as a low TEC for cobalt-containing materials. The family of Sr–Co–O system oxides with cubic perovskite crystal structure shows both high oxygen ionic and electronic conductivity. Zeng *et al.* [79] reported that scandium doping into the cobalt site in $SrCoO_{3-\delta}$ oxides effectively stabilized the cubic perovskite structure at a wide range of temperatures. It is likely that small amount of Sc^{3+} doping can effectively stabilize the cobalt ion at the lower valence state and/or the high spin state. Zhou *et al.* [80, 81] studied $SrSc_{0.2}Co_{0.8}O_{3-\delta}$ perovskite as a cathode material for IT-SOFCs. The material showed an almost linear thermal expansion from room temperature to 1000 °C in air with an average TEC of only 16.9×10^{-6} K^{-1}. The Sc doping not only dramatically reduced TEC but also resulted in extremely high oxygen vacancy concentrations in the lattice at low temperatures. The ASR is only 0.206 Ω cm^2 for $SrSc_{0.2}Co_{0.8}O_{3-\delta}$ at 550 °C. A peak power density as high as 564 mW cm^{-2} was obtained at 500 °C based on a 20 μm thick SDC electrolyte with $SrSc_{0.2}Co_{0.8}O_{3-\delta}$ cathode.

By introducing A-site-cation deficiency, the chemical and physical properties, including TEC, electrical conductivity, and oxygen nonstoichiometry, of the perovskites may also be altered. Kostogloudis and Ftikos [23] observed that the A-site-cation-deficient compound, $(La_{0.6}Sr_{0.4})_{1-x}Co_{0.2}Fe_{0.8}O_{3-\delta}$, had both lower TEC and lower electronic conductivity than the corresponding cation-stoichiometric perovskite ($La_{0.6}Sr_{0.4}Co_{0.2}Fe_{0.8}O_{3-\delta}$). Mineshige *et al.* [82] suggested that the reduced electronic conductivity could be attributed to the creation of additional oxygen vacancies. It was confirmed recently that additional oxygen vacancies are beneficial to the ORR [83]. The superior performance of A-site-cation-deficient LSCF cathodes was also reported by Doshi *et al.* [84], who found that the corresponding cathodic polarization resistance is as low as 0.1 Ω cm^2 at 500 °C. Zhou *et al.* [85] reported that the TEC of BSCF could be reduced by introducing A-site-cation deficiency with the formation of $(Ba_{0.5}Sr_{0.5})_{1-x}Co_{0.8}Fe_{0.2}O_{3-\delta}$ ($(BS)_{1-x}CF$) oxides. They found that the TEC is highly dependent on both the A-site-cation deficiency fraction and the selected temperature range. It decreased with increasing A-site-cation deficiency, especially in the 450–750 °C temperature range. A single SOFC with a $BS_{0.97}CF$ cathode exhibited peak power densities of 694 and 893 mW cm^{-2} at 600 and 650 °C, respectively, which are slightly lower than those of a similar cell with a cation-stoichiometric BSCF as the cathode.

In an alternative way, some cobalt-free perovskite oxide materials have been developed as the cathodes to decrease TEC. $La(Ni,Fe)O_{3-\delta}$ (LNF) has been considered as an attractive material for use as cathode, especially because of its high electronic conductivity, which reaches 580 S cm^{-1} at 800 °C for $x = 0.4$ [86]. Its TEC is also suitable to match that of the GDC, which favors the thermomechanical stability of the system. Another advantage of these LNF cathodes is their resistance against Cr poisoning, which can potentially diffuse from the cell interconnect material [87]. However, recent studies have claimed problems with its reactivity, with GDC forming poorly conducting phases at the electrolyte–cathode interface [88, 89]. Doping with strontium on the lanthanum site, $La_{1-y}Sr_yFe_{1-x}Ni_xO_{3-\delta}$, has been reported to increase the ionic conductivity by the introduction of oxygen vacancies in the system but has the detrimental effect of increasing the TEC. Cu doping in these systems has also been investigated, and fuel cells containing $LaNi_{0.2}Fe_{0.8-x}Cu_xO_3$ cathodes have been fabricated by Li and Zhu [90]. The materials were prepared by the coprecipitation method, and single-phase samples were obtained for $x < 0.15$. These fuel cells gave good peak power densities of 635 and 763 mW cm^{-2} at 580 and 650 °C, respectively. Hou *et al.* reported that molybdenum-doped $LaNiO_3$ ($LaNi_{1-x}Mo_xO_3$) showed acceptable electrochemical performance at 800 °C. The segregation of La_2MoO_6 from nominal $LaNi_{0.75}Mo_{0.25}O_{3-\delta}$ introduces La vacancies, which can partially reoxidize Ni(II) to give mixed valence in the Ni(III)–Ni(II) couple at 800 °C in air while leaving enough Mo(VI) to provide randomly distributed oxygen vacancies [91].

Another group of cobalt-free perovskites are based on $SrFeO_{3-\delta}$. Conventionally, rare earth elements, for example, La, have been used successfully to partially substitute Sr to restrain such a transformation represented, for example, by $(La,Sr)FeO_{3-\delta}$ (LSF) [92, 93]. This doped perovskite exhibits an extremely low TEC that matches well with SDC electrolytes. However, LSF does not deliver satisfactory electrochemical performance as a cathode in IT-SOFCs. Bi doping of $SrFeO_{3-\delta}$ results in the formation of a structure with high symmetry and extraordinary electrochemical performance for $Bi_{0.5}Sr_{0.5}FeO_{3-\delta}$ (BSF) due to the presence of a lone electron pair in Bi^{3+}, which is not available in La^{3+} [94–96]. BSF shows an activation energy (E_a) of 117 kJ mol^{-1} for ORR, which is relatively low with respect to other cobalt-free cathodes, for example, 183 kJ mol^{-1} for $La_{0.8}Sr_{0.2}FeO_{3-\delta}$ [97] and 142 kJ mol^{-1} for $GaBaFe_2O_{5+\delta}$ [98], and cobalt cathodes, for example, 164 kJ mol^{-1} for $La_{0.8}Sr_{0.2}CoO_{3-\delta}$ [97].

Even though some progress has been made with cobalt-free materials, they still suffer from unfavorable electrocatalysis on ORR and low electrical conductivity at reduced operating temperatures. The development of cobalt-free electrode materials with high electrical conductivity and high electrocatalytic activity for oxygen reduction is a big challenge to the whole community of SOFCs.

2.2.1.2 **Double Perovskites**

Layered perovskites have been extensively studied in the past because of their extraordinary magnetic and transport properties at low temperatures, in particular, layered cobaltites with the general formula $LnBaCo_2O_{5+\delta}$ (Ln = Gd, Pr, Y, La, etc.) have been the object of growing research activity in recent years [99–109]. These materials possess an ordered structure in which lanthanide and alkali earth ions occupy the A-site sublattice and the oxygen vacancies are localized into layers [104–110]. Figure 2.4 compares the lattice structure of a typical perovskite oxide and an A-site-cation-ordered double-perovskite oxide. The crystal structure of these oxides can be regarded as a layered crystal $A'A''B_2O_6$ by doubling the unit cell of the standard perovskite structure, consisting of consecutive layers of $[BO_2]$–$[A''O]$–$[BO_2]$–$[A'O_\delta]$ stacked along the *c*-axis [106]. The transformation of a cubic perovskite to such a layered structure reduces the oxygen bonding strength in the $A'O_\delta$ layer and provides disorder-free channels for ion motion [99], which remarkably enhances oxygen diffusivity and opens the possibility for developing a new class of materials suitable for application as LIT-SOFC cathodes.

Zhang *et al.* [110] systematically studied the phase structure/stability, oxygen content, electrical conductivity, and cathode performance of $LnBaCo_2O_{5+\delta}$ (Ln = La, Pr, Nd, Sm, Gd, and Y) oxides. Stability of the double-layered perovskite structure of $LnBaCo_2O_{5+\delta}$ oxides is closely related to the ionic radius of the Ln^{3+} cation. A stable layered structure was observed for Ln = Pr, Nd, Sm, and Gd, while it was in a metastable state for Ln = Y and La. The oxygen content and the nominal oxidation state of cobalt ions in the oxide increases with the ionic radius of Ln^{3+}. In addition, larger ionic radii for the Ln^{3+} cation translated into higher oxygen mobility in the LnO_δ layer. The oxygen permeability test shows that the rate of oxygen diffusion through thick $PrBaCo_2O_{5+\delta}$ membrane bulk is much slower than that in BSCF. This was attributed to the oxides' polycrystalline structure and to the fact that oxygen diffusion was limited to the LnO_δ layers only. However, attractive electrode performance was still observed for $LnBaCo_2O_{5+\delta}$, especially $PrBaCo_2O_{5+\delta}$, which

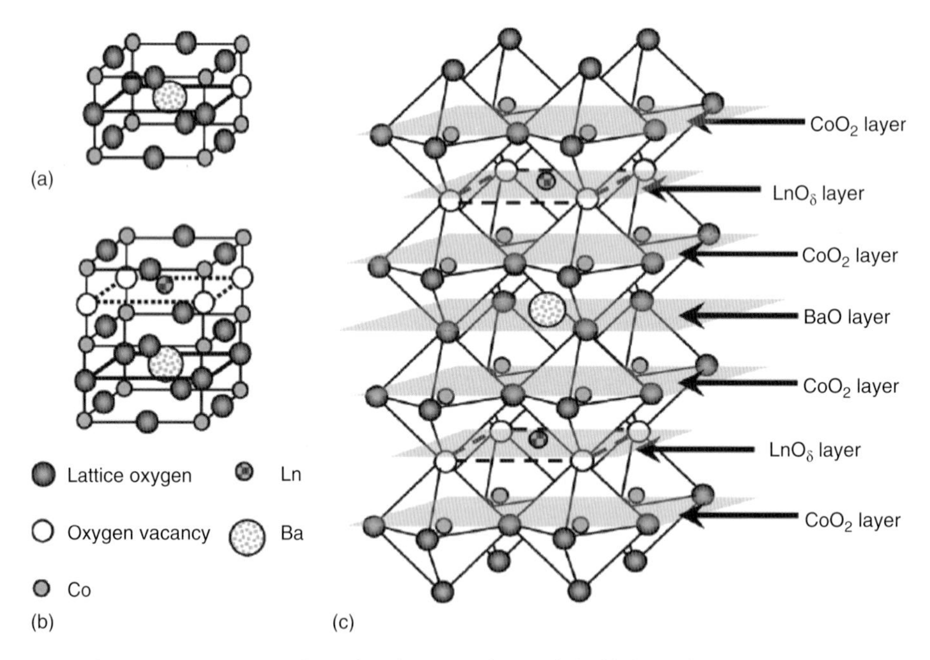

Figure 2.4 Structures of simple cubic perovskite and double-layered perovskite: (a) ABO_3 cubic perovskite, (b) $LnBaCoO_{5+\delta}$ double perovskite, and (c) layered structure. (Source: Reproduced with permission from Ref. [110].)

gave a relatively low ASR of 0.213 Ω cm^2 at 600 °C, because of the fast exchange kinetics across the oxides' surfaces. These results suggest that these layer-structured $LnBaCo_2O_{5+\delta}$ oxides have more promising application as materials of reduced-temperature SOFC electrodes rather than as materials of ceramic oxygen-separation membranes. Recent studies also demonstrated that layered cobaltites, with Ln = Gd [103, 105] and Pr [106], could be one of the aforementioned potential materials for LIT-SOFC applications due to their excellent oxygen transport properties at low temperatures, that is, high oxygen surface exchange coefficient and reasonable oxide ionic diffusivity, in combination with their high electronic conductivity. Tarancón *et al.* [105] obtained the oxygen surface exchange and oxygen tracer diffusion coefficients yielding optimum values for $GdBaCo_2O_{5+\delta}$ ($k^* = 2.8 \times 10^{-7}$ cm s^{-1} and $D^* = 4.8 \times 10^{-10}$ cm^2 s^{-1} at 575 °C) by $^{18}O/^{16}O$ isotope exchange depth profile (IEDP) method. Zhu *et al.* [111] reported that a fuel cell with a porous layer of $PrBaCo_2O_{5+\delta}$ deposited on a 42 µm thick SDC electrolyte provided a maximum power density of 583 mW cm^{-2} at 600 °C, using hydrogen as the fuel and stationary air as the oxidant. Chen *et al.* [112] showed that the phase reaction between $PrBaCo_2O_{5+\delta}$ and SDC is weak even at 1100 °C. It was found that the electrode's polarization resistance was mainly due to oxygen-ion transfer through the electrode–electrolyte interface and electron charge transfer over the electrode surface over the intermediate-temperature range 450–700 °C.

Kim and Manthiram [101] observed an increase in TEC with increase in the cation size of the rare earth metal in $LnBaCo_2O_{5+\delta}$. From the viewpoint of stability of SOFCs, they proposed samples with the intermediate rare earth metals, for example, Ln = Sm, as the best candidates. The ASR of a composite cathode (50 wt% $SmBaCo_2O_{5+\delta}$ and 50 wt% GDC) on a GDC electrolyte is only 0.05 Ω cm^2 at 700 °C. Moreover, by using this composite material, the TEC of the electrode was decreased to 12.5×10^{-6} K^{-1} as compared to the TEC of 20×10^{-6} K^{-1} for the single-phase $SmBaCo_2O_{5+\delta}$ [113].

Since the presence of Ba with a large ionic radius (1.60 Å) is beneficial to form large lattice spacing and higher freedom of oxygen ionic movement, higher Ba content in perovskite has been associated with enhanced ORR rate [114]. Most recently, Deng *et al.* [115] reported a B-site-ordered double perovskite with the A site fully occupied by Ba, for example, $Ba_2CoMo_{0.5}Nb_{0.5}O_{6-x}$ (BCMN) as the promising cathode for SOFCs operated at 700 °C. The Co concentration in BCMN, however, is not optimum, for example, only 50% on its B-site cation, resulting in its low electrical conductivity and electrocatalytic activity. Zhou *et al.* [116] reported on $Ba_2Bi_{0.1}Sc_{0.2}Co_{1.7}O_{6-x}$ (BBSC). High oxygen vacancy bulk diffusion and surface exchange rates, as well as high electrical conductivity, of BBSC translate into its extraordinary electrochemical performance at 600 °C (0.22 Ω cm^2) (Figure 2.5). BBSC is also attractive in terms of its low TEC (17.9×10^{-6} K^{-1}) due to the absence of Co^{4+}.

Despite these encouraging results, possible detrimental effects associated with the phase transition, as reported recently by Streule *et al.* [117, 118] for $PrBaCo_2O_{5+\delta'}$ at about 500 °C, could compromise the applicability of this, and other compounds of the same family, as cathodes for LIT-SOFCs. According to

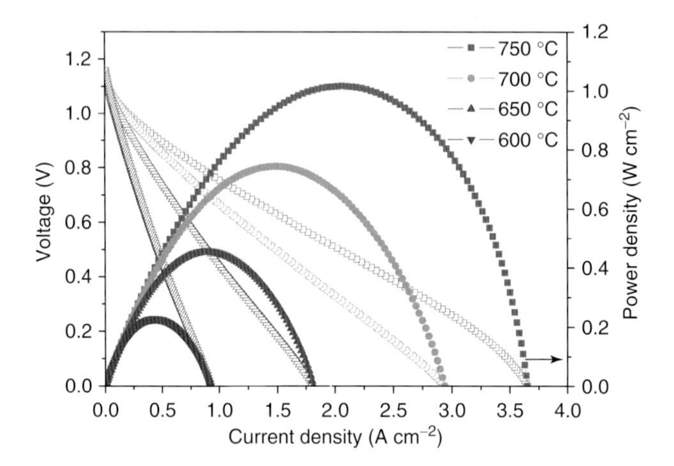

Figure 2.5 Cell voltage and power density as functions of current density. Performance obtained from a BBSC|SDC|YSZ|Ni–YSZ fuel cell. (Source: Reproduced with permission from Ref. [116].)

Streule *et al.*, apart from the extensively reported lattice distortion observed at low temperatures ($T \sim 75\,^\circ$C) [119, 120], PBC shows an order–disorder phase transition at 503 °C. This phase transition involves a rearrangement of oxygen vacancies from the low-temperature one-dimensional distribution alternating filled and empty chains of oxygen along the *a*-axis (orthorhombic symmetry, *Pmmm* space group) to the high-temperature two-dimensional distribution of vacancies in (001) layers, that is, in the PrO_δ plane (tetragonal symmetry, *P4/mmm* space group). Owing to the importance of oxygen transport properties for electrode applications, strong effects on the electrical and electrochemical performance would originate from the phase transition. Therefore, it is essential to dedicate some efforts to clarify possible problems associated with this particular feature. Tarancón *et al.* [121] demonstrated the significant effect of phase transition on high-temperature electrical properties of layered $GdBaCo_2O_{5+\delta}$ perovskite. The phase transition from orthorhombic to tetragonal symmetry takes place for values of $\delta < 0.45$, and an electronic conductivity decrease starts for values of $\delta < 0.25$. In addition, the electrochemical performance of $GdBaCo_2O_{5+\delta}$ decreases probably due to an excessive reduction of the oxygen content in the GdO_δ plane.

The phase transition from cubic to hexagonal structure was also observed in BBSC at 500–800 °C [122]. However, the impedance spectroscopy results of BBSC in symmetrical cell and three-electrode configurations at 700 °C show that the presence of hexagonal phase does not behave detrimentally toward the ORR performance. In addition, the application of cathodic polarization enables the regeneration of initial ORR performance and allows stable ORR performance over 72 h, most probably through cobalt reduction, which favors cubic structure formation and offsets the decomposition of cubic phase to hexagonal phase.

K_2NiF_4 Oxides Recently, there have been increased research interest in oxygen overstoichiometric $La_2NiO_{4+\delta}$-based compounds as novel mixed conductors for oxygen-separation membranes and cathodes of SOFCs [123, 124]. Those oxides have a K_2NiF_4-type structure and are usually formulated as $A_2BO_{4+\delta}$, which can be regarded as ABO_3 perovskite and AO rock-salt layers arranged one upon the other in the *c*-direction (Figure 2.6). There is sufficient space in the AO layer. This structure allows for the accommodation of oxygen overstoichiometry as oxygen interstitial species with negative charges, which are balanced through the oxidation of the B-site cations [125]. Previous studies have shown that these $A_2BO_{4+\delta}$ materials exhibit good characteristics in terms of electronic conductivity because of the mixed valence of B-site cation, the oxygen ionic transport properties due to the oxygen overstoichiometry, electrocatalysis for the oxygen reduction, and moderate thermal expansion properties [124–126]. The oxygen diffusion characteristics of the *c*-axis orientated $La_2NiO_{4+\delta}$ thin films (33–370 nm thick) grown on $SrTiO_3$ and

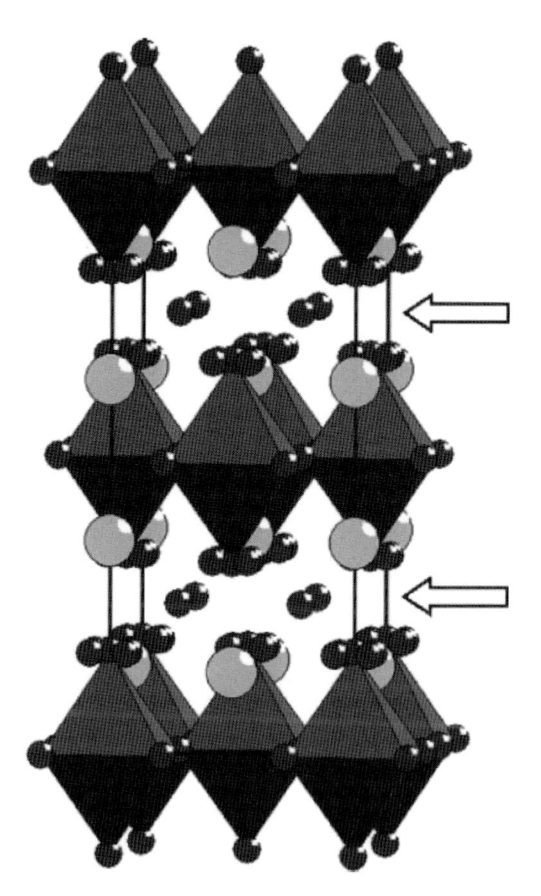

Figure 2.6 Crystal structure of $La_2NiO_{4+\delta}$. The arrows show layers of the interstitial oxygen. (Source: Reproduced with permission from Ref. [124].)

$NdGaO_3$ substrates have been examined by Burriel *et al.* [127]. Diffusion coefficients were found to increase as the film thickness increased, because of a reduction in the stress from the film substrate mismatch, although there was no effect on the surface exchange coefficient. Diffusion and surface exchange coefficients were shown to be higher along the *ab* plane by two orders of magnitude.

By partially replacing Ni with Co, the series of materials $La_2Ni_{1-x}Co_xO_{4+\delta}$, particularly the materials with high cobalt content, showed relatively high oxygen diffusion coefficient ($\sim 10^{-8}$ cm^2 s^{-1}) and oxygen surface exchange constant ($\sim 10^{-6}$ cm s^{-1}) over the temperature range 450–650 °C [125]. The high oxygen-ion diffusion coefficients and surface exchange constants are desirable properties for SOFC cathodes. In addition, low TECs of around $11.0–14.0 \times 10^{-6}$ K^{-1} match well with those of SOFC electrolytes [126]. It is reported that the doping of the A site of $La_2Ni_{1-x}Co_xO_{4+\delta}$ with Sr could further improve the electrode performance [128]. A single cell with $La_{1.2}Sr_{0.8}Co_{0.8}Ni_{0.2}O_{4+\delta}$–GDC as the cathode generated a peak power density of up to 350 mW cm^{-2} at 600 °C. In addition, the performance was fairly stable when a constant polarization voltage of 0.5 V was set for 36 h.

The effect of Cu doping on the electrode performance of the series $La_2Ni_{1-x}Cu_xO_{4+\delta}$ has also been examined by Aguadero *et al.* [129]. In terms of optimizing the synthesis of these nickelate systems, Weng *et al.* [130] have reported a new, efficient route for the synthesis of $La_{n+1}Ni_nO_{3n+1}$ ($n = 1, 2, 3, \ldots, N$). The method involves the heat treatment of nanosized cocrystallized precursors, which were prepared using a continuous hydrothermal flow system. The method led to high-quality samples, without the long heat treatments and multiple regrinding steps typically needed to prepare such systems.

Mazo *et al.* [131] reported that $La_{2-x}Sr_xCuO_{4-\delta}$ cuprates had high oxygen diffusivity at elevated temperatures (750–1000 K), among them $La_{1.7}Sr_{0.3}CuO_{4-\delta}$ has the highest oxygen diffusion coefficient. Another study indicated that $La_{2-x}Sr_xCuO_{4-\delta}$ could form different kinds of oxygen defects (interstitial oxygen or oxygen vacancy), depending on the Sr doping concentration, the oxygen partial pressure, and the preparation temperature [132]. At high Sr doping concentration, oxygen vacancy was formed, and the oxygen vacancy concentration increased with the Sr content. Li *et al.* [133] investigated the electrical properties of $La_{1.7}Sr_{0.3}CuO_{4-\delta}$ cathode and obtained a low ASR of 0.16 Ω cm^2 at 700 °C in air.

Further doping studies have examined varying the rare earth cation. Mauvy *et al.* [134] reported lower cathodic ASR values for praseodymium nickelate, $Pr_2NiO_{4+\delta}$. Currently, a $Pr_2NiO_{4+\delta}$–YSZ–$Pr_2NiO_{4+\delta}$ cell shows the best performance as a cathode layer, achieving an ASR of 0.5 Ω cm^2 at 610 °C [134]; however, $Pr_2NiO_{4+\delta}$ has been shown to react detrimentally with YSZ. The effect of doping on the Ni site in $Pr_2NiO_{4-\delta}$ was studied by Miyoshi *et al.* [135], who reported that the oxygen permeation rate was improved by doping with Cu and Mg. Neodymium nickelates have also been investigated, with reports suggesting that these show improved chemical stability compared to GDC and YSZ, since no reaction was found when using $Nd_{2-x}Sr_xNiO_{4-\delta}$ and $Nd_{1.95}NiO_{4-\delta}$, respectively, as electrodes [134, 136].

2.2.2
Cathodes on Stabilized Zirconia Electrolytes

Because of their excellent thermomechanical properties, high chemical stability, and pure ionic conductivity over a wide range of temperature and oxygen partial pressure of the atmosphere, yttria- and scandia-stabilized zirconia (YSZ, ScSZ) are still the most applied electrolyte of SOFCs up to now. ZrO_2 takes several lattice structures [137]. The doping of proper amount of yttrium or scandium into the ZrO_2 lattice can stabilize the conductive cubic fluorite phase, as well as increase the concentration of oxygen vacancies, and thus improve the oxygen ionic conductivity [138]. Although these electrolytes have insufficiently high ionic conductivity at temperatures lower than 600 °C, there are still considerable research interest in IT-SOFCs with stabilized zirconia electrolyte operating at 650–850 °C; at such a temperature range, the ohmic resistance is acceptable for a thin film electrolyte .

2.2.2.1 $La_{1-x}Sr_xMnO_3$-Based Perovskites

One practical problem during the development of cathode for SOFCs with stabilized zirconia electrolyte is the high reactivity between cathode and electrolyte at elevated temperature. Owing to their good chemical compatibility with the stabilized zirconia electrolyte, the $LaMnO_3$-based perovskites still receive considerable attention as the cathode of IT-SOFCs with stabilized zirconia electrolyte [139–141]. Among the various $LaMnO_3$-based materials, the Sr-doped lanthanum manganate LSM is the classical cathode material of high-temperature SOFCs because of its good properties in electrical conductivity, catalytic activity for oxygen reduction, thermal and chemical stability, and excellent mechanical properties at high temperatures [142]. However, LSM exhibits negligible ionic conductivities and the oxygen reduction is limited strictly to the TPB region, and thus a sharp increase in electrode polarization resistance is observed for pure LSM cathode. The electrode polarization resistance becomes so large that LSM is not practically applicable at temperatures lower than 700 °C [143–145].

Several ways have been tried to increase the low-temperature performance of LSM–based cathode. One general way is to add an ionic conducting second phase into LSM with the formation of a composite electrode. In such a manner, the active reaction sites successfully penetrate into the bulk of the electrode. The LSM–YSZ composite cathode was first reported by Kenjo and Nishiya [146]. After that, Wang *et al.* [147] fabricated anode-supported cells with LSM–YSZ cathodes and a peak power density as high as 0.8 W cm^{-2} was successfully demonstrated at 750 °C for a cell with an active cathode area of 4×4 cm^2 at a cell voltage of 0.7 V. The individual grain size of both LSM and YSZ is approximately 100 nm and the interface at cathode and electrolyte is well bonded. The percentage of contact line at cathode–electrolyte interface was measured to be 39% on average. By engineering the cathode through processing optimization, it is possible to create nanostructured cathodes, which are believed to effectively increase the TPB length, thereby enhancing the power output of the fuel cell. Song *et al.* [148] reported a dual-composite approach, in which both LSM and YSZ phases are placed on a

YSZ grain, that allows for the development of an optimal cathode microstructure with improved phase contiguity and interfacial coherence. A cell with such a well-controlled cathode microstructure delivered nearly constant power output over a period of 500 h, whereas a cathode prepared by conventional mechanical mixing technique undergoes significant degradation during the stability test due to thermochemically and electrochemically driven coarsening and shrinkage of the LSM phase.

Replacing YSZ in the LSM–YSZ composite with other materials with higher ionic conductivity may further improve the cathode performance. Perry Murray and Barnett [149] reported that at 700 °C, the interfacial polarization resistance is 7.28 Ω cm^2 for pure LSM cathode and 2.49 Ω cm^2 for an LSM–YSZ composite cathode, but only 1.06 Ω cm^2 for an LSM–GDC composite cathode. Because of the similar ionic conductivity between $Ce_{0.7}Bi_{0.3}O_2$ and GDC, the LSM–$Ce_{0.7}Bi_{0.3}O_2$ composite cathode demonstrated reasonably comparable electrode performance to LSM–GDC [150, 151]. One important concern about LSM-based composite electrode is how to avoid the interfacial reaction between the electrode components and between the electrode and electrolyte during the high-temperature fuel cell fabrication and operation processes. The LSM–LSGM composite cathode has also been investigated for the anode-supported SOFCs with YSZ thin film electrolyte [152]. However, noticeable solid-phase reactions between LSGM and YSZ electrolyte was observed above 1000 °C, introducing difficulty in firing the LSGM–LSM composite cathode onto the YSZ electrolyte without any interfacial reaction. Wang *et al.* [153] used $La_{0.8}Sr_{0.2}Mn_{1.1}O_3$–ScSZ–$CeO_2$ composite as the cathode for anode-supported SOFCs with thin YSZ electrolyte, and a maximum power density of 0.82 W cm^{-2} was reached at 650 °C.

2.2.2.2 **Doped $La_{0.8}Sr_{0.2}MnO_3$**

Besides the formation of a composite cathode by introducing a second ionic conducting phase, many attempts have also been made to introduce ionic conductivity into the electrode by compositional modification of the LSM electrode. Considerable efforts regarding LSM modification have been carried out on B-site substitution in LSM [142, 154–157]. Introducing cobalt ion into the B site of the LSM perovskite lattice formed new $La_{0.5}Sr_{0.5}Mn_{1-x}Co_xO_{3\pm\delta}$ perovskite oxides, and it was found that both the oxygen tracer diffusion coefficient and oxygen surface exchange constant increased because of the synergistic valence variations and nonstoichiometric defects induced into the perovskite lattice. On the other hand, it has been reported that lanthanum scandate is a mixed conductor with predominant oxide ion conduction at a wide range of pO_2, and the high polarizability of Sc^{3+} would make it easier for the oxide ions to be transported through the lattice [158–160]. From this point of view, scandium may be a promising dopant to increase the ion conductivity of LSM materials.

Gu *et al.* [161] synthesized perovskite-type $La_{0.8}Sr_{0.2}Mn_{1-x}Sc_xO_{3-\delta}$ (LSMSc) oxides and applied them as cathodes of SOFCs on a stabilized zirconia electrolyte. The introduction of Sc^{3+} into the B site of $La_{0.8}Sr_{0.2}MnO_{3-\delta}$ led to a decrease in the

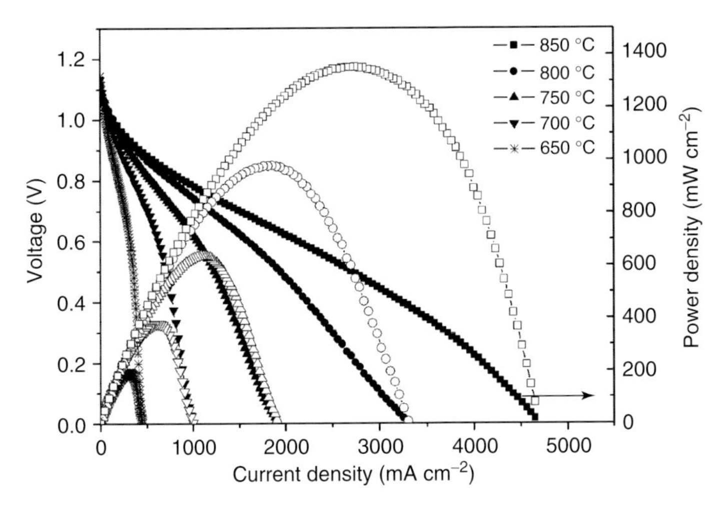

Figure 2.7 Cell voltage and power density as functions of current density. Performance obtained from a $La_{0.8}Sr_{0.2}Mn_{0.95}Sc_{0.05}O_{3-\delta}$|YSZ|Ni–YSZ fuel cell. (Source: Reproduced with permission from Ref. [162].)

oxides' TEC values and overall electrical conductivity. Among the various LSMSc oxides under study, $La_{0.8}Sr_{0.2}Mn_{0.95}Sc_{0.05}O_{3-\delta}$ demonstrated the lowest area-specific cathodic polarization resistance because of the suppressive effect of Sc^{3+} on surface SrO segregation and the optimization of the concentration of surface oxygen vacancies [162]. An anode-supported cell with a $La_{0.8}Sr_{0.2}Mn_{0.95}Sc_{0.05}O_{3-\delta}$ cathode demonstrated a peak power density of 650 mW cm^{-2} at 750 °C (Figure 2.7). The corresponding value for a similar cell with LSM cathode is only 200 mW cm^{-2} under identical experimental conditions. Zheng *et al.* [163] further introduced ScSZ as a second phase with the development of the LSMSc–ScSZ composite cathode. An anode-supported cell with LSMSc–ScSZ cathode delivered high power densities of ~1211 and 386 mW cm^{-2} at 800 and 650 °C, respectively, by applying hydrogen as the fuel and air as the oxidant. Yue *et al.* [164] also reported that cells with LSMSc-containing cathodes exhibit higher performance than those with the conventional LSM cathode, especially at lower temperatures.

2.2.2.3 **Cobalt-Containing Cathodes with a Buffering Layer**

As mentioned previously, among the various perovskite oxides, the cobalt-containing ones show the highest electronic and oxygen ionic conductivities, as well as the highest catalytic activity for oxygen electrochemical reduction. As a result, cobaltite perovskites should have much better low-temperature electrode performance than LSM on doped ceria electrolyte. However, cobalt-containing perovskites have high reactivity with stabilized zirconia electrolyte with the formation of $La_2Zr_2O_7$, $SrZrO_3$, or $BaZrO_3$ insulating the interfacial layer between the cathodes and the electrolyte layers [25, 165, 166], which block the charge transfer from the cathode to the electrolyte layers, resulting in a substantial increase in interfacial polarization resistance. In order to achieve a high electrode performance of cobaltite oxide on stabilized zirconia electrolyte, the interfacial reaction should be avoided. However, to obtain sufficient connectivity between cathode particles and adhesion of cathode layer to the electrolyte surface, a high-temperature firing process is typically required during the fuel cell fabrication. Most cobalt-containing cathode materials react readily with zirconia-based electrolytes at temperatures as low as 800 °C [25]. Duan *et al.* [166] demonstrated that BSCF readily reacts with YSZ at 900 °C. The key to achieve interfacial-reaction-free deposition of the cobaltite perovskite cathode onto the electrolyte surface is the development of the advanced fabrication technique that allows the deposition of the cathode layer onto the electrolyte surface below the initialization temperature at which interfacial reaction occurs.

To realize low-temperature fabrication, researchers tried to synthesize fine cobaltite perovskite powders with improved sintering activity by the advanced powder synthesis technique. By adopting a glycine–nitrate solution combustion technique, Lei *et al.* [167] successfully prepared an LSCF ultrafine powder with a specific surface area as high as 23 $m^2\ g^{-1}$, which shows high surface activity. Using this LSCF powder, they were able to fabricate LSCF cathode onto YSZ electrolyte at a temperature as low as 750 °C free of interfacial phase reaction. The corresponding cell demonstrated excellent power generation performance, with the peak power density greater than 1.0 W cm^{-2} and a power density of above 0.80 W cm^{-2} at 0.70 V at 700 °C. Until now, several other researchers have also succeeded in the synthesis of fine LSCF powder by advanced techniques, and in the fabrication of LSCF cathode onto YSZ electrolyte free of interfacial reaction; as a result, promising cell performance at reduced temperatures was demonstrated [168, 169].

However, thermodynamically, the phase reaction between cobaltite perovskite cathodes and stabilized zirconia electrolytes is unavoidable, the reduced fabrication temperature actually only suppresses such interfacial reaction kinetically. For practical application, the fuel cell may have a lifetime of couple of years, and in such a long operation period, stability is still a big concern. For example, by adopting an unsintered BSCF cathode directly on YSZ electrolyte, the cell performance degraded steadily during an operation period of 50 h at 800 °C. Later, it was found out that the interfacial reaction between BSCF and YSZ with the formation of $BaZrO_3$ insulator happened after the 50 h operation period at 800 °C [170].

As mentioned previously, most cobaltite perovskites have good chemical compatibility with doped ceria electrolyte. One strategy to avoid the interfacial reaction of cobaltite perovskite with stabilized zirconia is to apply a doped ceria interlayer to prevent direct contact between cobaltite perovskite cathode and stabilized zirconia electrolyte, therefore suppressing their potential phase reaction. Significant improvement in electrode performance was demonstrated by different researchers and different cobaltite perovskites, including LSCF, SSC, BSCF, and LSC [75, 171–174]. By applying BSCF as a cathode with a GDC interlayer, Duan *et al.* demonstrated an anode-supported fuel cell with the composition of Ni–YSZ|YSZ|GDC|BSCF that resulted in a maximum power density of 1.56 and 0.81 W cm^{-2} at 800 and 700 °C, respectively. The peak power density of the cell at 800 °C is only 0.44 W cm^{-2} for a similar cell by directly depositing the BSCF cathode on YSZ electrolyte [166], in which a serious interfacial reaction between the electrode and the cathode was detected. Besides its function as a separator in avoiding direct contact between the cathode and the electrolyte, the interlayer may also improve the surface exchange kinetics of the electrolyte. Tsai and Barnett [175] applied a doped ceria as the interlayer between LSM–YSZ cathode and YSZ electrolyte, and the electrode polarization resistance decreased by about 10 times, which is just the ratio of the surface exchange coefficient of SDC to YSZ.

Potentially, all the cobaltite perovskites with high oxygenic and electronic conductivity and large surface area can be applied as cathodes of stabilized zirconia electrolyte by adopting a doped ceria interlayer. However, the introduction of doped ceria also creates the risk for the interfacial reaction between doped ceria and stabilized zirconia [176, 177]. The reaction between them may form $(Zr,Ce)O_2$-based solid solutions, which have a very low electrical conductivity [178]. Therefore, the fabrication temperature should be specifically tailored.

It was found that the reaction between stabilized zirconia and doped ceria is negligent at temperatures below 1100 °C, but obvious at temperatures higher than 1300 °C [179]. A preferable fabrication technique should conduct at a temperature lower than that at which phase reaction occurs. Various fabrication techniques have been tried, including wet ceramic processes (spray coating, dip coating, screen printing) and physical vapor deposition (sputtering, electron beam evaporation, pulse laser) [175, 180–186]. The wet ceramic processes are relatively cheaper and cost-effective. However, they need a high-temperature sintering process for the formation of good connection between doped ceria particles and for the doped ceria layer to adhere well to the stabilized zirconia electrolyte. It is very difficult to manufacture a dense GDC layer by the wet ceramic process because of the poor sintering activity of the doped ceria powder and the absence of shrinkage of the electrolyte layer. Insufficient sintering temperature of the doped ceria layer produces an excessively porous layer, while too high sintering temperature may cause a reaction between the doped ceria and the stabilized electrolyte, which leads to low ionic conductivity of the electrolyte. Interestingly, it was found that the porous doped layer could not have an obvious effect on the electrode performance at a proper thickness [187]. Shiono *et al.* [188] demonstrated by EPMA (electron probe microanalysis) that the reaction between LSC cathode and SScZ electrolyte did not occur, even though the GDC interlayer used with a thickness of 3–6 μm was porous. In another study, Duan *et al.* [166] demonstrated that when the GDC layer is around 1 μm in thickness, the $BaZrO_3$ insulating phase was detected at the interface of BSCF and YSZ with the GDC layer at 1100–1200 °C, which means the porous GDC layer could not prevent the reaction between BSCF and YSZ. In other words, the diffusion of Ba cations from cathode through the porous GDC layer to the YSZ surface likely occurred in very thin porous GDC layers. Therefore, a proper control of the thickness as well as the firing temperature of the interlayer is necessary while adopting the wet ceramic process of the interlayer.

In another study, it was demonstrated that although the porous interlayer may have no obvious influence on the electrode performance, it may lead to an increase in the ohmic resistance of the electrolyte [186]. By fitting the resistance curves to an equivalent circuit, it is clearly seen that the polarization resistance was relatively unaffected and only the ohmic resistance of the fuel cell was significantly reduced when a fully dense SDC interlayer was used. Therefore, preparation of thin dense layer of doped ceria on YSZ electrolyte free of interfacial reaction is critical to achieve high performance, and in this aspect, physical vapor deposition techniques were found to be superior to the wet ceramic processes.

2.3 Cathodes on Proton-Conducting Electrolytes

Some perovskite oxides have proton conductivity under humidified atmosphere at elevated temperatures. The proton conduction is based on the existence of a proton defect in the oxide [189]. Because of the low activation energy required for proton conduction, potentially proton-conducting perovskites could have higher ionic conductivity than oxygen-ion-conducting electrolytes at lower temperatures. Indeed, some proton conductors, such as $BaCe_{0.8}Y_{0.2}O_3$, show even higher ionic conductivity than stabilized zirconia and doped ceria oxygen-ion-conducting electrolytes at temperatures $<600\,^{\circ}C$ [190, 191]. It means proton-conducting LIT-SOFCs may have higher power output than oxygen-ion-conducting LIT-SOFCs. Furthermore, SOFCs based on proton-conducting electrolytes (SOFC-H^+) have higher theoretical electromotive force and electrical efficiency and easier fuel management than SOFCs based on oxygen-ion-conducting electrolytes because of the production of water at the cathode side [192].

Although several oxides with high protonic conductivity at reduced temperatures have been developed, the progress of proton-conducting SOFCs is much retarded when compared to oxygen-ion-conducting SOFC-O^{2-}. A peak power density of >1.0 W cm^{-2} at intermediate temperatures of 600–800 °C is frequently reported in the literature for oxygen-ion-conducting SOFCs with thin film electrolytes, while the power density of proton-conducting SOFCs seldom exceeds 500 mW cm^{-2} [193–201]. The difference in their cathode behavior results in their different fuel cell performance. In a proton-conducting SOFC, hydrogen is oxidized at the anode to proton, which migrates selectively through the electrolyte to the cathode and undergoes a half-cell reaction with oxygen to produce water. The overall half reactions at the anode and cathode could be expressed as Eq. (2.2) and Eq. (2.3), respectively:

$$H_2 + 2O_O^{\times} \rightarrow 2OH_O^{\bullet} + 2e' \tag{2.2}$$

$$4OH_O^{\bullet} + O_2 + 4e' \rightarrow 2H_2O + 4O_O^{\times} \tag{2.3}$$

with water generated at cathode. The cathode reactions of SOFC-H^+ differ a lot with those of cells with oxygen-ion-conducting electrolyte (Eq. (2.1)).

In fact, Uchida *et al.* [202] have pointed out that the overpotential of the Pt cathode in a proton-conducting SOFC is not negligible at temperatures lower than 900 °C, whereas the overpotential of the anode is negligible. Therefore, in order to achieve high power density in proton-conducting LIT-SOFCs, besides the development of new electrolyte material with high protonic conductivity and the reduction of electrolyte thickness, the development of cathode with low overpotential is also of great importance. Up to now, many perovskite-type mixed ionic–electronic conductors, which perform well as cathodes of oxygen-ion-conducting SOFCs at reduced temperatures, have been investigated as potential cathodes of proton-conducting SOFCs.

2.3.1
Cobaltite

Iwahara *et al.* [203] examined several semiconducting oxides, including $LaCrO_3$, Sn-doped In_2O_3, $La_{0.4}Ca_{0.6}CoO_{3-\delta}$, and $La_{0.4}Sr_{0.6}CoO_{3-\delta}$, as potential cathodes for a SOFC using proton-conducting $SrCe_{0.95}Yb_{0.05}O_{3-\delta}$ electrolyte. Among these oxides, cobalt-containing perovskite-type oxides showed the best performance. Hibino *et al.* [194] used $Ba_{0.5}Pr_{0.5}CoO_3$ as a cathode for SOFCs with Y-doped $BaCeO_3$ electrolyte. The overpotential of the Pd-loaded FeO anode and the $Ba_{0.5}Pr_{0.5}CoO_3$ cathode at 600 °C is 25 and 53 mV, respectively, at 200 mA cm^{-2}, which are less than one-fourth of those of a Pt electrode.

$Sm_{0.5}Sr_{0.5}CoO_{3-\delta}$ was investigated as the cathode material for SOFCs with $BaCe_{0.8}Sm_{0.2}O_{3-\delta}$ electrolytes by Wu *et al.* [204]. To enlarge the TPB length, $BaCe_{0.8}Sm_{0.2}O_{3-\delta}$ was mixed with $Sm_{0.5}Sr_{0.5}CoO_{3-\delta}$ to form a composite cathode with the aim to create mixed electronic, protonic, and oxygen ionic conductivity in the cathode. The interfacial polarization resistance of the cell decreased with the $Sm_{0.5}Sr_{0.5}CoO_{3-\delta}$–$BaCe_{0.8}Sm_{0.2}O_{3-\delta}$ ratio, reached the lowest at 60 wt% of $Sm_{0.5}Sr_{0.5}CoO_{3-\delta}$ (about 0.67 Ω cm^2 at 600 °C), and then increased, whereas the bulk resistance of the fuel cell remains almost unchanged with the ratio. Moreover, the firing temperature of the cathodes had a great effect on the cell resistance, and high-performance cathode was obtained by firing at 1050 °C. A minimum interfacial polarization resistance and a maximum powder density of 0.21 Ω cm^2 and 0.24 W cm^{-2} were achieved at 700 °C, respectively. Kinetics of ORR on $Sm_{0.5}Sr_{0.5}CoO_{3-\delta}$-$BaCe_{0.8}Sm_{0.2}O_{3-\delta}$ composite cathode was recently studied by He *et al.* [205]. The results suggest that the migration of protons to TPBs and the surface diffusion of O_{ad}^- might be the limiting reaction steps for oxygen activation over the $Sm_{0.5}Sr_{0.5}CoO_{3-\delta}$-$BaCe_{0.8}Sm_{0.2}O_{3-\delta}$ composite cathodes in wet atmosphere, while the oxygen ions transfer into electrolyte, the reduction of O_{ad} to O_{ad}^-, and the surface diffusion of O_{ad}^- might be the limiting reactions for $Sm_{0.5}Sr_{0.5}CoO_{3-\delta}$-$BaCe_{0.8}Sm_{0.2}O_{3-\delta}$ composite cathode in dry atmosphere. Yang *et al.* [206] demonstrated that high power densities can be achieved in proton-conducting SOFCs using properly designed mixed proton–oxygen ion–electron conducting cathode derived from $Sm_{0.5}Sr_{0.5}CoO_{3-\delta}$ and $Ba(Zr_{0.1}Ce_{0.7}Y_{0.2})O_{3-\delta}$. The composite fired at 1000 °C offered unique transport properties and greatly increased the number of active sites, thus facilitating the electrochemical reactions involving H^+, O^{2-}, and e' or $h^{\bullet}$ on the entire cathode surfaces. The peak power densities were about 725, 598, 445, and 272 $mWcm^{-2}$ at 700, 650, 600, and 550 °C, respectively.

Lin *et al.* [207] evaluated the potential application of BSCF as a cathode for a proton-conducting SOFC based on $BaCe_{0.9}Y_{0.1}O_{2.95}$ electrolyte. Cation diffusion from $BaCe_{0.9}Y_{0.1}O_{2.95}$ to BSCF with the formation of a perovskite-type Ba^{2+}-enriched BSCF and a Ba^{2+}-deficient $BaCe_{0.9}Y_{0.1}O_{2.95}$ at a firing temperature as low as 900 °C was observed, and the higher the firing temperatures, the larger the deviations of the A/B ratio from unity for the perovskites. It was found that the impurity phases did not induce a significant change in the cathodic polarization resistance based on symmetric cell tests; however, the ohmic resistance of the cell obviously increased. Under optimized conditions, a maximum peak power density of ~550 and 100 mW cm^{-2} was reported at 700 and 400 °C, respectively, for the cell with the BSCF cathode fired at 950 °C and a 50 μm thick $BaCe_{0.9}Y_{0.1}O_{2.95}$ electrolyte. The results obtained by Lin *et al.* were much better than those reported by Peng *et al.* [208], who applied a $Ba_{0.5}Sr_{0.5}Co_{0.8}Fe_{0.2}O_{3-\delta}-BaCe_{0.9}Sm_{0.1}O_{2.95}$ composite oxide as the cathode for a proton-conducting $BaCe_{0.9}Sm_{0.1}O_{2.95}$ electrolyte and a firing temperature of 1100 °C for fixing the cathode layer to the electrolyte surface. They observed an area-specific polarization resistance as large as ~2.25 Ω cm^2 at 600 °C; for comparison, Lin *et al.* reported a value of only 0.5 Ω cm^2 under similar operating conditions. Lin *et al.* explained that the poor performance reported by Peng *et al.* could be related to the high firing temperature (1100 °C) and the composite cathode applied. They believed that in the composite cathode, the A-site-cation-deficient $BaCe_{0.9}Sm_{0.1}O_{2.95}$ was formed throughout the electrode layer, which could cover the BSCF surface and block the oxygen reduction over the cathode. In the case of the pure BSCF applied as the cathode layer, the phase reaction occurred only at the interface between the BSCF cathode and the $BaCe_{0.9}Y_{0.1}O_{2.95}$ electrolyte; therefore, the oxygen reduction properties of the cathode layer were not seriously affected.

2.3.2 Ferrite

Proton-conducting perovskites such as $BaCe_{0.8}Y_{0.2}O_3$ have a low TEC of less than 10×10^{-6} K^{-1}, while cobaltite perovskites typically have high TEC of larger than 20×10^{-6} K^{-1}. The large difference in TEC could result in a serious problem with the long-term operational stability. Although, in principle, the electrolyte's conductivity should be favorably high at reduced temperatures for proton conduction, the easy interfacial reaction occurring between cobaltite perovskite and proton-conducting perovskite electrolyte could result in a large ohmic drop in a thin film proton-conducting fuel cell with cobaltite perovskite-type cathode.

Several cobalt-free perovskite oxides have been reported as possible cathodes for fuel cells based on protonic electrolyte. Yamaura *et al.* [209] studied the cathodic polarization of strontium-doped lanthanum ferrite, $La_{1-x}Sr_xFeO_{3-\delta}$ (LSF), in the hydrogen–oxygen fuel cell using proton-conducting $SrCe_{0.95}Yb_{0.05}O_{3-\alpha}$ as the electrolyte. It is interesting that the overpotential of $La_{0.7}Sr_{0.3}FeO_{3-\delta}$ cathode was smaller than that of other perovskite-type oxides, such as $La_{0.7}Sr_{0.3}MnO_{3-\delta}$ and $La_{0.7}Sr_{0.3}CoO_{3-\delta}$, and platinum at 500–700 °C. The best cathodic performance was obtained for $La_{0.7}Sr_{0.3}FeO_{3-\delta}$ heat treated at 900 °C. The cathodic polarization resistance of $La_{0.7}Sr_{0.3}FeO_{3-\delta}$ is independent of oxygen partial pressure of the environment, whereas that of sputtered platinum was proportional to $pO_2^{1/4}$, indicating that the rate-determining step of the cathode reaction in $La_{0.7}Sr_{0.3}FeO_{3-\delta}$ differs from that in sputtered platinum.

Wang *et al.* [210] reported a novel cobalt-free $Ba_{0.5}Sr_{0.5}Zn_{0.2}Fe_{0.8}O_{3-\delta}$ perovskite-type mixed conductor for use in a high-temperature oxygen-permeable membrane. Later, Ding *et al.* [211] applied a $Ba_{0.5}Sr_{0.5}Zn_{0.2}Fe_{0.8}O_{3-\delta}-BaZr_{0.1}Ce_{0.7}Y_{0.2}O_{3-\delta}$ composite as the cathode of proton-conducting SOFCs with $BaZr_{0.1}Ce_{0.7}Y_{0.2}O_{3-\delta}$ electrolyte. An open-circuit potential of 1.015 V, a maximum power density of 486 mW cm^{-2}, and a low polarization resistance of the electrodes of only 0.08 $\Omega\ cm^2$ was achieved at 700 °C.

2.3.3 Bismuthate

As discussed, nowadays, most materials used for cathodes of proton-conducting SOFCs are mixed oxygen ionic and electronic conductors. The practical applications of the proton-conducting SOFCs are limited while using these materials as cathode most likely because of the limited active sites for oxygen reduction at the interface between the proton-conducting electrolyte and the oxygen-ion-conducting cathodes. Therefore, the composite cathode consisting of the cathode and electrolyte material was used to allow the simultaneous transport of proton, oxygen vacancy, and electronic defects, which effectively extend the active sites for oxygen reduction to a large extent and reduce the cathodic polarization resistance. It is expected that the material with mixed protonic and electronic conduction may improve the cathode performance dramatically.

Tao *et al.* [212] developed the single-phase proton-conducting $BaCe_{0.5}Bi_{0.5}O_{3-\delta}$ as the cathode for SOFC-H^+. Without addition of the electrolyte powder, the single-phase $BaCe_{0.5}Bi_{0.5}O_{3-\delta}$ showed promising cathode performance. A maximum power density of 321 mW cm^{-2} was obtained for the single cell, which is comparable with other composite cathodes reported recently under similar conditions.

2.4 Advanced Techniques in Cathode Fabrication

2.4.1 Wet Impregnation

With fabrication by wet impregnation, the composite is formed by first making a porous layer of the electrolyte material together with the electrolyte. The porous layer is used as the oxygen-ion-conducting channels for the electrode and the electronically conductive components are deposited into this layer in subsequent steps. The general procedure is shown diagrammatically in Figure 2.8. Synthesis

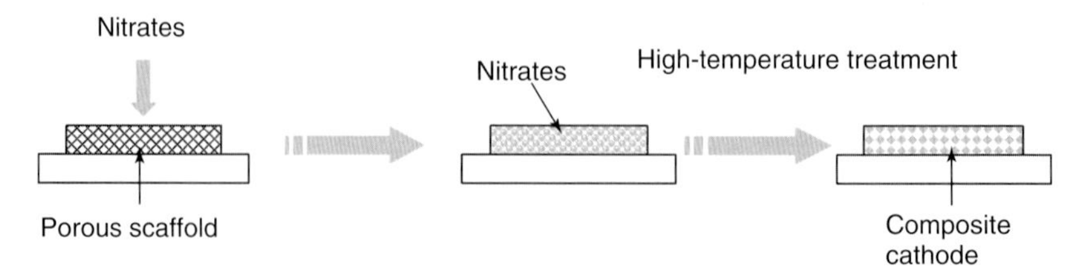

Figure 2.8 Schematic of a typical wet impregnation process for a composite cathode.

of composite electrodes by wet impregnation was initially developed to separate the calcination temperature of the YSZ from the sintering temperatures of the other electrode components. After that, other advantages have been observed for electrodes prepared in this way, such as improved microstructure of the cathode, matched TEC, and reduced cost of metal catalyst.

2.4.1.1 **Alleviated Phase Reaction**

Formation of undesired phases between cathode and electrolyte is an important issue to inhibit the enhancement of cathode performance. Although the LSM perovskites exhibit higher reactive stability than the $LaCoO_3$- and $LaFeO_3$-based perovskites, the reaction between YSZ and LSM can still occur during processing at high temperatures. The reaction layer, whether formed during processing or cell operation, is detrimental to fuel cell performance [213] because the conductivities of the reaction products are lower than those of the electrolyte and electrode materials [214–217]. In order to decrease the firing temperature for cathode fabrication, the wet impregnation technique is used as an effective way to add the second phase.

This technique also allowed $La_{0.6}Sr_{0.4}CoO_{3-\delta}$, $La_{0.8}Sr_{0.2}FeO_{3-\delta}$, $La_{0.8}Sr_{0.2}Co_{0.5}Fe_{0.5}O_{3-\delta}$, $LaNi_{0.6}Fe_{0.4}O_{3-\delta}$, and $La_{0.91}Sr_{0.09}Ni_{0.6}Fe_{0.4}O_{3-\delta}$ oxides to be used as the cathode on stabilized zirconia electrolyte [218–223]. The reason these materials were not widely applied is that they react rapidly with stabilized zirconia at 1000 °C, the minimum sintering temperature for YSZ, to form insulating $La_2Zr_2O_7$ and $SrZrO_3$ phases [15, 224]. By using the wet impregnation technique, the sintering temperature for cathode fabrication can be decreased to 700 °C. Therefore, it is possible to prepare these composite electrodes on stabilized zirconia electrolytes without any significant interfacial reaction between them. For example, Armstrong and Rich [219] demonstrated that a cell with a cathode composed of 30 vol% LSC in a YSZ scaffold exhibited a peak power density of 2.1 W cm^{-2} at 800 °C. Huang *et al.* [218] showed that the impedance of an LSC–YSZ electrode at 700 °C was 0.03 Ω cm^2.

2.4.1.2 **Optimized Microstructure**

In addition, this method can optimize the microstructure that would increase the length of TPB among the electrolyte, electrode, and oxygen gas phase, as shown in Figure 2.9 [225]. This diagram demonstrates that the YSZ within the composite can provide pathways for oxygen ions to migrate into the electrode. The distance

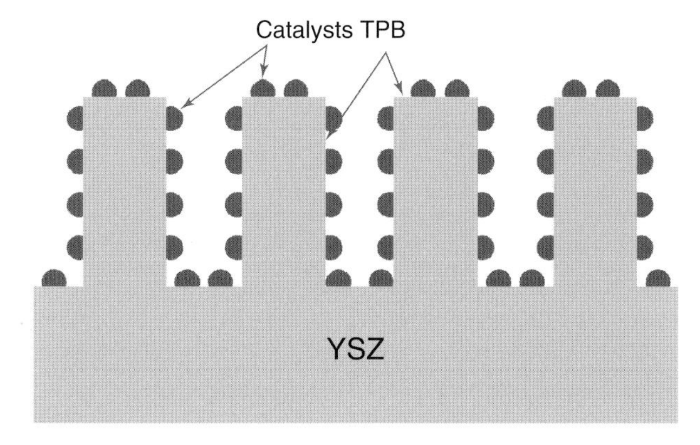

Figure 2.9 Schematic diagram of infiltrated YSZ cathode.

from the electrolyte interface, *h*, for which the YSZ pathways are effective in improving electrode performance, depends on the composition of the composite and the structure of the YSZ within the composite. Theoretical and experimental considerations suggest that the electrochemically active region typically extends 10–20 μm from the electrolyte into the electrode [226, 227].

The wet impregnation technique is also suitable for making nanostructured catalysts into various porous scaffold materials other than stabilized zirconia. Shah and Barnett [228] reported the lowest polarization resistance of 0.24 Ω cm^2 for $La_{0.6}Sr_{0.4}Co_{0.2}Fe_{0.8}O_{3-\delta}$-impregnated GDC at 600 °C. Jiang *et al.* [229] fabricated LSM cathodes impregnated with yttria-stabilized bismuth (YSB) oxide. YSB is chemically compatible with LSM at intermediate temperatures below 800 °C. Single cells with 25% of YSB loading generated a maximum power density of 300 mW cm^{-2} at temperatures as low as 600 °C. Wet impregnation was also used to fabricate cathode for SOFC-H$^+$. Nanosized $Sm_{0.5}Sr_{0.5}CoO_{3-\delta}$ was fabricated onto the inner face of porous $BaCe_{0.8}Sm_{0.2}O_{3-\delta}$ backbone to form a composite cathode. The lowest polarization resistance of the composite cathodes was about 0.21 Ω cm^2 at 600 °C, which was achieved with $BaCe_{0.8}Sm_{0.2}O_{3-\delta}$ backbone sintered at 1100 °C, $Sm_{0.5}Sr_{0.5}CoO_{3-\delta}$ layer fired at 800 °C, and $Sm_{0.5}Sr_{0.5}CoO_{3-\delta}$ loading of 55 wt% [230].

The impregnation technique allows for incorporation of additional nanostructured catalysts into the composite cathodes. When the additional phases are added, the performance of cathodes can be improved. Sholklapper *et al.* [231] prepared LSM–YSZ by a single-step impregnation method. By impregnating $Y_{0.2}Ce_{0.8}O_{1.9}$ (YDC) into LSM–YSZ as a third phase, the cell shows a drastic enhancement of power density at 0.7 V from 135 mW cm^{-2} before impregnation to 370 mW cm^{-2} after infiltration. Sholklapper *et al.* [232] also prepared a Ag–LSM–impregnated ScSZ cathode, which demonstrated an enhanced effectiveness in the cathode Ag metal catalyst, producing relatively stable cell power densities of 316 mW cm^{-2} at 0.7 V (and 467 mW cm^{-2} peak power at ~0.4 V) for over 500 h. Liang *et al.* [233] compared the performance of conventionally mixed LSM–YSZ, LSM–impregnated YSZ (LSM–YSZ), and Pd-impregnated LSM–YSZ (Pd+LSM–YSZ) cathodes. It is believed that Pd introduced in the form of nanosized particles facilitates the electrochemical reaction by promoting oxygen dissociation and diffusion processes as compared to other structured cathodes. The electrochemical performance of the impregnated cathodes has been significantly boosted due to the formation of nanosized LSM and Pd particles on the YSZ and LSM–YSZ substrates, respectively, and in turn, an increase in the area of the TPB, where the O_2 reduction reaction occurs, and power densities as high as 1.42 and 0.83 W cm^{-2} at 750 °C were achieved from single cells with the Pd+LSM–YSZ and LSM–YSZ cathodes, respectively, in contrast to 0.20 W cm^{-2} from the single cell with the conventional LSM–YSZ cathode.

2.4.1.3 Matched Thermal Expansion Coefficient

The TEC for conventional composites typically varies with the weighted-average TEC values of the individual components [234]. However, with composites formed by wet impregnation of a perovskite into porous YSZ, it was demonstrated that the TEC was closer to that of the YSZ matrix than to the weighted-average TEC [218]. This is because when the perovskite forms a coating over the YSZ, the YSZ still provides the mechanical properties of the composite. This, together with the need for less perovskite to achieve conductivity in the infiltrated composites, implies that wet-impregnated electrodes should exhibit an improved TEC match with the electrolyte compared to conventional electrodes.

2.4.1.4 Reduced Cost of Metal Catalyst

Noble metal is another candidate for cathode material because of its high catalytic reactivity for reduction/oxidation of oxygen. The composite cathode fabricated by mechanical mixing needs a large number of metal catalysts to be evolved to achieve a favorable performance. Pt–scandia-stabilized zirconia and PtAg–scandia-stabilized zirconia cermet cathodes need Pt of 40 and 19 mg cm^{-2}, respectively, for optimum electrode reactivity [235, 236]. A large initial investment is required to use these cermet cathodes, although the metal catalysts can be recycled. The high cost of metal catalysts would make the SOFCs too expensive to be practically used. Wet impregnation can reduce the loading of metal catalyst to a high extent, meanwhile maintain a high cathode performance. Liang *et al.* [237] showed that a high cathode performance was achieved on a Pd-impregnated YSZ composite. The Pd nanoparticles (20–80 nm) were uniformly distributed in the porous YSZ structure, and such nanostructured composite cathodes were highly active for ORR, with polarization resistances of 0.11 and 0.22 Ω cm^2 at 750 and 700 °C, respectively, and activation energy of 105 kJ mol^{-1} that is significantly lower than that of the conventional perovskite-based cathodes (130–201 kJ mol^{-1}). The loading of Pd is only ~1.4 mg cm^{-2}.

2.4.2 Surfactant-Assisted Assembly Approach

Electrodes have stringent requirements for use within an SOFC because of the high operating temperatures involved. These include stability in terms of chemical reactivity, phase, morphology, dimensionality, TEC, catalytic activity, electronic and ionic conductivity, and porosity [238]. Existing electrode materials are intrinsically dense having zero intragranular porosity at elevated temperatures and exhibiting low surface areas arising from intergranular necking produced through sintering processes. Porosity is a singular attribute, which controls not only the transport of gaseous oxidant to reactive sites but also the length of the TPB where charge transfer occurs for an electronically conducting electrode [239–242]. *TPB* is defined as the interface where the electronically conductive electrode meets both the YSZ electrolyte and the gaseous oxidant. Both mass transport (gaseous diffusion, adsorption processes, and surface diffusion) and charge-transfer processes at the TPB limit the efficiency of SOFCs. It is important to note that the porosity should not disrupt the percolation of electrons throughout the electrode microstructure.

Several researchers have attempted to improve porosity and enlarge the TPB by manipulating the electrode microstructure through traditional solid-state chemistry and material science techniques, which includes but is not limited to the impregnation of YSZ with noble metal salts and chemical deposition of electrode materials on YSZ substrates [243]. The common thread among these approaches involves enlarging the TPB by diminishing the dimensions of metal particles such as Pt in relation to YSZ grains. In essence, these materials are nanoscale or microscale versions of the bulk cermet electrode materials having a comparatively wide pore size distribution with low thermal stability.

Since the discovery of surfactant-templated mesoporous silicates [244, 245], there has been considerable interest in obtaining novel mesoporous transition-metal oxides, which might, for example, exhibit interesting catalytic, electrocatalytic, electrochromic, charge-transport, optical, and host–guest inclusion properties. In particular, mesoporous oxides of yttrium and zirconium have been synthesized using a variety of surfactant template molecules and transition-metal precursors [246–250]. At a higher length scale, a colloidal crystal of latex spheres has been used as a template for molding a periodic macroporous form of crystalline YSZ. However, the walls were composed of closely packed dense material and the yttrium content was quite low due to the limited solubility of the yttrium alkoxide precursor in zirconium alkoxide [251]. This methodology was well developed by Mamak and Ozin *et al.* [252–256] to prepare $La_{1-x}Sr_xMnO_3$–YSZ nanocomposites for use as SOFC cathodes [257]. This material with subhundred nanometer grain sizes for each phase is the first such nanocomposite where aqueous-based precursors of each component are incorporated in a single synthetic step. This approach utilizes the coassembly of an anionic yttrium–zirconium acetatoglycolate gel, cetyltrimethylammonium bromide (CTAB), as the cationic surfactant template and inorganic La, Mn, and Sr salts under alkaline aqueous conditions. The resulting as-synthesized product is an amorphous mesostructured organic–inorganic composite, which is transformed

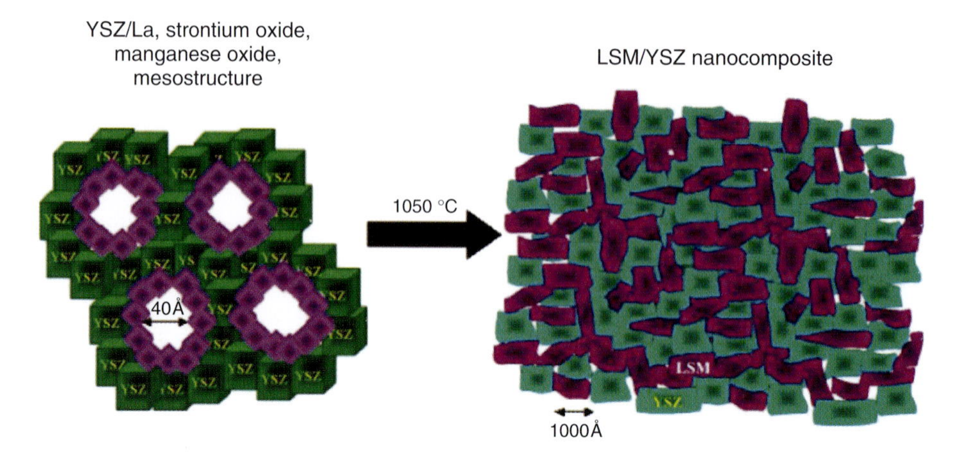

Figure 2.10 Structural evolution of meso-LSM–YSZ to nanocomposite-LSM–YSZ on sintering. The purple species in the mesostructure represent poorly crystalline La, Sr, and Mn oxides. (Source: Reproduced with permission from Ref. [257].)

to a mesoporous inorganic oxide with nanocrystalline YSZ walls on calcination. Calcination to temperatures above 600 °C leads to collapse of the mesopores followed by further crystallization of the nanocrystalline YSZ phase and a final crystallization of the LSM perovskite phase above 1000 °C. Structural evolution of meso-LSM–YSZ to nanocomposite-LSM–YSZ on sintering is shown in Figure 2.10. The nanoscale-phase separation created by CTAB in solution allows for the Y–Zr acetatoglycolate and La–Sr–Mn hydroxides to remain concentrated in different regions of the as-synthesized product so that each phase can later separately crystallize as cubic YSZ at 450 °C and as LSM perovskite above 800 °C. Impedance spectroscopy analysis of LSM–YSZ nanocomposite electrodes demonstrates a low polarization resistance of around 0.2 Ω cm^2 with an activation energy as low as 1.42 eV. Cathodic polarization studies show stable current densities over a 40 h test demonstration.

2.4.3 Spray Pyrolysis

Spray pyrolysis is a very versatile technique to obtain thin films of various materials and morphologies. Easy control of stoichiometry and simple experimental setup are the main advantages. The different microstructures needed for SOFC electrolytes (dense) and SOFC electrodes (porous) can be obtained using spray pyrolysis because the large number of parameters involved in the spray process helps fabricate thin films with very different microstructures. Beckel *et al.* [258–260] prepared $La_{0.6}Sr_{0.4}Co_{0.2}Fe_{0.8}O_{3-\delta}$ thin film cathodes by pressurized spray deposition. Hamedani *et al.* [261] prepared gradient porous LSM cathode film on YSZ electrolyte substrates using multiple-step ultrasonic spray pyrolysis technique. It was found that metal-organic precursors and organic solvent, which resulted in a homogenous crack-free deposition, have proven to be more satisfactory than aqueous solution.

Choy *et al.* [262] prepared LSM cathode using the flame-assisted spray pyrolysis method. The derived cathode yields a relatively low activation energy of 96.29 kJ mol^{-1} compared with the value reported by Chen *et al.* [263], which is 136.24 W cm^{-2} (LSM–YSZ–LSM) using the slurry painting technique.

The cathodes prepared by spray pyrolysis proved to be suitable for application in micro-solid oxide fuel cells (μ-SOFCs). The ~500 nm thick cathodes showed performance similar to traditional thick film LSCF cathodes. However, by using a new material composition $Ba_{0.25}La_{0.25}Sr_{0.5}Co_{0.8}Fe_{0.2}O_{3-\delta}$ or by modifying the microstructure, a significant improvement in performance was achieved. Reducing the grain size or introducing a thin dense cathode layer between the porous cathode and the electrolyte was a beneficial modification of the microstructure [260].

Hwang and coworkers [264–266] successfully prepared $La_{0.6}Sr_{0.4}Co_{0.2}Fe_{0.8}O_3$ and $Sm_{0.5}Sr_{0.5}CoO_3$ cathode films through an electrostatic-assisted ultrasonic spray pyrolysis (EAUSP) method. The results of impedance measurement indicate that the ASR values of the current $La_{0.6}Sr_{0.4}Co_{0.2}Fe_{0.8}O_3$ films and their activation energies are comparable to that obtained by conventional sample preparation routes. The $Sm_{0.5}Sr_{0.5}CoO_3$ film of a unique porous columnar structure was obtained using a deposition temperature of 400 °C, an applied voltage of 10 kV, and extended deposition times up to 600 s, rendering a minimum interfacial ASR of 0.035 Ω cm^2 at 700 °C in air [264].

2.5 Summary

Significant advances in single-phase cathode materials based on oxygen-ion-conducting or proton-ion-conducting electrolyte were achieved for LIT-SOFCs. The peak power densities of the fuel cells based on various cathode materials that have been employed in LIT-SOFCs are summarized in Table 2.1. Some cobalt-containing MIEC (mixed ionic–electronic conducting) oxides, such as perovskite-type BSCF and the layered perovskite $PrBaCoO_{3-\delta}$, showed high electrocatalytic activity for ORRs. However, these materials have stability problems in phase reaction and thermal expansion mismatching between the cathode and the electrolyte. Severe degradation of the surface oxygen exchange coefficient is observed for some $SrCoO_3$-based cathodes exposed to a CO_2-containing atmosphere. Accordingly, extremely pure air feed is required as the oxidant at the cathode when employing these cathodes at low operating temperatures, which would increase the operation cost of the fuel cell. Future work should be focused on improving the stability of single-phase cathode materials for IT-SOFCs.

Wet impregnation was verified to effectively improve microstructure of the cathode and compromise the disadvantages in phase reaction and thermal expansion matching. The disadvantage of wet impregnation is multiple fabricating processes

Table 2.1 Peak power densities (P_{max}) of the fuel cells based on various cathode materials.

Cathode	Electrolyte	T (°C)	P_{max} (mW cm^{-2})	References
$La_{0.8}Sr_{0.2}Mn_{1.1}O_3$–ScSZ–$CeO_2$	YSZ (15 μm)	650	820	[153]
$Sm_{0.5}Sr_{0.5}CoO_{3-\delta}$–SDC (10 wt%)	SDC (25 μm)	500	120	[56]
$SrNb_{0.1}Co_{0.9}O_{3-\delta}$	SDC (20 μm)	500	561	[72]
$SrSc_{0.2}Co_{0.8}O_{3-\delta}$	SDC (20 μm)	500	564	[80]
$PrBaCo_2O_{5+\delta}$	SDC (42 μm)	600	583	[111]
$Ba_2Bi_{0.1}Sc_{0.2}Co_{1.7}O_{6-x}$	SDC (5 μm)|YSZ (10 μm)	750	1016	[116]
$Bi_{0.5}Sr_{0.5}FeO_{3-\delta}$	SDC (2 μm)|YSZ (6 μm)	750	1090	[94]
$La_{1.2}Sr_{0.8}Co_{0.8}Ni_{0.2}O_{4+\delta}$–GDC	GDC (40 μm)	600	350	[128]
$La_{0.8}Sr_{0.2}Mn_{0.95}Sc_{0.05}O_{3-\delta}$	ScSZ (30 μm)	750	650	[162]
$Sm_{0.5}Sr_{0.5}CoO_{3-\delta}$–Ba$(Zr_{0.1}Ce_{0.7}Y_{0.2})O_{3-\delta}$	Ba$(Zr_{0.1}Ce_{0.7}Y_{0.2})O_{3-\delta}$ (65 μm)	650	598	[206]
$Ba_{0.5}Sr_{0.5}Co_{0.8}Fe_{0.2}O_{3-\delta}$	SDC (20 μm)	500	402	[62]
		600	1010	
	GDC (10 μm)	500	454	[63]
		600	1329	
	GDC (1 μm)|YSZ (15 μm)	700	810	[166]
	$BaCe_{0.9}Y_{0.1}O_{2.95}$ (50 μm)	700	550	[207]

that will raise the initiative cost of SOFCs. Advanced simple manufacture techniques, such as surfactant-assisted assembly and spray pyrolysis, are also beneficial in fabricating cathodes with optimized structure. Further research is required in such techniques to improve the cathode performance.

References

1. Minh, N.Q. (1995) Ceramic fuel cells. *J. Am. Ceram. Soc.*, **76** (3), 563–588.
2. Singhal, S.C. and Kendall, K. (2003) *High-Temperature Solid Oxide Fuel Cells: Fundamentals, Design and Applications*, Elsevier Science, Oxford.
3. Steele, B.C.H. (2001) Material science and engineering: the enabling technology for the commercialisation of fuel cell systems. *J. Mater. Sci.*, **36** (5), 1053–1068.
4. Brett, D.J.L., Atkinson, A., Brandon, N.P., and Skinnerd, S.J. (2008) Intermediate temperature solid oxide fuel cells. *Chem. Soc. Rev.*, **37** (8), 1568–1578.
5. Steele, B.C.H. (2000) Materials for IT-SOFC stacks: 35 years R&D: the inevitability of gradualness? *Solid State Ionics*, **134** (1–2), 3–20.
6. Steele, B.C.H. and Heinzel, A. (2001) Materials for fuel-cell technologies. *Nature*, **414** (6861), 345–352.
7. Tietz, F., Buchkremer, H.-P., and Stöver, D. (2002) Components manufacturing for solid oxide fuel cells. *Solid State Ionics*, **152–153**, 373–381.

8. Baumann, F.S. (2006) Oxygen reduction kinetics on mixed conducting SOFC model cathodes. PhD thesis. Max Planck Institute for Solid State Research, Stuttgart, Germany.
9. Stuart, B.A. (2004) Factors governing oxygen reduction in solid oxide fuel cell cathodes. *Chem. Rev.*, **104** (10), 4791–4844.
10. Schwarz, K. (2006) Materials design of solid electrolytes. *Proc. Natl. Acad. Sci. U.S.A.*, **103** (10), 3497.
11. Andersson, D.A., Simak, S.I., Skorodumova, N.V., Abrikosov, I.A., and Johansson, B. (2006) Optimization of ionic conductivity in doped ceria. *Proc. Natl. Acad. Sci. U.S.A.*, **103** (10), 3518–3521.
12. Yahiro, H., Eguchi, K., and Arai, H. (1989) Electrical properties and reducibilities of ceria-rare earth oxide systems and their application to solid oxide fuel cell. *Solid State Ionics*, **36** (1–2), 71–75.
13. Kharton, V.V., Marques, F.M.B., and Atkinson, A. (2004) Transport properties of solid oxide electrolyte ceramics: a brief review. *Solid State Ionics*, **174** (1–4), 135–149.
14. Tu, H.Y., Takeda, Y., Imanishi, N., and Yamamoto, O. (1997) $Ln_{1-x}Sr_xCoO_3(Ln = Sm, Dy)$ for the electrode of solid oxide fuel cells. *Solid State Ionics*, **100** (3–4), 283–288.
15. Godickemeier, M., Sasaki, K., Gauckler, L.J., and Reiss, I. (1996) Perovskite cathodes for solid oxide fuel cells based on ceria electrolytes. *Solid State Ionics*, **86–88**, 691–701.
16. Kawada, T., Masuda, K., Suzuki, J., Kaimai, A., Kawamura, K., Nigara, Y., Mizusaki, J., Yugami, H., Arashi, H., Sakai, N., and Yokokawa, H. (1999) Oxygen isotope exchange with a dense $La_{0.6}Sr_{0.4}CoO_{3-\delta}$ electrode on a $Ce_{0.9}Ca_{0.1}O_{1.9}$ electrolyte. *Solid State Ionics*, **121** (1–4), 271–279.
17. Adler, S.B. (1998) Mechanism and kinetics of oxygen reduction on porous $La_{1-x}Sr_xCoO_{3-\delta}$ electrodes. *Solid State Ionics*, **111** (1–2), 125–134.
18. Petrov, A.N., Kononchuk, O.F., Andreev, A.V., Cherepanov, V.A., and Kofstad, P. (1995) Crystal structure, electrical and magnetic properties of $La_{1-x}Sr_xCoO_{3-y}$. *Solid State Ionics*, **80** (3–4), 189–199.
19. Skinner, S.J. (2001) Recent advances in perovskite-type materials for solid oxide fuel cell cathodes. *Int. J. Inorg. Mater.*, **3** (2), 113–121.
20. Tai, L.W., Nasrallah, M.N., Anderson, H.U., Sparlin, D.M., and Sehlin, S.R. (1995) Structure and electrical-properties of $La_{1-x}Sr_xCo_{1-y}FeyO_3$. 2. The system $La_{1-x}Sr_xCo_{0.2}Fe_{0.8}O_3$. *Solid State Ionics*, **76** (3–4), 273–283.
21. Tai, L.W., Nasrallah, M.N., Anderson, H.U., Sparlin, D.M., and Sehlin, S.R. (1995) Structure and electrical-properties of $La_{1-x}Sr_xCo_{1-y}Fe_yO_3$. 1. The system $La_{0.8}Sr_{0.2}Co_{1-y}Fe_yO_3$. *Solid State Ionics*, **76** (3–4), 259–271.
22. Li, S.G., Jin, W., Xu, N., and Shi, J. (1999) Synthesis and oxygen permeation properties of $La_{0.2}Sr_{0.8}Co_{0.2}Fe_{0.8}O_{3-\delta}$ membranes. *Solid State Ionics*, **124** (1–2), 161–170.
23. Kostogloudis, G.C. and Ftikos, C. (1999) Properties of A-site-deficient $La_{0.6}Sr_{0.4}Co_{0.2}Fe_{0.8}O_{3-\delta}$-based perovskite. *Solid State Ionics*, **126** (1–2), 143–151.
24. Waller, D., Lane, J.A., Kilner, J.A., and Steele, B.C.H. (1996) The structure of and reaction of A-site deficient $La_{0.6}Sr_{0.4-x}Co_{0.2}Fe_{0.8}O_{3-\delta}$. *Mater. Lett.*, **27** (4–5), 225–228.
25. Tu, H.Y., Takeda, Y., Imanishi, N., and Yamamoto, O. (1999) $Ln_{0.4}Sr_{0.6}Co_{0.8}Fe_{0.2}O_{3-\delta}$ (Ln=La, Pr, Nd, Sm, Gd) for the electrode in solid oxide fuel. *Solid State Ionics*, **117** (3–4), 277–281.
26. Waller, D., Lane, J.A., Kilner, J.A., and Steele, B.C.H. (1996) The effect of thermal treatment on the resistance of LSCF electrodes on gadolinia doped ceria electrolytes. *Solid State Ionics*, **86–88**, 767–772.
27. Tai, L.W., Nasrallah, M.M., and Anderson, H.U. (1999) in *Proceedings of the third International Symposium on Solid Oxide Fuel Cells*, The Electrochemical Society Proceedings Series, (eds S.C. Singhal and H. Iwahara), The

Electrochemical Society, Pennington, NJ, p. 241.
28. Chen, C.C., Nasralla, M.M., Anderson, H.U., and Alim, M.A. (1995) Immittance response of $La_{0.6}Sr_{0.4}Co_{0.2}Fe_{0.8}O_3$ based electrochemical cells. *J. Electrochem. Soc.*, **142** (2), 491–496.
29. Chen, C.C. and Nasrallah, M.M. (1999) in *Proceedings of the third International Symposium on Solid Oxide Fuel Cells*, The Electrochemical Society Proceedings Series (eds H.U. Anderson, S.C. Singhal, and H. Iwahara), The Electrochemical Society, Pennington, NJ, p. 252.
30. Zhou, W., Shao, Z.P., Ran, R., Gu, H.X., Jin, W.Q., and Xu, N.P. (2008) LSCF nanopowder from cellulose–glycine-nitrate process and its application in intermediate-temperature solid-oxide fuel cells. *J. Am. Ceram. Soc.*, **91** (4), 1155–1162.
31. Bae, J.M. and Steele, B.C.H. (1998) Properties of $La_{0.6}Sr_{0.4}Co_{0.2}Fe_{0.8}O_{3-\delta}$ (LSCF) double layer cathodes on gadolinium-doped cerium oxide (CGO) electrolytes: I. Role of SiO_2. *Solid State Ionics*, **106** (3–4), 247–253.
32. Bae, J.M. and Steele, B.C.H. (1998) Properties of $La_{0.6}Sr_{0.4}Co_{0.2}Fe_{0.8}O_{3-x}$ (LSCF) double layer cathodes on gadolinium-doped cerium oxide (CGO) electrolytes: II. Role of oxygen exchange and diffusion. *Solid State Ionics*, **106** (3–4), 255–261.
33. Perry Murray, E., Sever, M.J., and Barnett, S.A. (2002) Electrochemical performance of $(La,Sr)(Co,Fe)O_3$–$(Ce,Gd)O_3$ composite cathodes. *Solid State Ionics*, **148** (1–2), 27–34.
34. Dusastre, V. and Kilner, J.A. (1999) Optimisation of composite cathodes for intermediate temperature SOFC applications. *Solid State Ionics*, **126** (1–2), 163–174.
35. Esquirol, A., Kilner, J., and Brandon, N. (2004) Oxygen transport in $La_{0.6}Sr_{0.4}Co_{0.2}Fe_{0.8}O_{3-\delta}/Ce_{0.8}Ge_{0.2}O_{2-x}$ composite cathode for IT-SOFCs. *Solid State Ionics*, **175** (1–4), 63–67.
36. Visco, S.J., Jacobson, C., and De Jonghe, L.C. (1997) in *Solid Oxide Fuel Cells V*, The Electrochemical Society Proceedings Series (eds U. Stimming, S.C. Singhal, H. Tagawa, and W. Lehnert), The Electrochemical Society, Pennington, NJ, p. 710.
37. Xia, C.R., Chen, F.L., and Liu, M.L. (2001) Reduced-temperature solid oxide fuel cells fabricated by screen printing. *Electrochem. Solid-State Lett.*, **4** (5), A52–A54.
38. Lee, S., Song, H.S., Hyun, S.H., Kim, J., and Moon, J. (2010) LSCF-SDC core-shell high-performance durable composite cathode. *J. Power Sources*, **195** (1), 118–123.
39. Ullmann, H., Trofimenko, N., Tietz, F., Stöver, D., and Ahmad-Khanlou, A. (2000) Correlation between thermal expansion and oxide ion transport in mixed conducting perovskite-type oxides for SOFC cathodes. *Solid State Ionics*, **138** (1–2), 79–90.
40. Petric, A., Huang, P., and Tietz, F. (2000) Evaluation of La-Sr-Co-Fe-O perovskites for solid oxide fuel cells and gas separation membranes. *Solid State Ionics*, **135** (1–4), 719–725.
41. Hashimoto, S., Fukuda, Y., Kuhn, M., Sato, K., Yashiro, K., and Mizusaki, J. (2011) Thermal and chemical lattice expansibility of $La_{0.6}Sr_{0.4}C_{o1-y}Fe_yO_{3-\delta}$ (y = 0.2, 0.4, 0.6 and 0.8) *Solid State Ionics*, **186** (1), 37–43.
42. Mogensen, M., Lindegaard, T., Hansen, U.R., and Mogensen, G. (1994) Physical properties of mixed conductor solid oxide fuel cell anodes of doped CeO_2. *J. Electrochem. Soc.*, **141** (8), 2122–2128.
43. Simner, S.P., Anderson, M.D., Templeton, J.W., and Stevenson, J.W. (2007) Silver-perovskite composite SOFC cathodes processed via mechanofusion. *J. Power Sources*, **168** (1), 236–239.
44. Christie, G.M., van Heuveln, F.H., and van Berkel, F.P.F. (1996) in *Proceedings of the 17th Risø International Symposium on Materials Science: High Temperature Electrochemistry: Ceramics and Metals* (eds F.W. Poulsen, N. Bonanos, S. Lidenoth, M. Morgensen, and B. Zochau-Christiansen), Risø National Laboratory, Roskilde, pp. 205–211.

45. Sahibzada, M., Benson, S.J., Rudkin, R.A., and Kilner, J.A. (1998) Pd-promoted $La_{0.6}Sr_{0.4}Fe_{0.8}Co_{0.2}O_3$ cathodes. *Solid State Ionics*, **113–115**, 285–290.
46. Hwang, H.J., Moon, J.W., Lee, S., and Lee, E.A. (2005) Electrochemical performance of LSCF-based composite cathodes for intermediate temperature SOFCs. *J. Power Sources*, **145** (2), 243–248.
47. Peña, M.A. and Fierro, J.L.G. (2001) Chemical structures and performance of perovskite oxides. *Chem. Rev.*, **101** (7), 1981–2018.
48. Orera, A. and Slater, P.R. (2010) New chemical systems for solid oxide fuel cells. *Chem. Mater.*, **22** (3), 675–690.
49. Takeda, Y., Ueno, H., Imanishi, N., Yamamoto, O., Sammes, N., and Phillipps, M.B. (1996) $Gd1-xSr_xCoO_3$ for the electrode of solid oxide fuel cells. *Solid State Ionics*, **86–88**, 1187–90.
50. Hibino, T., Hashimoto, A., Inoue, T., Tokuno, J.-I., Yoshida, S.-I., and Sano, M. (2000) A low-operating-temperature solid oxide fuel cell in hydrocarbon-air mixtures. *Science*, **288**, 2031–2033.
51. Ishihara, T., Honda, M., Shibayama, T., Minami, H., Nishiguchi, H., and Takita, Y. (1998) Intermediate temperature solid oxide fuel cells using a new $LaGaO_3$ based oxide ion conductor – I. Doped $SmCoO_3$ as a new cathode material. *J. Electrochem. Soc.*, **145** (9), 3177–3183.
52. Koyama, M., Wen, C., Masuyama, T., Otomo, J., Fukunaga, H., Yamada, K., Eguchi, K., and Takahashi, H. (2001) The mechanism of porous $Sm_{0.5}Sr_{0.5}CoO_3$ cathodes used in solid oxide fuel cells. *J. Electrochem. Soc.*, **148** (7), A795–A801.
53. Fukunaga, H., Koyama, M., Takahashi, N., Wen, C., and Yamada, K. (2000) Reaction model of dense $Sm_{0.5}Sr_{0.5}CoO_3$ as SOFC cathode. *Solid State Ionics*, **132** (3–4), 279–285.
54. Lv, H., Zhao, B.Y., Wu, Y.J., Sun, G., Chen, G., and Hu, K.A. (2007) Effect of B-site doping on $Sm_{0.5}Sr_{0.5}M_xCo_{1-x}O_{3-\delta}$ properties for IT-SOFC cathode material (M = Fe, Mn). *Mater. Res. Bull.*, **42** (12), 1999–2012.
55. Kim, S., Yang, Y.L., Jacobson, A.J., and Abeles, B. (1998) Diffusion and surface exchange coefficients in mixed ionic electronic conducting oxides from the pressure dependence of oxygen permeation. *Solid State Ionics*, **106** (3–4), 189–195.
56. Xia, C.R., Rauch, W., Chen, F., and Liu, M.L. (2002) $Sm_{0.5}Sr_{0.5}CoO_3$ cathodes for low-temperature SOFCs. *Solid State Ionics*, **149** (1–2), 11–19.
57. Shao, Z.P., Yang, W.S., Cong, Y., Dong, H., Tong, J.H., and Xiong, G.X. (2000) Investigation of the permeation behavior and stability of a $Ba_{0.5}Sr_{0.5}Co_{0.8}Fe_{0.2}O_{3-\delta}$ oxygen membrane. *J. Membr. Sci.*, **172** (1–2), 177–188.
58. Zeng, P.Y., Chen, Z.H., Zhou, W., Gu, H.X., Shao, Z.P., and Liu, S.M. (2007) Re-evaluation of $Ba_{0.5}Sr_{0.5}Co_{0.8}Fe_{0.2}O_{3-\delta}$ perovskite as oxygen semi-permeable membrane. *J. Membr. Sci.*, **291** (1–2), 148–156.
59. Shao, Z.P., Xiong, G.X., Dong, H., Yang, W.S., and Lin, L.W. (2001) Synthesis, oxygen permeation study and membrane performance of a $Ba_{0.5}Sr_{0.5}Co_{0.8}Fe_{0.2}O_{3-\delta}$ oxygen-permeable dense ceramic reactor for partial oxidation of methane to syngas. *Sep. Purif. Technol.*, **25** (1–3), 97–116.
60. Shao, Z.P., Dong, H., Xiong, G.X., Cong, Y., and Yang, W.S. (2001) Performance of a mixed-conducting ceramic membrane reactor with high oxygen permeability for methane conversion. *J. Membr. Sci.*, **183** (2), 181–192.
61. Shao, Z.P., Xiong, G.X., Tong, J.H., Dong, H., and Yang, W.S. (2001) Ba effect in doped $Sr(Co_{0.8}Fe_{0.2})O_{3-\delta}$ on the phase structure and oxygen permeation properties of the dense ceramic membranes. *Sep. Purif. Technol.*, **25** (1–3), 419–429.
62. Shao, Z.P. and Haile, S.M. (2004) A high-performance cathode for the next generation of solid-oxide fuel cells. *Nature*, **431** (7005), 170–173.

63. Liu, Q.L., Khor, K.A., and Chan, S.H. (2006) High-performance low-temperature solid oxide fuel cell with novel BSCF cathode. *J. Power Sources*, **161** (1), 123–128.

64. McIntosh, S., Vente, J.F., Haije, W.G., Blank, D.H.A., and Bouwmeester, H.J.M. (2006) Oxygen stoichiometry and chemical expansion of $Ba_{0.5}Sr_{0.5}Co_{0.8}Fe_{0.2}O_{3-\delta}$ measured by in situ neutron diffraction. *Chem. Mater.*, **18** (8), 2187–2193.

65. Baumann, F.S., Fleig, J., Habermeier, H.U., and Maier, J. (2006) $Ba_{0.5}Sr_{0.5}Co_{0.8}Fe_{0.2}O_{3-\delta}$ thin film microelectrodes investigated by impedance spectroscopy. *Solid State Ionics*, **177** (35–36), 3187–3191.

66. Baumann, F.S., Fleig, J., Habermeier, H.U., and Maier, J. (2006) Impedance spectroscopic study on well-defined $(La,Sr)(Co,Fe)O_{3-\delta}$ model electrodes. *Solid State Ionics*, **177** (11–12), 1071–1081.

67. Zhou, W., Ran, R., and Shao, Z.P. (2009) Progress in understanding and development of $Ba_{0.5}Sr_{0.5}Co_{0.8}Fe_{0.2}O_{3-\delta}$-based cathodes for intermediate-temperature solid-oxide fuel cells: a review. *J. Power Sources*, **192** (2), 231–246.

68. Zhou, W., Shao, Z.P., Ran, R., Zeng, P.Y., Gu, H.X., Jin, W.Q., and Xu, N.P. (2007) $Ba_{0.5}Sr_{0.5}Co_{0.8}Fe.2O_{3-\delta}$ + $LaCoO_3$ composite cathode for $Sm_{0.2}Ce_{0.8}O_{1.9}$-electrolyte based intermediate-temperature solid-oxide fuel cells. *J. Power Sources*, **168** (2), 330–337.

69. Aguadero, A., de la Calle, C., Alonso, J.A., Escudero, M.J., Fernández-Díaz, M.T., and Daza, L. (2007) Structural and electrical characterization of the novel $SrCo_{0.9}Sb_{0.1}O_{3-\delta}$ perovskite: evaluation as a solid oxide fuel cell cathode material. *Chem. Mater.*, **19** (26), 6437–6444.

70. Lin, B., Wang, S.L., Liu, H.L., Xie, K., Ding, H.P., Liu, M.F., and Meng, G.Y. (2009) $SrCo_{0.9}Sb_{0.1}O_{3-\delta}$ cubic perovskite as a novel cathode for intermediate-to-low temperature solid oxide fuel cells. *J. Alloys Compd.*, **472** (1–2), 556–558.

71. Aguadero, A., Pérez-Coll, D., de la Calle, C., Alonso, J.A., Escudero, M.J., and Daza, L. (2009) $SrC_{o1-x}Sb_xO_{3-\delta}$ perovskite oxides as cathode materials in solid oxide fuel cells. *J. Power Sources*, **192** (1), 132–137.

72. Zhou, W., Shao, Z.P., Ran, R., Jin, W.Q., and Xu, N.P. (2008) A novel efficient oxide electrode for electrocatalytic oxygen reduction at 400–600 °C. *Chem. Commun.*, 5791–5793.

73. Zhou, W., Jin, W.Q., Zhu, Z.H., and Shao, Z.P. (2010) Structural, electrical and electrochemical characterizations of $SrNb_{0.1}Co_{0.9}O_{3-\delta}$ as a cathode of solid oxide fuel cells operating below 600 °C. *Int. J. Hydrogen Energy*, **35** (3), 1356–1366.

74. Wei, B., Lü, Z., Li, S., Liu, Y., Liu, K., and Su, W.H. (2005) Thermal and electrical properties of new cathode material $Ba_{0.5}Sr_{0.5}Co_{0.8}Fe_{0.2}O_{3-\delta}$ for solid oxide fuel cells. *Electrochem. Solid-State Lett.*, **8** (8), A428–A431.

75. Lim, Y.H., Lee, J., Yoon, J.S., Kim, C.E., and Hwang, H.J. (2007) Electrochemical performance of $Ba_{0.5}Sr_{0.5}Co_xFe_{1-x}O_{3-\delta}$ ($x = 0.2–0.8$) cathode on a ScSZ electrolyte for intermediate temperature SOFCs. *J. Power Sources*, **171** (1), 79–85.

76. Kek, D., Panjan, P., and Wanzenberg, E. (2001) Electrical and microstructural investigations of cermet anode/YSZ thin film systems. *J. Eur. Ceram. Soc.*, **21** (10–11), 1861–1865.

77. Mori, M. and Sammes, N.M. (2000) Sintering and thermal expansion characterization of Al-doped and Co-doped lanthanum strontium chromites synthesized by the Pechini method. *Solid State Ionics*, **146** (3–4), 301–312.

78. Kharton, V.V., Kovalevsky, A.V., Tikhonovich, V.N., Naumovich, E.N., and Viskup, A.P. (1998) Mixed electronic and ionic conductivity of $LaCo(M)O_3$ (M=Ga, Cr, Fe or Ni): II. Oxygen permeation through Cr- and Ni-substituted $LaCoO_3$. *Solid State Ionics*, **110** (1–2), 53–60.

79. Zeng, P.Y., Ran, R., Chen, Z.H., Zhou, W., Gu, H.X., Shao, Z.P., and Liu, S.M. (2008) Efficient stabilization of cubic perovskite $SrCoO_{3-\delta}$ by B-site

low concentration scandium doping combined with sol–gel synthesis. *J. Alloys Compd.*, **455** (1–2), 465–470.

80. Zhou, W., Shao, Z.P., Ran, R., and Cai, R. (2008) Novel $SrSc_{0.2}Co_{0.8}O_{3-\delta}$ as a cathode material for low temperature solid-oxide fuel cell. *Electrochem. Commun.*, **10** (10), 1647–1651.
81. Zhou, W., An, B.M., Ran, R., and Shao, Z.P. (2009) Electrochemical performance of $SrSc_{0.2}Co_{0.8}O_{3-\delta}$ cathode on $Sm_{0.2}Ce_{0.8}O_{1.9}$ electrolyte for low temperature SOFCs. *J. Electrochem. Soc.*, **156** (8), B884–B890.
82. Mineshige, A., Izutsu, J., Nakamura, M., Nigaki, K., Abe, J., Kobune, M., Fujii, S., and Yazawa, T. (2005) Introduction of A-site deficiency into $La_{0.6}Sr_{0.4}Co_{0.2}Fe_{0.8}O_{3-\delta}$ and its effect on structure and conductivity. *Solid State Ionics*, **176** (11–12), 1145–1149.
83. Hansen, K.K. and Vels Hansen, K. (2007) A-site deficient $(La_{0.6}Sr_{0.4})_{1-s}Fe_{0.8}Co_{0.2}O_{3-\delta}$ perovskites as SOFC cathodes. *Solid State Ionics*, **178** (23–24), 1379–1384.
84. Doshi, R., Richard, V.L., Carter, J.D., Wang, X.P., and Krumpelt, M. (1999) Development of solid-oxide fuel cells that operate at 500 °C. *J. Electrochem. Soc.*, **146** (4), 1273–1278.
85. Zhou, W., Ran, R., Shao, Z.P., Jin, W.Q., and Xu, N.P. (2008) Evaluation of A-site cation-deficient $(Ba_{0.5}Sr_{0.5})_{1-x}Co_{0.8}Fe_{0.2}O_{3-\delta}$ $(x > 0)$ perovskite as a solid-oxide fuel cell cathode. *J. Power Sources*, **182** (1), 24–31.
86. Chiba, R., Yoshimura, F., and Sakurai, Y. (1999) An investigation of $LaNi_{1-x}Fe_xO_3$ as a cathode material for solid oxide fuel cells. *Solid State Ionics*, **124** (3–4), 281–288.
87. Komatsu, T., Arai, H., Chiba, R., Nozawa, K., Arakawa, M., and Sato, K. (2006) Cr poisoning suppression in solid oxide fuel cells using LaNi(Fe)O-3 electrodes. *Electrochem. Solid State Lett.*, **9** (1), A9–A12.
88. Millar, L., Taherparvar, H., Filkin, N., Slater, P., and Yeomans, J. (2008) Interaction of $(La_{1-x}Sr_x)_{1-y}MnO_3$–$Zr_{1-z}Y_zO_{2-\delta}$ cathodes and $LaNi_{0.6}Fe_{0.4}O_3$ current collecting layers for solid oxide fuel cell application. *Solid State Ionics*, **179** (19–20), 732–739.
89. Swierczek, K., Marzec, J., Palubiak, D., Zajac, W., and Molenda, J. (2006) LFN and LSCFN perovskites-structure and transport properties. *Solid State Ionics*, **177** (19–25), 1811–1817.
90. Li, S. and Zhu, B. (2009) Electrochemical performances of nanocomposite solid oxide fuel cells using nano-size material $LaNi_{0.2}Fe_{0.65}Cu_{0.15}O_3$ as cathode. *J. Nanosci. Nanotechnol.*, **9** (6), 3824–3827.
91. Hou, S., Alonso, J.A., Rajasekhara, S., Martinez-Lope, M.J., Fernandez-Diaz, M.T., and Goodenough, J.B. (2010) Defective Ni perovskites as cathode materials in intermediate-temperature solid-oxide fuel cells: a structure-properties correlation. *Chem. Mater.*, **22** (3), 1071–1079.
92. Simner, S.P., Bonnett, J.F., Canfield, N.L., Meinhardt, K.D., Sprenkle, V.L., and Stevenson, J.W. (2002) Optimized lanthanum ferrite-based cathodes for anode-supported SOFCs. *Electrochem. Solid-State Lett.*, **5** (7), A173–A175.
93. Simner, S.P., Shelton, J.P., Anderson, M.D., and Stevenson, J.W. (2003) Interaction between $La(Sr)FeO_3$ SOFC cathode and YSZ electrolyte. *Solid State Ionics*, **161** (1–2), 11–18.
94. Niu, Y.J., Zhou, W., Sunarso, J., Ge, L., Zhu, Z.H., and Shao, Z.P. (2010) High performance cobalt-free perovskite cathode for intermediate temperature solid oxide fuel cells. *J. Mater. Chem.*, **20** (43), 9619–9622.
95. Niu, Y.J., Sunarso, J., Liang, F.L., Zhou, W., Zhu, Z.H., and Shao, Z.P. (2011) A comparative study of oxygen reduction reaction on Bi- and La-doped $SrFeO_{3-\delta}$ perovskite cathodes. *J. Electrochem. Soc.*, **158** (2), B132–B138.
96. Niu, Y.J., Sunarso, J., Zhou, W., Liang, F.L., Ge, L., Zhu, Z.H., and Shao, Z.P. (2011) Evaluation and optimization of $Bi1\text{-}xSr_xFeO_{3-\delta}$ perovskites as cathodes of solid oxide fuel cells. *Int. J. Hydrogen Energy*, **36** (4), 3179–3186.
97. Ralph, J.M., Schoeler, A.C., and Krumpelt, M. (2001) Materials for

lower temperature solid oxide fuel cells. *J. Mater. Sci.*, **36** (5), 1161–1172.
98. Ding, H.P. and Xue, X.J. (2010) Cobalt-free layered perovskite $GdBaFe_2O_{5+x}$ as a novel cathode for intermediate temperature solid oxide fuel cells. *J. Power Sources*, **195** (15), 4718–4721.
99. Taskin, A.A., Lavrov, A.N., and Ando, Y. (2007) Fast oxygen diffusion in A-site ordered perovskites. *Prog. Solid State Chem.*, **35** (2–4), 481–490.
100. Kim, G., Wang, S., Jacobson, A.J., Yuan, Z., Donner, W., Chen, C.L., Reimus, L., Brodersen, P., and Mims, C.A. (2006) *Appl. Phys. Lett.*, **88** (2), 024103.
101. Kim, J.H. and Manthiram, A. (2008) $LnBaCo_{(2)}O_{(5+\delta)}$ oxides as cathodes for intermediate-temperature solid oxide fuel cells. *J. Electrochem. Soc.*, **155** (12), B385–B390.
102. Lin, B., Zhang, S.Q., Zhang, L.C., Bi, L., Ding, H.P., Liu, X.Q., Gao, J.F., and Meng, G.Y. (2008) Prontonic ceramic membrane fuel cells with layered $GdBaCo_2O_{5+x}$ cathode prepared by gel-casting and suspension spray. *J. Power Sources*, **177** (2), 330–333.
103. Taskin, A.A., Lavrov, A.N., and Ando, Y. (2005) Achieving fast oxygen diffusion in perovskites by cation ordering. *Appl. Phys. Lett.*, **86** (9), 091910.
104. Frontera, C., Caneiro, A., Carrillo, A.E., Oro-Sole, J., and Garcia-Munoz, J.L. (2005) Tailoring oxygen content on $PrBaCo_2O_{5+\delta}$ layered cobaltites. *Chem. Mater.*, **17** (22), 5439–5445.
105. Tarancón, A., Skinner, S.J., Chater, R.J., Hernández-Ramírez, F., and Kilner, J.A. (2007) Layered perovskites as promising cathodes for intermediate temperature solid oxide fuel cells. *J. Mater. Chem.*, **17** (30), 3175–3181.
106. Kim, G., Wang, S., Jacobson, A.J., Reimus, L., Brodersen, P., and Mims, C.A. (2007) Rapid oxygen ion diffusion and surface exchange kinetics in $PrBaCo_2O_{5+x}$ with a perovskite related structure and ordered A cations. *J. Mater. Chem.*, **17** (24), 2500–2505.
107. Li, N., Lu, Z., Wei, B., Huang, X.Q., Chen, K.F., Zhang, Y.H., and Su, W.H. (2008) Characterization of $GdBaCo_2O_{5+\delta}$ cathode for IT-SOFCs. *J. Alloys Compd.*, **454** (1–2), 274–279.
108. Tarancón, A., Morata, A., Dezanneau, G., Skinner, S.J., Kilner, J.A., Estrade, S., Hernandez-Ramirez, F., Peiro, F., and Morante, J.R. (2007) $GdBaCo_2O_{5+x}$ layered perovskite as an intermediate temperature solid oxide fuel cell cathode. *J. Power Sources*, **174** (1), 255–263.
109. Chang, A.M., Skinner, S.J., and Kilner, J.A. (2006) Electrical properties of $GdBaCo_2O_{5+x}$ for ITSOFC applications. *Solid State Ionics*, **177** (19–25), 2009–2011.
110. Zhang, K., Ge, L., Ran, R., Shao, Z.P., and Liu, S.M. (2008) Synthesis, characterization and evaluation of cation-ordered $LnBaCo_2O_{5+\delta}$ as materials of oxygen permeation membranes and cathodes of SOFCs. *Acta Mater.*, **56** (17), 4876–4889.
111. Zhu, C.J., Liu, X.M., Yi, C.S., Yan, D.T., and Su, W.H. (2008) Electrochemical performance of $PrBaCo_2O_{5+\delta}$ layered perovskite as an intermediate-temperature solid oxide fuel cell cathode. *J. Power Sources*, **185** (1), 193–196.
112. Chen, D.J., Ran, R., Zhang, K., Wang, J., and Shao, Z.P. (2009) Intermediate-temperature electrochemical performance of a polycrystalline $PrBaCo_2O_{5+\delta}$ cathode on samarium-doped ceria electrolyte. *J. Power Sources*, **188** (1), 96–105.
113. Kim, J.H., Kim, Y.M., Connor, P.A., Irvine, J.T.S., Bae, J., and Zhou, W.Z. (2009) Structural, thermal and electrochemical properties of layered perovskite $SmBaCo_2O_{5+\delta}$, a potential cathode material for intermediate-temperature solid oxide fuel cells. *J. Power Sources*, **194** (2), 704–711.
114. Zhu, X.F., Wang, H.H., and Yang, W.S. (2004) Novel cobalt-free oxygen permeable membrane. *Chem. Commun.*, 1130–1131.
115. Deng, Z.Q., Smit, J.P., Niu, H.J., Evans, G., Li, M.R., Xu, Z.L., Claridge, J.B., and Rosseinsky, M.J. (2009) *Chem. Mater.*, **21** (21), 5154–5162.
116. Zhou, W., Sunarso, J., Chen, Z.G., Ge, L., Motuzas, J., Zou, J., Wang,

G.X., Julbe, A., and Zhu, Z.H. (2011) Novel B-site ordered double perovskite $Ba_2Bi_{0.1}Sc_{0.2}Co_{1.7}O_{6-x}$ for highly efficient oxygen reduction reaction. *Energy Environ. Sci.*, **4** (3), 872–875.

117. Streule, S., Podlensyak, A., Sheptyakov, D., Pomjakushina, E., Stingaciu, M., Conder, K., Medarde, M., Patrakeev, M.V., Leonidov, I.A., Kozhevnikov, V.L., and Mesot, J. (2006) High-temperature order–disorder transition and polaronic conductivity in $PrBaCo_2O_{5.48}$. *Phys. Rev. B*, **73** (9), 94203.

118. Streule, S., Podlensyak, A., Pomjakushina, E., Conder, K., Sheptyakov, D., Medarde, M., and Mesot, J. (2006) Oxygen order–disorder phase transition in $PrBaCo_2O_{5.48}$ at high temperature. *Physica B*, **378–380**, 539–540.

119. Maignan, A., Caignaert, V., Raveau, B., Khomskii, D., and Sawatzky, G. (2004) Thermoelectric power of $HoBaCo_2O_{5.5}$: possible evidence of the spin blockade in cobaltites. *Phys. Rev. Lett.*, **93** (2), 26401.

120. Frontera, C., García-Muñoz, J.L., Llobet, A., Mañosa, L., and Aranda, M.A.G. (2003) Selective spin-state and metal-insulator transitions in $GdBaCo_2O_{5.5}$. *J. Solid State Chem.*, **171** (1–2), 349–352.

121. Tarancón, A., Marrero-López, D., Peña-Martínez, J., Ruiz-Morales, J.C., and Núñez, P. (2008) Effect of phase transition on high-temperature electrical properties of $GdBaCo_2O_{5+x}$ layered perovskite. *Solid State Ionics*, **179** (17–18), 611–618.

122. Zhou, W., Sunarso, J., Motuzas, J., Liang, F.L., Chen, Z.G., Ge, L., Liu, S.M., Julbe, A., and Zhu, Z.H. (2011) Deactivation and regeneration of oxygen reduction reactivity on double perovskite $Ba_2Bi_{0.1}Sc_{0.2}Co_{1.7}O_{6-x}$ cathode for intermediate temperature solid oxide fuel cells. *Chem. Mater.*, **23** (6), 1618–1624.

123. Ishikawa, K., Kondo, S., Okano, H., Suzuki, S., and Suzuki, Y. (1987) Nonstoichiometry and electrical-resistivity in 2 mixed metal-oxides, La_2NiO_{4-x} and $LaSrNiO_{4-x}$. *Bull. Chem. Soc. Jpn.*, **60** (4), 1295–1298.

124. Kharton, V.V., Kovalevsky, A.V., Avdeev, M., Tsipis, E.V., Patrakeev, M.V., Yaremchenko, A.A., Naumovich, E.N., and Frade, J.R. (2007) Chemically induced expansion of $La_2Ni_{O4+\delta}$-based materials. *Chem. Mater.*, **19** (8), 2027–2933.

125. Munnings, C.N., Skinner, S.J., Amow, G., Whitfield, P.S., and Davidson, I.J. (2005) Oxygen transport in the $La_2Ni_{1-x}Co_xO_{4+\delta}$ system. *Solid State Ionics*, **176** (23–24), 1895–1901.

126. Al Daroukh, M., Vashook, V.V., Ullmann, H., Tietz, F., and Arual Raj, I. (2003) Oxides of the AMO_3 and $A2MO_4$-type: structural stability, electrical conductivity and thermal expansion. *Solid State Ionics*, **158** (1–2), 141–150.

127. Burriel, M., Garcia, G., Santiso, J., Kilner, J.A., Chater, R.J., and Skinner, S.J. (2008) Anisotropic oxygen diffusion properties in epitaxial thin films of $La_2NiO_{4+\delta}$. *J. Mater. Chem.*, **18** (4), 416–422.

128. Zhao, F., Wang, X.F., Wang, Z.Y., Peng, R.R., and Xia, C.R. (2008) K2NiF4 type $La_{2-x}Sr_xCo_{0.8}Ni_{0.2}O_{4+\delta}$ as the cathodes for solid oxide fuel cells. *Solid State Ionics*, **179** (27–32), 1450–1453.

129. Aguadero, A., Alonso, J.A., Escudero, M.J., and Daza, L. (2008) Evaluation of the $La_2Ni_{1-x}CuxO_{4+\delta}$ system as SOFC cathode material with 8YSZ and LSGM as electrolytes. *Solid State Ionics*, **179** (11–12), 393–400.

130. Weng, X.L., Boldrin, P., Abrahams, I., Skinner, S.J., Kellici, S., and Darr, J.A. (2008) Direct syntheses of $Lan+1NinO_{3n+1}$ phases (n = 1, 2, 3 and infinity) from nanosized co-crystallites. *J. Solid State Chem.*, **181** (5), 1123.

131. Mazo, G.N. and Savvin, S.N. (2004) The molecular dynamics study of oxygen mobility in $La_{2-x}Sr_xCuO_{4-\delta}$. *Solid State Ionics*, **175** (1–4), 371–374.

132. Kanai, H., Mizusaki, J., Tagawa, H., Hoshiyama, S., Hirano, K., Fujita, K., Tezuka, M., and Hashimoto, T. (1997) Defect chemistry of $La_{2-x}Sr_xCuO_{4-\delta}$:

oxygen nonstoichiometry and thermodynamic stability. *J. Solid State Chem.*, **131** (1), 150–159.

133. Li, Q., Zhao, H., Huo, L.H., Sun, L.P., Cheng, X.L., and Grenier, J.C. (2007) Electrode properties of Sr doped La_2CuO_4 as new cathode material for intermediate-temperature SOFCs. *Electrochem. Commun.*, **9** (7), 1508–1512.
134. Mauvy, F., Lalanne, C., Bassat, J.M., Grenier, J.C., Zhao, H., Huo, L.H., and Stevens, P. (2006) Electrode properties of $Ln_2NiO_{4+\delta}$ (Ln = La, Nd, Pr). *J. Electrochem. Soc.*, **153** (8), A1547–A1553.
135. Miyoshi, S., Furuno, T., Sangoanruang, O., Matsumoto, H., and Ishihara, T. (2007) Mixed conductivity and oxygen permeability of doped Pr_2NiO_4-based oxides. *J. Electrochem. Soc.*, **154** (1), B57–B62.
136. Sun, L.P., Li, Q., Zhao, H., Huo, L.H., and Grenier, J.C. (2008) Preparation and electrochemical properties of Sr-doped Nd_2NiO_4 cathode materials for intermediate-temperature solid oxide fuel cells. *J. Power Sources*, **183** (1), 43–48.
137. Fergus, J.W. (2006) Electrolytes for solid oxide fuel cells. *J. Power Sources*, **162** (1), 30–40.
138. Goodenough, J.B. (2003) Oxide-ion electrolytes. *Annu. Rev. Mater. Res.*, **33**, 91–128.
139. Jiang, S.P. (2008) Development of lanthanum strontium manganite perovskite cathode materials of solid oxide fuel cells: a review. *J. Mater. Sci.*, **43** (21), 6799–6833.
140. Huijsmans, J.P.P. (2001) Ceramics in solid oxide fuel cells. *Curr. Opin. Solid State Mater. Sci.*, **5**, 317–323.
141. Van Herle, J., McEvov, A.J., and Ravindranathan Thampi, K. (1996) A study on the $La_{1-x}Sr_xMnO_{3-\delta}$ oxygen cathode. *Electrochim. Acta*, **41** (9), 1447–1454.
142. Carter, S., Selcuk, A., Chater, R.J., Kajda, J., Kilner, J.A., and Steele, B.C.H. (1992) Oxygen transport in selected nonstoichiometric perovskite-structure oxides. *Solid State Ionics*, **53–56**, 597–605.
143. Jiang, S.P. (2002) A comparison of O_2 reduction reactions on porous $(La,Sr)MnO_3$ and $(La,Sr)(Co,Fe)O_3$ electrodes. *Solid State Ionics*, **146** (1–2), 1–22.
144. Yasuda, I., Ogasawara, K., Hishinuma, M., Kawada, T., and Dokiya, M. (1996) Oxygen tracer diffusion coefficient of (La, Sr)$MnO_{3\pm\delta}$. *Solid State Ionics*, **86–88**, 1197–1201.
145. Jiang, S.P. (2003) Issues on development of $(La,Sr)MnO_3$ cathode for solid oxide fuel cells. *J. Power Sources*, **124**, 390–402.
146. Kenjo, T. and Nishiya, M. (1992) $LaMnO_3$ air cathodes containing ZrO_2 electrolyte for high temperature solid oxide fuel cells. *Solid State Ionics*, **57** (3–4), 295–302.
147. Wang, W.G., Liu, Y.L., Barfod, R., Schougaard, S.B., Gordes, P., Ramousse, S., Hendriksen, P.V., and Mogensen, M. (2005) Nanostructured lanthanum manganate composite cathode. *Electrochem. Solid-State Lett.*, **8** (12), A619–A621.
148. Song, H.S., Hyun, S.H., Kim, J., Lee, H.W., and Moon, J. (2008) A nanocomposite material for highly durable solid oxide fuel cell cathodes. *J. Mater. Chem.*, **18** (10), 1087–1092.
149. Perry Murray, E. and Barnett, S.A. (2001) $(La,Sr)MnO_3$-$(Ce,Gd)O_{2-x}$ composite cathodes for solid oxide fuel cells. *Solid State Ionics*, **143** (3–4), 265–273.
150. Zhao, H., Feng, S., and Xu, W. (2000) A soft chemistry route for the synthesis of nano solid electrolytes $Ce_{1-x}Bi_xO_{2-x/2}$. *Mater. Res. Bull*, **35** (14–15), 2379–2386.
151. Zhao, H., Huo, L., and Gao, S. (2004) Electrochemical properties of LSM–CBO composite cathode. *J. Power Sources*, **125** (2), 149–154.
152. Armstrong, T.J. and Virkar, A.V. (2002) Performance of solid oxide fuel cells with LSGM-LSM composite cathodes. *J. Electrochem. Soc.*, **149** (12), A1565–A1571.
153. Wang, Z.W., Cheng, M.J., Dong, Y.L., Zhang, M., and Zhang, H.M. (2005) Investigation of LSM1.1-ScSZ composite cathodes for anode-supported solid

oxide fuel cells. *Solid State Ionics*, **176** (35–36), 2555–2561.

154. Pai, M.R., Wani, B.N., Sreedhar, B., Singh, S., and Gupta, N.M. (2006) Catalytic and redox properties of nano-sized $La_{0.8}Sr_{0.2}Mn_{1-x}Fe_xO_{3-\delta}$ mixed oxides synthesized by different routes. *J. Mol. Catal. A: Chem.*, **246** (1–2), 128–135.
155. Porta, P., De Rossi, S., Faticanti, M., Minelli, G., Pettiti, I., Lisi, L., and Turco, M. (1999) Perovskite-type oxides I. Structural, magnetic, and morphological properties of $LaMn_{1-x}Cu_xO_3$ and $LaCo_{1-x}Cu_xO_3$ solid solutions with large surface area. *J. Solid State Chem.*, **146** (2), 291–304.
156. De Souza, R.A. and Kilner, J.A. (1998) Oxygen transport in $La_{1-x}Sr_xMn_{1-y}CoyO_{3\pm\delta}$ perovskites: part I. Oxygen tracer diffusion. *Solid State Ionics*, **106** (3–4), 175–187.
157. De Souza, R.A. and Kilner, J.A. (1999) Oxygen transport in $La_{1-x}Sr_xMn_{1-y}CoyO_{3\pm\delta}$ perovskites: part II. Oxygen surface exchange. *Solid State Ionics*, **126** (1–2), 153–161.
158. Lybye, D. and Bonanos, N. (1999) Proton and oxide ion conductivity of doped $LaScO_3$. *Solid State Ionics*, **125** (1–4), 339–344.
159. Lybye, D., Poulsen, F., and Mogensen, M. (2000) Conductivity of A- and B-site doped $LaAlO_3$, $LaGaO_3$, $LaScO_3$ and $LaInO_3$ perovskites. *Solid State Ionics*, **128** (1–4), 91–103.
160. Nomura, K., Takeuchi, T., Tanase, S., Kageyama, H., Tanimoto, K., and Miyazaki, Y. (2002) Proton conduction in $(La_{0.9}Sr_{0.1})MIIIO_{3-\delta}$ (MIII=Sc, In, and Lu) perovskites. *Solid State Ionics*, **154–155**, 647–652.
161. Gu, H.X., Zheng, Y., Ran, R., Shao, Z.P., Jin, W.Q., Xu, N.P., and Ahn, J.M. (2008) Synthesis and assessment of $La_{0.8}Sr_{0.2}ScyMn_{1-y}O_{3-\delta}$ as cathodes for solid-oxide fuel cells on scandium-stabilized zirconia electrolyte. *J. Power Sources*, **183** (2), 471–478.
162. Zheng, Y., Ran, R., and Shao, Z.P. (2008) Activation and deactivation kinetics of oxygen reduction over a $La_{0.8}Sr_{0.2}Sc_{0.1}Mn_{0.9}O_3$ cathode. *J. Phys. Chem. C*, **112** (47), 18690–18700.
163. Zheng, Y., Ran, R., Gu, H.X., Cai, R., and Shao, Z.P. (2008) Characterization and optimization of $La_{0.8}Sr_{0.2}Sc_{0.1}Mn_{0.9}O_{3-\delta}$ based composite electrodes for intermediate-temperature solid-oxide fuel cells. *J. Power Sources*, **185** (2), 641–648.
164. Yue, X.L., Yan, A.Y., Zhang, M., Liu, L., Dong, Y.L., and Cheng, M.J. (2008) Investigation on scandium-doped manganate $La_{0.8}Sr_{0.2}Mn_{1-x}Sc_xO_{3-\delta}$ cathode for intermediate temperature solid oxide fuel cells. *J. Power Sources*, **185** (2), 691–697.
165. Yamamoto, O., Takeda, Y., Kanno, R., and Noda, M. (1987) Perovskite-type oxides as oxygen electrodes for high temperature oxide fuel cells. *Solid State Ionics*, **22** (2–3), 241–246.
166. Duan, Z.S., Yang, M., Yan, A.Y., Hou, Z.F., Dong, Y.L., Chong, Y., Cheng, M.J., and Yang, W.S. (2006) $Ba_{0.5}Sr_{0.5}Co_{0.8}Fe_{0.2}O_{3-\delta}$ as a cathode for IT-SOFCs with a GDC interlayer. *J. Power Sources*, **160** (1), 57–64.
167. Lei, Z., Zhu, Q.S., and Zhao, L. (2005) Low temperature processing of interlayer-free $La_{0.6}Sr_{0.4}Co_{0.2}Fe_{0.8}O_{3-\delta}$ cathodes for intermediate temperature solid oxide fuel cells. *J. Power Sources*, **161** (2), 1169–1175.
168. Murata, K., Fukui, T., Abe, H., Naito, M., and Nogi, K. (2005) Morphology control of $La(Sr)Fe(Co)O_3$–a cathodes for IT-SOFCs. *J. Power Sources*, **145** (2), 257–261.
169. Lee, S., Song, H.S., Hyun, S.H., Kim, J., and Moon, J. (2009) Interlayer-free nanostructured $La_{0.58}Sr_{0.4}Co_{0.2}Fe_{0.8}O_{3-\delta}$ cathode on scandium stabilized zirconia electrolyte for intermediate-temperature solid oxide fuel cells. *J. Power Sources*, **187** (1), 74–79.
170. Kim, Y.M., Kim-Lohsoontorn, P., and Bae, J. (2010) Effect of unsintered gadolinium-doped ceria buffer layer on performance of metal-supported solid oxide fuel cells using unsintered barium strontium cobalt ferrite cathode. *J. Power Sources*, **195** (19), 6420–6427.
171. Charojrochkul, S., Choy, K.L., and Steele, B.C.H. (1999) Cathode/electrolyte systems for solid oxide

fuel cells fabricated using flame assisted vapour deposition technique. *Solid State Ionics*, **121** (1–4), 107–113.

172. Rossignol, C., Ralph, J.M., Bae, J.M., and Vaughey, J.T. (2004) $Ln_{1-x}Sr_xCoO_3$ (Ln=Gd, Pr) as a cathode for intermediate-temperature solid oxide fuel cells. *Solid State Ionics*, **175** (1–4), 59–61.
173. Gong, Y.H., Ji, W.J., Zhang, L., Li, M., Xie, B., Wang, H.Q., Jiang, Y.S., and Song, Y.Z. (2011) Low temperature deposited $(Ce,Gd)O_{2-x}$ interlayer for $La_{0.6}Sr_{0.4}Co_{0.2}Fe_{0.8}O_3$ cathode based solid oxide fuel cell. *J. Power Sources*, **196** (5), 2768–2772.
174. Shiono, M., Kobayashi, K., Nguyen, T.L., Hosoda, K., Kato, T., Ota, K., and Dokiya, M. (2004) Effect of CeO_2 interlayer on ZrO_2 electrolyte/$La(Sr)CoO_3$ cathode for low-temperature SOFCs. *Solid State Ionics*, **170** (1–2), 1–7.
175. Tsai, T. and Barnett, S.A. (1997) Increased solid-oxide fuel cell power density using interfacial ceria layers. *Solid State Ionics*, **98** (3–4), 191–196.
176. Tsoga, A., Gupta, A., Naoumidis, A., and Nikolopoulos, P. (2000) *Acta Mater.*, **48**, 4709.
177. Nguyen, T.L., Kobayashi, K., Honda, T., and Iimura, Y. (2004) Preparation and evaluation of doped ceria interlayer on supported stabilized zirconia electrolyte SOFCs by wet ceramic processes. *Solid State Ionics*, **174** (1–4), 163–174.
178. Tsoga, A., Naoumidis, A., and Stover, D. (2000) Total electrical conductivity and defect structure of ZrO_2-CeO_2-Y_2O_3-Gd_2O_3 solid solutions. *Solid State Ionics*, **135** (1–4), 403–409.
179. Martínez-Amesti, A., Larrañaga, A., Rodríguez-Martínez, L.M., Nó, L., Pizarro, J.L., Laresgoiti, A., and Arriortua, I. (2009) Chemical compatibility between YSZ and SDC sintered at different atmospheres for SOFC applications. *J. Power Sources*, **192** (1), 151–157.
180. Fonseca, F.C., Uhlenbruck, S., Nedéléc, R., and Buchkremer, H.P. (2010) Properties of bias-assisted sputtered gadolinia-doped ceria interlayers for solid oxide fuel cells. *J. Power Sources*, **195** (6), 1599–1604.
181. Tsai, T.P., Perry, E., and Barnett, S. (1997) Low-temperature solid-oxide fuel cells utilizing thin bilayer electrolytes. *J. Electrochem. Soc.*, **144** (5), L130–L132.
182. Brahim, C., Ringuede, A., Gourba, E., Cassir, M., Billard, A., and Briois, P. (2006) Electrical properties of thin bilayered YSZ/GDC SOFC electrolyte elaborated by sputtering. *J. Power Sources*, **156** (1), 45–49.
183. Wang, D.F., Wang, J.X., He, C.R., Tao, Y.K., Xu, C., and Wang, W.G. (2010) Preparation of a $Gd_{0.1}Ce_{0.9}O_{2-\delta}$ interlayer for intermediate-temperature solid oxide fuel cells by spray coating. *J. Alloys Compd.*, **505** (1), 118–124.
184. Nguyen, T.L., Kato, T., Nozaki, K., Honda, T., Negishi, A., Kato, K., and Iimura, Y. (2006) Application of $(Sm_{0.5}Sr_{0.5})$ CoO_3 as a cathode material to $(Zr,Sc)O_2$ electrolyte with ceria-based interlayers for reduced-temperature operation SOFCs. *J. Electrochem. Soc.*, **153** (7), A1310–A1316.
185. Matsuda, M., Hosomi, T., Murata, K., Fukui, T., and Miyake, M. (2007) Fabrication of bilayered YSZ/SDC electrolyte film by electrophoretic deposition for reduced-temperature operating anode-supported SOFC. *J. Power Sources*, **165** (1), 102–107.
186. Lu, Z.G., Zhou, X.D., Fisher, D., Templeton, J., Stevenson, J., Wu, N.J., and Ignatiev, A. (2010) Enhanced performance of an anode-supported YSZ thin electrolyte fuel cell with a laser-deposited $Sm_{0.2}Ce_{0.8}O_{1.9}$ interlayer. *Electrochem. Commun.*, **12** (2), 179–182.
187. Nguyena, T.L., Kobayashi, K., Hondaa, T., Iimuraa, Y., Katoa, K., Neghisia, A., Nozakia, K., Tapperoa, F., Sasakib, K., Shirahamab, H., Otac, K., Dokiyab, M., and Katoa, T. (2004) Preparation and evaluation of doped ceria interlayer on supported stabilized zirconia electrolyte SOFCs by wet ceramic processes. *Solid State Ionics*, **174** (1–4), 163–174.
188. Shiono, M., Kobayashi, K., Nguyen, T.L., Hosoda, K., Kato, T., Ota, K., and Dokiya, M. (2004) Effect of CeO_2 interlayer on ZrO_2 electrolyte/$La(Sr)CoO_3$

cathode for low-temperature SOFCs. *Solid State Ionics*, **170** (1–2), 1–7.
189. Kreuer, K.D. (2003) Proton-conducting oxides. *Annu. Rev. Mater. Res.*, **33**, 333–359.
190. Zuo, C.D., Zha, S.W., Liu, M.L., Hatano, M., and Uchiyama, M. (2006) $Ba(Zr_{0.1}Ce_{0.7}Y_{0.2})O_{3-\delta}$ as an electrolyte for low-temperature solid-oxide fuel cells. *Adv. Mater.*, **18** (24), 3318–3320.
191. Suksamai, W. and Metcalfe, I.S. (2007) Measurement of proton and oxide ion fluxes in a working Y-doped $BaCeO_3$ SOFC. *Solid State Ionics*, **178** (7–10), 627–634.
192. Amsak, W., Assabumrungrat, S., Douglas, P.L., Laosiripojana, N., and Charojrochkul, S. (2006) Theoretical performance analysis of ethanol-fuelled solid oxide fuel cells with different electrolytes. *Chem. Eng. J.*, **119** (1), 11–18.
193. Iwahara, H., Uchida, H., and Ogaki, K. (1988) Proton conduction in sintered oxides based on $BaCeO_3$. *J. Electrochem. Soc.*, **135** (2), 529–533.
194. Hibino, T., Hashimoto, A., Suzuki, M., and Sano, M. (2002) A solid oxide fuel cell using Y-doped $BaCeO_3$ with Pd-loaded FeO anode and $Ba_{0.5}Pr_{0.5}CoO_3$ cathode at low temperatures. *J. Electrochem. Soc.*, **149** (11), A1503–A1508.
195. Hirabayashi, D., Tomita, A., Brito, M.E., Hibino, T., Harada, U., Nagao, M., and Sano, M. (2004) Solid oxide fuel cells operating without using an anode material. *Solid State Ionics*, **168** (1–2), 23–29.
196. Maffei, N., Pelletier, L., and Mctarlan, A. (2004) Performance characteristics of Gd-doped barium cerate-based fuel cells. *J. Power Sources*, **136** (1), 24–29.
197. Ito, N., Iijima, M., Kimura, K., and Iguchi, S. (2005) New intermediate temperature fuel cell with ultra-thin proton conductor electrolyte. *J. Power Sources*, **152**, 200–203.
198. Pelletier, L., McFarlan, A., and Maffei, N. (2005) Ammonia fuel cell using doped barium cerate proton conducting solid electrolytes. *J. Power Sources*, **145** (2), 262–265.
199. Tomita, A., Hibino, T., and Sano, M. (2005) Surface modification of a doped $BaCeO_3$ to function as an electrolyte and as an anode for SOFCs. *Electrochem. Solid-State Lett.*, **8** (7), A333–A336.
200. Tomita, A., Tsunekawa, K., Hibino, T., Teranishi, S., Tachi, Y., and Sano, M. (2006) Chemical and redox stabilities of a solid oxide fuel cell with $BaCe_{0.8}Y_{0.2}O_{3-\alpha}$ functioning as an electrolyte and as an anode. *Solid State Ionics*, **177** (33–34), 2951–2956.
201. Akimune, Y., Matsuo, K., Higashiyama, H., Honda, K., Yamanaka, M., Uchiyama, M., and Hatano, M. (2007) Nano-Ag particles for electrodes in a yttria-doped $BaCeO_3$ protonic conductor. *Solid State Ionics*, **178** (7–10), 575–579.
202. Uchida, H., Tanaka, S., and Iwahara, H. (1985) Polarization at Pt electrodes of a fuel-cell with a high temperature type proton conductive solid electrolyte. *J. Appl. Electrochem.*, **15** (1), 93–97.
203. Iwahara, H., Yajima, T., Hibino, T., and Ushida, H. (1993) Performance of solid oxide fuel-cell using proton and oxide-ion mixed conductors based on $BaCe_{1-x}Sm_xO_{3-\delta}$. *J. Electrochem. Soc.*, **140** (6), 1687–1691.
204. Wu, T.Z., Peng, R.R., and Xia, C.R. (2008) $Sm_{0.5}Sr_{0.5}CoO_{3-\delta}$-$BaCe_{0.8}Sm_{0.2}O_{3-\delta}$ composite cathodes for proton-conducting solid oxide fuel cells. *Solid State Ionics*, **179** (27–32), 1505–1508.
205. He, F., Wu, T.Z., Peng, R.R., and Xia, C.R. (2009) Cathode reaction models and performance analysis of $Sm_{0.5}Sr_{0.5}CoO_{3-\delta}$-$BaCe_{0.8}Sm_{0.2}O_{3-\delta}$ composite cathode for solid oxide fuel cells with proton conducting electrolyte. *J. Power Sources*, **194** (1), 263–268.
206. Yang, L., Zuo, C.D., Wang, S.Z., Cheng, Z., and Liu, M.L. (2008) A novel composite cathode for low-temperature SOFCs based on oxide proton conductors. *Adv. Mater.*, **20** (17), 3280–3283.
207. Lin, Y., Ran, R., Zheng, Y., Shao, Z.P., Jin, W.Q., Xu, N.P., and Ahn, J. (2008) Evaluation of $Ba_{0.5}Sr_{0.5}Co_{0.8}Fe_{0.2}O_{3-\delta}$

as a potential cathode for an anode-supported proton-conducting solid-oxide fuel cell. *J. Power Sources*, **180** (1), 15–22.

208. Peng, R.R., Wu, Y., Yang, L.Z., and Mao, Z.Q. (2006) Electrochemical properties of intermediate-temperature SOFCs based on proton conducting Sm-doped $BaCeO_3$ electrolyte thin film. *Solid State Ionics*, **177** (3–4), 389–393.

209. Yamaura, H., Ikuta, T., Yahiro, H., and Okada, G. (2005) Cathodic polarization of strontium-doped lanthanum ferrite in proton-conducting solid oxide fuel cell. *Solid State Ionics*, **176** (3–4), 269–274.

210. Wang, H., Tablet, C., Feldhoff, A., and Caro, J. (2005) *Adv. Mater.*, **17** (14), 1785–1788.

211. Ding, H.P., Lin, B., Liu, X.Q., and Meng, G.Y. (2008) High performance protonic ceramic membrane fuel cells (PCMFCs) with $Ba_{0.5}Sr_{0.5}Zn_{0.2}Fe_{0.8}O_{3-\delta}$ perovskite cathode. *Electrochem. Commun.*, **10** (9), 1388–1391.

212. Tao, Z.T., Bi, L., Yan, L.T., Sun, W.P., Zhu, Z.W., Peng, R.R., and Liu, W. (2009) A novel single phase cathode material for a proton-conducting SOFC. *Electrochem. Commun.*, **11** (3), 688–690.

213. Badwal, S.P.S. (2001) Stability of solid oxide fuel cell components. *Solid State Ionics*, **143** (1), 39–46.

214. Brugnoni, C., Ducati, U., and Scagliotti, M. (1995) SOFC cathode/electrolyte interface. Part I: reactivity between $La_{0.85}Sr_{0.15}MnO_3$ and $ZrO_2.Y_2O_3$. *Solid State Ionics*, **76** (3–4), 177–182.

215. Chiodelli, G. and Scagliotti, M. (1994) Electrical characterization of lanthanum zirconate reaction layers by impedance spectroscopy. *Solid State Ionics*, **73** (3–4), 265–271.

216. Lee, H.Y. and Oh, S.M. (1996) Origin of cathodic degradation and new phase formation at the $La_{0.9}Sr_{0.1}MnO_3$/YSZ interface. *Solid State Ionics*, **90** (1–4), 133–140.

217. Mitterdorfer, A. and Gauckler, L.J. (1998) $La_2Zr_2O_7$ formation and oxygen reduction kinetics of the $La_{0.85}Sr_{0.15}MnyO_3$, O_2(g)YSZ system. *Solid State Ionics*, **111** (3–4), 185–218.

218. Huang, Y., Ahn, K., Vohs, J.M., and Gorte, R.J. (2004) Characterization of Sr-Doped $LaCoO_3$.YSZ composites prepared by impregnation methods. *J. Electrochem. Soc.*, **151**, A1592–A1597.

219. Armstrong, T.J. and Rich, J.G. (2006) Anode-supported solid oxide fuel cells with $La_{0.6}Sr_{0.4}CoO_{3-\delta}$-$Zr_{0.84}Y_{0.16}O_{2-\delta}$ composite cathodes fabricated by an infiltration method. *J. Electrochem. Soc.*, **153**, A515–A520.

220. Huang, Y.Y., Vohs, J.M., and Gorte, R.J. (2004) Fabrication of Sr-doped $LaFeO_3$ YSZ composite cathodes. *J. Electrochem. Soc.*, **151** (10), A646–A651.

221. Wang, W., Gross, M.D., Vohs, J.M., and Gorte, R.J. (2007) The stability of LSF-YSZ electrodes prepared by infiltration. *J. Electrochem. Soc.*, **154** (5), B439–B445.

222. Chen, J., Liang, F.L., Liu, L.N., Jiang, S.P., Chi, B., Pu, J., and Li, J. (2008) Nano-structured (La, Sr)(Co, Fe)O_3 + YSZ composite cathodes for intermediate temperature solid oxide fuel cells. *J. Power Sources*, **183** (2), 586–589.

223. Lee, S., Bevilacqua, M., Fornasiero, P., Vohs, J.M., and Gorte, R.J. (2009) Solid oxide fuel cell cathodes prepared by infiltration of $LaNi_{0.6}Fe_{0.4}O_3$ and $La_{0.91}Sr_{0.09}Ni_{0.6}Fe_{0.4}O_3$ in porous yttria-stabilized zirconia. *J. Power Sources*, **193** (2), 747–753.

224. Horita, T., Yamaji, K., Sakai, N., Yokokawa, H., Weber, A., and Ivers-Tiffee, E. (2001) Oxygen reduction mechanism at porous $La_{1-x}Sr_xCoO_{3-\delta}$ cathodes/$La_{0.8}Sr_{0.2}Ga_{0.8}Mg_{0.2}O_{2.8}$ electrolyte interface for solid oxide fuel cells. *Electrochim. Acta*, **46** (12), 1837–1845.

225. Vohs, J.M. and Gorte, R.J. (2009) High-performance SOFC cathodes prepared by infiltration. *Adv. Mater.*, **21** (9), 943–956.

226. Zhao, F. and Virkar, A.V. (2005) Dependence of polarization in anode-supported solid oxide fuel cells on various cell parameters. *J. Power Sources*, **141** (1), 79–95.

227. Virkar, V., Chen, J., Tanner, C.W., and Kim, J.W. (2000) The role of electrode microstructure on activation and concentration polarizations in solid oxide fuel cells. *Solid State Ionics*, **131** (1–2), 189–198.

228. Shah, M. and Barnett, S.A. (2008) Solid oxide fuel cell cathodes by infiltration of $La_{0.6}Sr_{0.4}Co_{0.2}Fe_{0.8}O_{3-\delta}$ into Gd-doped ceria. *Solid State Ionics*, **179** (35–36), 2059–2064.

229. Jiang, Z.Y., Zhang, L., Feng, K., and Xia, C.R. (2008) Nanoscale bismuth oxide impregnated $(La,Sr)MnO_3$ cathodes for intermediate-temperature solid oxide fuel cells. *J. Power Sources*, **185** (1), 40–48.

230. Wu, T.Z., Zhao, Y.Q., Peng, R.R., and Xia, C.R. (2009) Nano-sized $Sm_{0.5}Sr_{0.5}CoO_{3-\delta}$ as the cathode for solid oxide fuel cells with proton-conducting electrolytes of $BaCe_{0.8}Sm_{0.2}O_{2.9}$. *Electrochim. Acta*, **54** (21), 4888–4892.

231. Sholklapper, T.Z., Kurokawa, H., Jacobson, C.P., Visco, S.J., and De Jonghe, L.C. (2007) Nanostructured solid oxide fuel cell electrodes. *Nano Lett.*, **7** (7), 2136–2141.

232. Sholklapper, T.Z., Radmilovic, V., Jacobson, C.P., Visco, S.J., and De Jonghe, L.C. (2008) Nanocomposite Ag-LSM solid oxide fuel cell electrodes. *J. Power Sources*, **175** (1), 206–210.

233. Liang, F.L., Chen, J., Cheng, J.L., Jiang, S.P., He, T.M., Pu, J., and Li, J. (2008) Novel nano-structured Pd+yttrium doped ZrO_2 cathodes for intermediate temperature solid oxide fuel cells. *Electrochem. Commun.*, **10** (1), 42–46.

234. Clemmer, R.M.C. and Corbin, S.F. (2004) Influence of porous composite microstructure on the processing and properties of solid oxide fuel cell anodes. *Solid State Ionics*, **166** (3–4), 251–259.

235. Sasaki, K., Tamura, J., and Dokiya, M. (2001) Pt-cermet cathode for reduced temperature SOFCs. *Solid State Ionics*, **144** (3–4), 223–232.

236. Sasaki, K., Tamura, J., and Dokiya, M. (2001) Noble metal alloy-Zr(Sc)O_2 cermet cathode for reduced-temperature SOFCs. *Solid State Ionics*, **144** (3–4), 233–240.

237. Liang, F.L., Chen, J., Jiang, S.P., Pu, J., Chi, B., and Li, J. (2009) High performance solid oxide fuel cells with electrocatalytically enhanced (La, Sr)MnO_3 cathodes. *Electrochem. Commun.*, **11** (5), 1048–1051.

238. Takahashi, T. and Minh, N.Q. (1995) *Science and Technology of Ceramic Fuel Cells*, Elsevier, New York.

239. Verweij, H. (1998) Nanocrystalline and nanoporous ceramics. *Adv. Mater.*, **10** (17), 1483–1486.

240. Ziehfreund, A., Simon, U., and Maier, W.F. (1996) Oxygen ion conductivity of platinum-impregnated stabilized zirconia in bulk and microporous materials. *Adv. Mater.*, **8** (5), 424–427.

241. van Berkel, F.P.F., van Heuveln, F.H., and Huijsmans, J.P.P. (1994) Characterization of solid oxide fuel cell electrodes by impedance spectroscopy and I-V characteristics. *Solid State Ionics*, **72** (2), 240–247.

242. Steele, B.C.H. (1997) Behaviour of porous cathodes in high temperature fuel cells. *Solid State Ionics*, **94** (1–4), 239–248.

243. Shiga, H., Okubo, T., and Sadakata, M. (1996) Preparation of nanostructured platinum/yttria-stabilized zirconia cermet by the sol–gel method. *Ind. Eng. Chem. Res.*, **35** (12), 4479–4486.

244. Kresge, C.T., Leonowicz, M.E., Roth, W.J., Vartuli, J.C., and Beck, J.S. (1992) Ordered mesoporous molecular sieves synthesized by a liquid-crystal template mechanism. *Nature*, **359** (6397), 710–712.

245. Beck, J.S., Vartuli, J.C., Roth, W.J., Leonowicz, M.E., Kresge, C.T., Schmitt, K.D., Chu, C.T., Olson, D.H., Sheppard, E.W., McCullen, S.B., Higgins, J.B., and Schlenker, J.L. (1992) A new family of mesoporous molecular sieves prepared with liquid crystal templates. *J. Am. Chem. Soc.*, **114** (27), 10834–10843.

246. Wong, M.S. and Ying, J.Y. (1998) Amphiphilic templating of mesostructured zirconium oxide. *Chem. Mater.*, **10** (8), 2067–2077.

247. Antonelli, D.M. (1999) Synthesis and mechanistic studies of sulfated meso- and microporous zirconias with chelating carboxylate surfactants. *Adv. Mater.*, **11** (6), 487–492.

248. Antonelli, D.M. and Ying, J.Y. (1996) Synthesis of a stable hexagonally packed mesoporous niobium oxide molecular sieve through a novel ligand-assisted templating mechanism. *Angew. Chem. Int. Ed.*, **35** (4), 426–430.

249. Sun, T. and Ying, J.Y. (1997) Synthesis of microporous transition-metal-oxide molecular sieves by a supramolecular templating mechanism. *Nature*, **389** (6652), 704–706.

250. Kim, A., Bruinsma, P., Chen, Y., Wang, L., and Liu, J. (1997) Amphoteric surfactant templating route for mesoporous zirconia. *Chem. Comm.*, **2**, 161–162.

251. Holland, B.T., Blanford, C.F., Do, T., and Stein, A. (1999) Synthesis of highly ordered, three-dimensional, macroporous structures of amorphous or crystalline inorganic oxides, phosphates, and hybrid composites. *Chem. Mater.*, **11** (3), 795–805.

252. Mamak, M., Coombs, N., and Ozin, G.A. (2001) Electroactive mesoporous yttria stabilized zirconia containing platinum or nickel oxide nanoclusters: a new class of solid oxide fuel cell electrode materials. *Adv. Funct. Mater.*, **11** (1), 59–63.

253. Mamak, M., Coombs, N., and Ozin, G.A. (2001) Mesoporous nickel-yttria-zirconia fuel cell materials. *Chem. Mater.*, **13** (10), 3564–3570.

254. Mamak, M., Coombs, N., and Ozin, G.A. (2000) Self-assembling solid oxide fuel cell materials: mesoporous yttria-zirconia and metal-yttria-zirconia solid solutions. *J. Am. Chem. Soc.*, **122** (37), 8932–8939.

255. Mamak, M., Coombs, N., and Ozin, G.A. (2000) Mesoporous yttria-zirconia and metal-yttria-zirconia solid solutions for fuel cells. *Adv. Mater.*, **12** (3), 198–202.

256. Mamak, M., Coombs, N., and Ozin, G.A. (2002) Practical solid oxide fuel cells with anodes derived from self-assembled mesoporous-NiO-YSZ. *Chem. Commun.*, **20**, 2300–2301.

257. Mamak, M., Métraux, G.S., Petrov, S., Coombs, N., Ozin, G.A., and Green, M.A. (2003) Lanthanum strontium manganite/yttria-stabilized zirconia nanocomposites derived from a surfactant assisted, co-assembled mesoporous phase. *J. Am. Chem. Soc.*, **125** (17), 5161–5175.

258. Beckel, D., Dubach, A., Studart, A.R., and Gauckler, L.J. (2006) Spray pyrolysis of $La_{0.6}Sr_{0.4}Co_{0.2}Fe_{0.8}O_{3-\delta}$ thin film cathodes. *J. Electroceram.*, **16** (3), 221–228.

259. Beckel, D., Dubach, A., Grundy, A.N., Infortuna, A., and Gauckler, L.J. (2008) Solid-state dewetting of $La_{0.6}Sr_{0.4}Co_{0.2}Fe_{0.8}O_{3-\delta}$ thin films during annealing. *J. Eur. Ceram. Soc.*, **28** (1), 49–60.

260. Beckel, D., Muecke, U.P., Gyger, T., Florey, G., Infortuna, A., and Gauckler, L.J. (2007) Electrochemical performance of LSCF based thin film cathodes prepared by spray pyrolysis. *Solid State Ionics*, **178** (5–6), 407–415.

261. Hamedani, H.A., Dahmen, K.H., Li, D., Peydaye-Saheli, H., Garmestani, H., and Khaleel, M. (2008) Fabrication of gradient porous LSM cathode by optimizing deposition parameters in ultrasonic spray pyrolysis. *Mater. Sci. Eng., B*, **153** (1–3), 1–9.

262. Choy, K.L., Charojrochkul, S., and Steele, B.C.H. (1997) Fabrication of cathode for solid oxide fuel cells using flame assisted vapour deposition technique. *Solid State Ionics*, **96** (1–2), 49–54.

263. Chen, C.C., Nasrallah, M.M., Anderson, H.U., and Honolulu, H.I. (1993) in *Proceedings of the 3rd International Symposium on Solid Oxide Fuel Cells, May, 1993*, The Electrochemical Society Proceedings Series (eds S.C. Singhal and H. Iwahara), The Electrochemical Society, Pennington, NJ, p. 598.

264. Chang, C.L. and Hwang, B.H. (2008) Microstructure and electrochemical characterization of $Sm_{0.5}Sr_{0.5}CoO_3$ films as SOFC cathode prepared by the electrostatic-assisted ultrasonic spray pyrolysis method. *Int. J. Appl. Ceram. Technol.*, **5** (6), 582–588.

265. Chang, C.L., Hsu, C.S., and Hwang, B.H. (2008) Unique porous thick $Sm_{0.5}Sr_{0.5}CoO_3$ solid oxide fuel cell cathode films prepared by spray pyrolysis. *J. Power Sources*, **179** (2), 734–738.

266. Chen, J.C. and Hwang, B.H. (2008) Microstructure and properties of the Ni-CGO composite anodes preparedby the electrostatic-assisted ultrasonic spray pyrolysis method. *J. Am. Ceram. Soc.*, **91** (1), 97–102.

3
Oxide Ion-Conducting Materials for Electrolytes

Tatsumi Ishihara

3.1
Introduction

The electrolytes for solid oxide fuel cells (SOFCs) must be stable in both reducing and oxidizing environments and must have sufficiently high ionic conductivity ($>\log(\sigma\ (\mathrm{S\,cm^{-1}})) = -2$), with low electronic conductivity at the cell operation temperature. Present SOFCs have extensively used stabilized zirconia with fluorite structure, especially yttria-stabilized zirconia, as the electrolyte. Another important requirement for the electrolyte material is compatibility with cell component materials, namely, low reactivity and similar thermal expansion coefficient, sufficient mechanical strength, and easy handling property.

Although oxide ion conductors are now widely used as the electrolytes in SOFCs, as discussed in this chapter, high-temperature proton conductors are also drawing attention as the electrolytes in SOFCs. Since protons have the smallest ionic size among elements, they exhibit high mobility and also small activation energy for conduction. Therefore, if the ionic conductivity is simply compared, proton-conducting materials show much higher ionic conductivity in low-temperature regions and are suitable as the electrolyte of fuel cells operating at low temperatures [1, 2]. For fuel cells using proton-conducting electrolytes, a polymer ion conductor such as Nafion is used; however, because of insufficient stability, there are still many issues to overcome in polymer electrolyte membrane fuel cells (PEMFCs). In contrast, ceramic proton conductors, typically, $AZrO_3$ or $ACeO_3$ ($A = Sr$ or Ba), are highly stable and intermediate-temperature operable; hence, no expensive noble metal is required for electrolytes. The mechanism of proton in ceramics is explained by the following equation:

$$H_2O + V_O^{\bullet\bullet} \rightarrow O_O^{\times} + 2H_i^{\bullet} \tag{3.1}$$

Recently, extremely high power density at intermediate temperatures was reported in the cells using $Sn_2P_2O_7$, a new family of proton conductors [3, 4]. However, similar to polymer electrolyte fuel cells, hydrogen is required as fuel in SOFCs using proton-conducting ceramic electrolytes, albeit not like the highly pure H_2 in PEMFCs. Therefore, considering the practical applications, at present,

Materials for High-Temperature Fuel Cells, First Edition. Edited by San Ping Jiang and Yushan Yan.

oxide ion conductors are most reasonably accepted as the electrolyte in SOFCs. By using oxide ion conductors as the electrolyte, hydrocarbon can be directly used as fuel, and thus, it is possible to use various fuels, which is one of the advantages of SOFCs using oxide ion conductors. This could realize a simple fuel handling process (e.g., balance of plant, BOP) for SOFCs.

Oxide ion conductors generally conduct ions through oxygen vacancy, which could be introduced by a dopant. A typical case is stabilized ZrO_2 and the oxygen vacancy, $V_O^{\bullet\bullet}$, that is introduced as in the following equation, similar to Y doping:

$$Y_2O_3 \rightarrow 2Y'_{Zr} + 3O_O^{\times} + V_O^{\bullet\bullet} \tag{3.2}$$

Oxide ion conductivity was first reported as early as in 1899 in ZrO_2 with 15 wt% Y_2O_3 (stabilized zirconia denoted as YSZ) by Nernst [5], and so, the history of oxide ion conductors is longer than a century. In the history of oxide ion conductors, fluoride structure oxides consisting of tetravalent cation (ZrO_2, CeO_2, ThO_2, etc.) have been widely studied. In this chapter, fundamentals of oxide ion conductivity and basic properties of oxide ion conductors are introduced from the viewpoint of materials, and improved conductivity related to thin films is also introduced. There are several good reviews already published and detailed information is available from those reviews [6–8].

3.2 Oxide Ion Conductivity in Metal Oxide

3.2.1 Fluorite Oxides

In the history of oxide ion conductor development, fluoride structure oxides consisting of tetravalent cations have been widely studied. Figure 3.1 shows the fluorite structure. Fluorite-type structure is a face-centered cubic arrangement of cations, with anions occupying all the tetrahedral sites. It has a large number of octahedral interstitial free volume sites. Thus, this structure is a rather open

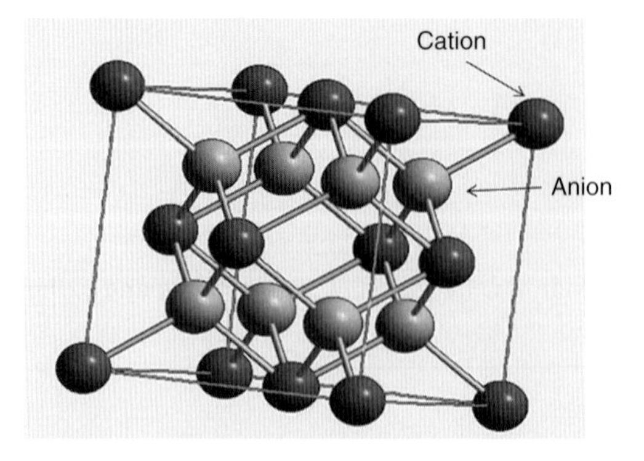

Figure 3.1 Fluorite structure of a typical oxide ion conductor.

structure and rapid ion diffusion is achieved. Oxide ion conductivity has a strong relationship between oxygen vacancy formation and its diffusivity. Therefore, in the previous open literature, there are many reports on fluorite oxide suggesting that ionic size of the dopant is highly important for determining the oxide ion conductivity. In this section, the effects of dopant on oxide ion conductivity are briefly introduced, in particular, for stabilized ZrO_2 and CeO_2, which is the typical electrolyte for SOFCs.

3.2.1.1 Stabilized ZrO_2

ZrO_2 is the most widely used electrolyte material for the current SOFCs and has a long history. ZrO_2 typically has three crystal structures: monoclinic, tetragonal (>1443 K), and cubic (>2643 K) [9]. In pure ZrO_2, the cubic phase is stable only at high temperatures, and at room temperature, and the monoclinic phase is the stable one from chemical equilibrium. ZrO_2 is principally classified as an electrical insulator. To achieve high oxide ion conductivity, introduction of oxygen vacancy is essential, as discussed earlier, and so substitution of lower valence cation is generally performed for ZrO_2. In case of Y-doped ZrO_2, formation of oxygen vacancy is as shown in Eq. (3.2). Here, high-temperature stable phase of cubic structure is stabilized down to room temperature by substitution of the lattice position with lower valence cation, and so oxygen-vacancy-introduced ZrO_2 is called *stabilized ZrO_2*, which includes the tetragonal (partially stabilized) and cubic (fully stabilized) phases. For the ZrO_2–Y_2O_3 system, several equilibrium phase diagrams have been reported [9, 10]. Addition of Y_2O_3 to ZrO_2 reduces the tetragonal/monoclinic transformation temperature. In the composition range from 0 to 2.5 mol% Y_2O_3, the tetragonal solid solution will transform on cooling to the monoclinic phase. At higher Y_2O_3 content, a mixture of nontransformable tetragonal and cubic solid solution exists. Further increase in Y_2O_3 content results in a homogenous cubic solid solution. The minimum amount required to fully stabilize the cubic phase of ZrO_2 is about 8–10 mol% at 1273 K. Other ZrO_2–M_2O_3 systems, where M is Y, Sc, Nd, Sm, or Gd, have also shown stabilized solution at a certain range. The minimum amount of dopant necessary to stabilize ZrO_2 in cubic structure is close to the composition that gives the highest conductivity (8 mol% Y_2O_3, 10 mol% Sc_2O_3, 15 mol% Nd_2O_3, 10 mol% Sm_2O_3, and 10 mol% Gd_2O_3). This is discussed later in relation to the cluster formation of oxygen vacancy and dopant.

The concentration of the vacancies is given simply by the electron neutrality condition. In this case, therefore, $2\left[Y'_{Zr}\right] = \left[V_O^{\bullet\bullet}\right]$ for Y_2O_3-stabilized ZrO_2. On the other hand, it is well known that the ionic conductivity, σ, can be expressed as

$$\sigma = en\mu \tag{3.3}$$

where n, is the number of mobile oxide ion vacancies; μ, its mobility; and e, the charge (in case of oxide ion, $e=2$). By introduction of oxygen vacancy, oxide ion conductivity occurs in cubic-phase ZrO_2, suggesting that the vacancy concentration is linearly dependent on the dopant level. However, this is not true and the higher dopant concentration just leads to the formation of vacancy and dopant cluster resulting in decreased oxide ion conductivity.

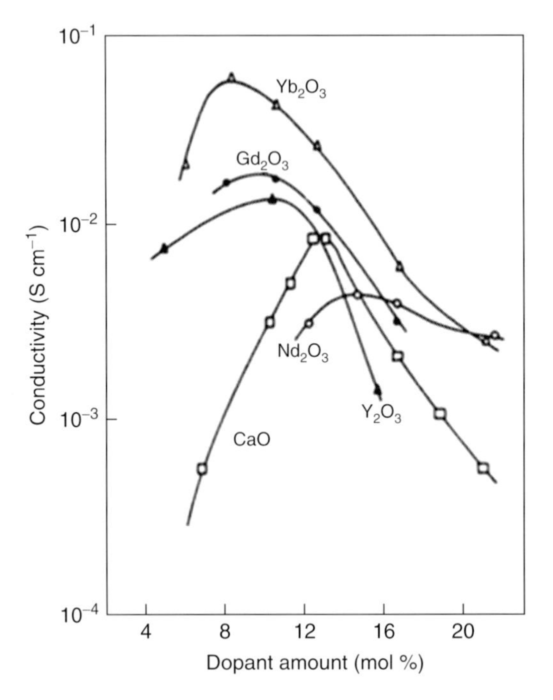

Figure 3.2 Oxide ion conductivity of stabilized ZrO_2 at 1353 K as a function of the dopant amount [11].

The conductivity of stabilized ZrO_2 varies with dopant concentration. As shown in Figure 3.2, the conductivities of doped zirconia show a maximum at a specific concentration of the dopant [11]. It is obvious that the electrical conductivity in ZrO_2 is strongly dependent on the dopant element and its concentration. In the small doping amount, conductivity monotonically increases with increasing dopant amount, which is expected from the theory and evidently, introduced defects behave as a point defect. Therefore, conductivity is mainly determined by the amount of oxygen vacancy, namely, the amount of dopant. On the other hand, conductivity as well as activation energy for conduction are strongly affected by the dopant ionic size, which is reported by Arachi *et al.* [12] for the ZrO_2–Ln_2O_3 (Ln = lanthanide) system.

Figure 3.3 shows the maximum oxide ionic conductivity and apparent enthalpy of oxide ion transport as a function of the ionic size of the dopant. It is evident that the conductivity increased with decreasing ionic size of doped cations. Explanation for this conductivity behavior is based on structural effects. The content of dopant with the highest conductivity in the ZrO_2–Ln_2O_3 system decreases with increasing dopant ionic radius. The dopants, Dy^{3+} and Gd^{3+}, with larger ionic radius show a limiting value of 8 mol%. The dopant Sc^{3+}, which has the closest ionic radius to the host ion, Zr^{4+}, shows the highest conductivity and the highest dopant content at which cluster formation starts. Scandia-doped zirconia is quite attractive as the electrolyte for SOFCs, especially for the intermediate-temperature (873–1073 K)

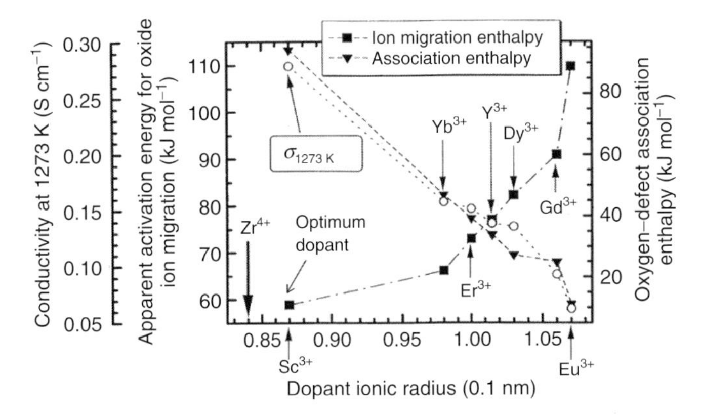

Figure 3.3 Maximum conductivities of doped zirconia at a specific concentration of the dopant [12].

SOFCs, and as-sintered sample shows the conductivity of 0.3 S cm^{-1} at 1273 K. However, the degradation in conductivity is more clearly observed for Sc-stabilized ZrO_2 due to annealing effects, which are discussed later.

Although high oxide ion conductivity is exhibited by Sc_2O_3-stabilized ZrO_2, Y_2O_3-stabilized ZrO_2 is most popularly used as the electrolyte in SOFCs because of its stability and the high cost of Sc_2O_3. In general, the ionic conductivity of cubic-stabilized ZrO_2 is independent of oxygen partial pressure over several orders of magnitude. Typical pO_2 dependence of stabilized ZrO_2 is shown in Figure 3.4.

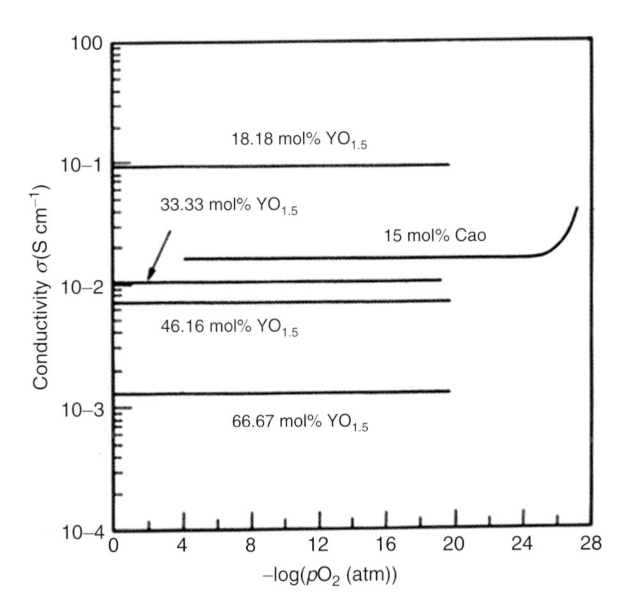

Figure 3.4 Typical pO_2 dependence of stabilized ZrO_2 at 1273 K [13].

Conductivity is independent of oxygen partial pressure ($1 < pO_2 < 10^{-20}$ atm), suggesting that the ionic transport number is close to unity in a wide pO_2 range [13]. Therefore, high transport number of oxide ion is one reason for using Y_2O_3-stabilized ZrO_2 as the electrolyte in SOFCs. Although oxide ion conductivity is dominant in stabilized ZrO_2, electronic conduction cannot be absolutely zero, and this partial electronic conductivity will determine the theoretical conversion efficiency of SOFCs. The ionic (σ_i) and electronic conductivity (σ_e or σ_h) are reported for $Zr_{0.84}Y_{0.16}O_2$ as a function of temperature and oxygen partial pressure. The following equations are reported for ionic and electronic conductivity [14].

$$\sigma_i \left(\text{S cm}^{-1}\right) = 1.63 \times 10^2 \exp\left(-0.79\ kT^{-1}\right) \tag{3.4}$$

$$\sigma_e \left(\text{S cm}^{-1}\right) = 1.31 \times 10^7 \exp\left(-3.88\ kT^{-1}\right) pO_2{}^{-1/4} \tag{3.5}$$

$$\sigma_h \left(\text{S cm}^{-1}\right) = 2.35 \times 10^2 \exp\left(-1.67\ kT^{-1}\right) pO_2{}^{1/4} \tag{3.6}$$

In a typical SOFC operation, oxygen partial pressure varies from 0.21 to 10^{-21} atm, the electronic conductivity (both electron and hole) is negligible compared to the ionic conductivity. At very low pO_2, the electrical conductivity becomes significant and the total conductivity starts to increase with decreasing pO_2. The oxygen partial pressure, at which electronic conductivity becomes significant, is higher at higher temperatures. In a highly reducing atmosphere, the increase in conductivity occurs mainly in the bulk material. The grain-boundary conductivity varies little with atmosphere [15].

Another reason for YSZ's popularity as the choice of electrolyte of SOFCs is its high mechanical strength. At room temperature, YSZ (8 mol% Y_2O_3) typically has a bending strength of about 300–400 MPa and a fracture toughness of about 3 MN $m^{-3/2}$. Even at elevated temperatures, YSZ shows sufficient mechanical property for electrolytes. A mean strength of about 280 MPa at 1173 K and a bending strength of about 225 MPa at 1273 K are reported for YSZ [16]. Therefore, the mechanical strength of YSZ is sufficiently high for electrolytes of SOFCs; however, attempts have been made for improving the mechanical strength of YSZ using additives. It is reported that the toughness of YSZ is much improved by addition of partially stabilized ZrO_2, Al_2O_3, and MgO [16, 17]. The composition of YSZ with 20 wt% Al_2O_3 has a bending strength of 33 kgf mm^{-2} or 323 MPa (compared with 24 kgf mm^{-2} or 235 MPa for YSZ) and an ionic conductivity of about 0.10 S cm^{-1} at 1273 K (compared with 0.12 S cm^{-1} for YSZ) [18].

Because of high toughness and strength, tetragonal ZrO_2 polycrystal (t-ZrO_2) has also been proposed as the SOFC electrolyte material in spite of its slightly low oxide ionic conductivity compared with that of fully stabilized ZrO_2 [19, 20]. For comparison, the toughness of YSZ ranges between 1 and 3 MPa $m^{1/2}$, whereas that of t-ZrO_2 ranges between 6 and 9 MPa $m^{1/2}$. At temperatures below 873 K, the electrical conductivity of t-ZrO_2 is greater than that of YSZ. This suggests the possibility of using t-ZrO_2 as an electrolyte in SOFCs operated at low temperatures. However, there are two main concerns regarding SOFCs based on t-ZrO_2 electrolytes: mechanical intensity and aging effect, namely, degradation of oxide ionic conductivity with time.

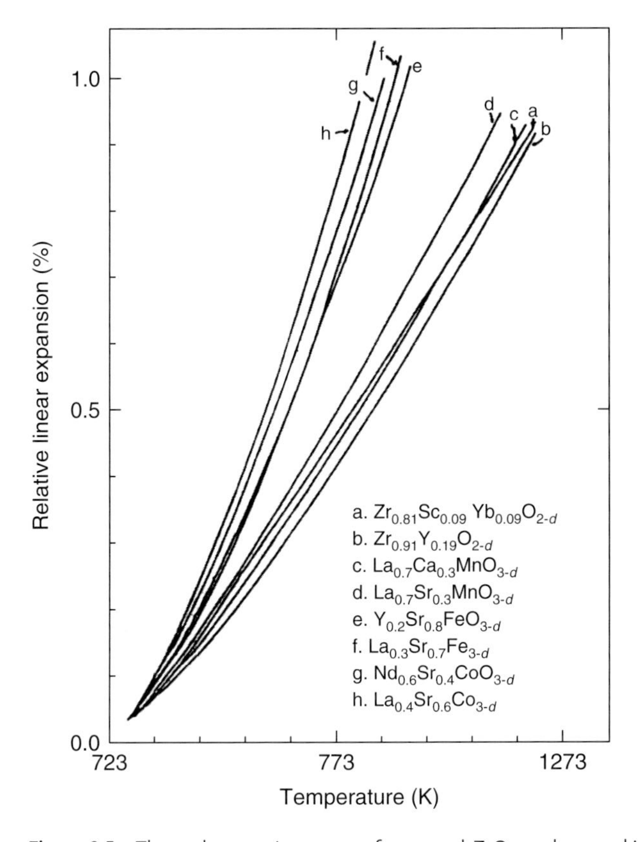

Figure 3.5 Thermal expansion curves for several ZrO_2 and perovskite cathode oxides [21].

Thermal expansion property is also an important factor for the electrolyte of SOFCs. Since a small difference in the thermal expansion coefficient of cell components in SOFCs produces large stresses during fabrication and operation, matching thermal expansion in the cell components is also important. Figure 3.5 shows thermal expansion curves for several stabilized ZrO_2 and perovskite oxide materials, which is the typical cathode material for SOFCs [21]. It is evident that thermal expansion of ZrO_2 (typically $10.8 \times 10^{-6}\,K^{-1}$ for 8 mol% Y_2O_3–ZrO_2) is smaller than that of a typical perovskite cathode, Ni anode, and interconnector. Therefore, for cathode, $LaMnO_3$-based oxide is widely used owing to its similar thermal expansion coefficient and low reactivity. On the other hand, for anode, addition of YSZ to Ni is generally performed to adjust thermal expansion and decrease overpotential.

3.2.1.2 **Doped CeO_2**

Doped ceria has been suggested as an alternative electrolyte for the low-temperature SOFCs. Detailed reviews of the electrical conductivity and conduction mechanism in ceria-based electrolytes have been also presented by Mogensen *et al.* [22] and

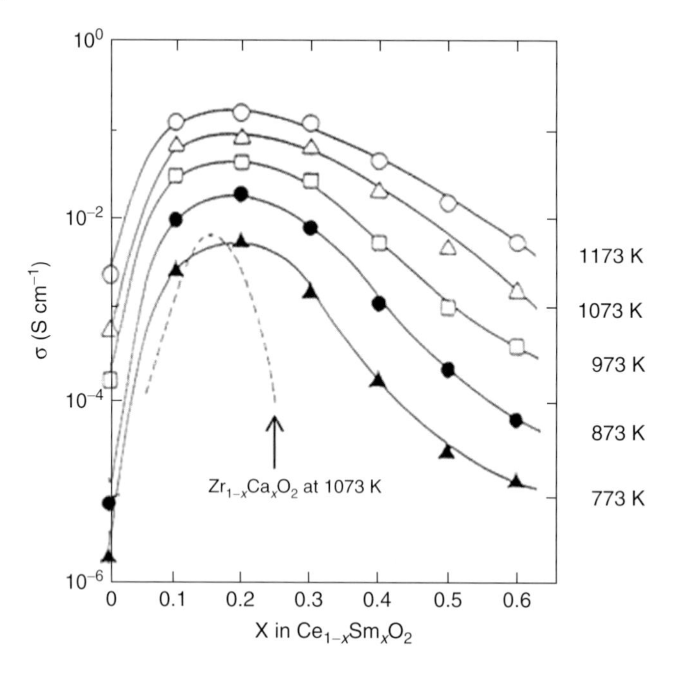

Figure 3.6 A typical concentration dependence of the electrical conductivity in the $CeO_2-La_2O_3$ system [25].

Steele [23]. Ceria possesses the same fluorite structure as stabilized zirconia. In a manner similar to ZrO_2, mobile oxygen vacancies are introduced by substituting Ce^{4+} with trivalent rare earth ions. The conductivity of doped ceria systems depends on the kind of dopant and its concentration, which is similar to the case of stabilized ZrO_2. Recent studies on oxide ion conductivity reveal that the clusterization of doped cations and vacancy occurs at dopant concentrations much smaller than that considered previously [24].

A typical concentration dependence of electrical conductivity in the $CeO_2-Sm_2O_3$ system reported by Yahiro *et al.* [25] is shown in Figure 3.6. The maximum conductivity is also observed for each dopant, which is similar with that of stabilized ZrO_2 and exists at around 10 mol% of Sm_2O_3. It is suggested that the conductivity of the $CeO_2-Ln_2O_3$ system depends on the dopant's ionic radius, and its relation is summarized in Figure 3.7 [26]. The binding energy calculated by Butler *et al.* [27] shows a close relationship to the conductivity as shown in this figure; the dopant with low binding energy exhibits high conductivity. The systems $CeO_2-Gd_2O_3$ and $CeO_2-Sm_2O_3$ show ionic conductivity as high as 5×10^{-3} S cm^{-1} at 773 K, and this conductivity value corresponds to 0.2 Ω cm^2 ohmic loss for an electrolyte of 10 μm thickness. Thus, these systems are quite attractive for electrolytes in low-temperature SOFCs and have been extensively studied.

Ceria-based oxide ion conductors have pure ionic conductivity at high oxygen partial pressures. At lower oxygen partial pressuresthat are prevalent on the anode side of SOFCs, the materials become partially reduced. This leads to an

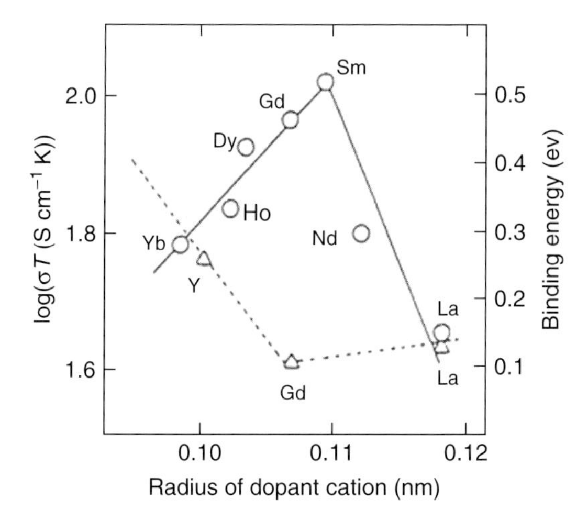

Figure 3.7 Conductivity of $Ce_{0.8}Sm_{0.2}O_2$ as a function of ionic size of pO_2 [26].

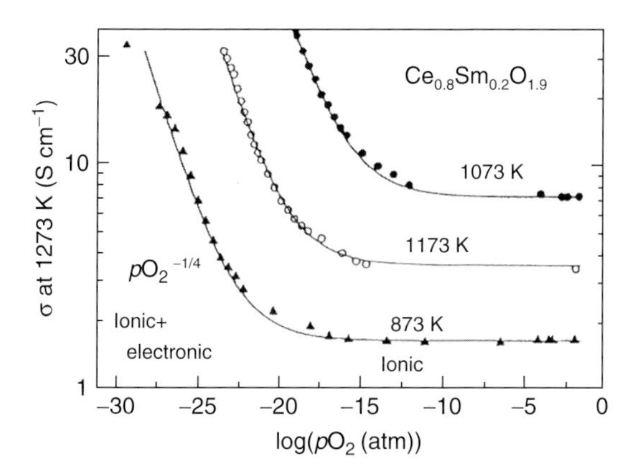

Figure 3.8 Total conductivity of $Ce_{0.8}Sm_{0.2}O_{1.9-\delta}$ as a function of oxygen partial pressure [26].

electronic conductivity in a large volume fraction of the electrolyte extending from the anode side [28]. When the fuel cell is constructed with such an electrolyte, electronic current flows through the electrolyte even at open circuit, and the terminal voltage is somewhat lower than the theoretical value. In Figure 3.8, total electrical conductivities (ionic and electronic) of $Ce_{0.8}Sm_{0.2}O_{1.9-\delta}$ are shown as a function of oxygen partial pressure [26]. At a high pO_2 range, conductivity is independent of oxygen partial pressure, suggesting pure oxide ion conductivity; however, in a reducing atmosphere, apparently, conductivity increases with pO_2 with $pO_2^{-1/4}$ dependency, suggesting that the free electron is formed by reduction of Ce^{4+} to Ce^{3+}. When CeO_2 is used as the electrolyte in SOFCs, this partial electronic conduction in fuel atmosphere causes several problems, in particular, decreased cell efficiency, which is discussed later. Reduction of Ce^{4+} to Ce^{3+} also results in volume expansion and so large stress occurs in CeO_2 electrolyte during SOFC operation, that is, large pO_2 gradient through electrolyte membrane, and in some cases, fracture of electrolyte occurs when CeO_2-based oxide is used as the electrolyte. Godickemeier and Gaucker [29] analyzed the efficiency of cells with $Ce_{0.8}Sm_{0.2}O_{1.9}$ by considering the electronic conduction and reported that the maximum efficiency based on Gibbs free energy was 50% at 1073 K and 60% at 873 K. Therefore, SOFCs with $Ce_{0.8}Sm_{0.2}O_{1.9}$ should be operated at temperatures lower than 873 K. However, to apply doped CeO_2 in SOFCs operating at such low temperatures, film electrolyte is required because of its decreased oxide ion conductivity. Partial electronic conduction becomes more significant in a film sample for application of SOFC electrolyte because of the enlarged pO_2 gradient per unit length.

Ceria-based electrolytes could be used in SOFCs operated at 823 K or lower. To operate at higher temperatures, a dual-layer electrolyte, with a thin YSZ layer on doped ceria, has been proposed [30]. However, the interdiffusion at the YSZ–GDC (Gd-doped ceria) interface could be an issue for practical applications in SOFCs, and cells using the YSZ–GDC bilayer electrolyte show much lower power density than that expected from film thickness.

In any case, as discussed, in order to achieve high oxide ion conductivity, design of the dopant and its concentration is highly important. Although, at present, YSZ is widely used as the electrolyte in SOFCs, usage of further higher oxide ion conductivity is strongly required for the SOFC operation temperature. However, in case of the tetravalent oxides with fluorite structure, up to now, higher oxide ion conductivity is due to lower chemical stability in reducing atmosphere; thus, an alternative to YSZ is quite limited, which includes Sc_2O_3–ZrO_2 doped with 1 mol% CeO_2 [31] or Sm_2O_3- or Gd_2O_3-doped CeO_2. Similar to fluoride oxide, perovskite oxide also has a large free volume and so, for a long period, the oxide ion conductivity in perovskite oxide has also been in need of an alternative to YSZ as the electrolyte for SOFCs.

3.2.2 Perovskite Oxide

Although the oxide with perovskite structure is anticipated to be a superior oxide ion conductor, typical perovskite oxides such as $LaCoO_3$ and $LaFeO_3$ are well known as famous mixed electronic and oxide ionic conductors, which can be used for the cathode of SOFCs. Therefore, these mixed conducting perovskite oxides can be a promising material group as the cathode catalysts of SOFCs or oxygen-permeating membranes. Now, the large majority of perovskite oxides exhibiting oxide ion conduction is classified as mixed conductors, which show both electronic and oxide ionic conduction and cannot be used as electrolytes in SOFCs. Diffusivity of oxide ion in typical perovskite is shown in Figure 3.9 [32]. Some perovskite oxide show fast diffusivity of oxide ion and can be used as the oxide ion-conducting electrolyte. However, such perovskite oxides show much larger electronic or mainly hole

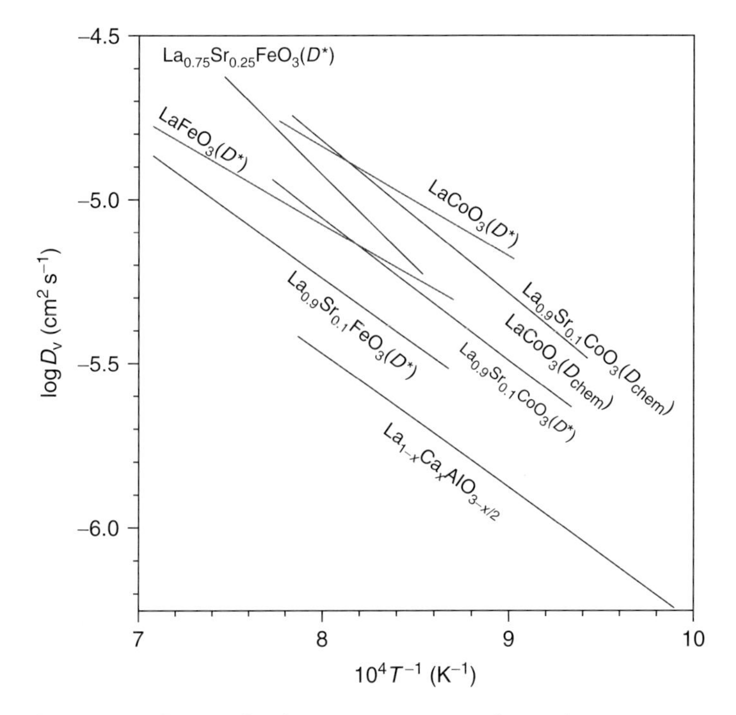

Figure 3.9 Diffusivity of oxide ion in various perovskite oxides [32].

conduction and so are only considered as an active electrode. However, recently, a few perovskite oxides showed high transport number of oxide ion. In this section, oxide ion conductors with perovskite oxide, in particular, $LaGaO_3$, are reviewed.

Takahashi and Iwahara [33] have done pioneering works on perovskite oxide ion conductor. They reported fast oxide ion conductivity and reasonably high transport number of oxide ion in Ti- and Al-based perovskite oxide, and it is evident that Al- or Mg-doped $CaTiO_3$ exhibits high conductivity but it is still lower than that of YSZ. Takahashi and Iwahara [33] investigated the oxide ion conductivity in $CaTiO_3$ in detail. Although the high transport number of oxide ion is exhibited by $CaTi_{0.95}Mg_{0.05}O_3$ at intermediate temperatures, Ca-doped $LaAlO_3$ is another attractive candidate as an oxide ion conductor because no electronic conduction appears in reducing atmosphere and its transport number is higher than 0.9 over the entire temperature range.

After the report by Takahashi and Iwahara [33], many researchers have investigated the oxide ion conductivity in $LaAlO_3$-based oxide. However, the reported oxide ion conductors with perovskite structure exhibited lower ionic conductivity than that of Y_2O_3–ZrO_2 [34]. In the conventional study on perovskite oxide of ABO_3, it was widely believed that the electric or dielectric property is determined by B-site cations. However, a migrating oxide ion has to pass through the triangular space consisting of two large A-site and one small B-site cations in the crystal lattice.

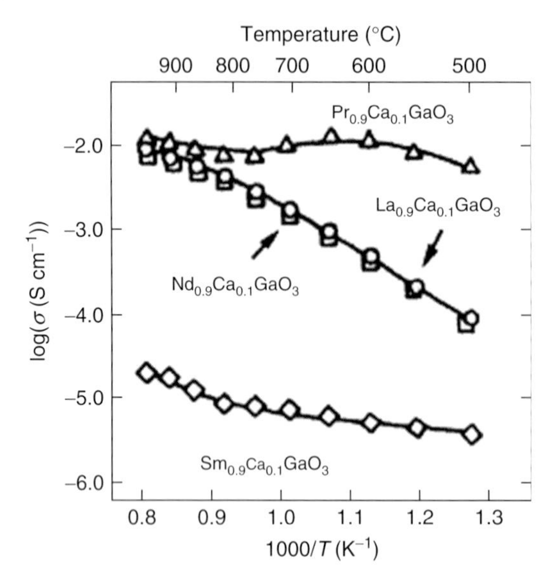

Figure 3.10 Arrhenius plots of the electrical conductivity of Ca-doped $LnGaO_3$ (Ln = La, Pr, Nd, Sm) [34].

Therefore, the ionic size of the A-site cation seems to influence oxide ion conductivity greatly. Effects of A-site cation on the oxide ion conductivity in $LnAlO_3$-based perovskite oxide have been reported. Figure 3.10 shows the electrical conductivity in Ca-doped $LnGaO_3$-based oxide [34]. Electrical conductivity of Ga-based perovskite oxide increased with increasing ionic size of A-site cations. This suggests that the larger unit volume of lattice is important for high oxide ion conductivity because of larger free volume. Therefore, doping larger cation for B site is also important. Although high oxide ion conductivity was reported in $Nd_{0.9}Ca_{0.1}Al_{0.5}Ga_{0.5}O_3$, it was still lower than that of YSZ. However, of great interest is the higher oxide ion conductivity reported in $LaGaO_3$-based perovskite oxide. Another type of perovskite oxide ion conductor is $LaScO_3$, which is also reported as a high-temperature proton conductor [35]. Figure 3.11 shows the pO_2 dependence of four different perovskite oxides at similar composition [36]. In spite of similar composition, the oxide ion conductivity is much different, that is, higher in $LaGaO_3$; however, $LaAlO_3$, $LaScO_3$, and $LaInO_3$ show lower oxide ion conduction, with hole conduction at higher pO_2 range [37]. A similar study was performed by Nomura *et al.* and the order of oxide ion conductivity in these perovskite oxides is not simply explained by free volume size, but by size matching of dopant, in particular, Mg to B-site cations. Mg is too big as a dopant for the B site of those four perovskite oxides; however, it is the closest one to Ga in size [37]. Comparison of the defect association energy is also required, as discussed later; however, it is evident that $LaGaO_3$ is a promising oxide ion conductor over a wide pO_2 range [38].

In the following section, the oxide ion conductivity in $LaGaO_3$ perovskite oxide is briefly introduced. The high oxide ion conductivity in $LaGaO_3$-based perovskite,

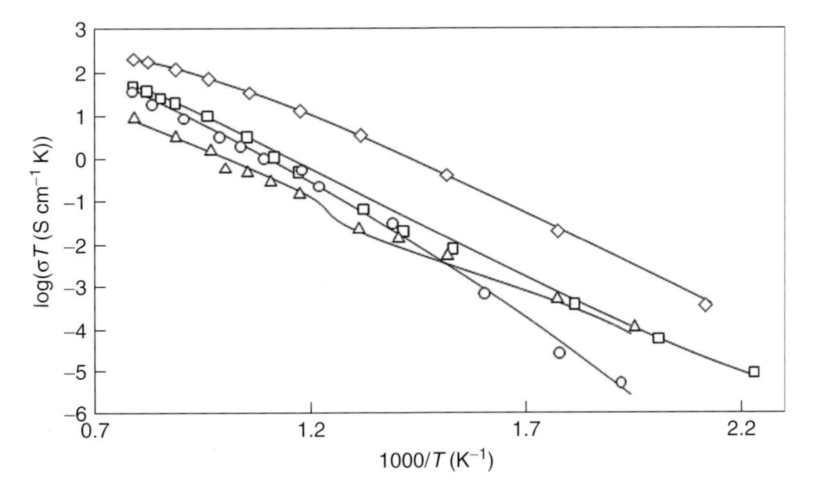

Figure 3.11 Electrical conductivity of four different perovskite oxides at similar composition in air [36]: $La_{0.9}Sr_{0.1}Ga_{0.9}Mg_{0.1}O_3$, (♦); $La_{0.9}Sr_{0.1}Sc_{0.9}Mg_{0.1}O_3$, (□); $La_{0.9}Sr_{0.1}Al_{0.9}Mg_{0.1}O_3$, (○); and $La_{0.9}Sr_{0.1}In_{0.9}Mg_{0.1}O_3$, (Δ).

which is the first pure oxide ion conductor, was reported in 1994 [39]. The high oxide ion conductivity in this oxide is achieved by double doping of lower valence cation into A and B sites of perovskite oxide, ABO_3. It is obvious that the oxide ion conductivity strongly depends on the cations for the A site, which is similar to the case of Al-based oxide, and the highest conductivity is achieved in $LaGaO_3$, which is also the largest unit lattice volume in Ga-based perovskite. The electrical conductivity of Ga-based perovskite oxides is almost independent of the oxygen partial pressure in the range from 1 to 10^{-21} atm. Therefore, it is expected that the oxide ion conduction will be dominant in all Ga-based perovskite oxides.

Doping a lower valence cation generally forms oxygen vacancies due to the electric neutrality; the oxide ion conductivity will increase with increasing amount of oxygen vacancies. Therefore, doping alkaline earth cations to La sites was investigated and the oxide ion conductivity obtained is shown in Figure 3.12 [39]. The electrical conductivity in $LaGaO_3$ depends strongly on the alkaline earth cations doped at La sites and is increased in the following order: Sr > Ba > Ca. Therefore, strontium, the ionic size of which is almost the same as that of La^{3+}, is the most suitable dopant for La sites in $LaGaO_3$. This may also be explained by the formation of local stress in the lattice due to a mismatch in ionic size. Theoretically, increasing the amount of Sr will increase the amount of oxygen vacancy and hence the oxide ion conductivity. However, solid solubility of Sr into La sites of $LaGaO_3$ is poor and it is observed that the secondary phases, $SrGaO_3$ or La_4SrO_7, form when the amount of Sr becomes higher than 10 mol%. Therefore, the amount of oxygen vacancy introduced by La-site doping is not large.

It is found that doping Mg is highly effective for increasing the conductivity since additional oxide ion vacancies are formed. The oxide ion conductivity is further increased by increasing the amount of doped Mg and it attained the maximum

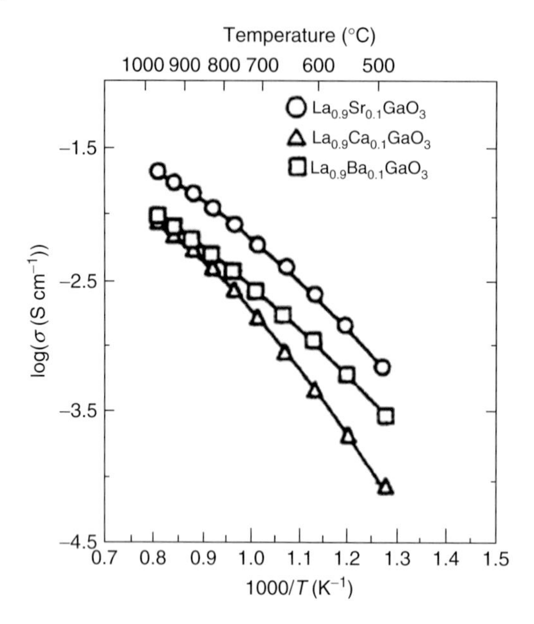

Figure 3.12 Effects of alkaline earth cations in the La site on the oxide ion conductivity of $LaGaO_3$ [39].

at 20 mol% Mg doped for Ga sites. This enlargement in the limit of Sr solid solution in La sites was also reported by Majewski *et al.* [40]. This seems to be due to the enlarged crystal lattice. In any case, the highest oxide ion conductivity in $LaGaO_3$-based oxide is reported at the composition of $La_{0.8}Sr_{0.2}Ga_{0.8}Mg_{0.2}O_3$ [41].

As this oxide consists of four elements, the optimum composition is slightly varied with researchers. Oxide ion conductivity in $LaGaO_3$-based oxide was investigated by several groups [42, 43], and various cations were examined as the dopant for $LaGaO_3$-based oxides. Huang and Petric [43] investigated the oxide ion conductivity in various compositions and expressed the oxide ion conductivity in contour maps [44], as shown in Figure 3.13, in which the optimum composition reported by two other groups are also shown. Huang and Goodenough [45] reported that the highest oxide ion conductivity was obtained at the composition of $La_{0.8}Sr_{0.2}Ga_{0.85}Mg_{0.15}O_3$. On the other hand, Huang *et al.* [44, 45] and Huang and Goodenough [44, 45] reported the optimized composition in $La_{1-X}Sr_XGa_{1-Y}Mg_YO_3$, where $X = 0.2$, $Y = 0.17$. However, the optimized composition among the three groups is close to each other and exists at $Y = 0.15–0.2$ in $La_{0.8}Sr_{0.2}Ga_{1-Y}Mg_YO_3$. Difference may arise from the uniformity of composition and also grain size.

Figure 3.14 shows the comparison of oxide ion conductivity of double-doped $LaGaO_3$ with that of the conventional oxide ion conductors [16]. It is obvious that the oxide ion conductivity in $La_{0.8}Sr_{0.2}Ga_{0.8}Mg_{0.2}O_3$ is higher than the typical conductivity of ZrO_2- or CeO_2-based oxides and somewhat lower than those of Bi_2O_3-based oxides. It is well known that n-type semiconductors are dominant in CeO_2- or Bi_2O_3-based oxides in a reducing atmosphere, and furthermore, thermal

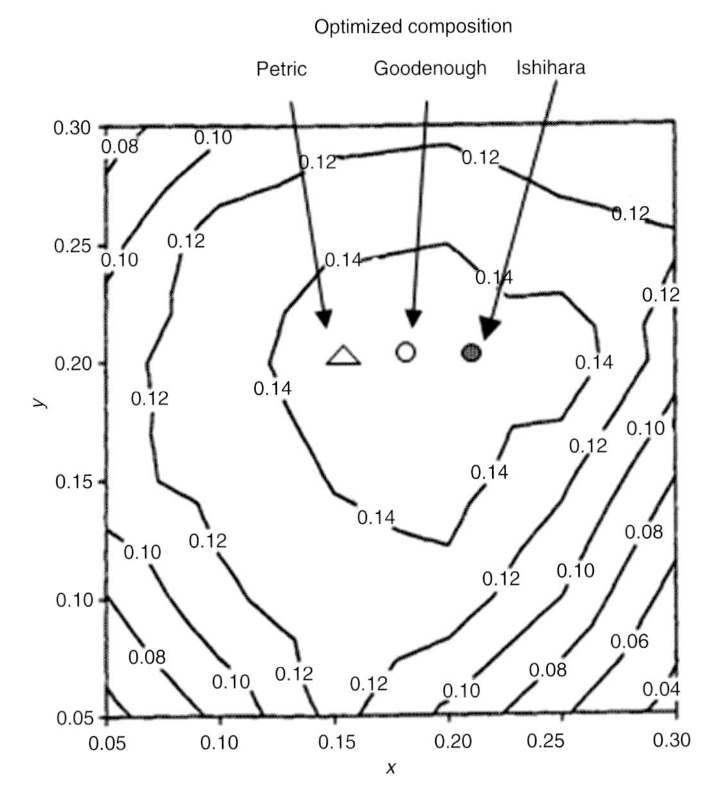

Figure 3.13 Contour plot of the conductivity in $La_{1-X}Sr_XGa_{1-Y}Mg_YO_3$ at 1073 K [43, 44].

stability is not satisfactory in Bi_2O_3-based oxides. In contrast, $La_{0.8}Sr_{0.2}Ga_{0.8}Mg_{0.2}O_3$ exhibits fully ionic conduction at pO_2 from 10^{-20} to 1 atm. Therefore, double-doped $LaGaO_3$ perovskite oxide shows great promise for the solid electrolyte of fuel cell and oxygen sensor.

The diffusivity in oxide ions in $(La_{0.8}Sr_{0.2})(Ga_{0.8}Mg_{0.2})O_{2.8}$ (LSGM) was further studied with ^{18}O tracer diffusion measurements [46]. When compared with fluoride oxide, LSGM exhibits larger diffusion coefficient that could originate from the high mobility of the oxide ions (Table 3.1). Therefore, it is considered that the perovskite structure has a large free volume in lattice and this could allow the high diffusivity of oxide ion, resulting in high oxide ion conductivity.

Recently, visualization of oxygen transport in perovskite, in particular, $LaGaO_3$ perovskite is simulated based on quantum chemistry [47]. In case of perovskite oxide, the migrating ion must pass through the opening of a triangle defined by two A-site (La^{3+}) ions and one B-site ion. Owing to the geometrical reason for lattice relaxation, it is suggested that there is a small deviation from the direct path for vacancy migration, as illustrated schematically in Figure 3.15. The calculations reveal a curved route around the octahedron edge, with the saddle point away from the adjacent B-site cation. Therefore, because of the large free volume, high

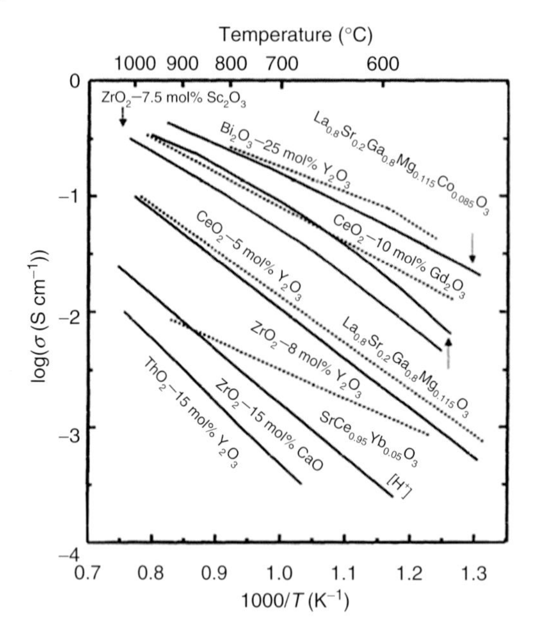

Figure 3.14 Comparison of oxide ion conductivity of $LaGaO_3$ doped with Sr, Mg, and Co with that of the fluorite oxide ion conductor in N_2 atmosphere.

Table 3.1 Comparison of mobility of oxide ion in selected fluorite and LSGM oxide at 1073 K.

	D_t ($cm^2\ s^{-1}$)	E_a (eV)	δ	$[V_O^{\cdot\cdot}]$ (cm^{-3})	D ($cm^2\ s^{-1}$)	μ ($cm^2\ V^{-1}\ s^{-1}$)
$Zr_{0.81}Y_{0.19}O_{2-\delta}$	6.2×10^{-8}	1.0	0.10	2.95×10^{21}	1.31×10^{-6}	1.41×10^{-5}
$Zr_{0.858}Ca_{0.142}O_{2-\delta}$	7.54×10^{-9}	1.53	0.142	4.19×10^{21}	1.06×10^{-7}	1.15×10^{-6}
$Zr_{0.85}Ca_{0.15}O_{2-\delta}$	1.87×10^{-8}	1.22	0.15	4.43×10^{21}	2.49×10^{-7}	2.69×10^{-6}
$Ce_{0.9}Gd_{0.1}O_{2-\delta}$	2.70×10^{-8}	0.9	0.05	1.26×10^{21}	1.08×10^{-6}	1.17×10^{-5}
$La_{0.9}Sr_{0.1}Ga_{0.8}Mg_{0.2}O_{3-\delta}$	3.24×10^{-7}	0.74	0.15	2.53×10^{21}	6.4×10^{-6}	6.93×10^{-5}
$La_{0.8}Sr_{0.2}Ga_{0.8}Mg_{0.2}O_{3-\delta}$	4.13×10^{-7}	0.63	0.20	3.34×10^{21}	6.12×10^{-6}	6.62×10^{-5}
$La_{0.8}Sr_{0.2}Ga_{0.8}Mg_{0.125}Co_{0.085}O_{3-\delta}$	4.50×10^{-7}	0.42	0.1645	2.78×10^{21}	8.21×10^{-6}	8.89×10^{-5}

D_t, tracer diffusion coefficient; D, self-diffusion coefficient.

mobility of oxide ions is achieved in perovskite oxide, which is the origin of high conductivity in this oxide.

3.2.3
Perovskite-Related Oxide

Oxide ion conductivity in the oxide with a perovskite-related structure is also reported, in particular, $Ba_2In_2O_5$-based oxide is popularly studied. In this section,

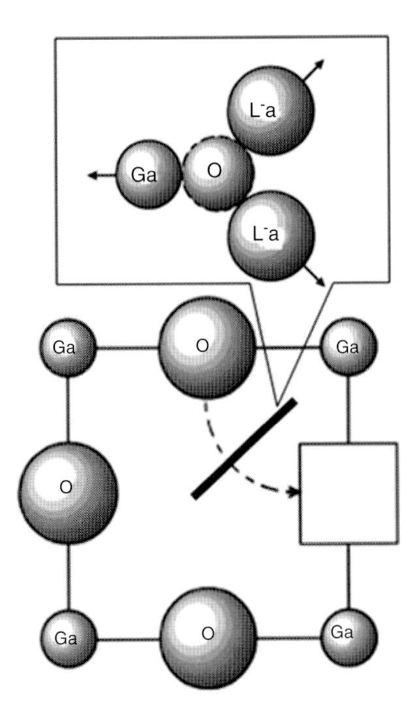

Figure 3.15 Calculated path for oxygen vacancy migration [47].

oxide ion conductivities in doped-$Ba_2In_2O_5$ and K_2NiO_4 structure oxides are introduced, particularly from dopant effects. Brownmillerite structure ($A_2B_2O_5$) is a perovskite-related structure in which one-sixth of the oxygen in the unit cell is originally deficient. In this oxide, oxygen vacancy is ordered in the [101] direction at low temperatures but is disordered at higher temperatures. The high oxide ion conductivity in $Ba_2In_2O_5$-based oxide was first reported by Goodenough *et al.* [48] in 1990, which is shown in Figure 3.16. Oxide ion conductivity in this $Ba_2In_2O_5$ disordered phases is higher than that of YSZ and so many researchers have studied dopant effects on this oxide [49, 50]. Similar to ZrO_2, high-temperature cubic phase can be stabilized to lower temperature by doping aliovalent cations. Goodenough *et al.* reported that Zr for the In site is effective, and Yao *et al.* reported that Ga is effective for stabilizing the high-temperature phase. As shown in Figure 3.16, conductivity in low-temperature range is much improved by stabilizing the high-temperature phase, and the discontinuity in the Arrhenius plot of conductivity on $Ba_2In_2O_5$ disappears by stabilizing the high-temperature cubic phase. However, high temperature conductivity is neither improved nor slightly decreased by the dopant. Therefore, conductivity of doped $Ba_2In_2O_5$ is almost the same as that of YSZ.

Kakinuma *et al.* [51] reported the effects of La^{3+} on oxide ion conductivity in the Ba site in $Ba_2In_2O_5$. In the conventional oxide ion conductors, introduction of oxygen vacancy by doping lower valence cation is essential; however, in case of $Ba_2In_2O_5$, doping the higher valence cation of La^{3+} to the Ba site is effective in

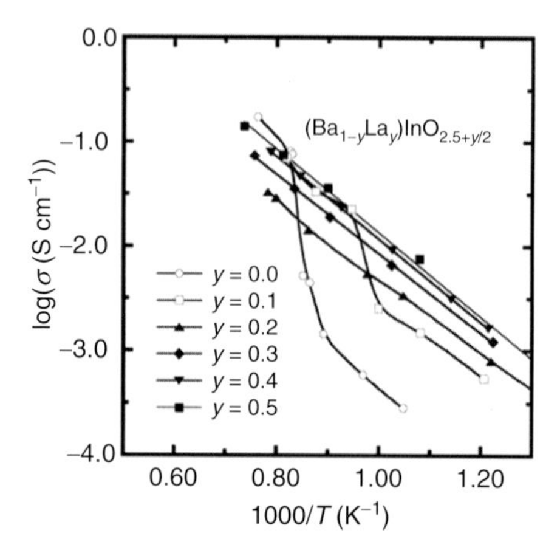

Figure 3.16 Oxide ion conductivity in $Ba_2In_2O_5$-based oxide [48].

increasing the oxide ion conductivity. Figure 3.17 shows oxide ion conductivity as a function of oxygen content. It is interesting that oxide ion conductivity almost monotonically increases with increasing oxygen content. This can be explained by the disordered oxygen vacancy structure stabilized by introduction of excess oxygen. In addition, introduction of Sr^{2+}, which decreases the unit lattice volume, is also effective in increasing the oxide ion conductivity of this oxide. In spite of no change in oxygen content, substitution of Sr for Ba site in $Ba_2In_2O_5$ is effective in increasing oxide ion conductivity. Here, free volume is also an important factor for high oxide ion conductivity, which is also suggested for perovskite oxide. The highest oxide ion conductivity is reported for $(Ba_{0.5}Sr_{0.2}La_{0.3})InO_{2.85}$.

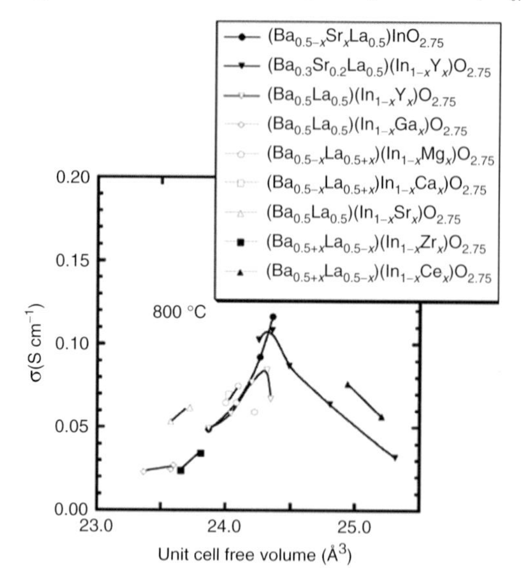

Figure 3.17 Oxide ion conductivity as a function of oxygen content in $Ba_2In_2O_5$ [51].

Application of Sr- and La-doped $Ba_2In_2O_5$ to the electrolyte of SOFCs has also been reported. Their open circuit potential is 0.93 V, which is slightly lower than that of the theoretical value. This is explained by hole conduction in the high pO_2 range. However, the maximum power density is almost 0.6 W cm^{-2} at 1073 K, which is a reasonably high power density. This power density implies that $Ba_2In_2O_5$ perovskite-related oxide is also interesting as the electrolyte in SOFCs.

Another type of defect perovskite for oxide ion conductor is K_2NiF_4, in which perovskite oxide block (ABO_3) is connected with rock salt one (AO), which has a large free volume for oxide ion conduction. A typical case of this defect perovskite is Sr_2TiO_4, and the oxide ion conductivity of Ti-based oxide has been reported [52]. However, introduction of oxygen vacancy or interstitial oxygen into rock salt block is rather difficult and the observed oxide ion conductivity in large part of K_2NiF_4 structure oxide is low. But, recently, high oxygen diffusivity was reported for Pr_2NiO_4 doped with Cu and Ga, although high hole conduction was simultaneously observed [53, 54]. By doping Cu and Ga, interstitial oxygen is introduced into rock salt block, thus oxide ion conductivity in this oxide is mainly based on interstitial oxygen, but not oxygen vacancy. Figure 3.18 shows the nuclear density distribution on the (100) plane of the mixed conductor $(Pr_{0.9}La_{0.1})_2(Ni_{0.74}Cu_{0.21}Ga_{0.05})O_4$ at

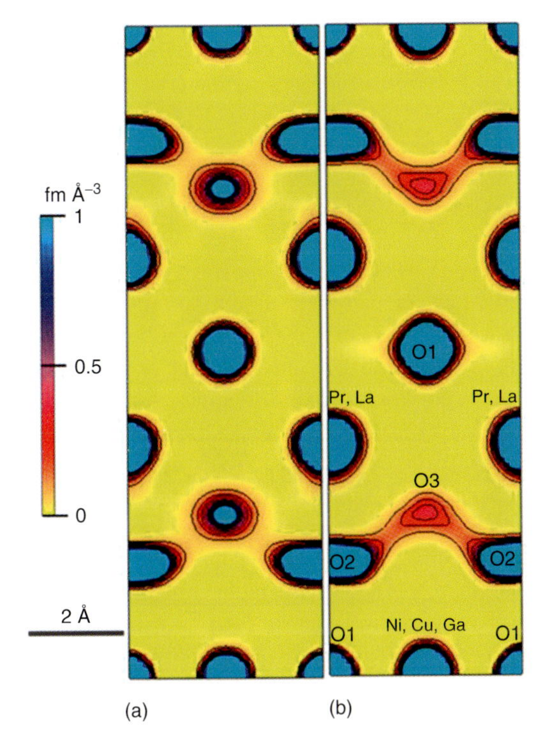

Figure 3.18 Nuclear density distribution on the (100) plane of the mixed conductor $(Pr_{0.9}La_{0.1})_2(Ni_{0.74}Cu_{0.21}Ga_{0.05})O_4$ at (a) 879.6 K (606.6 °C) and (b) 1288.6 K (1015.6 °C). Contour lines from 0.1 to 1.0 by the step of 0.1 fm $Å^{-3}$ [53, 54].

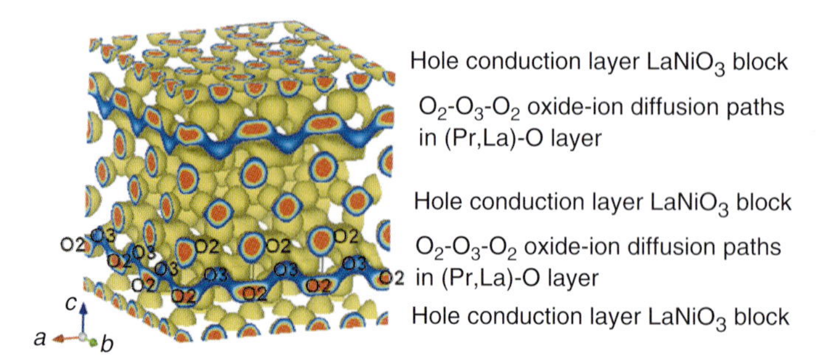

Figure 3.19 Three-dimensional image of oxide ion and hole conduction route in Pr_2NiO_4.

(a) 883.6 K and (b) 1288.6 K. Contour lines from 0.1 to 1.0 by the step of 0.1 fm $Å^{-3}$ are shown in the figure. Apparently, at high temperatures, interstitial oxygen is introduced into the O_3 position, which is the interstitial position in rock salt block and then oxide ion is conducted through this position. On the other hand, perovskite block has a high density of electron and it is suggested that hole is mainly conducted in perovskite block. Therefore, K_2NiF_4 is highly unisotropic for oxide ion and hole conduction and different conduction routes for ion and electronic charge carrier, as shown in Figure 3.19. The estimated oxide ion conductivity in Pr_2NiO_4 from oxygen permeation rate is higher than that of $LaGaO_3$-based oxide, suggesting that defect perovskite oxide is also an interesting material group from the viewpoint of oxide ion conductors.

3.2.4
New Class of Oxide Ion-Conducting Oxide

To a large degree, the known fast oxide ion conductors possess either cubic or pseudocubic crystal lattice. Even $LaGaO_3$ or perovskite is not an exception as it has a pseudocubic perovskite structure. Therefore, it is generally believed that a high symmetry in crystal lattice is an essential requirement for fast oxide ion conduction. So far, there have been no reports in the literature for a notable oxide ion conductivity in noncubic structured oxides. Among a few exceptions, the oxide ion conductivity in hexagonal apatite oxide of $La_{10}Si_6O_{27}$ and $Nd_{10}Si_6O_{27}$ reported by Nakayama and Sakamoto [55] and Nakayama [56] is highly interesting. Figure 3.20 shows the comparison of oxide ion conductivity in $La_{10}Si_6O_{27}$ with that of doped bismuth oxides and zirconia. The electrical conductivity values of this oxide at temperatures higher than 873 K are not high enough compared with that of the conventional fast oxide ion conductors such as 8 mol% Y_2O_3-stabilized ZrO_2. However, at lower temperatures, $La_{10}Si_6O_{27}$ exhibits higher oxide ion conductivity than the conventional oxide ion conductors. Sansom *et al.* [57] studied the relationship between crystal structure and oxide ion conductivity in this oxide. The refinement crystal structure of the $La_{10}Si_6O_{26}$ belongs to a hexagonal space group *P*3 (no. 147), with $a = b = 972.48$ pm, $c = 718.95$ pm. The refinement structure suggests that the $La_{10}Si_6O_{26}$ has a unique structure

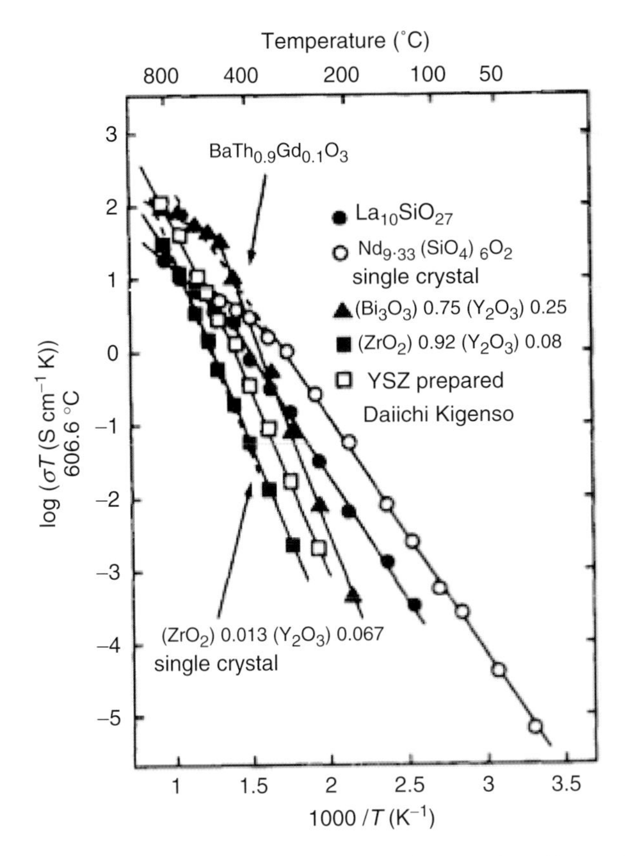

Figure 3.20 Comparison of oxide ion conductivity of $La_{10}Si_6O_{27}$ with that of doped bismuth oxides and zirconia [55].

in channel oxygen sites and the high oxygen ion conductivity could be assigned to this disorder in the channel site [57]. Islam also studied oxygen diffusion route in this hexagonal apatite oxide, and as shown in Figure 3.21, relatively high oxide ion conductivity is assigned to an interstitial oxygen; however, a quite unique diffusion route the so-called snake like transport route is pointed out [58].

In an analog of $La_{10}Si_6O_{10}$ hexagonal apatite, La_2GeO_5 is reported as a high oxide ion conductor [59]. Figure 3.22 shows the crystal structure of La_2GeO_5 phase, and the difference between this structure and that of apatite is just the tilting angle of the GeO_4 tetragonal pyramid. In La_2GeO_5, as shown in Figure 3.22a, lanthanum and oxygen have a straight arrangement along the [111] direction resulting in the more straight transport route of oxide ions. The conductivity increases with increasing amount of La deficiency and the maximum value is attained at $X = 0.39$ in $La_{2-X}GeO_5$. Oxide ion transport number in La_2GeO_5-based oxide is also estimated to be unity from measured electromotive forces of H_2-O_2 and N_2-O_2 gas concentration cells. Figure 3.23 shows the comparison of the

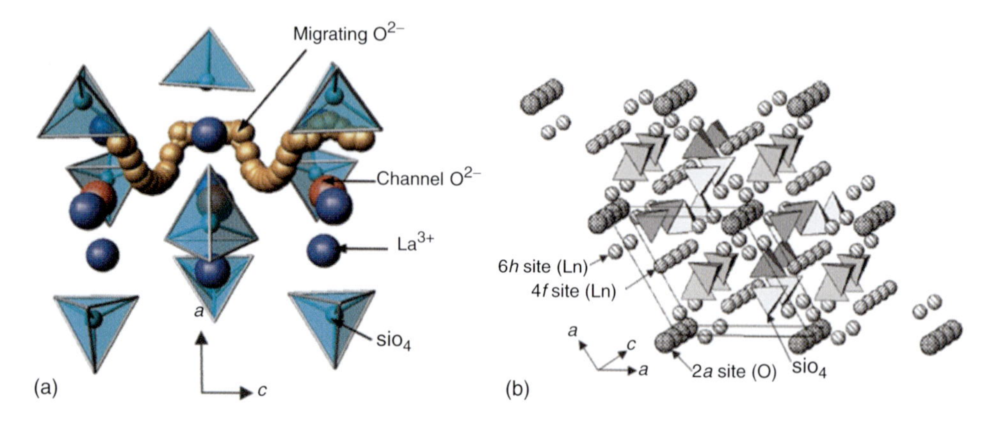

Figure 3.21 (a) Oxygen diffusion route in a hexagonal apatite oxide and (b) unit lattice [57].

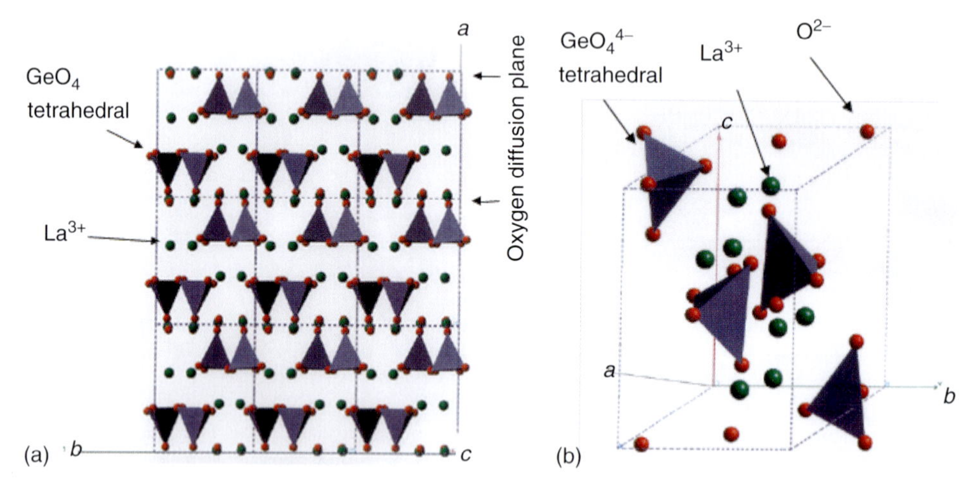

Figure 3.22 Crystal structure of La_2GeO_5 phase in (a) [111] direction and (b) unit lattice [59].

oxide ion conductivity of $La_{1.61}GeO_5$ with that of fluorite and perovskite-structured oxides. The comparison clearly reveals that the oxide ion conductivity of $La_{1.61}GeO_5$ is much higher than that of Y_2O_3-stabilized ZrO_2 and almost the same as that of $Gd_{0.15}Ce_{0.85}O_2$ or $La_{0.9}Sr_{0.1}Ga_{0.8}Mg_{0.2}O_3$ at temperatures above 973 K. However, at low temperatures, the oxide ion conductivity of $La_{1.61}GeO_5$ becomes much lower because of the change in activation energy. Change in the slope of the Arrhenius plot can be explained by the crystal structure change in a short range, which corresponds to the order–disorder change in oxygen vacancy structure. Although there are many other requirements for the SOFC electrolyte, this high value of electrical conductivity at high temperatures is attractive.

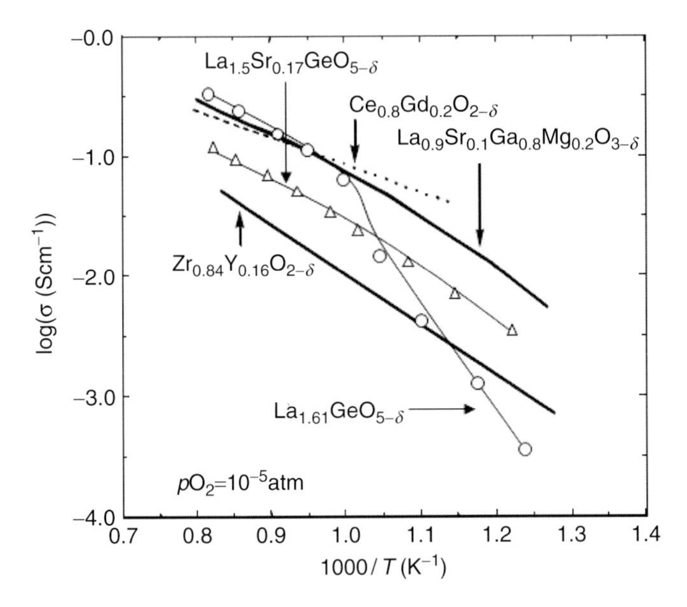

Figure 3.23 Comparison of the oxide ion conductivity of $La_{1.61}GeO_5$ with that of fluorite- and perovskite-structured oxides [59].

A bismuth-based oxide, the so-called BIMEVOX (bismuth metal vanadium oxide), was also reported as a high oxide ion conductor with noncubic structure; however, this oxide exhibits whole oxide ionic conductivity only in a limited pO_2 range [60] and is not interesting as the electrolyte of SOFCs because of its high electronic conductivity in reducing atmosphere. To use this property, double layer of SDC (samarium-doped ceria)–Bi_2O_3-based oxide is studied as the electrolyte in SOFCs [61]. By preventing the reduction of Bi_2O_3 by the SDC layer set in the fuel side, reasonably high open circuit voltage (OCV) and large power density were achieved, as shown in Figure 3.24. However, it is anticipated that the counterdiffusion of cations and also low melting point of Bi_2O_3 may still cause problems for durability of SOFCs.

Oxide ion conductivity in β-phase $La_2Mo_2O_9$ was reported in 2000 by Lacorre *et al.* [62] (Figure 3.25). Similar to $Ba_2In_2O_5$, this oxide also shows large jump in conductivity around 973 K, at which temperature, α-phase is transferred to β-phase, a more disordered structure. In this oxide, much higher electronic conduction was reported like Bi_2O_3 and it was easily reduced. Therefore, its application as SOFC electrolyte is highly difficult. However, by doping lower valence cation, in particular, W for Mo site or Ca, Sr for La site, stability against reduction becomes little better [63]. It was reported that the highest conductivity in this series is achieved at $La_2Mo_{1.7}W_{0.3}O_9$ and the conductivity is around $\log(\sigma\ (S\,cm^{-1})) = -0.8$ at 1048 K. Although chemical stability in reducing atmosphere is improved by dopant and also conductivity increases, higher electrical conductivity in reducing atmosphere is still not enough for an SOFC electrolyte.

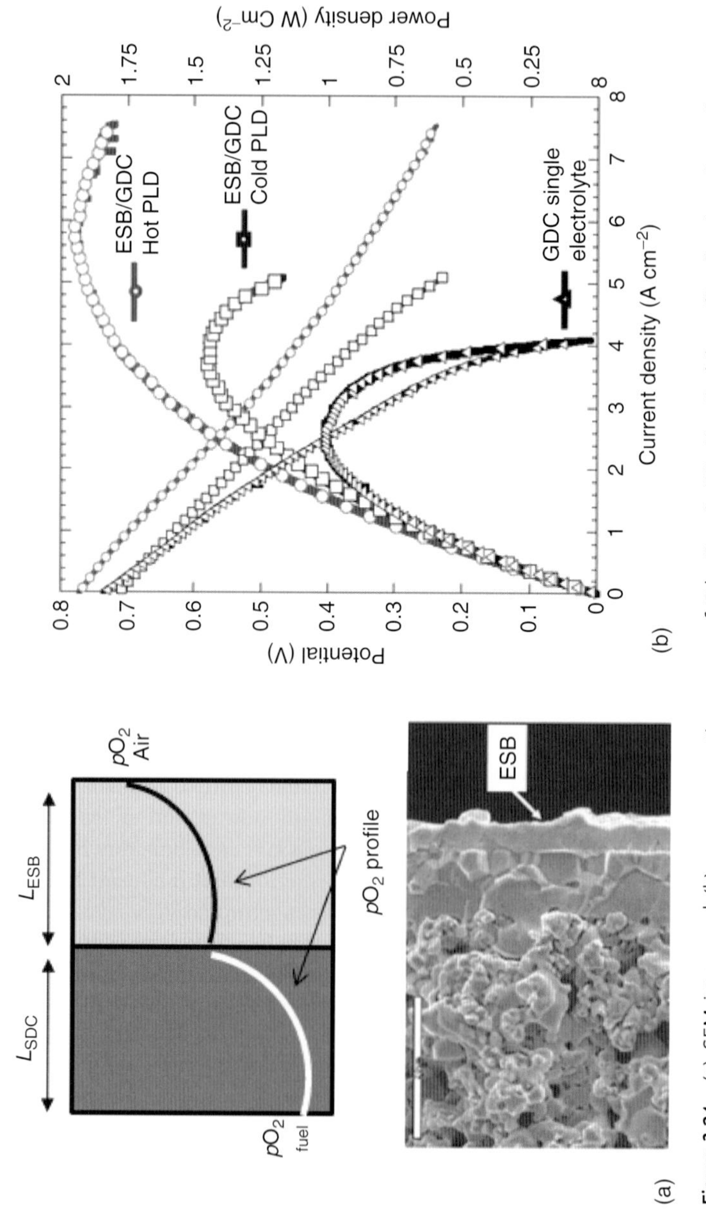

Figure 3.24 (a) SEM image and (b) power generation curves of $Gd_{0.2}Ce_{0.8}O_2/Bi_{1.6}Er_{0.4}O_3$ bilayer film for electrolytes of SOFCs [61].

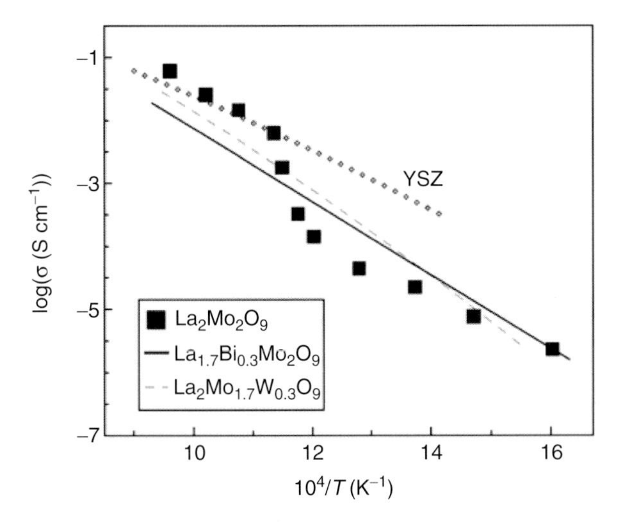

Figure 3.25 Oxide ion conductivity in $La_2Mo_2O_7$ and doped $La_2Mo_2O_7$ [62].

3.3 Electrolyte Efficiency

Main charge carrier in electrolytes for SOFCs is oxide ion; however, electron or hole, a minor charge carrier is also conducting a small part of charge even under operation conditions. Since the concentration of the minor carriers (electrons and/or holes) determines the chemical leakage of oxygen when an oxide ion conductor is used as the electrolyte in SOFCs [64], the analysis of the performance of electron and hole is an important subject for the electrolyte materials. Partial electronic conduction is commonly analyzed by the ion blocking method, the so-called Wagner polarization method. Partial electronic conductivity is the sum of electronic and hole contribution to the total conductivity and each conductivity is proportional to a carrier density. Therefore, the total electronic conductivity can be expressed as follows:

$$\sigma = \frac{ILF}{RT} = \sigma_{\mathrm{n}} + \sigma_{\mathrm{p}} = \sigma_{\mathrm{n}}^{0}\left\{1 - \exp\left(\frac{FE\,(L)}{RT}\right)\right\} + \sigma_{\mathrm{p}}^{0}\left\{\exp\left(\frac{FE\,(L)}{RT}\right) - 1\right\} \tag{3.7}$$

Here, I, L, $E(L)$, F, R, and T represent current, length of the sample, applied voltage, the Faraday constant, the gas constant, and temperature, respectively. When the hole conduction is dominant, the second term in the equation is dominant and the current increases exponentially with applied potential. Since the predominant charge carriers change from holes to electrons with decreasing pO_2 and p–n transition occurs at intermediate pO_2, the current, I, shows a typical "S"-shaped curve against potential, $E(L)$. Figure 3.26 shows the I–$E(L)$ curve observed in Ni-doped LSGM as an example by the ion blocking method [65]. Differential of the observed current with respect to potential, which corresponds to the pO_2 in the sample, gives the dependence of the partial electronic conductivity on pO_2.

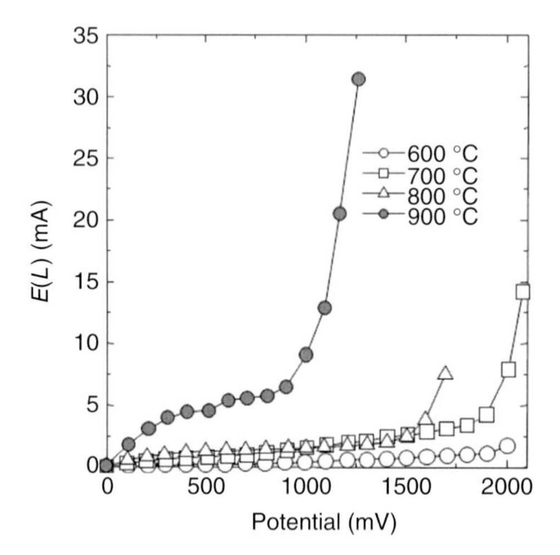

Figure 3.26 *I–E(L)* curve observed in Ni-doped LSGM as an example of the ion blocking method [65].

Determination of hole and electron conductivity and transport numbers of oxide ion in $LaGaO_3$-based oxides were performed by the polarization method by Baker *et al.* [66], Yamaji *et al.* [67], and Kim and Yoo. [68]. Kim and Yoo [68] reported that pO_2 dependence of hole and electron conductivity is proportional to $pO_2{}^{1/4}$ and $pO_2{}^{-1/4}$, respectively, and obeys the Hebb–Wagner theory.

$$\sigma\ (\mathrm{el}) = \sigma^0_{\mathrm{electron}} pO_2{}^{1/4} + \sigma^0_{\mathrm{hole}} pO_2{}^{1/4} \tag{3.8}$$

The results of the polarization method clearly indicate that $LaGaO_3$-based oxide exhibits almost pure oxide ion conductivity over a wide range of oxygen partial pressure ($10^5 > pO_2 > 10^{-25}$ atm). Compared with CeO_2-based oxides or Bi_2O_3, this is a major advantage of the $LaGaO_3$-based oxides and the redox stability is comparable to that of ZrO_2-based oxide. Consequently, $LaGaO_3$-based oxides are highly promising as an electrolyte for SOFCs, which is similar to YSZ. Kim and Yoo [68] also investigated the temperature dependence of hole and electronic conductivity in Mg-doped gallate, $La_{0.9}Sr_{0.1}Ga_{0.8}Mg_{0.2}O_3$, with the polarization method. Figure 3.27 shows the evaluated boundaries of the electrolytic domain of various oxide ion conductors in plots of log(pO_2 (atm)) against reciprocal temperature. The lower boundary of the electrolytic domain (defined as $t_{\mathrm{ion}} > 0.99$) for LSGM is 10^{-23} atm at 1273 K. This pressure is even lower than that of CaO-stabilized ZrO_2 and YSZ, which is also plotted in Figure 3.27. Consequently, it is clear that the electrolytic domain of LSGM or YSZ covers the pO_2 range required for the operation of SOFCs.

These electrical properties determine the energy conversion losses by an electrolyte. This can be expressed in terms of the loss (or lowering) from the theoretical conversion based on the Gibbs energy. In Figure 3.28a, those effects are given as a function of products of current density (J^{ex}) and electrolyte thickness (L).

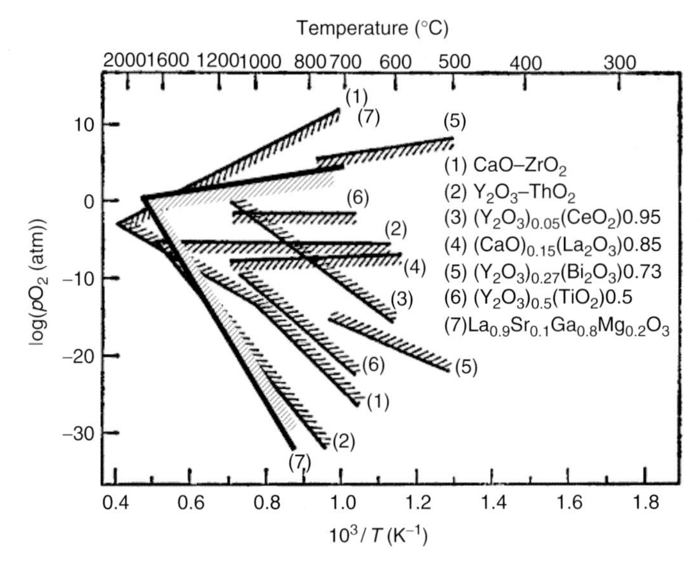

Figure 3.27 Evaluated boundaries of the electrolytic domain of various oxide ion conductors plotted in the axis of log(pO_2 (atm)) against reciprocal temperature.

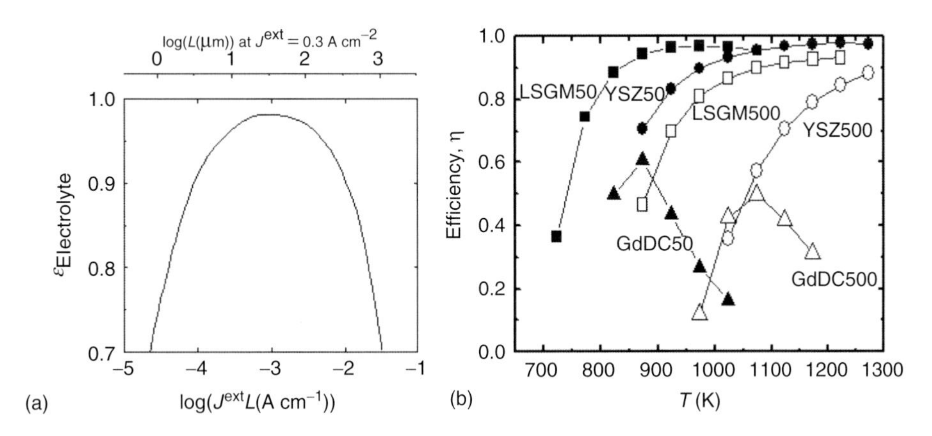

Figure 3.28 Gibbs-energy-based conversion loss occurring in electrolytes due to the Joule effects and shorting effects (a) as a function of products of current density and thickness at a given temperature (1273 K) and (b) as a function of temperature for a given current density (0.3 A cm^{-2}) with the parameter thickness [69].

In a large $J^{ex}L$ region, the lowering in efficiency, $1-\eta$(electrolyte), decreases with increasing $J^{ex}L$. This is the Joule effect. On the other hand, even in a small $J^{ex}L$ region, η(electrolyte) decreases with decreasing $J^{ex}L$. This is the shorting effect due to the electronic conduction. In this region, oxide ions are transported and undergo electrochemical-related reactions without generating electricity. This can be regarded as a normal chemical reaction of fuel with permeated oxygen gas (oxide ion and holes), or in other words, as an electrochemical short-circuiting condition. By combining the Joule effect and shorting effect, the decrease in the Gibbs energy conversion efficiency can be characterized as the behavior that exhibits the maximum point [64, 69].

A similar maximum behavior can be observed even for temperature dependence when the thickness of the electrolyte and the current density are fixed as shown in Figure 3.28b. In this figure, three electrolytes are compared with each other, in which the electronic contributions of YSZ are small so that the efficiency lowering is very small over a wide temperature range. For LSGM, the high efficient region extends to lower temperatures than those for YSZ because the oxide ionic conductivity of LSGM is higher, hence the Joule effect is smaller. For GDC, the efficiency at high temperatures is quite low due to the large contribution of the electron, which is also expecting the limited electrolyte domain.

3.4
Strain Effects on Oxide Ion Conductivity

Although ionic conductivity is not high enough, if a thin film could be prepared, then a reasonably small resistance of the electrolyte can be achieved when the materials are used as the electrolyte in SOFCs. Therefore, thin film ion conductor is widely adopted as the electrolyte in SOFCs. For example, in the case of YSZ, various methods ranging from physical to chemical have been proposed, and conventional wet process such as the slurry coating or tape casting method are popularly used for their cost and for mass production. Generally, the thickness of the film used for electrolytes is around a few tens of micrometers because it is reliable. However, in order to achieve a high power density at intermediate temperatures, an extremely thin film of around a few hundreds of nanometers in thickness is used, in particular, for micro-SOFCs.

Recently, there has been much interest on nanosize effects in ionic conductivity [70]. "Nanosize effects" mean the enhancement in conductivity when the size of grain or film thickness becomes nanometer size. Sata *et al.* [71] reported for the first time that an increase in the interface density in multilayered calcium fluoride and barium fluoride (CaF_2–BaF_2) strongly enhanced the ionic conductivity along the interfacial direction, particularly when the film thickness ranged between 20 and 100 nm, as compared with either bulk calcium fluoride or bulk barium fluoride. Maier *et al.* attributes this conductivity enhancement to the presence of space charge regions at the interfaces. The mechanisms of such increase in ionic conductivity are still under discussion and not clear yet. On the other hand, in case of an ultrathin film of oxide ion conductors, misfit in the crystal lattice of the film and the substrate seems to cause a residual stress, as schematically shown in Figure 3.29 [72], resulting in a large influence on conductivity.

Kosacki *et al.* [73, 74] deposited epitaxial YSZ thin films by pulsed laser deposition on (001)-oriented single crystals of MgO. They reported high ionic conductivity for YSZ films thinner than 60 nm. In particular, when the YSZ thickness decreased

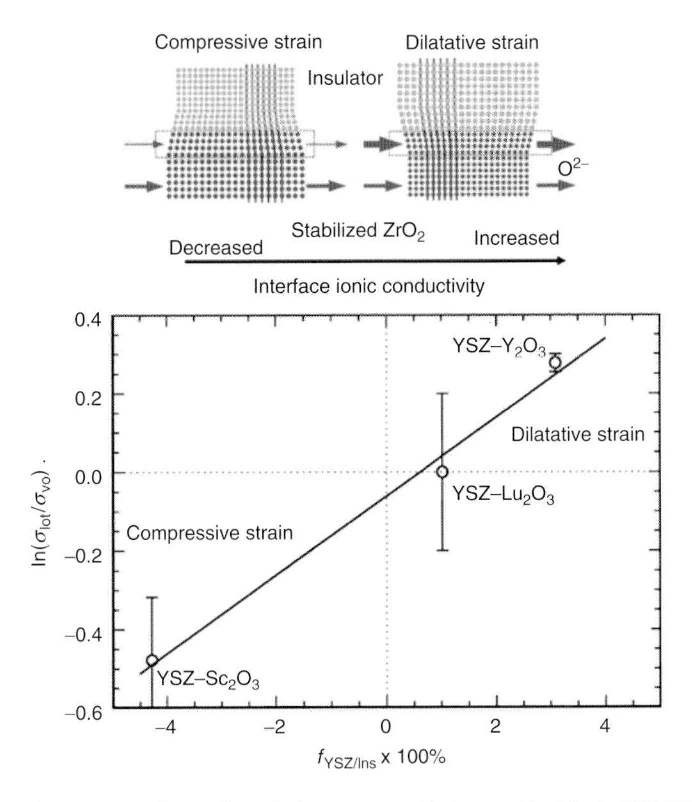

Figure 3.29 Effects of residual stress on oxide ion conductivity in YSZ [72].

from 60 to 15 nm, the conductivity at 673 K increased about 150 times. The increase was attributed to a highly conductive layer at the YSZ–MgO interface, about 1.6 nm thick, given the enhancement in conductivity occurred only in films thinner than 60 nm. The estimated interfacial conductivity is higher than the YSZ bulk conductivity by more than three orders of magnitude. The activation energy decreased from 1.09 to 0.62 eV when the YSZ thickness was reduced from 60 to 15 nm. The lattice mismatch at the YSZ–MgO interface is very large (18%), and thus, it is expected to be compensated not only by the elastic strain but also by the formation of misfit dislocations. Similar enhancement in oxide ion conductivity in ZrO_2- or CeO_2-based oxide film was reported by several groups [75–80]. The authors also studied the effects of film thickness of La_2GeO_5-based oxide film and the oxide ion conductivity, as shown in Figure 3.30, as a function of film thickness [81]. Like in ZrO_2- or CeO_2-based oxide, oxide ion conductivity increases with decreasing film thickness. However, when compareding with ZrO_2 or CeO_2, the length of the highly conducting phase is much larger in the La_2GeO_5 phase, which might be related to the higher mobility of oxide ion in this oxide than that in La_2GeO_5, as discussed earlier.

Improvement in oxide ion conductivity by applying tensile strain is also simulated based on quantum calculation [82]. As shown in Figure 3.29, oxide ion conductivity

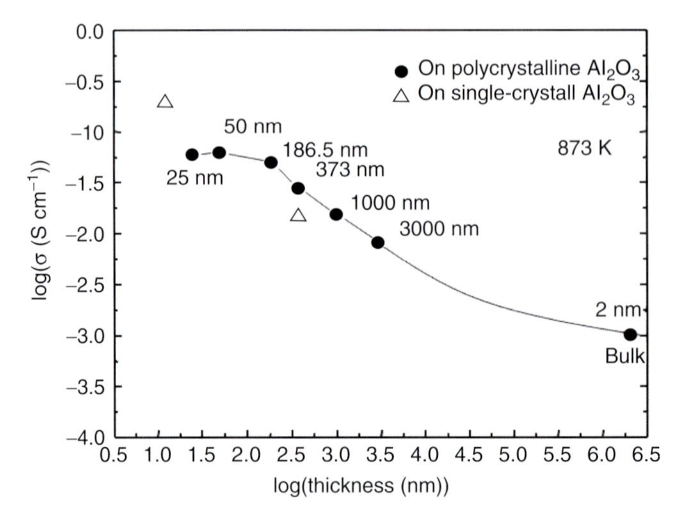

Figure 3.30 Electrical conductivity in La_2GeO_5 film as a function of thickness [81].

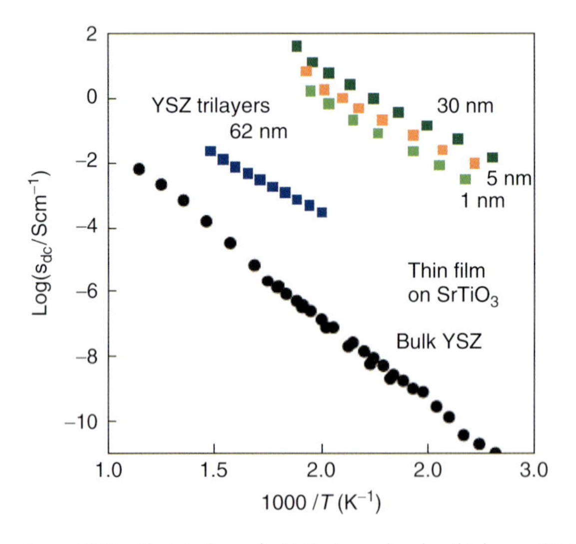

Figure 3.31 Electrical conductivity in molecular thickness YSZ film on $SrTiO_3$ [84].

is improved by 2 or 2.5 orders of magnitude with increasing tensile strain. This increase is caused by decreased activation energy for oxygen transport because of expanding space for hopping of oxygen. Large increase in conductivity (eight orders of magnitude) is reported for YSZ molecular film sandwiched between $SrTiO_3$ [83]; however, the improvement in conductivity cannot be repeated by other groups, as shown in Figure 3.31, and furthermore, the improved conductivity is not due to ionic conduction but electronic conduction. Therefore, Kilner [84] concluded that the improved conductivity by nanosize thickness film is mainly due to the improved electronic conductivity. The main reason for these differences in nanosize effects on ionic conductivity can be due to the lack of ionic transport number measurement because the film is usually deposited on the dense substrate. Therefore, diffusion of oxygen tracer or at least pO_2 dependence measurement is required. It thus seems likely that the unusual improvement in oxide ion conductivity is achieved on an ion-conducting electrolyte film, and this could be highly effective for increasing the power density of SOFCs at intermediate temperatures.

In contrast to a film with nanoscale effects, it is also reported that nanocrystalline CeO_2 shows drastically decreased oxide ion conductivity because of a negative charge at the grain boundary [85]. Therefore, the improved ionic conductivity seems to be due to the interface property of the oxide ion conductor and the substrate. The improved or depressed conductivity reported for YSZ and doped CeO_2 shows the importance of understanding the relationship between the electrolyte's microstructure, including its degree of crystallinity, grain size, strain within grains, dopant segregation, and space charge, and its electrical properties. As a result, the properties of thin films are rather different from those of bulk material. No common explanation is so far to be found in the literature and this might be related to the microstructure of the sample. Since the internal resistance of oxide electrolytes is generally the main reason for the potential drop in SOFCs, an unusual increase in oxide ion conductivity by nanosize effects is highly useful for decreasing the operating temperature of SOFCs, especially important for micro-SOFCs.

In summary, in addition to the intrinsic material property, microstructure of the conductor, including the grain size, grain boundary, and local strain, is also an important parameter for oxide ion conductors.

3.5 Degradation in Conductivity

Recently, the durability of SOFCs has gained much attention, and so change in oxide ion conductivity in an electrolyte also strongly influences the durability of power density of the cell. Change in oxide ion conductivity in electrolytes is briefly introduced in this section. Two reasons are considered for decrease in oxide ion conductivity: change in crystal structure, the so-called annealing effects, and diffusion of doped cation into or from electrolytes. Reactivity between electrolyte and electrode is generally considered as the main reason for decrease in electrolytes. For example, stabilized ZrO_2 reacts with La in the cathode and the resulting product $La_2Zr_2O_7$ is highly resistive, resulting in decreased power density [86, 87]. Although formation of $La_2Zr_2O_7$ is observed at a temperature higher than 1373 K, formation of a small amount of $La_2Zr_2O_7$ is observed between ZrO_2 electrolyte and $LaMnO_3$-based perovskite cathode during long-term operation. Therefore, preventing reaction between cell components is important. On the other hand, in case of stabilized ZrO_2, decrease in conductivity is observed by changing the crystal phase. This phenomenon is called *annealing effects*. This change in conductivity mainly involves gradual enhancement on the blocking effects as a result of the segregation of oversaturated impurities at the boundaries. For partially saturated

ZrO_2, the conductivity aging effect is considered to be due to the precipitation of the tetragonal phase from the cubic matrix. The time dependence of the resistivity of polycrystalline YSZ has been measured and is given by the following general equation [88]:

$$\gamma\,(t) = A - B_1 \exp\left(-K_1 t\right) - B_2 \exp\left(-K_2 t\right) \tag{3.9}$$

where γ is the resistivity, t is time, and A, B_1, K_1, B_2, and K_2 are positive constants, with subscripts 1 and 2 designating relations to the bulk and grain-boundary resistance, respectively. Annealing effects are more significantly observed for Sc-stabilized ZrO_2 (SSZ). ZrO_2 with 8 mol% Sc_2O_3 exhibited a significant decrease in conductivity with time at 1273 K [89]. The conductivity of 0.3 $S\,cm^{-1}$ at 1273 K (just after sintering) decreases to 0.12 $S\,cm^{-1}$ after 1000 h at 1273 K. This conductivity value is comparable to that of ZrO_2 with 9 mol% Y_2O_3 after aging for 1000 h. On the other hand, ZrO_2 with 11 mol% Sc_2O_3 showed no aging effect by annealing at 1273 K for more than 6000 h. However, ZrO_2 with 11 mol% Sc_2O_3 shows a phase transition from the rhombohedral structure (low-temperature phase) to the cubic structure (high-temperature phase) at 873 K, with an accompanying small volume change due to the phase transition. The cubic phase is stabilized at room temperature by the addition of a small amount of CeO_2 [90] and Al_2O_3 [91]. The conductivity of SSZ with CeO_2 and Al_2O_3 is slightly lower than that of the undoped SSZ. Similar aging effects have been observed in other zirconia-based oxide ion conductors. In Table 3.2, the conductivity changes in the ZrO_2–M_2O_3 system by annealing are summarized [92]. Comparing the counterdiffusion of cations between electrolyte and electrode, such conductivity aging effects are not the significant reason for decreases in the power density of the cell.

Table 3.2 Electrical conductivity, bending strength, and thermal expansion coefficient of zirconia-based electrolytes [92].

Electrolyte	Conductivity at 1273 K ($S\,cm^{-1}$)		Bending strength (MPa)	Thermal expansion coefficient, 1 K^{-1} x 10^{-6}
	As-sintered	Annealing		
ZrO_2–3 mol% Y_2O_3	0.059	0.050	1200	10.8
ZrO_2–3 mol% Yb_2O_3	0.063	0.09	—	—
ZrO_2–2.9 mol% Sc_2O_3	0.090	0.063	—	—
ZrO_2–8 mol% Y_2O_3	0.13	0.09	230	10.5
ZrO_2–9 mol% Y_2O_3	0.13	0.12	—	—
ZrO_2–8 mol% Yb_2O_3	0.20	0.15	—	—
ZrO_2–10 mol% Yb_2O_3	0.15	0.15	—	—
ZrO_2–8 mol% Sc_2O_3	0.30	0.12	270	10.7
ZrO_2–11 mol% Sc_2O_3	0.30	0.30	255	10.0
ZrO_2–11 mol% Sc_2O_3 -1 wt% Al_2O_3	0.26	0.26	250	—

3.6 Concluding Remarks

In this chapter, the current status of the promising oxide ion conductors as the electrolyte in SOFCs and their related properties are introduced. Most oxide ion conductors have a cubic structure; however, recently, nonfluorite oxide ion conductors, such as cubic perovskite or defect perovskite, and apatite-related oxides are reported as a new oxide ion conductor. Some of the noncubic oxide conductors, such as $La_{10}Si_6O_{27}$ or La_2GeO_5, are highly interesting. Therefore, in the near future, there is a high possibility that new noncubic oxides exhibiting extremely fast ion conductivity will be found. At present, the electrolyte for high-temperature SOFCs, which operate at around 1273 K, seems to be Y_2O_3-stabilized ZrO_2; however, the electrolyte for the intermediate temperature has not been determined yet. The most promising candidate of electrolyte is considered to be $LaGaO_3$-based oxide. On the other hand, if the operating temperature can decrease to the temperature range 673–873 K, development of SOFCs for practical use could be accelerated significantly because of improved reliability, durability, and easiness of operation of the system as power generator. When compared with the proton-conducting electrolytes, one of the great advantages of SOFCs using oxide ion-conducting electrolytes is the direct usage of hydrocarbon for fuel. Therefore, a simple reforming system is considered for SOFC systems using oxide ion conductors as the electrolyte. Another interesting phenomenon the so-called nanoionic effects has also been reported. Although significant improvement in oxide ion conductivity has not yet been possible, this may suggest that the control of microstructure of the ionic conductors is highly important for high oxide ion conductivity. The day when SOFCs are operable at low-temperature ranges is not far in the future. Oxide ion conductor film with few nm thickness has an extremely high potential as the electrolyte for SOFC for a small size combined heat and power generating system but the large size power generators which set distributively.

References

1. Iwahara, H. (1998) *Solid State Ionics*, **28–30**, 573.
2. Norby, T. (1999) *Solid State Ionics*,break **125**, 1.
3. Heo, P., Nagao, M., Kamiya, T., Sano, M., Tomita, A., and Hibino, T. (2007) *J. Electrochem. Soc.*, **154**, B63.
4. Jin, Y.C., Shen, Y.B., and Hibino, T. (2010) *J. Mater. Chem.*, **20**, 6214.
5. Nernst, W. (1899) *Z. Elektrochem.*, **6**, 41.
6. Ishihara, T., Sammes, N.M., and Yamamoto, O. (2003) in *High Temperature Solid Oxide Fuel Cells: Fundamentals, Design and Applications* (eds S.C. Singhal and K. Kendall), Elsevier, Oxford, pp. 83–117.
7. Malavasi, L., Fisherb, C.A.J., and Islam, M.S. (2010) *Chem. Soc. Rev.*, **39**, 4370.
8. Khartona, V.V., Marquesa, F.M.B., and Atkinson, A. (2004) *Solid State Ionics*, **174**, 135.
9. Scott, H.G. (1975) *J. Mater. Sci.*, **10**, 1527.
10. Stubican, V.S. (1988) in *Science and Technology of Zirconia III* (eds S. Somiya, N. Yamamoto, and H. Yanagida), American Ceramic Society, Columbus, OH, p. 71.

11. Baumard, J.F. and Abelard, P. (1984) in *Science and Technology of Zirconia II* (eds N. Claussen, M. Ruhle, and A.H. Heuer), American Ceramic Society, Columbus, OH, p. 555.
12. Arachi, Y., Sakai, H., Yamamoto, O., Takeda, Y., and Imanishi, N. (1999) *Solid State Ionics*, **121**, 133.
13. Subbarao, E.C. (1981) in *Science and Technology of Zirconia* (eds A.H. Heuer and L.W. Hobbs), American Ceramic Society, Columbus, OH, p. 1.
14. Park, J.H. and Blumenthal, R.N. (1989) *J. Electrochem. Soc.*, **136**, 2867.
15. Ovenston, A. (1992) *Solid State Ionics*, **58**, 221.
16. Minh, N.Q. and Takahashi, T. (eds) (1995) *Science and Technology of Ceramic Fuel Cells*, Elsevier, Amsterdam, p. 90.
17. Heussner, K.H. and Claussen, N. (1989) *J. Eur. Ceram. Soc.*, **5**, 193.
18. Ishizaki, F., Yoshida, T., and Sakurada, S. (1989) Proceedings of the International Symposium on Solid Oxide Fuel Cells, Nagoya, The Electrochemical Society, p. 172.
19. Evans, A., Bieberle-Hütter, A., Bonderer, L.J., Stuckenholz, S., and Gauckler, L.J. (2011) *J. Power Sources*, **196**, 10069.
20. Weppner, W. (1992) *Solid State Ionics*, **52**, 15.
21. Minh, N.Q. and Takahashi, T. (eds) (1995) *Science and Technology of Ceramic Fuel Cells*, Elsevier, Amsterdam, p. 89.
22. Mogensen, M., Sammes, N.M., and Tompsett, G.A. (2000) *Solid State Ionics*, **129**, 63.
23. Steele, B.C.H. (2000) *Solid State Ionics*, **129**, 95.
24. Navrotsky, A., Simoncic, P., Yokokawa, H., Chen, W., and Lee, T. (2007) *Faraday Discuss.*, **134**, 171.
25. Yahiro, H., Eguchi, Y., Eguchi, K., and Arai, H. (1988) *J. Appl. Electrochem.*, **18**, 527.
26. Yahiro, H., Eguchi, K., and Arai, H. (1989) *Solid State Ionics*, **36**, 71.
27. Butler, V., Catlow, C.R.A., Fender, B.E.F., and Harding, J.H. (1983) *Solid State Ionics*, **8**, 109.
28. Tuller, H.L. and Nowick, A.S. (1975) *J. Electrochem. Soc.*, **122**, 255.
29. Godickemeier, M. and Gaucker, L.J. (1998) *J. Electrochem. Soc.*, **145**, 414.
30. (a) Tsai, T., Perry, E., and Barnett, S. (1997) *J. Electrochem. Soc.*, **144**, L130. (b) Kwon, T.H., Lee, T.W., and Yoo, H.I. (2011) *Solid State Ionics*, **195**, 25.
31. Omar, S., Najib, W.B., and Bonanos, N. (2011) *Solid State Ionics*, **189**, 100.
32. Kilner, J.A. (2000) *Solid State Ionics*, **129**, 13.
33. Takahashi, T. and Iwahara, H. (1971) *Energy Convers.*, **11**, 105–111.
34. Ishihara, T., Matsuda, H., and Takita, Y. (1994) *J. Electrochem. Soc.*, **141**, 3444.
35. Nomura, K., Takeuchi, T., Kamo, S., Kageyama, H., and Miyazaki, Y. (2004) *Solid State Ionics*, **175**, 553.
36. Lybye, D., Poulsen, F.W., and Mogensen, M. (2000) *Solid State Ionics*, **128**, 91.
37. Nomura, K. and Tanase, S. (1997) *Solid State Ionics*, **98**, 229.
38. Mogensen, M., Lybye, D., Bonanos, N., Hendriksen, P.V., and Poulsen, F.W. (2004) *Solid State Ionics*, **174**, 279.
39. Ishihara, T., Matsuda, H., and Takita, Y. (1994) *J. Am. Chem. Soc.*, **116**, 3801.
40. Majewski, P., Rozumek, M., and Aldinger, F. (2001) *J. Alloys Compd.*, **329**, 253–258.
41. Ishihara, T., Matsuda, H., and Takita, Y. (1995) *Solid State Ionics*, **79**, 147.
42. Feng, M. and Goodenough, J.B. (1994) *Eur. J. Solid State Inorg. Chem.*, **31**, 663.
43. Huang, P.N. and Petric, P. (1996) *J. Electrochem. Soc.*, **143**, 1644.
44. Huang, K., Tichy, R., and Goodenough, J.B. (1998) *J. Am. Ceram. Soc.*, **81**, 2565.
45. Huang, K. and Goodenough, J.B. (2000) *J. Alloy. Compd*, **303–304**, 454.
46. Ishihara, T., Kilner, J.A., Honda, M., and Takita, Y. (1997) *J. Am. Chem. Soc.*, **119**, 2747.
47. Ishihara, T., Shibayama, T., Honda, M., Nishiguchi, H., and Takita, Y. (1999) *Chem. Commun.*, **1227**.
48. Goodenough, J.B., Ruiz-Diaz, J.E., and Zhen, Y.S. (1990) *Solid State Ionics*, **44**, 21–31.
49. Yao, T., Uchimoto, Y., Kinuhata, M., Inagaki, T., and Yoshida, H. (2000) *Solid State Ionics*, **132**, 189.
50. Kharton, V.V., Marques, F.M.B., and Atkinson, A. (2004) *Solid State Ionics*, **174**, 135.

51. Kakinuma, K., Yamamura, H., and Atake, T. (2005) *Defect Diffus. Forum*, **159**, 242–244.
52. Sirikanda, N., Matsumoto, H., and Ishihara, T. (2010) *Solid State Ionics*, **181**, 315.
53. Ishihara, T., Sirikanda, N., Nakashima, K., Miyoshi, S., and Matsumoto, H. (2010) *J. Electrochem. Soc.*, **157**, B141.
54. Yashima, M., Enoki, M., Wakita, T., Ali, R., Matsushita, T., Izumi, F., and Ishihara, T. (2008) *J. Am. Chem. Soc.*, **130**, 2762.
55. Nakayama, S. and Sakamoto, M. (1998) *J. Eur. Ceram. Soc.*, **18**, 1413.
56. Nakayama, S. (1999) *Mater. Integr*, **12** (4), 57.
57. Sansom, J.E., Richings, D., and Slater, P.R. (2001) *Solid State Ionics*, **139**, 205.
58. Tolchard, J.R., Islam, M.S., and Slater, P.R. (2003) *J. Mater. Chem.*, **13**, 1956.
59. Ishihara, T., Arikawa, H., Akbay, T., Nishiguchi, H., and Takita, Y. (2001) *J. Am. Chem. Soc.*, **123**, 203.
60. Abraham, F., Boivin, J.C., Mairesse, G., and Nowogrocki, G. (1990) *Solid State Ionics*, **40–41**, 934.
61. Ahn, J.S., Camaratta, M.A., Pergolesi, D., Lee, K.T., Yoon, H., Lee, B.W., Jung, D.W., Traversa, E., and Wachsmana, E.D. (2010) *J. Electrochem. Soc.*, **157**, B376.
62. Lacorre, P., Goutenoire, F., Bohnke, O., Retoux, R., and Laligant, Y. (2000) *Nature*, **404**, 856.
63. (a) Tealdi, C., Malavasi, L., Ritter, C., Flor, G., and Costa, G. (2008) *J. Solid State Chem.*, **181**, 603. (b) Tealdi, C., Chiodelli, G., Flor, G., and Leonardi, S. (2010) *Solid State Ionics*, **181**, 1456.
64. Yokokawa, H., Sakai, N., Horita, T., and Yamaji, K. (2001) *Fuel Cells*, **1**, 117.
65. Ishihara, T., Ishikawa, S., Hosoi, K., Nishiguchi, H., and Takita, Y. (2004) *Solid State Ionics*, **175**, 319.
66. Baker, R.T., Gharbage, B., and Marques, F.M.B. (1997) *J. Electrochem. Soc.*, **144**, 3130.
67. Yamaji, K., Horita, T., Ishikawa, M., Sakai, N., Yokokawa, H., and Dokiya, M. (1997) Solid oxide fuel cells V. *Proc. Electrochem. Soc.*, **97** (18), 1041.
68. Kim, J.H. and Yoo, H.I. (2001) *Solid State Ionics*, **140**, 105.
69. Yokokawa, H., Sakai, N., Horita, T., Yamaji, K., and Brito, M.E. (2005) Solid oxide electrolytes for high temperature fuel cells. *Electrochemistry*, **73**, 20–30.
70. Maier, J. (2004) *Physical Chemistry of Ionic Materials. Ions and Electrons in Solids*, John Wiley & Sons, Ltd, Chichester.
71. Sata, N., Eberman, K., Eberl, K., and Maier, J. (2000) *Nature*, **408**, 946.
72. Fabbri, E., Pergolesi, D., and Traversa, E. (2010) *Sci. Technol. Adv. Mater.*, **11**, 054503.
73. Kosacki, I., Rouleau, C.M., Becher, P.E., Bentley, J., and Lowndes, D.H. (2004) *Electrochem. Solid-State Lett.*, **7**, A459.
74. Kosacki, I., Rouleau, C.M., Becher, P.F., Bentley, J., and Lowndes, D.H. (2005) *Solid State Ionics.*, **176**, 1319.
75. Sillassen, M., Eklund, P., Pryds, N., Johnson, E., Helmersson, U., and Bøttiger, J. (2010) *Adv. Funct. Mater.*, **20**, 2071.
76. Guo, X., Vasco, E., Mi, S., Szot, K., Wachsman, E., and Waser, R. (2005) *Acta Mater.*, **53**, 5161.
77. Karthikeyan, A., Chang, C.H., and Ramanathan, S. (2006) *Appl. Phys. Lett.*, **89**, 183116.
78. Korte, C., Peters, A., Janek, J., Hesse, D., and Zakharov, N. (2008) *Phys. Chem. Chem. Phys.*, **10**, 4623.
79. Azad, S., Marina, O.A., and Wang, C.M. (2005) *Appl. Phys. Lett.*, **86**, 131906.
80. Rupp, J.L.M., Infortuna, A., and Gauckler, L.J. (2007) *J. Am. Ceram. Soc.*, **90**, 1792.
81. Yan, J.W., Matsumoto, H., and Ishihara, T. (2005) *Electrochem. Solid-State Lett.*, **8**, A607.
82. Kushima, A. and Yildiz, B. (2010) *J. Mater. Chem.*, **20**, 4809.
83. Garcia-Barriocanal, J., Rivera-Calzada, A., Varela, M., Sefrioui, Z., Iborra, E., Leon, C., Pennycook, S.J., and Santamaria, J. (2008) *Science*, **321**, 676.
84. Kilner, J.A. (2008) *Nat. Mater.*, **7**, 838.
85. Kim, S.T. and Maier, J. (2002) *J. Electrochem. Soc.*, **149**, J73.
86. van Roosmalen, J.A.M. and Cordfunke, E.H.P. (1992) *Solid State Ionics*, **52**, 303.

87. Clausen, C., Bagger, C., Bilde-Sorensen, J.B., and Horsewell, A. (1994) *Solid State Ionics*, **70/71**, 59.
88. Moghadam, F.K. and Stevenson, D.A. (1982) *J. Am. Ceram. Soc.*, **65**, 213.
89. Yamamoto, O., Arachi, Y., Takeda, Y., Imanishi, N., Mizutani, Y., Kawai, M., and Nakamura, Y. (1995) *Solid State Ionics*, **79**, 137.
90. Arachi, Y., Ashai, T., Yamamoto, O., Takeda, Y., Imanishi, N., Kawada, K., and Tamakoshi, C. (2001) *J. Electrochem. Soc.*, **148**, A520.
91. Mizutani, Y., Kawai, M., Nomura, K., Nakamura, Y., and Yamamoto, O. (1997) in *Proceedings of the 5th International Symposium on Solid Oxide Fuel Cells* (eds U. Stimming, S.C. Singhal, H. Tagawa, and W. Lehnert), The Electrochemical Society, p. 37.
92. Ishihara, T., Sammes, N., and Yamamoto, O. (2003) in *High Temperature Solid Oxide Fuel Cell, Fundamentals, Design and Applications* (eds S.C. Singhal and K. Kendal), Elsevier, Oxford, p. 91.

4
Proton-Conducting Materials as Electrolytes for Solid Oxide Fuel Cells

Rong Lan and Shanwen Tao

4.1
Introduction

Solid oxide fuel cells (SOFCs) are electrochemical devices that can convert chemical energy into electricity at a very high efficiency [1–3]. The key materials for SOFCs are cathode, electrolyte, and anode. As for electrolyte materials, both O^{2-} and H^+ ion conducting materials can be used in SOFCs. Most SOFC technologies of today are based on O^{2-} ion conducting materials, typically yttria-stabilized zirconia (YSZ) [4, 5], gadolinia- or samaria-doped ceria (SDC) [6], and Sr- and Mg-doped $LaGaO_3$ [7] (see also Chapter 3). However, proton-conducting materials can also be used as electrolytes in SOFCs [8]. SOFCs based on oxygen-ion-conducting electrolytes are normally called *SOFC-O*, while those based on proton-conducting electrolytes are called *SOFC-H*. The use of proton-conducting materials can have some advantages such as water is generated at the cathode side, preventing fuel dilution at the anode and formation of NO_x or SO_x can be avoided when ammonia or H_2S is used as the fuel. Although there are excellent reviews on proton-conducting materials [9–15], this chapter provides an introduction on proton-conducting materials and their applications in SOFCs.

4.2
The Principle of Proton-Conducting Oxides

Proton conduction in solid oxides was discovered in $SrCe_{0.95}Yb_{0.05}O_{3-\delta}$, $SrCe_{0.95}Mg_{0.05}O_{3-\delta}$, $SrCe_{0.95}Sc_{0.05}O_{3-\delta}$, and $SrCe_{0.9}Sc_{0.1}O_{3-\delta}$ by Iwahara *et al.* in 1980 [16, 17]. The proton conduction of doped $SrCeO_3$ was confirmed using a steam concentration cell (Figure 4.1). The researchers believed that proton conduction in these oxides may be related to electron holes and proposed the possible proton conduction mechanism:

$$H_2O + 2h^+ \left(\text{oxides}\right) \rightarrow 2H^+ \left(\text{oxides}\right) + \frac{1}{2}O_2 \tag{4.1}$$

Materials for High-Temperature Fuel Cells, First Edition. Edited by San Ping Jiang and Yushan Yan.

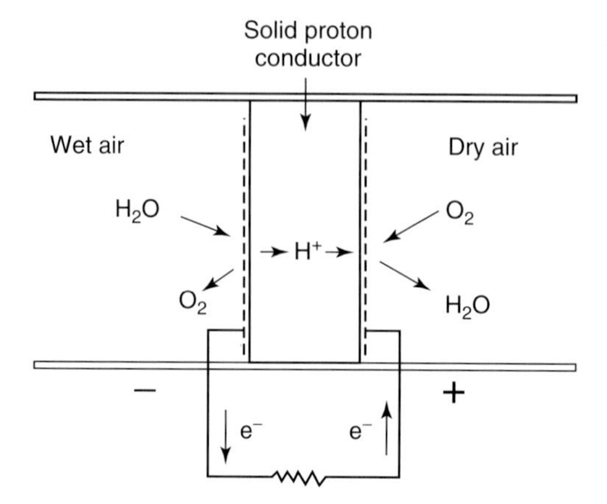

Figure 4.1 First confirmation of proton conduction through the concept of steam concentration cell. (Source: Reproduced with permission from Ref. [16].)

Later, Iwahara *et al.* [18] proposed that water may interact with oxygen vacancies to form protons,

$$H_2O(g) + V_O^{\bullet\bullet} \rightarrow 2H^+ + O_O^\times \tag{4.2}$$

However, it is believed that the formed proton may be associated with an oxygen ion. It has been recognized that water may interact with oxygen vacancies and lattice to form proton defects $(OH_O^\bullet)$ according to the following reaction [9]:

$$H_2O(g) + V_O^{\bullet\bullet} + O_O^\times \rightarrow 2OH_O^\bullet \tag{4.3}$$

The proton may jump to a neighboring oxygen ion through the rotational diffusion of a proton around an oxygen ion. In general, proton conduction in oxides is very much related to the presence of oxygen vacancies.

Figure 4.2 illustrates the relation of ionic conduction of a typical proton-conducting oxide. According to Eq. (4.3), most proton-conducting oxides uptake water at intermediate temperatures, typically between 200 and 500 °C [19]. The

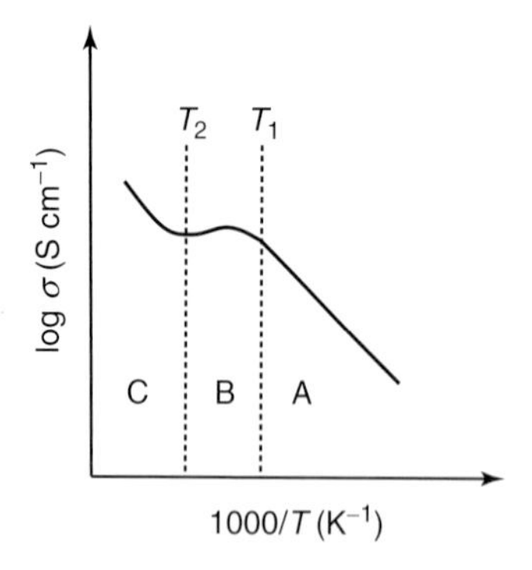

Figure 4.2 Typical conduction curve of a proton-conducting oxide in a humidified atmosphere.

oxygen ion vacancies of $Sr_3CaZr_{1-x}Ta_{1+y}O_{8.5-x/2}$ are filled by O–H groups by exposing the sample to a wet 5% H_2/Ar atmosphere at intermediate temperatures. At temperatures below T_1, proton defects are dominant in the oxides; therefore, the materials perform almost pure protonic conduction in zone A (Figure 4.2). At temperatures above T_1, water starts to dissociate from the lattice

$$2OH_O^{\bullet} \rightarrow H_2O\,(g) + V_O^{\bullet\bullet} + O_O^{\times} \tag{4.4}$$

This is a reversible process of reaction (4.3). Although the concentration of charge carrier (proton defects) decreases at temperatures above T_1, the mobility of proton defects increases at increased temperature; therefore, the total ionic conductivity still increases at a certain temperature range. At temperatures above T_2 (zone A), almost all the water molecules associated with oxygen vacancies dissociate, and therefore, the oxides dominate by oxygen ion conduction. In zone B, at a temperature between T_1 and T_2, the material exhibits mixed H^+/O^{2-} ionic conduction. In proton-conducting oxides, protons are the dominant mobile ions at lower temperatures (normally below 600–700 °C) in doped perovskite oxides such as $BeCeO_3$ and $BaZrO_3$. At higher temperatures, these oxides become mixed H^+/O^{2-} ion conductors rather than pure O^{2-} ion conductors [20].

It should be noted that proton conduction is also related to oxygen partial pressure. Most $BeCeO_3$- and $BaZrO_3$-based proton-conducting oxides exhibit mixed ionic–electronic conduction in an oxygen-rich atmosphere, while in a reducing atmosphere, the electronic conductivity is very low [21].

Since the discovery of proton conduction in doped $SrCeO_3$, many other proton-conducting materials have been investigated, which are described below.

4.3 Proton-Conducting Materials for Solid Oxide Fuel Cells

4.3.1 $BaCeO_3$- and $BaZrO_3$-Based Proton-Conducting Oxides

After the discovery of proton conduction in doped $SrCeO_3$ by Iwahara in 1980, there were some other early reports on proton-conducting materials based on doped $SrCeO_3$ [16, 22–27]. In 1985, Virkar and Maiti [28] reported the ionic conduction of pure and Y_2O_3-doped $BaCeO_3$ and they believed that the conductivity of these oxides becomes predominantly ionic at $pO_2 < 10^{-6}$ atm at a temperature of 600–1000 °C. In 1987, Mitsui *et al.* [29] first reported proton conduction in $BaCe_{0.95}Yb_{0.05}O_{3-\delta}$ and $BaZr_{0.95}Yb_{0.05}O_{3-\delta}$, and it was found that the conductivity of $BaCe_{0.95}Yb_{0.05}O_{3-\delta}$ in wet air reached 10^{-2} S cm^{-1} at 800 °C, which is about one order of magnitude higher than that of $SrCe_{0.95}Yb_{0.05}O_{3-\delta}$. The typical ionic conductivity of rare-earth-element-doped $SrCeO_3$ and $BaCeO_3$ is shown in Figure 4.3 [15].

As seen in Figure 4.3, the ionic conductivity of doped $BaCeO_3$ is much higher than that of doped $SrCeO_3$. One possible reason for this is that the lattice parameters of $BaCeO_3$ are much larger than those of $SrCeO_3$, which gives rise to more space

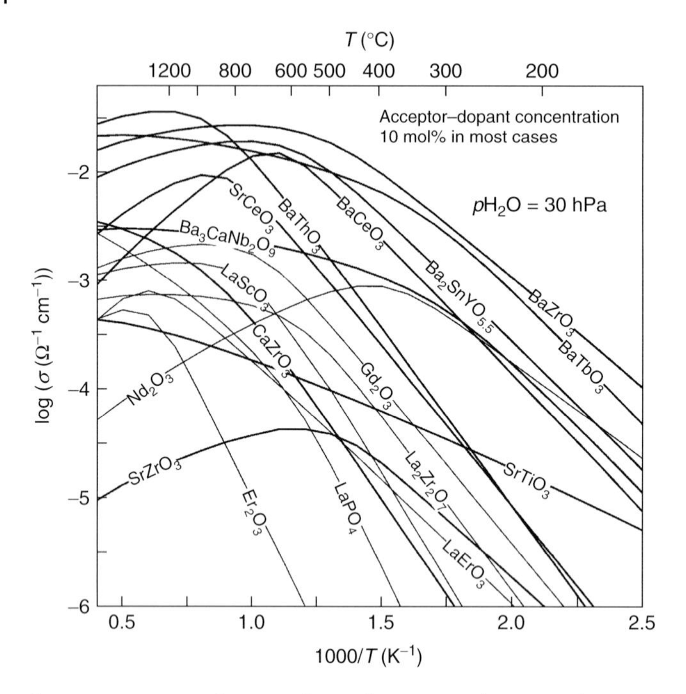

Figure 4.3 Ionic conductivity of typical proton-conducting oxides. (Source: Reproduced with permission from Ref. [15].)

or free volume for protons to migrate smoothly in the lattice. However, large lattice also means longer or weaker metal–oxygen bonds, which make the lattice less stable. It has been reported that $BaCeO_3$-based materials can react with H_2O and CO_2 to form $Ba(OH)_2$ and $BaCO_3$, respectively [30–33].

$$BaCeO_3 + H_2O \rightarrow Ba(OH)_2 + CeO_2 \tag{4.5}$$

$$BaCeO_3 + CO_2 \rightarrow BaCO_3 + CeO_2 \tag{4.6}$$

Not only cerates but also some perovskite zirconates such as doped $BaZrO_3$ and $SrZrO_3$ exhibit proton conduction [34–37]. The initially reported proton conductivity of Y_2O_3-doped $SrZrO_3$ and $BaZrO_3$ is below 10^{-2} S cm^{-1} at 1000 °C [34]. However, Kreuer [15] reported that the bulk conductivity of Y_2O_3-doped $BaZrO_3$ is comparable to that of doped $BaCeO_3$. The reported low total conductivity is due to the presence of a huge grain boundary resistance [34, 38]. Irvine *et al.* attempted to reduce the grain boundary resistance of $BaZr_{0.9}Y_{0.1}O_{2.95}$ starting from a core–shell modification of its grain with a thin film of $BaCe_{0.9}Y_{0.1}O_{2.95}$. Impedance spectroscopy performed on the material in wet 5% H_2/Ar atmosphere showed a significant reduction in both bulk and grain boundary resistances of the initial barium zirconate when modified with a thin film of barium cerate; however, the total conductivity of the core–shell proton-conducting oxides is still not high enough to be used as electrolytes in SOFCs [39]. Tao and Irvine [33, 40] found that introduction of some ZnO can significantly reduce the grain boundary resistance of Y_2O_3-doped $BaZrO_3$ at high temperatures, but again the total conductivity was lower than ZnO-free Y_2O_3-doped $BaZrO_3$. Recently, Traversa *et al.* reported the fabrication of grain-boundary-free $BaZr_{0.8}Y_{0.2}O_{3-\delta}$ (BZY) proton-conducting electrolyte thin films by pulsed laser deposition (PLD) on different single-crystal substrates. Highly textured, epitaxially oriented BZY films were obtained on (100)-oriented MgO substrates, showing the largest proton conductivity ever reported for BZY samples, that is, 0.11 S cm^{-1} at 500 °C, indicating that PLD could be an efficient way to fabricate BZY electrolyte for SOFCs [41]. It should be noted that the highly conductive BZY film was deposited on a single-crystal substrate, but this may be challenging on a real polycrystalline SOFC anode or cathode. $BaZrO_3$-based proton-conducting materials are stable in H_2O and CO_2 [42, 43]; however, normally, a temperature of 1700 °C is required to sinter Y_2O_3-doped $BaZrO_3$ [38, 44]. Haile and Iwahara reported that the substitution of Ce in $BaCeO_3$ by Zr can stabilize $BaCeO_3$ [42, 43], but normally, these oxides need to be sintered to dense at 1550 °C. Tao and Irvine found that the addition of ZnO as a sintering aid can effectively reduce the sintering temperature of BZY from 1700 to 1325 °C [33, 40, 45]. A similar effect was also reported independently by Babilo and Haile [46]. It was reported that the total ionic conductivity of $BaCe_{0.5}Zr_{0.3}Y_{0.16}Zn_{0.04}O_{3-\delta}$ (BCZYZn) exceeds 10 mS cm^{-1} at 600 °C, which is a potential stable and easily sintered proton-conducting oxide for SOFCs [33]. A complex oxide such as $BaCe_{0.7}Zr_{0.1}Y_{0.2}O_{3-\delta}$ exhibits high proton conductivity and is stable in air. It can also be another good electrolyte for SOFCs [47].

Recently, it was reported that the oxide $BaZr_{0.1}Ce_{0.7}Y_{0.1}Yb_{0.1}O_{3-\delta}$ (BZCYYb) exhibits a conductivity of 1.18×10^{-2} S cm^{-1} at 600 °C in wet O_2, although some p-type electronic conduction might be present at high pO_2 [40, 48]. Excellent fuel cell performance was also observed, which is described below. Importantly, sulfur-tolerant and anticoking anodes for SOFCs have been reported, with the integration of a mixed H^+/e' conducting material such as BZCYYb at the anode of SOFC-H [48].

4.3.2
Other Perovskite-Related Proton-Conducting Oxides

Not only cerates and zirconates but also some other oxides such as $Ba_2SnYO_{5.5}$ also exhibit proton conduction. $Ba_2YSnO_{5.5}$ shows good chemical durability with respect to carbonate formation; however, structural instability becomes apparent in a high grain boundary impedance, during the collapse of the structure on reduction, and in the macrocracking induced by lattice strain resulting from hydration. The bulk conductivity is about 1.5×10^{-3} S cm^{-1} at 400 °C, which is lower than that for Y-doped $BaCeO_3$ [49]. Protonic conduction of both stoichiometric and nonstoichiometric mixed perovskite ceramics with the general formula $A_2(B'_{1+x}B''_{1-x})O_{6-\delta}$ (where A = Sr^{2+} and Ba^{2+}; B′ = Ga^{3+}, Gd^{3+}, and Nd^{3+}; B″ = Nb^{5+} and Ta^{5+}; and $x = 0$–0.2) has been investigated [50], but the proton conductivity is not high enough. It was observed that $Sr_3Sr_{1.5}Nb_{1.5}O_{9-\delta}$ exhibits initially high total conductivity that decreases after annealing at 1100 °C for 6 weeks [51]. It was also reported that the double perovskite oxide $Ba_2Ca_{1.18}Nb_{1.92}O_6$ (BCN18) exhibits a proton conductivity of 1.5×10^{-3} S cm^{-1} at 400 °C in wet Ar/O_2 and is stable in CO_2 and H_2O [52]. The proton conductivity of $Sr_3CaZr_{0.5}Ta_{1.5}O_{8.75}$ was also investigated, with a bulk conductivity in wet H_2/Ar of 5.15×10^{-4} S cm^{-1} at 500 °C [53]. The reported proton conductivity of these oxides is not high enough for use as electrolyte in SOFCs. The ionic conductivities of typical proton-conducting oxides are shown in Figure 4.3.

4.3.3
Niobate- and Tantalate-Based Proton-Conducting Oxides

Norby and Haugsrud [54, 55] reported that rare-earth ortho-niobates and ortho-tantalates exhibit significant proton conduction. For $La_{0.99}Ca_{0.01}NbO_4$ (LCN), a maximum conductivity of 10^{-3} S cm^{-1} at 800 °C was observed [54, 56]. The conductivity is also related to the lanthanide element. The proton conductivity of $Gd_{0.95}Ca_{0.05}NbO_4$ is 2.0×10^{-4} S cm^{-1}, with a total conductivity 6.0×10^{-4} S cm^{-1} at 800 °C [57]. The proton conductivity of lanthanum tantalates is generally lower than that for niobates. For example, the conductivity of $La_{0.99}Ca_{0.01}TaO_4$ is only 5.0×10^{-4} S cm^{-1} at 800 °C [58], which is only one-fifth of that of the niobate analog. The performance of a fuel cell based on these materials is not good, mainly due to the relatively low proton conductivity of the electrolyte [13].

4.3.4
Proton Conduction in Typical O^{2-} Ion Conducting Materials

According to Eq. (4.3), the migration of protons in an oxide is connected to oxygen ions. From this point of view, the movement of oxygen ions may accompany proton migration.

Yokokawa *et al.* investigated the relationship between thermodynamic, electronic hole, and proton properties in doped ceria through thermodynamic analysis. It was found that the rather high concentrations of holes and protons in doped ceria, compared with YSZ, are closely related to the less stabilized dopants in ceria. It has been predicted from the bulk thermodynamic treatment on protons and holes that their concentration in doped ceria increases with decreasing temperature, approaching the saturation level [59]. This indicates that doped ceria may exhibit more significant proton conduction than YSZ. Nigara *et al.* [60, 61] found that hydrogen permeated the bulk of the fluorite type $(CeO_2)_{0.9}(CaO)_{0.1}$ at 1075–1800 K, indicating that $(CeO_2)_{0.9}(CaO)_{0.1}$ is an electron–proton mixed conductor under the H_2–N_2 atmosphere at high temperatures.

Power generation has been reported at temperatures as low as the room temperature using dense bulk nanostructured YSZ and SDC as electrolytes. This behavior is observed only when the material is nanostructured (grain size similar to 15 nm). Open-circuit electromotive force (EMF) (up to 180 mV for YSZ and 400 mV for SDC) and closed circuit currents (similar to 6 nA for YSZ and 30 nA for SDC) were measured from water concentration cells, indicating proton conduction within the nanostructured oxides. Proton conduction has been confirmed in both

YSZ and SDC at room temperature [62]. The proton conductivity of SDC and YSZ is about 10^{-8}–10^{-7} S cm^{-1} at room temperature [63]. It was also found that proton conduction in nanocrystalline solid electrolytes takes place exclusively along the grain boundaries [64, 65]. However, proton conduction was also observed in $Ce_{0.9}Gd_{0.1}O_{2-\delta}$ in a dense sample with grain sizes in the range 91–252 nm [66].

Ammonia has been successfully synthesized from H_2 and N_2 at atmospheric pressure based on an electrochemical cell using $Ce_{0.8}M_{0.2}O_{2-\delta}$ (M = La, Y, Gd, Sm) as electrolyte, indicating that these oxides exhibit proton conduction [67].

$La_{0.9}Sr_{0.1}Ga_{0.8}Mg_{0.2}O_{3-\delta}$ (LSGM) is widely accepted as a O^{2-} ion conducting oxide [7]; however, the proton conduction in LSGM ceramics has been discovered in a hydrogen-containing atmosphere. In a hydrogen atmosphere, LSGM was found to be an excellent proton conductor, with a proton conductivity of 1.4×10^{-2} to 1.4×10^{-1} S cm^{-1} and a proton transport number of $t(H^+) > 0.99$, which are comparable to those of the proton-conducting $BaCeO_3$-based oxides and $Ba_3Ca_{1.18}Nb_{1.82}O_{9-\delta}$. In a water-vapor-containing air atmosphere, the specimen was found to be a mixed-ion (H^+/O^{2-}) conductor, and the proton and oxide-ion transport numbers were about 0.05–0.20 and 0.95–0.80, respectively, whereas in dry oxygen-containing atmospheres, the specimen was a pure oxide-ion conductor [68]. Proton conduction was further confirmed through electrochemical synthesis of ammonia [69, 70].

$Ba_2In_2O_5$ is a typical oxygen ion conductor due to the presence of oxygen vacancies; however, it is also a proton conductor at a lower temperature. The conductivity in wet air is much higher than in dry air at a temperature below 600 °C [71]. The stability of $Ba_2In_2O_5$-based material in H_2 is not good above 480 °C due to the reduction of indium [72]. This may limit the application of $Ba_2In_2O_5$-based materials as the electrolyte for SOFCs.

Kendrick *et al.* reported that the $La_{1-x}Ba_{1+x}GaO_{4-x/2}$ oxide exhibits both proton and oxide-ion conduction. It was found that O^{2-} conduction proceeds via a cooperative "cog-wheel"-type process involving the breaking and re-formation of Ga_2O_7 units, whereas the rate-limiting step for proton conduction is intratetrahedron proton transfer. Both mechanisms are unusual for ceramic oxide materials, and similar cooperative processes may be important in related systems containing tetrahedral moieties [73]. In this type of materials, protons and oxygen ions are in different pathways.

4.3.5
Other Proton-Conducting Materials

It has been reported that the Ca-doped La_6WO_{12} exhibits mixed ionic and electronic conductivities; n- and p-type electronic conduction mechanisms predominate at high temperatures under reducing and oxidizing conditions, respectively. Protons are the major ionic charge carriers under wet conditions and predominate in conductivity below 750 °C. The maximum in proton conductivity is observed for $La_6W_6O_{12}$ with values reaching 3×10^{-3} S cm^{-1} at approximately 800 °C. The high proton conductivity for the undoped material is explained by assuming interaction between water vapor and the intrinsic anti-Frenkel oxygen vacancies [74, 75]. It was found that some phosphates, such as the Ca- and Sr-doped $LaPO_4$, also exhibit reasonable proton conduction but the conductivity is lower than 10^{-3} S cm^{-1} at 900 °C, which is not high enough for use as electrolyte for SOFCs [76].

4.4
Solid Oxide Fuel Cells Based on Proton-Conducting Electrolytes

To the best of our knowledge, Iwahara *et al.* first reported the application of the proton-conducting oxide $SrCe_{0.95}Yb_{0.05}O_{3-\delta}$ as electrolyte for SOFCs. Pt or Ni was

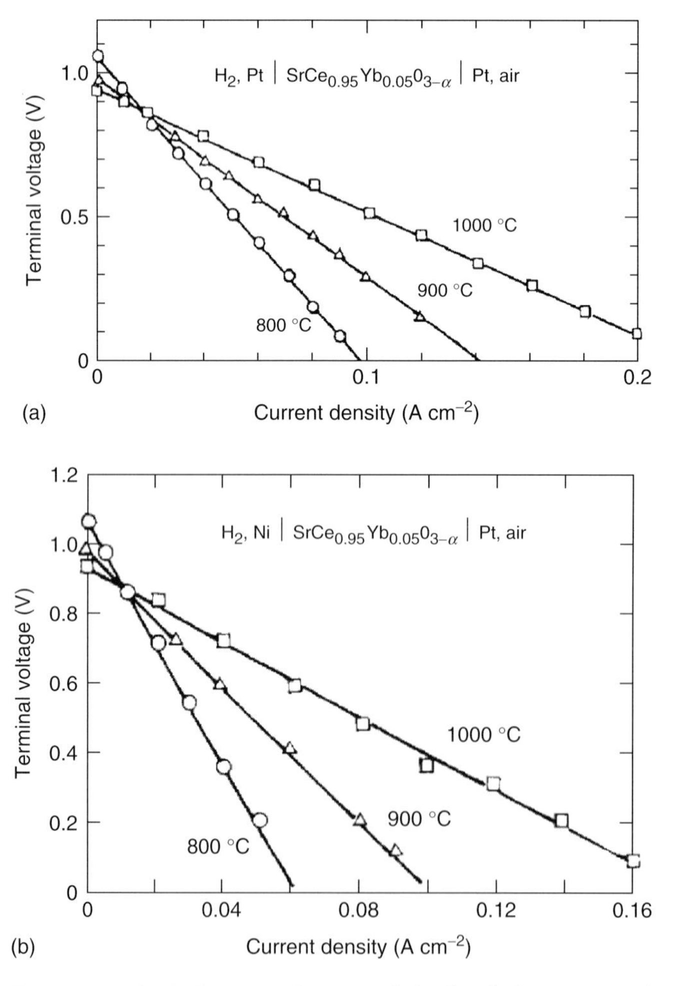

Figure 4.4 (a,b) Performance of a H_2/air fuel cell with the proton-conducting electrolyte $SrCe_{0.95}Yb_{0.05}O_{3-\delta}$ (electrolyte thickness, 0.5 mm). (Source: Reproduced with permission from Ref. [25].)

used as the anode and Pt or $La_{0.4}Sr_{0.6}CoO_{3-\delta}$ as the cathode. A power density of 50 mW cm^{-2} was obtained at 1000 °C when Pt was used as both electrodes (Figure 4.4) [25].

At the anode, hydrogen loses electrons to form protons:

$$H_2 - 2e' \rightarrow 2H^+ \tag{4.7}$$

The formed protons are transferred to the cathode to react with the oxygen in air and yield water. The reaction at cathode is

$$2H^+ + \frac{1}{2}O_2 + 2e' \rightarrow H_2O \tag{4.8}$$

The overall reaction for a hydrogen fuel cell based on a proton-conducting electrolyte is

$$H_2 + \frac{1}{2}O_2 \rightarrow 6H_2O \tag{4.9}$$

Recently, a good review paper was published on SOFCs based on proton-conducting oxides [77].

Conventional SOFCs contain electrolyte and two electrodes. It was also reported that anode-free SOFCs were fabricated based on the proton-conducting electrolyte $BaCe_{0.76}Y_{0.20}Pr_{0.04}O_{3-\delta}$. The anode was spontaneously formed by reduction of the electrolyte surface under reducing gas conditions because it showed high mixed protonic–electronic conduction. A power density of 140 mW cm^{-2} at 800 °C for a H_2/air cell was observed (Figure 4.5) [78]. Although no anode material was used in this cell, gold mesh was used as current collector and may partially play the role as anode.

Single-chamber SOFC is an interesting type of fuel cell. It was first developed based on an oxygen-ion-conducting electrolyte $Ce_{0.8}Sm_{0.2}O_{1.9}$ [79]. The proton-conducting oxide $BaCe_{0.8}Y_{0.2}O_{3-\delta}$ was also used to fabricate the first single-chamber SOFC. The fuel cell consisted of Pt/$BaCe_{0.8}Y_{0.2}O_{3-\delta}$/Au, in which two electrodes were exposed to the same mixture of CH_4 and air. The working mechanism of the fuel cell was clarified to be based on the difference in catalytic activity for the partial oxidation of methane between two electrode materials: Pt catalyzes the partial oxidation of methane to form hydrogen and carbon monoxide, while Au is inactive to this reaction. Therefore, Pt acts as a fuel electrode, while Au acts as an oxygen electrode at which electrochemical reduction of oxygen takes place on discharging the cell. A maximum power density of 160 mW cm^{-2} at 950 °C has been achieved when a mixture of CH_4/O_2 (=2 : 1) was fed into the cell (Figure 4.6) [80].

The stable easily sintered BCZYZn is an ideal electrolyte for SOFCs due to the lower sintering temperature. BCZYZn can be sintered to dense at 1250 °C when it is prepared by a soft-chemical method [81]. An open-circuit voltage (OCV) of 1.009 V and a maximal power density of 350 mW cm^{-2} were achieved at 700 °C (Figure 4.7). $SrCo_{0.9}Sb_{0.1}O_{3-\delta}$ (SCS) cubic perovskite cathode was also applied to an SOFC based on dense BCZYZn electrolyte. An open-circuit potential of 0.987 V, a maximum power density of 364 mW cm^{-2}, and a low polarization resistance of the electrodes of 0.07 Ω cm^2 was achieved at 700 °C [82]. When the $LaSr_3Co_{15}Fe_{15}O_{10}$

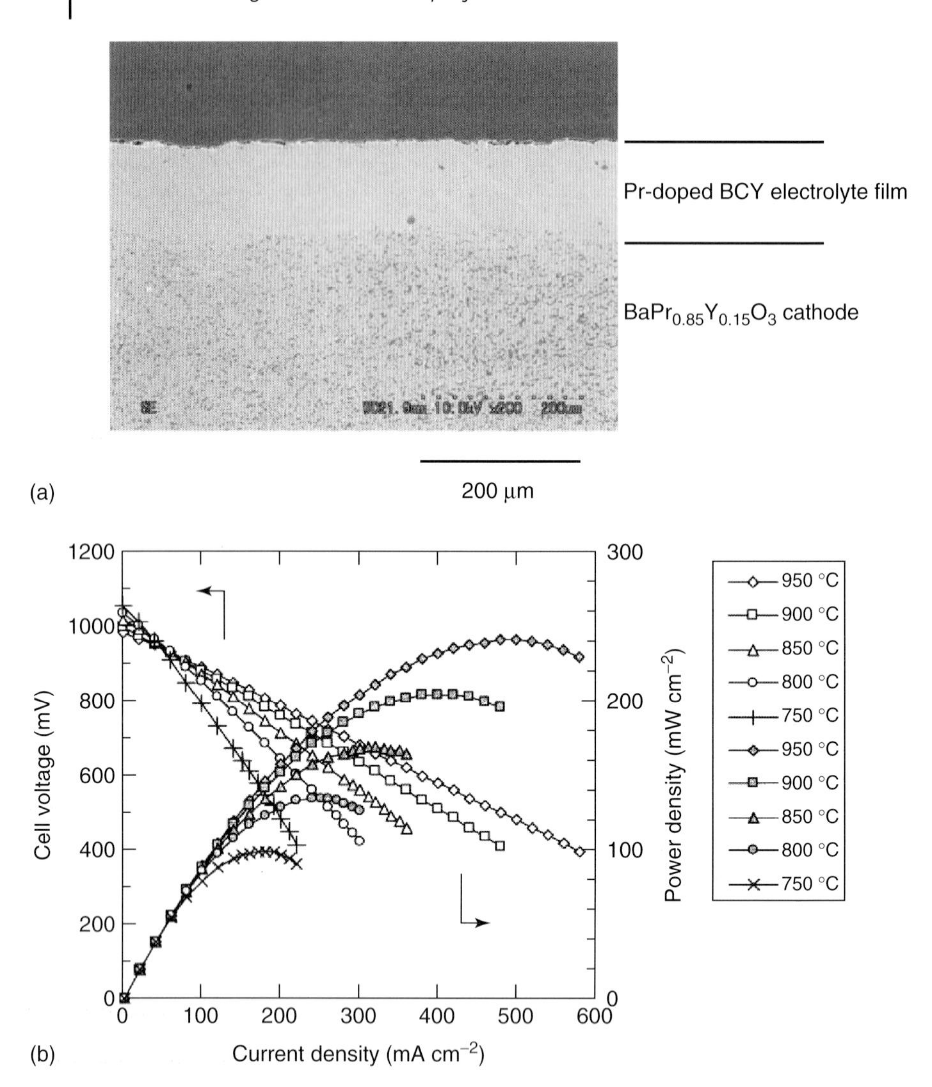

Figure 4.5 (a) SEM image of anode-free solid oxide fuel cells and (b) cell voltage and power density versus current density of B-type cell using thin Pr-doped BCY electrolyte film with wet hydrogen between 750 and 950 °C. (Source: Reproduced with permission from Ref. [78].)

(LSCF)–BCZYZn composite cathode was applied, the open current voltage and maximum power density reached 1.00 V and 247 mW cm^{-2}, respectively, at 650 °C [83]. Similar performance was achieved when $Ba_{0.5}Sr_{0.5}Zn_{0.2}Fe_{0.8}O_3$ was used as the cathode [84]. The proton conductor $BaZr_{0.1}Ce_{0.7}Y_{0.2}O_{3-\delta}$ (BZCY7) was also reported as a good electrolyte for SOFCs, with a power density of 270 mW cm^{-2} at 700 °C for a H_2/air cell [47].

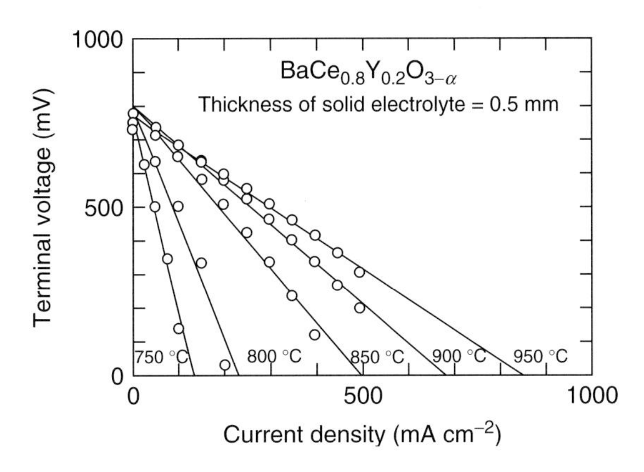

Figure 4.6 Fuel cell performance of a single-chamber solid oxide fuel cell based on the proton-conducting electrolyte $BaCe_{0.8}Y_{0.2}O_{3-\delta}$, with Pt as anode, Au as cathode, and a CH_4/O_2 ratio of 2 : 1. (Source: Reproduced with permission from Ref. [80].)

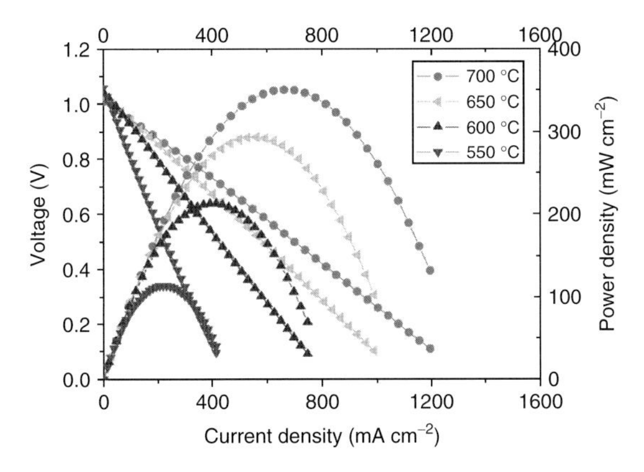

Figure 4.7 H_2/air fuel cell performance of a solid oxide fuel cell based on $BaCe_{0.5}Zr_{0.3}Y_{0.16}Zn_{0.04}O_{3-\delta}$ electrolyte. (Source: Reproduced with permission from Ref. [81].)

Excellent fuel cell performance was achieved based on BZCYYb electrolyte. At 750 °C, power densities of 560 and 1100 mW cm^{-2} were achieved when dry C_3H_8 and H_2 were used as the fuel, respectively [48]. When $PrBaCo_2O_{5+\delta}$ (PBCO) was used as cathode and BZCYYb as electrolyte, an OCV of 0.983 V and a maximal power density of 490 mW cm^{-2} were achieved at 700 °C for a H_2/air cell [85]. Reasonably high OCV was observed when a mixed H^+/e' conducting oxide BZCYYb was used as the electrolyte, indicating that the transport number for protons is quite high in this material.

So far, the best fuel cell performance for SOFCs based on proton-conducting electrolytes was reported in 2005 by Ito *et al.* A new approach for intermediate-temperature fuel cells using an ultrathin proton conductor $BaCe_{0.8}Y_{0.2}O_{3-\delta}$ coated

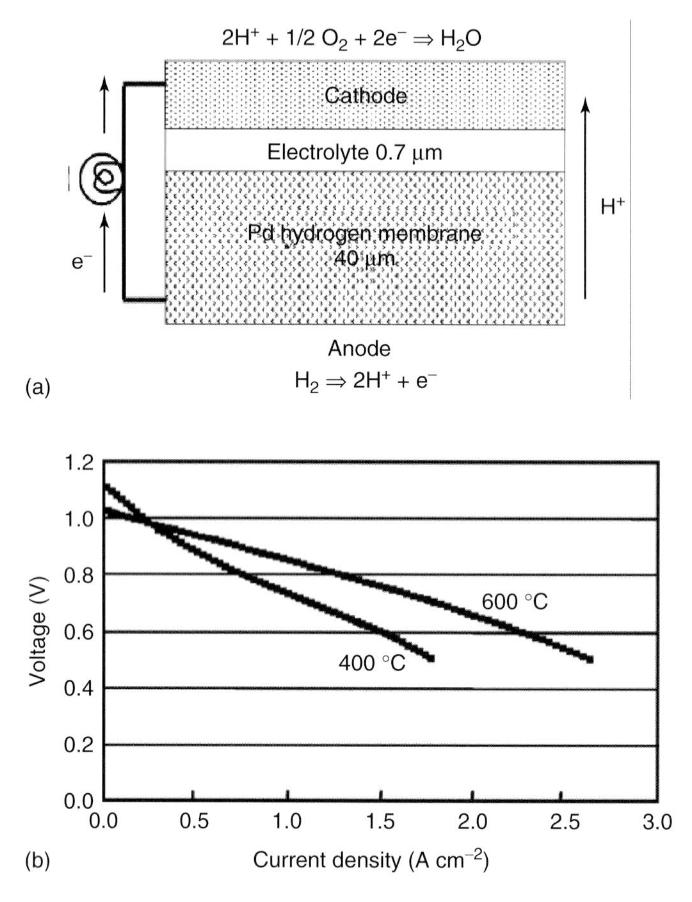

Figure 4.8 (a) Schematic diagram of an SOFC fabricated on Pd substrate and (b) performance of H_2/air fuel cell at 400 and 600 °C. (Source: Reproduced with permission from Ref. [86].)

on the solid Pd film anode. The electrolyte layer, which is 0.7 μm in thickness, was coated on a Pd film by PLD. The power density was 900 and 1400 mW cm^{-2} at the operating temperature of 400 and 600 °C, respectively (Figure 4.8) [86]; however, long-term stability could be an issue, as the electrolyte was too thin, let alone the stability of $BaCe_{0.8}Y_{0.2}O_{3-\delta}$ in CO_2 and H_2O. When $BaCe_{0.8}Y_{0.2}O_{3-\delta}$ was replaced by stable $BaZr_{0.8}Y_{0.2}O_{3-\delta}$ (thickness, 1 μm), with Pd thin film anode and Pt cathode, the power density was only 9.1 mW cm^{-2} at 400 °C due to the high electrode polarization resistance, which could be related to the electrolyte–electrode interfaces [87]. Although a high proton conductivity of 0.11 S cm^{-1} at 500 °C was observed for grain-boundary-free $BaZr_{0.8}Y_{0.2}O_{3-\delta}$ thin film deposited on (100)-oriented MgO substrates by the PLD method [41], it might face a similar problem with electrolyte–electrode interfaces when fabricating a real cell. The power density of H_2/air fuel cell based on $BaZr_{0.8}Y_{0.2}O_{3-\delta}$ electrolyte was 136 mW cm^{-2} at 400 °C when it is prepared by the PLD method and Pt was used as both anode and cathode [88]. To date, the performance of fuel cells based on $BaZr_{0.8}Y_{0.2}O_{3-\delta}$ electrolyte is not as high as those of $BaCeO_3$-based electrolytes [89].

It has been reported that an SOFC using Ni–$LaNbO_4$ as anode substrate (thickness, ~1200 μm), $La_{0.995}Sr_{0.005}NbO_4$ as electrolyte (thickness, ~30 μm), and $La_{1-x}Sr_xMnO_3$ as cathode (thickness, ~250 μm) has been fabricated. The OCV after reduction of the anode reached a maximum of approximately 830 mV at 800 °C, which is about ~85% of the theoretical Nernst voltage between wet 5% H_2/Ar and wet air. A power density of 1.32 mW cm^{-2} has been achieved at 800 °C when wet 5% H_2/Ar was used as the fuel [13]. A slightly better performance for a cell with similar cell construction was reported by Meng *et al.* A cell with LCN thin electrolyte (20 μm), NiO–$La_{0.5}Ce_{0.5}O_{1.75}$ (NiO–LDC) anode, and $(La_{0.8}Sr_{0.2})_{0.9}MnO_{3-\delta}$-$La_{0.5}Ce_{0.5}O_{1.75}$ (LSM–LDC) composite cathode was fabricated. The OCV and maximum power density, respectively, reached 0.98 V and 65 mW cm^{-2} at 800 °C when H_2 was fed as the fuel [90]. The difference in cell performance may be due to the electrode compositions and the fabrication methods. The lower power density is related to the low proton conductivity of the electrolyte. From this point of view, the known lanthanide niobates and tantalates are not conductive enough for use as electrolytes for SOFCs.

Not only H_2 and hydrocarbons, but also ammonia, can be used as the fuel for SOFCs based on proton-conducting electrolytes. To avoid the formation of nitric oxide, doped $BaCeO_3$ and $BaZrO_3$ based on a proton-conducting electrolyte is an ideal solution for direct ammonia fuel cells [91–96].

In 2003, Coors reported the use of hydrocarbons as the fuel in a proton-conducting SOFC. Proton conduction implies that water vapor is produced at the cathode, where it is swept away by air, rather than at the anode as in an SOFC based on O^{2-} ion conducting electrolytes, where it dilutes the fuel. Since carbon dioxide is the only exhaust gas produced, higher fuel utilization is possible. Ambipolar steam permeation from the cathode to the anode provides the steam for direct re-forming of hydrocarbons, so external steam injection is not required. Therefore, high thermodynamic efficiency is achieved and coking is not a problem [97]. $BaCe_{0.9}Y_{0.1}O_{3-\delta}$ was used as electrolyte, and power densities of 42 and 22 mW cm^{-2} at 793 °C were observed when H_2 and CH_4 were used as the fuel, respectively.

Research on ammonia SOFCs has been reviewed by Ni *et al* and Lan *et al.* [98, 99]. It has been reported that H_2 and NH_3 exhibit similar OCV and power density for an SOFC based on $BaCe_{0.9}Nd_{0.1}O_{3-\delta}$ electrolyte. At 700 °C, the power densities were 335 and 315 mW cm^{-2} for H_2/air and NH_3/air cells, respectively [96]. High-performance direct ammonia SOFCs based on oxygen-conducting $Ce_{0.8}Sm_{0.2}O_{1.9}$ electrolyte with a working temperature of 550–650 °C was also reported by Meng *et al.* [100]. The power density of direct ammonia fuel cells is just slightly lower than that of hydrogen fuel cells [100]. It is expected that ammonia fuel cells with

Table 4.1 Comparison of fuel cell performances on SOFCs based on proton-conducting electrolytes.

Electrolyte	Electrolyte preparation method	Electrolyte thickness	Cathode	Anode	Open-circuit voltage	Power density	References
$SrCe_{0.95}Yb_{0.05}O_{3-\delta}$	Sintering at 1500 °C for 10 h	500 μm	Pt	Pt	1.05 V at 800 °C (H_2/air)	25 mW cm^{-2} at 800 °C (H_2/air)	[25]
$BaCe_{0.9}Y_{0.1}O_{3-\delta}$	Not specified	460 μm	Porous Pt	Ni (1 μm thick)	1.01 V for H_2/air; 0.96 V for CH_4/air at 793 °C	42 mW cm^{-2} for H_2/air; 22 mW cm^{-2} for CH_4/air at 793 °C	[97]
$BaCe_{0.76}Y_{0.20}Pr_{0.04}O_{3-\delta}$	Sintering at 1650 °C for 6 h	500 μm	$BaPr_{0.85}Y_{0.15}O_3$	None (Au mesh as anode current collector)	1.02 V for H_2/air at 800 °C	140 mW cm^{-2} at 800 °C	[78]
$BaCe_{0.8}Y_{0.2}O_{3-\delta}$	Sintering at 1650 °C for 10 h	500 μm	Au	Pt	0.8 V at 800 °C for mixed fuel (CH_4/O_2 = 2 : 1)	52 mW cm^{-2} at 800 °C	[80]
BCZYZn	Sintering at 1250 °C for 5 h	20 μm	$GdBa_{0.5}Sr_{0.5}Co_2O_{5+\delta}$	NiO–BCZYZn (65 : 35 w/o)	1.05 V for H_2/air at 700 °C	350 mW cm^{-2} at 700 °C	[81]
BZCY7	Sintering at 1350 °C for 6 h	65 μm	$BaCe_{0.4}Pr_{0.4}Y_{0.2}O_3$	NiO–BZCY7	1.0 V for H_2/air at 700 °C	270 mW cm^{-2} at 700 °C	[47]

BZCYYb	Sintering at 1400 °C for 5 h	10 µm	LSCF-BZCYYb	NiO–BZCYYb	1.0 V for H_2/air; 0.89 V for C_3H_8/air at 750 °C	1100 mW cm^{-2} for H_2/air; 560 mW cm^{-2} for C_3H_8/air at 750 °C	[48]
$BaCe_{0.8}Y_{0.2}O_3$	PLD	0.7 µm	Perovskite cathode (not specified)	Pd thin film (40 µm)	1.1 V at 400 °C; 1.0 V at 600 °C for H_2/air	900 mW cm^{-2} at 400 °C; 1400 mW cm^{-2} at 600 °C for H_2/air	[86]
$BaZr_{0.8}Y_{0.2}O_3$	PLD	1 µm	Pt	Pd thin film (400 nm)	1.0 V at 400 °C	9.1 mW cm^{-2} for H_2/air at 400 °C	[87]
LCN	Sintering at 1400 °C for 5 h	20 µm	LSM-LDC at 50 : 50	LCN-LDC	0.98 V for H_2/air at 800 °C	65 mW cm^{-2} at 800 °C	[90]
BCNO	Sintering at 1400 °C for 5 h	20 µm	$La_{0.5}Sr_{0.5}CoO_{3-\delta}$	NiO–NCNO	0.951 V for H_2/air; 0.95 V for NH_3/air at 700 °C	355 mW cm^{-2} for H_2/air; 315 mW cm^{-2} for NH_3/air at 700 °C	[96]

BCZYZn, $BaCe_{0.5}Zr_{0.3}Y_{0.16}Zn_{0.04}O_{3-\delta}$; BZCY7, $BaZr_{0.1}Ce_{0.7}Y_{0.2}O_{3-\delta}$; BCZYYb, $BaCe_{0.7}Zr_{0.1}Y_{0.1}Yb_{0.1}O_{3-\delta}$; LSCF, $La_{0.6}Sr_{0.4}Co_{0.2}Fe_{0.8}O_{3-\delta}$; PLD, pulsed laser deposition; LCN, $La_{0.99}Ca_{0.01}NbO_4$; LSM, $(La_{0.8}Sr_{0.2})_{0.9}MnO_{3-\delta}$; LDC, $La_{0.5}Ce_{0.5}O_{1.75}$; BCNO, $BaCe_{0.9}Nd_{0.1}O_{3-\delta}$.

high power densities can be achieved for SOFCs based on proton-conducting electrolytes.

H_2S has been reported as the fuel in an SOFC based on YSZ electrolyte. It may have the risk of formation of SO_x [101]. If an SOFC-H is used, toxic H_2S can be converted into valued sulfur while electricity is generated. Typical proton conductor $BaCe_{1-x}Y_xO_{3-\delta}$ is not stable in H_2S atmosphere [102]. The $YSZ-K_3PO_4-Ca_3(PO_4)_2$ composite exhibits proton conduction and has been proposed as electrolyte for H_2S SOFCs [102]. The typical fuel cell performances based on proton-conducting electrolytes are listed in Table 4.1.

4.5 Electrode Materials and Anode Reactions for SOFCs Based on Proton-Conducting Electrolytes

The electrode materials for SOFCs based on oxygen-ion-conducting electrolyte have been investigated in detail. The typical anode is a metal-oxide cermet such

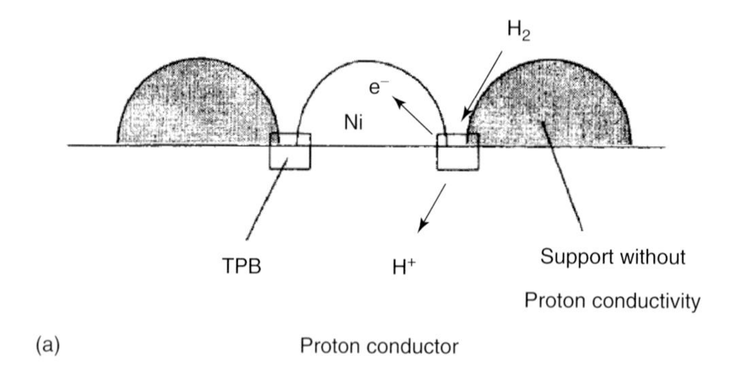

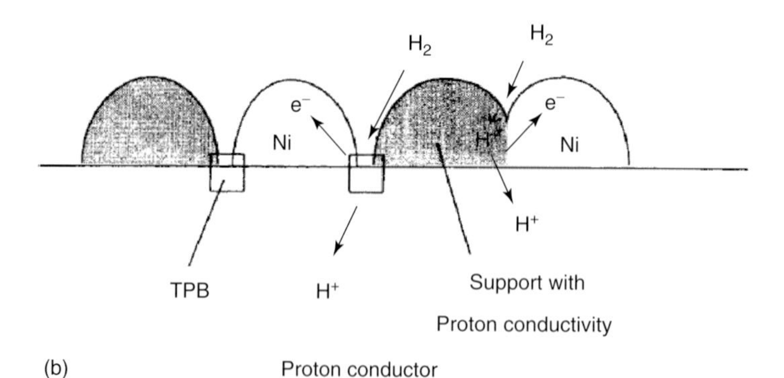

Figure 4.9 Schematic diagram of the anode surface (a) without and (b) with a proton-conducting support for a proton-conducting fuel cell. (Source: Reproduced with permission from Ref. [106].)

as Ni–YSZ or a redox-stable anode such as $(La_{0.75}Sr_{0.25})Cr_{0.5}Mn_{0.5}O_{3-\delta}$ [103, 104]. Perovskite oxides, such as lanthanum strontium manganite, are commonly used as cathode materials for SOFCs [105]. In principle, the materials used for SOFCs based on O^{2-} ion conducting electrolytes can also be adopted for SOFCs based on proton-conducting electrolytes. For SOFCs based on proton-conducting electrolytes, the proton conduction in the anode is very important to extend the reaction at electrodes from triple phase boundary to the whole of the electrodes (Figure 4.9) [106]. Therefore, some proton-conducting electrolyte materials are normally mixed with metal or conducting oxides to form a composite anode or cathode in order to reduce polarization resistance and improve power density. It should be noted that in this reaction, hydrogen is commonly used as the fuel for SOFCs. The reaction mechanism has been described in Eq. (4.7), Eq. (4.8), and Eq. (4.9). Unlike SOFCs based on oxygen-ion-conducting electrolytes, water is formed at cathode for fuel cells based on proton-conducting electrolytes, and therefore, the fuel is not diluted.

Hydrocarbons such as methane can be used as the fuel for this type of SOFCs. The cathode reaction is the same; however, the anode reaction is more complicated.

If dry methane is used as the fuel, it would be expected that coking may happen at the anode,

$$CH_4 \rightarrow C + 2H_2 \tag{4.10}$$

However, no coking was observed when dry methane was used as the fuel [97]. The reaction mechanism of a proton-conducting SOFC when methane was used as the fuel has been described by Kreuer (Figure 4.10) [15].

At the anode, water is formed from proton defects formed through Eq. (4.11):

$$4OH_O^{\bullet} \rightarrow 2H_2O + 2V_O^{\bullet\bullet} + 2O_O^{\times} \tag{4.11}$$

The formed water will re-form methane,

$$CH_4 + 2H_2O \rightarrow CO_2 + 4H_2 \tag{4.12}$$

The formed H_2 will interact with lattice oxygen to form proton defects:

$$4H_2 + 8O_O^{\times} \rightarrow 8OH_O^{\bullet} + 8e' \tag{4.13}$$

The overall reaction at anode is

$$CH_4 + 6O_O^{\times} \rightarrow CO_2 + 2V_O^{\bullet\bullet} + 4OH_O^{\bullet} + 8e' \tag{4.14}$$

At the cathode, O_2 is reduced by both proton defects and oxygen vacancies

$$4OH_O^{\bullet} + O_2 + 4e' \rightarrow 4O_O^{\times} + 2H_2O \tag{4.15}$$

$$2V_O^{\bullet\bullet} + O_2 + 4e' \rightarrow 2O_O^{\times} \tag{4.16}$$

The overall reaction at cathode is

$$4OH_O^{\bullet} + 2V_O^{\bullet\bullet} + 2O_2 + 8e' \rightarrow 6O_O^{\times} + 2H_2O \tag{4.17}$$

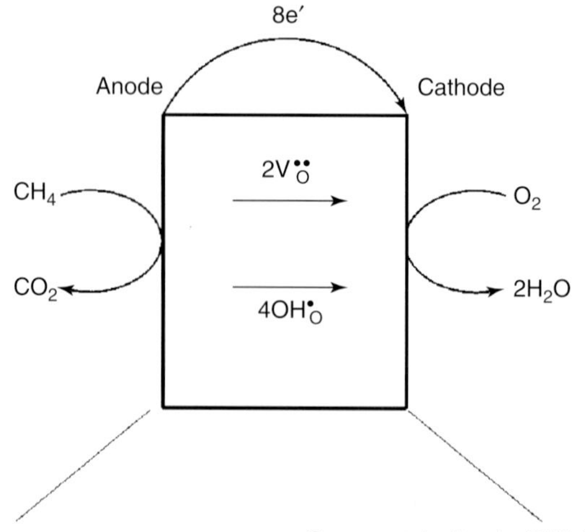

Water formation :

$4OH_O^{\bullet} + 2H_2O + 2V_O^{\bullet\bullet} + 2O_O^{x}$

Steam reforming :

$CH_4 + 2H_2O = CO_2 + 4H_2$

Electrochemical hydrogen oxidation :

$4H_2 + 8O_O^{x} = 8OH_O^{\bullet} + 8e'$

Oxygen reduction by $OH_O^{\bullet} / V_O^{\bullet\bullet}$

$4OH_O^{\bullet} + O2 + 4e' = 4O_O^{x} + 2H_2O$

$2V_O^{\bullet\bullet} + O_2 + 4e' = 2O_O^{x} + 2H_2O$

$CH_4 + 6O_O^{x} = CO_2 + 2V_O^{\bullet\bullet} + 4OH_O^{\bullet} + 8e'$

$4OH_O^{\bullet} + 2V_O^{\bullet\bullet} + 2O_2 + 8e' = 6O_O^{x} + 2H_2O$

$CH_4 + 2O_2 = CO_2 + 2H_2O$

(a)

$$\tilde{D}_{H_2O} = \frac{(2-x)D_{OH_O^{\bullet}}D_{V_O^{\bullet\bullet}}}{x\,D_{OH_O^{\bullet}} + 2(1-x)D_{V_O^{\bullet\bullet}}}$$

with

$$x = \frac{[OH_O^{\bullet}]}{[A]} \text{ (degree of hydration)}$$

$2V_O^{\bullet\bullet}$

$2H_2O$

$CO_2 + 4H_2$

$4OH_O^{\bullet}$

(b)

Figure 4.10 Self-regulated methane reformation and overall cell reaction in an SOFC operating with an electrolyte with protonic defects and oxide-ion vacancies as mobile charge carriers: (a) the overall cell reaction and (b) the steam reforming step. (Source: Reproduced with permission from Ref. [15].)

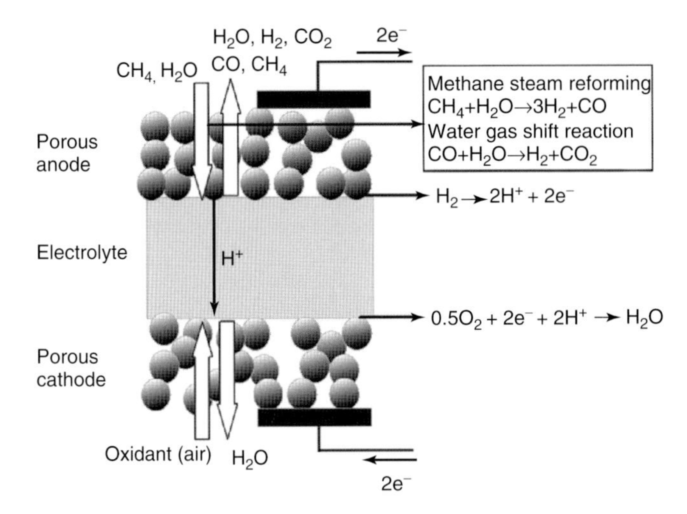

Figure 4.11 Schematic diagram of an SOFC with internal re-formation of methane. (Source: Reproduced with permission from Ref. [107].)

Combining Eq. (4.9) and Eq. (4.12), the overall reaction is

$$CH_4 + 2O_2 \rightarrow CO_2 + 2H_2O \tag{4.18}$$

H_2 can also be formed through internal steam reforming (Figure 4.11) [107]. The formed H_2 is then used as the fuel for SOFC-H.

When ammonia is used as the fuel in an SOFC based on proton-conducting electrolytes, at the anode,

$$2NH_3 - 6e' \rightarrow N_2 + 6H^+ \tag{4.19}$$

The cathode reaction is the same as Eq. (4.17). The overall reaction for an ammonia fuel cell based on proton-conducting electrolytes is

$$2NH_3 + \frac{3}{2}O_2 \rightarrow N_2 + 3H_2O \tag{4.20}$$

If the operating temperature of SOFC-H is high enough, ammonia may thermally decompose into H_2 and N_2, and the obtained H_2 is used as the fuel for fuel cells. The reaction mechanism of an ammonia fuel cell based on a proton-conducting electrolyte is shown in Figure 4.12 [98]. Formation of toxic NO_x can be avoided when a proton-conducting electrolyte is used for ammonia SOFCs. Similar mechanism will be applied to SOFC-H when H_2S is used as the fuel.

4.6 Conclusion

SOFCs based on O^{2-} ion conducting electrolytes have been investigated in detail; however, the development of SOFCs based on proton-conducting electrolytes is

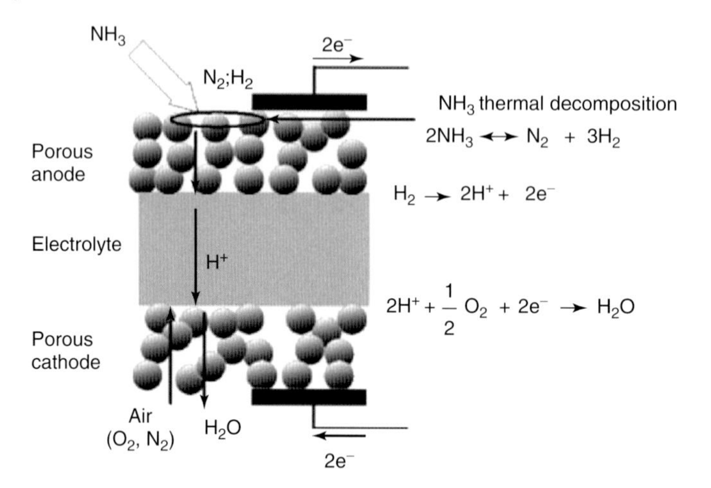

Figure 4.12 Diagram of a NH_3/air fuel cell based on proton-conducting electrolytes. (Source: Reproduced with permission from Ref. [98].)

lagging behind. The excellent performance of SOFCs based on BZCYYb [48] indicates that this type could be as good as SOFCs based on YSZ, LSGM, and gadolinium doped ceria (CGO) electrolytes. The use of new technology such as PLD makes it feasible to deposit dense thin film with ~1 μm thickness, to be used for intermediate-temperature (400–600 °C) operation. The mechanic strength and cracking of this thin film could be an issue for long-time operation. For some fuels such as ammonia, hydrogen sulfide, and hydrocarbons, SOFCs based on proton-conducting electrolytes have obvious advantages over those based on O^{2-} ion conducting electrolytes without risking formation of toxic NO_x and SO_x.

Electrode materials use for SOFCs based on O^{2-} ion conducting electrolyte can also be adopted for SOFCs based on proton-conducting electrolytes, but the reaction mechanism for SOFC-H is different from that of SOFC-O. The development of SOFC-H is still at an early stage, and further investigation on compatibility and performance is necessary. It is believed that SOFCs based on proton-conducting electrolytes will be as good as those based on O^{2-} ion conducting electrolytes.

References

1. Steele, B.C.H. and Heinzel, A. (2001) Materials for fuel-cell technologies. *Nature*, **414**, 345–352.
2. Minh, N.Q. and Takahashi, T. (1995) *Science and Technology of Ceramic Fuel Cells*, Vol. 9, Elsevier, Amsterdam.
3. Singhal, S.C. and Kendall, K. (2003) *High Temperature Solid Oxide Fuel Cells: Fundamentals, Design, and Applications*, Elsevier, Oxford.
4. Ruiz-Morales, J.C., Canales-Vazquez, J., Savaniu, C., Marrero-Lopez, D., Zhou, W.Z., and Irvine, J.T.S. (2006) Disruption of extended defects in solid oxide fuel cell anodes for methane oxidation. *Nature*, **439**, 568–571.
5. Zhang, L., Jiang, S.P., Wang, W., and Zhang, Y.J. (2007) NiO/YSZ, anode-supported, thin-electrolyte, solid oxide fuel cells fabricated by

gel casting. *J. Power Sources*, **170**, 55–60.

6. Oishi, N., Atkinson, A., Brandon, N.P., Kilner, J.A., and Steele, B.C.H. (2005) Fabrication of an anode-supported gadolinium-doped ceria solid oxide fuel cell and its operation at 550 °C. *J. Am. Ceram. Soc.*, **88**, 1394–1396.
7. Ishihara, T., Matsuda, H., and Takita, Y. (1994) Doped $LAGAO_3$ perovskite-type oxide as a new oxide ionic conductor. *J. Am. Chem. Soc.*, **116**, 3801–3803.
8. Xie, K., Yan, R.Q., Jiang, Y.Z., Liu, X.Q., and Meng, G.Y. (2008) A simple and easy one-step fabrication of thin $BaZr_{0.1}Ce_{0.7}Y_{0.2}O_{3-\delta}$ electrolyte membrane for solid oxide fuel cells. *J. Membr. Sci.*, **325**, 6–10.
9. Kreuer, K.D. (1996) Proton conductivity: materials and applications. *Chem. Mater.*, **8**, 610–641.
10. Haile, S.M. (2003) Fuel cell materials and components. *Acta Mater.*, **51**, 5981–6000.
11. Orera, A. and Slater, P. (2009) New chemical systems for solid oxide fuel cells. *Chem. Mater.*, **22**, 675–690.
12. Fabbri, E., Pergolesi, D., and Traversa, E. (2010) Materials challenges toward proton-conducting oxide fuel cells: a critical review. *Chem. Soc. Rev.*, **39**, 4355–4369.
13. Magraso, A., Fontaine, M.L., Larring, Y., Bredesen, R., Syvertsen, G.E., Lein, H.L., Grande, T., Huse, M., Strandbakke, R., Haugsrud, R., and Norby, T. (2011) Development of proton conducting SOFCs based on $LaNbO_4$ electrolyte – status in Norway. *Fuel Cells*, **11**, 17–25.
14. Malavasi, L., Fisher, C.A.J., and Islam, M.S. (2010) Oxide-ion and proton conducting electrolyte materials for clean energy applications: structural and mechanistic features. *Chem. Soc. Rev.*, **39**, 4370–4387.
15. Kreuer, K.D. (2003) Proton-conducting oxides. *Annu. Rev. Mater. Res.*, **33**, 333–359.
16. Iwahara, H., Esaka, T., Uchida, H., and Maeda, N. (1981) Proton conduction in sintered oxides and its application to steam electrolysis for hydrogen production. *Solid State Ionics*, **3–4**, 359–363.
17. Takahashi, T. and Iwahara, H. (1980) Solid-state ionics – protonic conduction in perovskite type oxide solid-solutions. *Rev. Chim. Minerale*, **17**, 243–253.
18. Uchida, H., Yoshikawa, H., and Iwahara, H. (1989) Formation of protons in $SrCeO_3$-based proton conducting oxides. Part I. Gas evolution and absorption in doped $SrCeO_3$ at high temperature. *Solid State Ionics*, **34**, 103–110.
19. Irvine, J.T.S., Corcoran, D.J.D., Lashtabeg, A., and Walton, J.C. (2002) Incorporation of molecular species into the vacancies of perovskite oxides. *Solid State Ionics*, **154**, 447–453.
20. Ma, G.L., Shimura, T., and Iwahara, H. (1999) Simultaneous doping with La^{3+} and Y^{3+} for Ba^{2+}- and Ce^{4+}-sites in $BaCeO_3$ and the ionic conduction. *Solid State Ionics*, **120**, 51–60.
21. Liu, M.L., Hu, H.X., and Rauch, W. (1997) Ionic and electronic transport in $BaCe_{0.8}Gd_{0.2}O_3$ solid electrolyte, in *Proceedings of the First International Symposium on Ceramic Membranes*, Electrochemical Society Series, Vol. **95** (eds H.U. Anderson, A.C. Khandkar, and M. Liu), The Electrochemical Society, Pennington, NJ, pp. 192–220.
22. Uchida, H., Maeda, N., and Iwahara, H. (1982) Steam concentration cell using a high-temperature type proton conductive solid electrolyte. *J. Appl. Electrochem.*, **12**, 645–651.
23. Iwahara, H., Uchida, H., and Maeda, N. (1982) High-temperature fuel and steam electrolysis cells using proton conductive solid electrolytes. *J. Power Sources*, **7**, 293–301.
24. Uchida, H., Maeda, N., and Iwahara, H. (1983) Relation between proton and hole conduction in $SrCeO_3$-based solid electrolytes under water-containing atmospheres at high-temperatures. *Solid State Ionics*, **11**, 117–124.
25. Iwahara, H., Uchida, H., and Tanaka, S. (1983) High-temperature type proton conductor based on $SrCeO_3$ and its application to solid electrolyte fuel cells. *Solid State Ionics*, **9–10**, 1021–1025.

26. Iwahara, H., Esaka, T., Uchida, H., Yamauchi, T., and Ogaki, K. (1986) High-temperature type protonic conductor based on $SrCeO_3$ and its application to the extraction of hydrogen gas. *Solid State Ionics*, **18–19**, 1003–1007.
27. Ishigaki, T., Yamauchi, S., Kishio, K., Fueki, K., and Iwahara, H. (1986) Dissolution of deuterium into proton conductor $SrCe_{0.95}Yb_{0.05}O_{3-\delta}$. *Solid State Ionics*, **21**, 239–41.
28. Virkar, A.N. and Maiti, H.S. (1985) Oxygen ion conduction in pure and yttria-doped barium cerate. *J. Power Sources*, **14**, 295–303.
29. Mitsui, A., Miyayama, M., and Yanagida, H. (1987) Evaluation of the activation-energy for proton conduction in perovskite-type oxides. *Solid State Ionics*, **22**, 213–217.
30. Kreuer, K.D. (1997) On the development of proton conducting materials for technological applications. *Solid State Ionics*, **97**, 1–15.
31. Chen, F.L., Sorensen, O.T., Meng, G.Y., and Peng, D.K. (1997) Chemical stability study of $BaCe_{0.9}Nd_{0.1}O_{3-\delta}$ high-temperature proton-conducting ceramic. *J. Mater. Chem.*, **7**, 481–485.
32. Tanner, C.W. and Virkar, A.V. (1996) Instability of $BaCeO_3$ in H2O-containing atmospheres. *J. Electrochem. Soc.*, **143**, 1386–1389.
33. Tao, S.W. and Irvine, J.T.S. (2006) A stable, easily sintered proton-conducting oxide electrolyte for moderate-temperature fuel cells and electrolyzers. *Adv. Mater.*, **18**, 1581.
34. Iwahara, H., Yajima, T., Hibino, T., Ozaki, K., and Suzuki, H. (1993) Protonic conduction in calcium, strontium and barium zirconates. *Solid State Ionics*, **61**, 65–69.
35. Yajima, T., Kazeoka, H., Yogo, T., and Iwahara, H. (1991) Proton conduction in sintered oxides based on $CaZrO_3$. *Solid State Ionics*, **47**, 271–275.
36. Yajima, T., Suzuki, H., Yogo, T., and Iwahara, H. (1992) Protonic conduction in $SrZrO_3$-based oxides. *Solid State Ionics*, **51**, 101–107.
37. Hibino, T., Mizutani, K., Yajima, T., and Iwahara, H. (1992) Evaluation of proton conductivity in $SrCeO_3$, $BaCeO_3$, $CaZrO_3$ and $SrZrO_3$ by temperature programmed desorption method. *Solid State Ionics*, **57**, 303–306.
38. Bohn, H.G. and Schober, T. (2000) Electrical conductivity of the high-temperature proton conductor $BaZr_{0.9}Y_{0.1}O_{2.95}$. *J. Am. Ceram. Soc.*, **83**, 768–772.
39. Savaniu, C.D., Canales-Vazquez, J., and Irvine, J.T.S. (2005) Investigation of proton conducting $BaZr_{0.9}Y_{0.1}O_{2.95}$: $BaCe_{0.9}Y_{0.1}O_{2.95}$ core-shell structures. *J. Mater. Chem.*, **15**, 598–604.
40. Tao, S.W. and Irvine, J.T.S. (2007) Conductivity studies of dense yttrium-doped $BaZrO_3$ sintered at 1325 °C. *J. Solid State Chem.*, **180**, 3493–3503.
41. Pergolesi, D., Fabbri, E., D'Epifanio, A., Di Bartolomeo, E., Tebano, A., Sanna, S., Licoccia, S., Balestrino, G., and Traversa, E. (2010) High proton conduction in grain-boundary-free yttrium-doped barium zirconate films grown by pulsed laser deposition. *Nat. Mater.*, **9**, 846–852.
42. Ryu, K.H. and Haile, S.M. (1999) Chemical stability and proton conductivity of doped $BaCeO_3$-$BaZrO_3$ solid solutions. *Solid State Ionics*, **125**, 355–367.
43. Katahira, K., Kohchi, Y., Shimura, T., and Iwahara, H. (2000) Protonic conduction in Zr-substituted $BaCeO_3$. *Solid State Ionics*, **138**, 91–98.
44. Kreuer, K.D., Adams, S., Munch, W., Fuchs, A., Klock, U., and Maier, J. (2001) Proton conducting alkaline earth zirconates and titanates for high drain electrochemical applications. *Solid State Ionics*, **145**, 295–306.
45. Irvine, J.T.S., Tao, S.W., Savaniu, C.D., and A.K. Azad (2004) Patent No. GB20040006818 20040326; GB20040027329 20041214; WO2005GB01169 20050324.
46. Babilo, P. and Haile, S.M. (2005) Enhanced sintering of yttrium doped barium zirconate by addition of ZnO. *J. Am. Ceram. Soc.*, **88**, 2362–2368.
47. Zuo, C.D., Zha, S.W., Liu, M.L., Hatano, M., and Uchiyama, M. (2006) $BaZr_{0.1}Ce_{0.7}Y_{0.2}O_{3-\delta}$ as an electrolyte

for low-temperature solid-oxide fuel cells. *Adv. Mater.*, **18**, 3318–3320.
48. Yang, L., Wang, S.Z., Blinn, K., Liu, M.F., Liu, Z., Cheng, Z., and Liu, M.L. (2009) Enhanced sulfur and coking tolerance of a xixed Ion conductor for SOFCs: $BaZr_{0.1}Ce_{0.7}Y_{0.2-x}Yb_xO_{3-\delta}$. *Science*, **326**, 126–129.
49. Murugaraj, P., Kreuer, K.D., He, T., Schober, T., and Maier, J. (1997) High proton conductivity in barium yttrium stannate $Ba_2YSnO_{5.5}$. *Solid State Ionics*, **98**, 1–6.
50. Liang, K.C. and Nowick, A.S. (1993) High-temperature protonic conduction in mixed perovskite ceramics. *Solid State Ionics*, **61**, 77–81.
51. Glockner, R., Neiman, A., Larring, Y., and Norby, T. (1999) Protons in $Sr_3(Sr_{1+x}Nb_{2-x})O_{9-3x/2}$ perovskite. *Solid State Ionics*, **125**, 369–376.
52. Bohn, H.G., Schober, T., Mono, T., and Schilling, W. (1999) The high temperature proton conductor $Ba_3Ca_{1.18}Nb_{1.82}O_{3-\delta}$. I. Electrical conductivity. *Solid State Ionics*, **117**, 219–228.
53. Savaniu, C. and Irvine, J.T.S. (2003) $Sr_3Ca_{1-x}ZnxZr_{0.5}Ta_{1.5}O_{8.75}$: a study of the influence of the B-site dopant nature upon protonic conduction. *Solid State Ionics*, **162**, 105–113.
54. Haugsrud, R. and Norby, T. (2006) Proton conduction in rare-earth ortho-niobates and ortho-tantalates. *Nat. Mater.*, **5**, 193–196.
55. Haugsrud, R. and Norby, T. (2006) High-temperature proton conductivity in acceptor-doped $LaNbO_4$. *Solid State Ionics*, **177**, 1129–1135.
56. Mokkelbost, T., Kaus, I., Haugsrud, R., Norby, T., Grande, T., and Einarsrud, M.A. (2008) High-temperature proton-conducting lanthanum ortho-niobate-based materials. Part II: sintering properties and solubility of alkaline earth oxides. *J. Am. Ceram. Soc.*, **91**, 879–886.
57. Haugsrud, R., Ballesteros, B., Lira-Cantu, M., and Norby, T. (2006) Ionic and electronic conductivity of 5% Ca-doped $GdNbO_4$. *J. Electrochem. Soc.*, **153**, J87–J90.
58. Haugsrud, R. and Norby, T. (2007) High-temperature proton conductivity in acceptor-substituted rare-earth ortho-tantalates, $LnTaO_4$. *J Am Ceram Soc*, **90**, 1116–1121.
59. Yokokawa, H., Horita, T., Sakai, N., Yamaji, K., Brito, M.E., Xiong, Y.P., and Kishimoto, H. (2006) Ceria: Relation among thermodynamic, electronic hole and proton properties. *Solid State Ionics*, **177**, 1705–1714.
60. Nigara, Y., Mizusaki, J., Kawamura, K., Kawada, T., and Ishigame, M. (1998) Hydrogen permeability in $(CeO_2)0.9(CaO)_{0.1}$ at high temperatures. *Solid State Ionics*, **113**, 347–354.
61. Nigara, Y., Yashiro, K., Kawada, T., and Mizusaki, J. (2001) The atomic hydrogen permeability in $(CeO_2)_{0.85}(CaO)_{0.15}$ at high temperatures. *Solid State Ionics*, **145**, 365–370.
62. Kim, S., Anselmi-Tambtirini, U., Park, H.J., Martin, M., and Munir, Z.A. (2008) Unprecedented room-temperature electrical power generation using nanoscale fluorite-structured oxide electrolytes. *Adv. Mater.*, **20**, 556.
63. Avila-Paredes, H.J., Barrera-Calva, E., Anderson, H.U., De Souza, R.A., Martin, M., Munir, Z.A., and Kim, S. (2010) Room-temperature protonic conduction in nanocrystalline films of yttria-stabilized zirconia. *J. Mater. Chem.*, **20**, 6235–6238.
64. Avila-Paredes, H.J., Chen, C.T., Wang, S.Z., De Souza, R.A., Martin, M., Munir, Z., and Kim, S. (2010) Grain boundaries in dense nanocrystalline ceria ceramics: exclusive pathways for proton conduction at room temperature. *J. Mater. Chem.*, **20**, 10110–10112.
65. Kim, S., Avila-Paredes, H.J., Wang, S.Z., Chen, C.T., De Souza, R.A., Martin, M., and Munir, Z.A. (2009) On the conduction pathway for protons in nanocrystalline yttria-stabilized zirconia. *Phys. Chem. Chem. Phys.*, **11**, 3035–3038.
66. Ruiz-Trejo, E. and Kilner, J.A. (2009) Possible proton conduction in $Ce_{0.9}Gd_{0.1}O_{2-\delta}$ nanoceramics. *J. Appl. Electrochem.*, **39**, 523–528.
67. Liu, R.Q., Xie, Y.H., Wang, J.D., Li, Z.J., and Wang, B.H. (2006) Synthesis

of ammonia at atmospheric pressure with $Ce_{0.8}M_{0.2}O_{2-\delta}$ (M=La, Y, Gd, Sm) and their proton conduction at intermediate temperature. *Solid State Ion.*, **177**, 73–76.

68. Ma, G.L., Zhang, F., Zhu, J.L., and Meng, G.Y. (2006) Proton conduction in $La_{0.9}Sr_{0.1}Ga_{0.8}Mg_{0.2}O_{3-\delta}$. *Chem. Mater.*, **18**, 6006–6011.
69. Chen, C., Wang, W.B., and Ma, G.L. (2009) Proton conduction in $La_{0.9}M_{0.1}Ga_{0.8}Mg_{0.2}O_{3-\delta}$ at intermediate temperature and its application to synthesis of ammonia at atmospheric pressure. *Acta. Chim. Sinica*, **67**, 623–628.
70. Chen, C. and Ma, G.L. (2008) Preparation, proton conduction, and application in ammonia synthesis at atmospheric pressure of $La_{0.9}Ba_{0.1}Ga_{1-x}MgxO_{3-\delta}$. *J. Mater. Sci.*, **43**, 5109–5114.
71. Zhang, G.B. and Smyth, D.M. (1995) Protonic conduction in $Ba_2In_2O_5$. *Solid State Ion.*, **82**, 153–160.
72. Jankovic, J., Wilkinson, D.P., and Hui, R. (2011) Proton conductivity and stability of $Ba_2In_2O_5$ in hydrogen containing atmospheres. *J. Electrochem. Soc.*, **158**, B61–B68.
73. Kendrick, E., Kendrick, J., Knight, K.S., Islam, M.S., and Slater, P.R. (2007) Cooperative mechanisms of fast-ion conduction in gallium-based oxides with tetrahedral moieties. *Nat. Mater.*, **6**, 871–875.
74. Haugsrud, R. and Kjolseth, C. (2008) Effects of protons and acceptor substitution on the electrical conductivity of La_6WO_{12}. *J. Phys. Chem. Solids*, **69**, 1758–1765.
75. Haugsrud, R. (2007) Defects and transport properties in Ln_6WO_{12} (Ln = La, Nd, Gd, Er). *Solid State Ionics.*, **178**, 555–560.
76. Norby, T. and Christiansen, N. (1995) Proton conduction in Ca-substituted and Sr-substituted $LaPO_4$. *Solid State Ionics.*, **77**, 240–243.
77. Lefebvre-Joud, F., Gauthier, G., and Mougin, J. (2009) Current status of proton-conducting solid oxide fuel cells development. *J. Appl. Electrochem.*, **39**, 535–543.
78. Hirabayashi, D., Tomita, A., Brito, M.E., Hibino, T., Harada, U., Nagao, M., and Sano, M. (2004) Solid oxide fuel cells operating without using an anode material. *Solid State Ionics*, **168**, 23–29.
79. Hibino, T., Hashimoto, A., Inoue, T., Tokuno, J., Yoshida, S., and Sano, M. (2000) A low-operating-temperature solid oxide fuel cell in hydrocarbon-air mixtures. *Science*, **288**, 2031–2033.
80. Asano, K., Hibino, T., and Iwahara, H. (1995) A novel solid oxide fuel cell system using the partial oxidation of methane. *J. Electrochem. Soc.*, **142**, 3241–3245.
81. Zhang, X.L., Jin, M.F., and Sheng, J.M. (2010) Layered $GdBa_{0.5}Sr_{0.5}Co_2O_{5+\delta}$ delta as a cathode for proton-conducting solid oxide fuel cells with stable $BaCe_{0.5}Zr_{0.3}Y_{0.16}Zn_{0.04}O_{3-\delta}$ delta electrolyte. *J. Alloy. Compd.*, **496**, 241–243.
82. Lin, B., Dong, Y.C., Wang, S.L., Fang, D.R., Ding, H.P., Zhang, X.Z., Liu, X.Q., and Meng, G.Y. (2009) Stable, easily sintered $BaCe_{0.5}Zr_{0.3}Y_{0.16}Zn_{0.04}O_{3-\delta}$ electrolyte-based proton-conducting solid oxide fuel cells by gel-casting and suspension spray. *J. Alloy. Compd.*, **478**, 590–593.
83. Zhang, S.Q., Bi, L., Zhang, L., Tao, Z.T., Sun, W.P., Wang, H.Q., and Liu, W. (2009) Stable $BaCe_{0.5}Zr_{0.3}Y_{0.16}Zn_{0.04}O_{3-\delta}$ thin membrane prepared by in situ tape casting for proton-conducting solid oxide fuel cells. *J. Power Sources*, **188**, 343–346.
84. Lin, B., Hu, M.J., Ma, J.J., Jiang, Y.Z., Tao, S.W., and Meng, G.Y. (2008) Stable, easily sintered $BaCe_{0.5}Zr_{0.3}Y_{0.16}Zn_{0.04}O_{3-\delta}$ electrolyte-based protonic ceramic membrane fuel cells with $Ba_{0.5}Sr_{0.5}Zn_{0.2}Fe_{0.8}O_{3-\delta}$ perovskite cathode. *J. Power Sources*, **183**, 479–484.
85. Ding, H.P., Xie, Y.Y., and Xue, X.J. (2011) Electrochemical performance of $BaZr_{0.1}Ce_{0.7}Y_{0.1}Yb_{0.1}O_{3-\delta}$ electrolyte based proton-conducting SOFC solid oxide fuel cell with layered perovskite $PrBaCo_2O_{5+\delta}$ cathode. *J. Power Sources*, **196**, 2602–2607.

86. Ito, N., Iijima, M., Kimura, K., and Iguchi, S. (2005) New intermediate temperature fuel cell with ultra-thin proton conductor electrolyte. *J. Power Sources*, **152**, 200–203.

87. Kang, S., Heo, P., Lee, Y.H., Ha, J., Chang, I., and Cha, S.W. (2011) Low intermediate temperature ceramic fuel cell with Y-doped $BaZrO_3$ electrolyte and thin film Pd anode on porous substrate. *Electrochem. Commun.*, **13**, 374–377.

88. Shim, J.H., Park, J.S., An, J., Gur, T.M., Kang, S., and Prinz, F.B. (2009) Intermediate-temperature ceramic fuel cells with thin film yttrium-doped barium zirconate electrolytes. *Chem. Mater.*, **21**, 3290–3296.

89. Bi, L., Fabbri, E., Sun, Z.Q., and Traversa, E.S. (2011) Interactive anodic powders improve densification and electrochemical properties of $BaZr_{0.8}Y_{0.2}O_{3-\delta}$ electrolyte films for anode-supported solid oxide fuel cells. *Energy. Environ. Sci.*, **4**, 1352–1357.

90. Lin, B., Wang, S.L., Liu, X.Q., and Meng, G.Y. (2009) Stable proton-conducting Ca-doped $LaNbO_4$ thin electrolyte-based protonic ceramic membrane fuel cells by in situ screen printing. *J. Alloy. Compd.*, **478**, 355–357.

91. Maffei, N., Pelletier, L., Charland, J., and McFarlan, A. (2005) An intermediate temperature direct ammonia fuel cell using a proton conducting electrolyte. *J. Power Sources*, **140**, 264–267.

92. Zhang, L.M. and Yang, W.S. (2008) Direct ammonia solid oxide fuel cell based on thin proton-conducting electrolyte. *J. Power Sources*, **179**, 92–95.

93. Ni, M., Leung, D.Y.C., and Leung, M.K.H. (2008) Electrochemical modeling of ammonia-fed solid oxide fuel cells based on proton conducting electrolyte. *J. Power Sources*, **183**, 687–692.

94. Maffei, N., Pelletier, L., Charland, J.P., and McFarlan, A. (2007) A direct ammonia fuel cell using barium cerate proton conducting electrolyte doped with gadolinium and praseodymium. *Fuel Cells*, **7**, 323–328.

95. Xie, K., Yan, R.Q., Chen, X.R., Wang, S.L., Jiang, Y.Z., Liu, X.Q., and Meng, G.Y. (2009) A stable and easily sintering $BaCeO_3$-based proton-conductive electrolyte. *J. Alloy. Compd.*, **473**, 323–329.

96. Xie, K., Ma, Q.L., Lin, B., Jiang, Y.Z., Gao, J.F., Liu, X.Q., and Meng, G.Y. (2007) An ammonia fuelled SOFC with a $BaCe_{0.9}Nd_{0.1}O_{3-\delta}$ thin electrolyte prepared with a suspension spray. *J. Power Sources*, **170**, 38–41.

97. Coors, W.G. (2003) Protonic ceramic fuel cells for high-efficiency operation with methane. *J. Power Sources*, **118**, 150–156.

98. Ni, M., Leung, M.K.H., and Leung, D.Y.C. (2009) Ammonia-fed solid oxide fuel cells for power generation-A review. *Int. J. Energy Res.*, **33**, 943–59.

99. Lan, R., Irvine, J.T.S., and Tao, S.W. (2012) Ammonia and related chemicals as potential indirect hydrogen storage materials. *Int. J. Hydrogen Energy*, **37**, 1482–1494.

100. Meng, G.Y., Jiang, C.R., Ma, J.J., Ma, Q.L., and Liu, X.Q. (2007) Comparative study on the performance of a SDC-based SOFC fueled by ammonia and hydrogen. *J. Power Sources*, **173**, 189–193.

101. Pujare, N.U., Semkow, K.W., and Sammells, A.F. (1987) A direct H_2s/Air solid oxide fuel-cell. *J. Electrochem. Soc.*, **134**, 2639–2640.

102. Chen, H., Xu, Z.R., Peng, C., Shi, Z.C., Luo, J.L., Sanger, A., and Chuang, K.T. (2010) Proton conductive YSZ-phosphate composite electrolyte for H(2)S SOFC. *Ceram. Int.*, **36**, 2163–2167.

103. Atkinson, A., Barnett, S., Gorte, R.J., Irvine, J.T.S., McEvoy, A.J., Mogensen, M., Singhal, S.C., and Vohs, J. (2004) Advanced anodes for high-temperature fuel cells. *Nat. Mater.*, **3**, 17–27.

104. Cowin, P.I., Petit, C.T.G., Lan, R., Irvine, J.T.S., and Tao, S.W. (2011) Recent progress in the development of anode materials for solid oxide fuel cells. *Adv. Energy. Mater.*, **1**, 314–332.

105. Jiang, S.P. (2008) Development of lanthanum strontium manganite perovskite cathode materials of solid oxide fuel cells: a review. *J. Mater. Sci.*, **43**, 6799–6833.

106. Tao, S.W., Wu, Q.Y., Peng, D.K., and Meng, G.Y. (2000) Electrode materials for intermediate temperature proton-conducting fuel cells. *J. Appl. Electrochem.*, **30**, 153–157.

107. Ni, M., Leung, D.Y.C., and Leung, M.K.H. (2008) Modeling of methane fed solid oxide fuel cells: comparison between proton conducting electrolyte and oxygen ion conducting electrolyte. *J. Power Sources*, **183**, 133–142.

5
Metallic Interconnect Materials of Solid Oxide Fuel Cells

Li Jian, Hua Bin, and Zhang Wenying

5.1
Introduction

A solid oxide fuel cell (SOFC) is an electrochemical device that directly converts the chemical energy of fossil, biomass, or other hydrocarbon fuels into electricity without involving the processes of combustion and mechanical motion, which makes SOFC a promising power-generation technology with the advantages of high efficiency, low emission, and fuel flexibility. The interconnect, as shown in Figure 5.1 of an external manifolding design, is a key component of SOFC stacks, which separates fuel and oxidant gases, distributes reactive gases to electrodes, and electrically connects adjacent cells. Furthermore, in some stack designs, the interconnect is also a structural element employed to maintain the mechanical integrity of the stack and to offer mechanical contact surfaces for gas path sealing [1–3]. Recent progress on thin film electrolyte fabrication and high-performance cell materials has made it possible to reduce the operating temperature of SOFCs from the conventional 1000 °C to an intermediate temperature range 600–800 °C without compromising on cell power density and durability in a planar design. The lowered operating temperature of SOFCs allows the use of metallic alloys in interconnects. In recent years, more and more attention has been paid to develop proper metallic materials for the interconnect application [4–9].

As the metallic interconnects of intermediate-temperature solid oxide fuel cells (IT-SOFCs) are operated at temperatures in the range 600–800 °C, their material faces challenges arising from the high working temperature and stringent dual atmospheres (oxidizing on the cathode side and reducing on the anode side), which necessitates their meeting the following requirements [4, 5, 10–13].

1) **Close thermal expansion match to the cell components.** Since the thermal expansion coefficient (TEC) of the cell components (electrolyte and electrodes) varies in the range $10–13 \times 10^{-6}\ K^{-1}$, it is preferred that the TEC of interconnect materials matches with this range at operating temperatures to avoid failures caused by the generation of excessive thermal stresses during thermal cycles.

Materials for High-Temperature Fuel Cells, First Edition. Edited by San Ping Jiang and Yushan Yan.

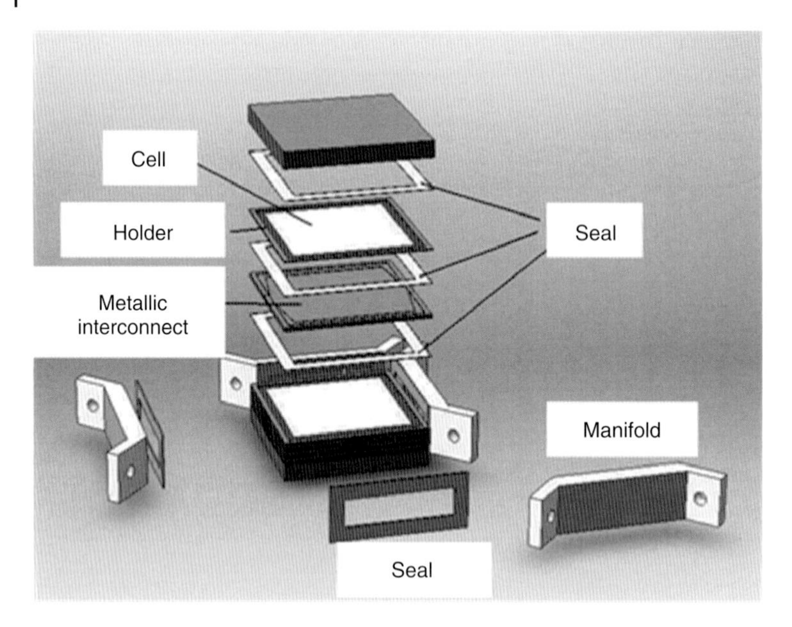

Figure 5.1 The schematic design of planar solid oxide fuel cells.

2) **Sufficient oxidation resistance and electrical conductivity**. The interconnect is exposed simultaneously to a cathode atmosphere (usually air; the oxygen partial pressure pO_2 is 0.21 atm) on one side and an anode atmosphere (fuel; typical pO_2 is between 10^{-18} and 10^{-8} atm) on the other, and oxidation occurs inevitably to form an oxide scale on the surface of the interconnect, increasing its ohmic resistance. To assure a robust stack performance, $0.1\,\Omega\,cm^2$ is the accepted maximum of the area-specific resistance (ASR) for the interconnect. Ionic conductivity of interconnect materials should be negligibly low.
3) **Tolerable chemical compatibility**. The interconnect is usually in contact with the anode, cathode, and seal components; chemical reaction or interdiffusion between the elements in the interconnect and adjacent components may degrade the performance of the electrodes by forming high-resistance phases and compromise the function of the seal by disintegrating the sealing surfaces. Hence, chemical compatibility of interconnect materials with adjacent components under both reducing and oxidizing atmospheres during SOFC operation is important for a durable stack performance.
4) **Adequate gastightness**. The interconnect connects the anode and cathode of adjacent cells; therefore, the material is required to be impermeable to the reactant gases to avoid their direct mixing during cell operation. It is known from the Nernst equation that a slight variation in partial pressure of oxygen or hydrogen, resulting from interconnect leakage, will lead to a noticeable change in the open circuit voltage and, in turn, the cell performance. On the other hand, direct combustion of the fuel and oxidant gases will significantly increase the local temperature, damaging the interconnect and the cells.

5) **High thermal conductivity**. As a structural component in the stack, the interconnect is expected to be able to dissipate excessive heat generated by the exothermic electrode reactions to maintain a stable and uniform temperature distribution in the stack; furthermore, the interconnect is employed as a "heat exchanger" to balance the heat generated in the cathode and the anode of a planar cell, especially in the case of on-cell fuel re-forming, where an endothermic reaction occurs. Conventionally, a thermal conductivity greater than $20\,W\,m^{-1}\,K^{-1}$ for interconnect materials is desired.
6) **Excellent sulfidation and carburization resistances**. For metallic interconnect materials, sulfidation or carburization may occur during the operation of an SOFC stack fueled by sulfur- or carbon-bearing fuels, which increases the ASR of metallic interconnects by forming sulfides or carbides and disintegrates the contact between the interconnect and the electrodes by spallation of the formed scales and metal dusting.
7) **Acceptable elevated-temperature mechanical property**. This requirement is of special relevance to planar SOFCs where the interconnect serves as a structural support. Basically, metallic interconnects are not subjected to significantly high mechanical loads during operation; therefore, the minimum requirement of the elevated-temperature mechanical property is sufficient high temperature strength and creep resistance for maintaining the structural rigidity of the interconnect in long-term operations.
8) **Low material and fabrication costs**. The cost of the interconnect covers a considerable portion of that of SOFC stacks, which suggests that cost-effective materials and interconnect fabrication methods are of critical importance for cost reduction of SOFC stacks, and in turn for commercialization of SOFC technology.

In order to meet these requirements, oxidation-resistant alloys are considered the candidate materials. Such alloys usually contain different amounts of Al, Si, or Cr as the alloying elements to form a protective oxide layer by the preferential oxidation of Al to Al_2O_3, Si to SiO_2, or Cr to Cr_2O_3 on exposure to SOFC environments at operating temperatures. Al_2O_3- and SiO_2-forming alloys are less interesting for SOFC interconnect applications because of the low electrical conductivity of the thermally grown oxide scales, which leads to an unacceptably high ASR, even though they can provide superior oxidation resistance. Since an oxide scale is inevitably formed on the surface of the alloys under IT-SOFC operation conditions, the ideal solution is to develop metallic alloys that have sufficient oxidation resistance and produce less than $0.1\,\Omega\,cm^2$ ASR over the expected lifetime (40 000 h) of IT-SOFCs. This requires the alloys to form a relatively dense oxide scale that is well adhered to the substrate, chemically stable, and homogenous with sufficiently high electronic conductivity and significantly low growth rate, while they meet the rest of the requirements mentioned earlier. Therefore, Cr_2O_3-forming alloys, especially the ferritic Fe–Cr and Ni–Cr alloys, have attracted much attention as promising metallic interconnect materials because of their closely matched TEC to that of the cell components and the thermally grown Cr_2O_3, with a combination of relatively low growth rate and acceptably high conductivity [4, 10, 14].

In this chapter, recent studies on metallic interconnect materials for IT-SOFCs are summarized in terms of oxidation behavior, electrical conductivity, surface modification, and new alloy development as a complement to several review articles on metallic SOFC interconnects published in the past decade [4–7, 15].

5.2
Oxidation Behaviors of Candidate Alloys

Three types of Cr_2O_3-forming heat-resistant alloys have been under investigation as the metallic interconnect materials: Cr-, Ni-, and Fe-based alloys. Cr in the alloy is preferentially oxidized in the atmosphere of IT-SOFCs to form a dense and well-adherent Cr_2O_3 layer, which provides oxidation resistance by significantly slowing down the outward diffusion of metal cations from the alloy and inward diffusion of oxygen from the reactant gases. Cr-based alloys, for example, the oxide-dispersion-strengthened alloy $Cr_5Fe_1Y_2O_3$, have been the choice for metallic interconnects of high-temperature (~900 °C) SOFCs. The alloy has a typical TEC of $11.8 \times 10^{-6}\ K^{-1}$ over 20–1000 °C, well matching that of the cell components, and demonstrates excellent oxidation resistance and high-temperature mechanical strength. However, the problem of long-term stability is yet to be solved because the oxidation resistance is inadequate for operation above 800 °C, especially in the presence of water and/or CO. The evaporation of Cr from the alloy followed by deposition on the cathode can result in rapid degradation of the electrochemical performance of cells. In addition, it is difficult and costly to fabricate and the data on the oxidation resistance at lower temperatures are not yet available [16–20]. Therefore, Cr-based alloys are not considered for application as metallic interconnects in IT-SOFCs.

Ni-based alloys are well known for their excellent oxidation/corrosion resistance, so they are considered as potential candidates for metallic interconnect applications. Owing to the nature of their austenitic microstructure, Ni-based alloys usually have a relatively high TEC, typically in the range $14.0–19.0 \times 10^{-6}\ K^{-1}$ from room temperature to 800 °C; on the other hand, they are much more expensive than the ferritic type Fe–Cr alloys [8, 21–23]. Hence, Ni-based alloys are not favored as interconnect materials, and less attention has been paid to some special Ni-based alloys. Haynes 230, a Ni-based alloy with a TEC of $15.2 \times 10^{-6}\ K^{-1}$ between 25 and 800 °C, has been studied as a potential candidate for metallic interconnect materials for IT-SOFCs [24–32].

Fe-based alloys, especially ferritic Fe–Cr stainless steels, exhibit a TEC typically in the range $12.0–13.0 \times 10^{-6}\ K^{-1}$ from room temperature to 800 °C, which is better matched to that of the cell components. This is one of the primary reasons for Fe–Cr ferritic stainless steels to be applied as the interconnects in SOFC stacks. In addition, Fe–Cr ferritic stainless steels usually possess relatively high oxidation resistance and robust mechanical properties, and are of low cost. Thus, in recent years, several kinds of Fe–Cr ferritic alloys such as SUS 430 and Crofer 22 APU, with TEC around $12.0 \times 10^{-6}\ K^{-1}$ and Cr content in the range 16–22 wt%, have been under extensive study as candidate materials for the metallic interconnects, although their oxidation resistance under IT-SOFC environments is lower than that of the Ni-based alloys [33–51].

An oxide scale is inevitably formed on the surface of the alloys under IT-SOFC operating conditions. ASR of less than $0.1\,\Omega\,cm^2$ in 40 000 h of expected stack life requires the metallic interconnects to possess significantly high electrical conductivity and adequate oxidation resistance to prevent the formation of overgrown thick oxide scales and subsequent spallation due to thermal stresses induced by thermal expansion mismatch between the scale and the substrate alloy [52–54]. The oxidation of the alloys considered for metallic interconnect applications is usually a diffusion-controlled process, which obeys the parabolic law [55–58] of Wagner's theory, expressed as

$$\Delta W^2 = K_p t \tag{5.1}$$

or

$$X^2 = K_p t \tag{5.2}$$

where ΔW is the specific area weight gain of the sample, X is the thickness of the thermally grown oxide scale, t is the oxidation time, and K_p ($g^2\,cm^{-4}\,s^{-1}$ or $\mu m^2\,s^{-1}$) is the parabolic rate constant or oxidation rate. For a description of oxidation resistance, oxidation kinetics according to Eq. (5.1) or Eq. (5.2) is usually used by plotting the specific area weight gain or scale thickness against the oxidation time at a certain oxidation temperature. The slope of the curve represents the parabolic rate constant K_p, which is primarily dependent on alloying composition and is also affected by thermal history and surface conditions, and so on. In the case of oxidation of Ni–Cr and Fe–Cr alloys [37, 40], where the oxidation is controlled by outward diffusion of cations through the well-developed Cr_2O_3 layer, the oxidation kinetics is basically in agreement with the diffusion-controlled parabolic growth relationship; a typical oxidation kinetics curve is shown in Figure 5.2 [59], either with a constant or a changed oxidation rate. On the basis of the oxidation kinetics, long-term oxidation behavior and the possibility of the use of the alloy as an interconnect material can be predicted by extrapolation. The transition in oxidation rate with oxidation time can be attributed to the formation of a specific oxide in the scale [28, 37].

5.2.1 Oxidation in Cathode Atmosphere

When exposed to the cathode atmosphere, typically air, of IT-SOFCs in the intermediate-temperature range, Cr in the alloy is preferentially oxidized initially to form a dense protective layer of Cr_2O_3 on the alloy surface. As mentioned, the scale growth is dominated by the outward diffusion of cations across the Cr_2O_3 layer, following the parabolic law. Yang *et al.* [8] and Fergus [5] summarized the parabolic rate constants of Fe- and Ni-based alloys at different temperatures. The Ni–Cr alloys usually demonstrate a lower growth rate, that is, higher oxidation resistance, than the Fe–Cr ferritic stainless steels. Also, alloys with higher Cr content are more

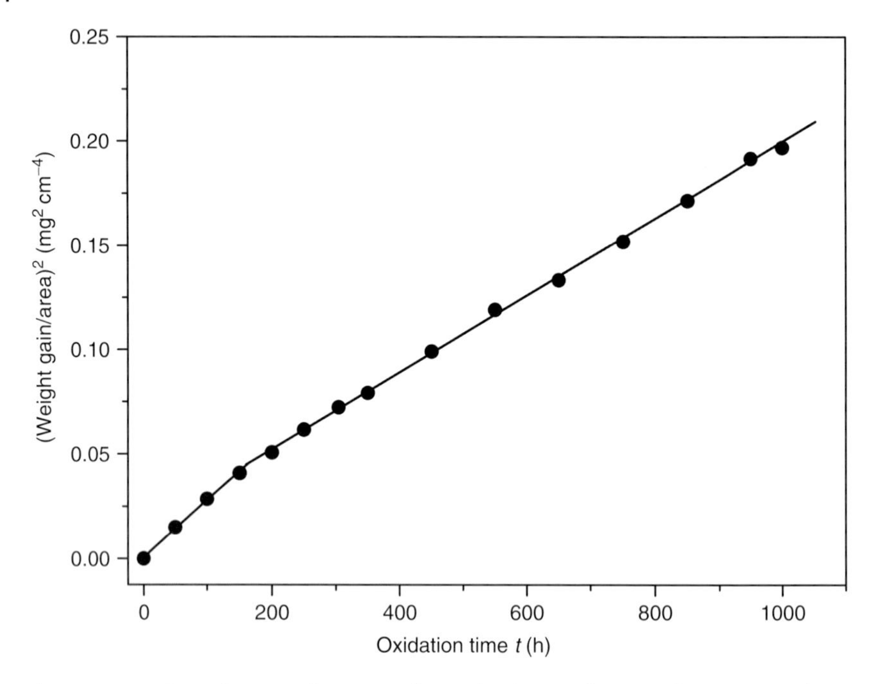

Figure 5.2 Oxidation kinetics of a Fe–Cr alloy oxidized at 750 °C in air for up to 1000 h. (Source: Reproduced with permission from Ref. [59].)

oxidation resistant than those with lower Cr content, while the percentage of the other alloying elements remains the same. The scale growth rate of Fe–Cr based alloys is in the range 10^{-14}–10^{-12} g^2 cm^{-4} s^{-1} at 800 °C in air.

The growth rate, microstructure, and composition of the thermally grown oxide scale are dependent on the alloy composition. The addition of a few tenths of a percentage of Mn to the Fe–Cr or Ni–Cr alloys leads to the formation of cubic Mn–Cr spinel; the typical morphology of the spinel microstructure is shown in Figure 5.3. As the outward diffusion rate of Mn ions in Cr_2O_3 is faster than that of Cr in Cr_2O_3 [60–62], the thermally grown oxide scale formed on Fe–Cr–Mn or Ni–Cr–Mn alloys exhibits a duplex structure with Mn–Cr spinel on top of the dense Cr_2O_3 sublayer, as typically shown in Figure 5.4. Such a duplex oxide scale was observed frequently in Fe–Cr–Mn and Ni–Cr–Mn alloys, such as SUS 430, Crofer 22 APU, and Haynes 230 [28, 32, 37, 41]. Fe implantation of carbon on the alloy surface can increase Mn diffusion and promote the formation of Mn–Cr spinel oxide [63]. Generally, Cr_2O_3 and Mn–Cr spinel are the main phases in the oxide scale of the candidate alloys for metallic interconnects exposed to IT-SOFC operating environments. Because the electrical conductivity of Mn–Cr spinel is orders of magnitude higher than that of Cr_2O_3, the formation of such a Mn–Cr spinel layer can result in a higher electrical conductivity of the scale than that obtained with only Cr_2O_3 [6, 38, 64–66]. In addition, the thickness of the alloy

Figure 5.3 The surface morphology of Crofer 22 APU oxidized at 750 °C in air for 300 h.

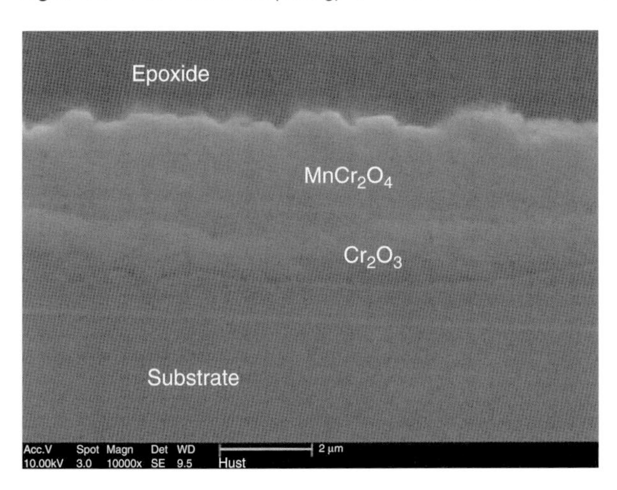

Figure 5.4 The duplex scale of SUS 430 alloy oxidized at 750 °C in air for 500 h. (Source: Reproduced with permission from Ref. [37].)

affects the oxidation rate; the thinner the alloy, the lower the oxidation resistance under the same oxidation condition. In thin sheet alloys, depletion of Cr occurs underneath the chromium oxide, which becomes unstable; consequently, Fe is oxidized. Such accelerated oxidation behavior of thin sheet alloys can be improved by reducing Mn, increasing Cr, and adding W [67].

In the case of oxidation with single parabolic rate constant K_p, the growth of the oxide scale can be well predicted according to the parabolic law, and the degree of oxidation of the alloys in 40 000 h can be evaluated. However, recent studies indicate that the Cr_2O_3-forming Fe (or Ni)–Cr–Mn alloys do not always follow the constant rate kinetics and that multistage oxidation is observed, each of which follows Wagner's parabolic law with different rate constants [28, 37]. Figure 5.5 [28] shows the oxidation kinetics of Haynes 230 alloy oxidized in air at 800 °C for up to 1000 h; three straight lines are obtained and the slopes of the first and the third stages have almost the same value, which is lower than that of the slope of the second stage. This phenomenon is a result of the faster diffusion of Mn ions in Cr_2O_3 and Mn deficiency and subsequent recovery in the substrate adjacent to the alloy–scale interface [28, 37]. As reported by Caplan *et al.* [68] in Fe–Cr–Mn alloys, the Mn rapidly diffuses from the alloy to the surface to form a Mn-rich oxide scale, causing severe localized Mn depletion until the Mn flux is replenished. The underlying alloy adjacent to the Mn-rich oxide scale has a transient composition that approximates to that of the Fe–Cr alloy, resulting in Cr_2O_3 formation. Thus, the first slow oxidation stage is related to the growth of the Cr_2O_3 layer, controlled by sluggish Cr ion diffusion through the dense Cr_2O_3 scale. The faster second stage is a result of rapid diffusion of Mn ions passing the established Cr_2O_3 scale to form Mn–Cr spinel on top of the Cr_2O_3 layer. Owing to the limitation of Mn content in the Haynes 230 alloy, the fast growth of Mn–Cr spinel may be interrupted as the requirement for a continuous supply of Mn flux is met. Such Mn depletion may cease the fast growth of Mn–Cr spinel, and then the oxidation kinetics reaches the third stage, controlled by Cr diffusion as in the first stage. Marasco and Young [69]

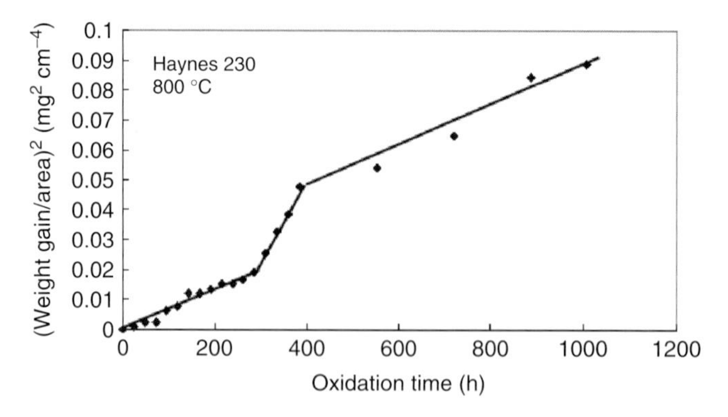

Figure 5.5 The oxidation kinetics of Haynes 230 alloy oxidized at 800 °C in air for up to 1000 h. (Source: Reproduced with permission from Ref. [28].)

and Cox *et al.* [70] proposed a similar mechanism to explain multistage oxidation in the formation of M_3O_4-type (M = Fe, Cr, or Mn) and M_2O_3-type oxides. The formation of M_3O_4 is preceded by the growth of M_2O_3, and the growth rate of M_3O_4 is considered an order of magnitude greater than that of M_2O_3. As required, the metallic interconnects are expected to function for 40 000 h with an ASR lower than 0.1 $\Omega\,cm^2$; therefore, it is important for the oxidation evaluation of candidate alloys to last longer to fully understand their oxidation kinetics.

As known, grain boundaries are a rapid diffusion path of Fe migration, and the diffusion of cations along the grain boundary is expected to affect the growth of oxide scales [61, 71–75]. As shown in Figure 5.6a, a relatively fast diffusion of elements is observed at the grain boundaries of Crofer 22 APU alloy oxidized at 750 °C in air for 1000 h, forming larger sized oxides that decorate the grain boundaries and lower the oxidation resistance and contact electrical conduction. By engineering the grain boundary, such as precipitation of the Laves phase at grain boundaries [76, 77], the accelerated grain boundary oxidation can be effectively depressed and a uniform oxide scale surface can be obtained, as shown in Figure 5.6b. In addition, the Laves phase can preferentially consume Si presented in the alloy substrate to avoid the formation of electrically resistive Si-rich oxide subscales and improve electrical performance [78]. However, the Laves phase does not completely block grain boundary diffusion and remove Si from the matrix. Furthermore, addition of rare-earth elements/reactive elements (REs) to the alloy can improve its oxidation resistance by forming RE oxides at grain boundaries [79–83]. In general, controlling elemental diffusion along grain boundaries is an effective way to form a thin and compact oxide scale that is needed for the metallic interconnect application.

5.2.2 Oxidation in Anode Atmosphere

The oxidation behavior of metallic interconnects in the oxidant atmosphere (air) has been widely investigated; however, it is less understood in the anode fuel atmosphere of IT-SOFCs, such as wet H_2 or re-formed gases typically with an oxygen partial pressure of 10^{-18} to 10^{-8} atm. Compared to the cathode side, the anode environment is more complex, particularly when a hydrocarbon fuel is used. The presence of water, H_2, and carbon makes metallic interconnects susceptible to various forms of corrosion. Even though the oxygen partial pressure on the anode side is significantly lower than that on the cathode side, oxides such as Cr_2O_3 and Mn–Cr spinel are still thermodynamically stable. There are some values of the scale growth rate in moist hydrogen reported in the literature; however, they are orders of magnitude different. Brylewski *et al.* [84] reported a parabolic rate constant, K_p, of $3.8 \times 10^{-6}\,\mu m^2\,s^{-1}$ for SUS 430 alloy oxidized in wet hydrogen at 800 °C, similar to that oxidized in air (and $3.0 \times 10^{-6}\,\mu m^2\,s^{-1}$). But for alloy 446 and Fe22CrMn, the weight gain after oxidation at 800 °C in H_2/H_2O for 250 h is only about half of that in air under the same conditions [85, 86]. The phenomenon of faster oxidation in H_2/H_2O than in air was also observed in Ni-based alloys at temperatures between 700 and 1100 °C [24, 25]. It is generally considered that the

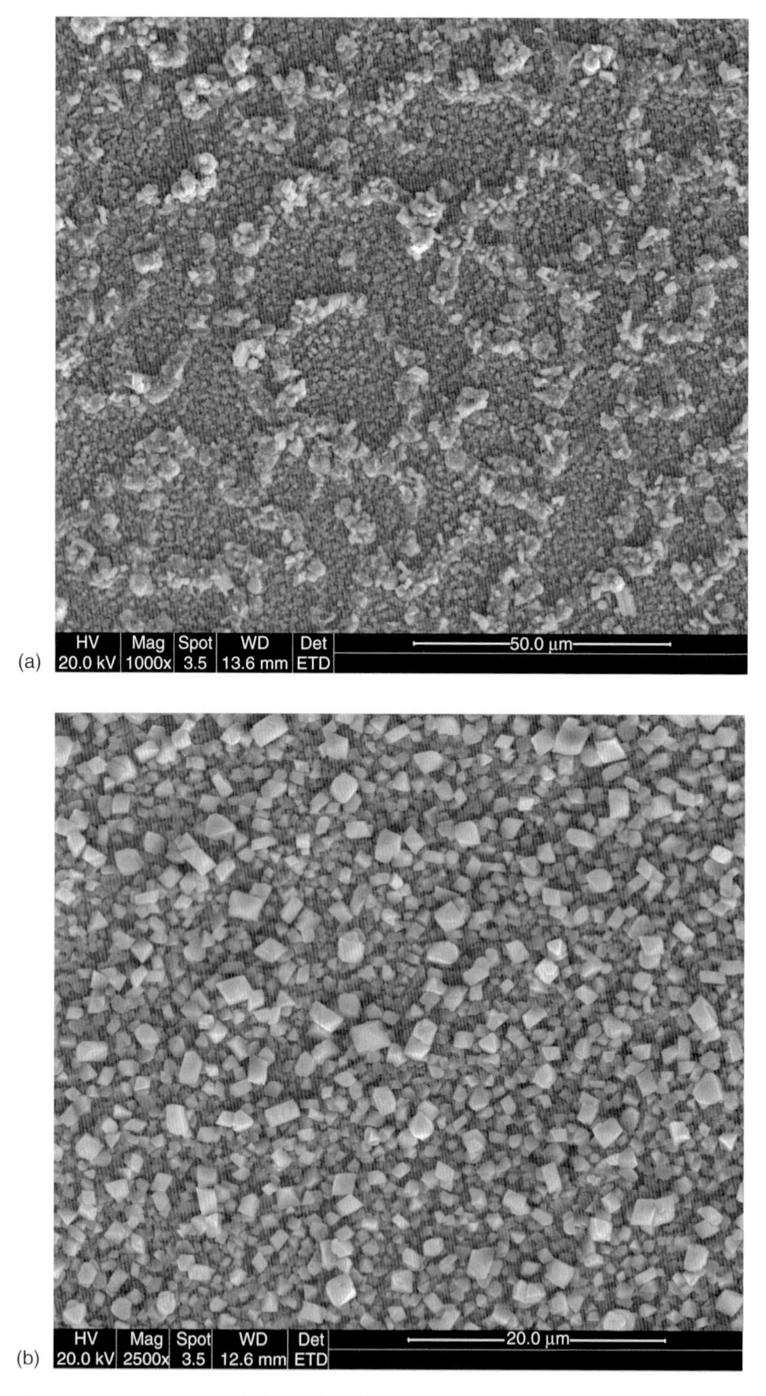

Figure 5.6 Surface morphology of Crofer 22 APU (a) and Crofer 22 H (b) oxidized at 750 °C in air for up to 1000 h.

anode atmosphere is more corrosive than the cathode atmosphere in IT-SOFCs. The reason for this remains unclear so far.

XRD (X-ray diffraction) analysis indicates that the thermally grown oxide scale of ferritic Fe–Cr alloys formed in wet hydrogen at intermediate temperatures is usually composed of the same major phases as those found in the scale formed in air. Cr_2O_3 and Mn–Cr spinel in a duplex scale structure are still the major oxidation products. However, the surface morphology and property of the scale can be significantly different. It has been found [33, 47, 52, 84, 87] that the top spinel phase in the scale is generally needle shaped in wet hydrogen, as shown in Figure 5.7, whereas that formed in air is prism shaped (Figure 5.3). In the case of the ZMG 232 alloy, the oxide scale formed in H_2/H_2O atmosphere consists of the top spinel phase and the bottom Cr_2O_3 phase, with a thin layer of SiO_2 adhered to the substrate alloy, and high concentration of Al is observed at the oxide scale–alloy interface [33, 52, 87]. Reduction of Si and Al content in Fe–Cr alloys results in thinner oxide scale and insulating layer of SiO_2 and/or Al_2O_3, improving the electrical conductivity of the scale [88]. In addition, in the H_2/H_2O atmosphere, MnO and Fe-rich oxides are frequently observed in Fe–Cr (Crofer 22 APU) and Ni–Cr (Haynes 230) alloys [89] and grain boundary oxidation is enhanced [76, 90]. The formation of Mn- and/or Fe-rich oxides in high-water-vapor environment is not appreciated, as it results in the absence of the protective Cr_2O_3 layer and scale spallation under the stresses generated. Alloys with higher Cr content are less sensitive to the catastrophic oxidation.

Compared with Fe-based alloys, Ni-based alloys, even those with a relatively low Cr percentage, usually exhibit higher resistance to oxidation and scale spallation [91–96]. Besides the dominant phases of Mn–Cr spinel and Cr_2O_3, there is another

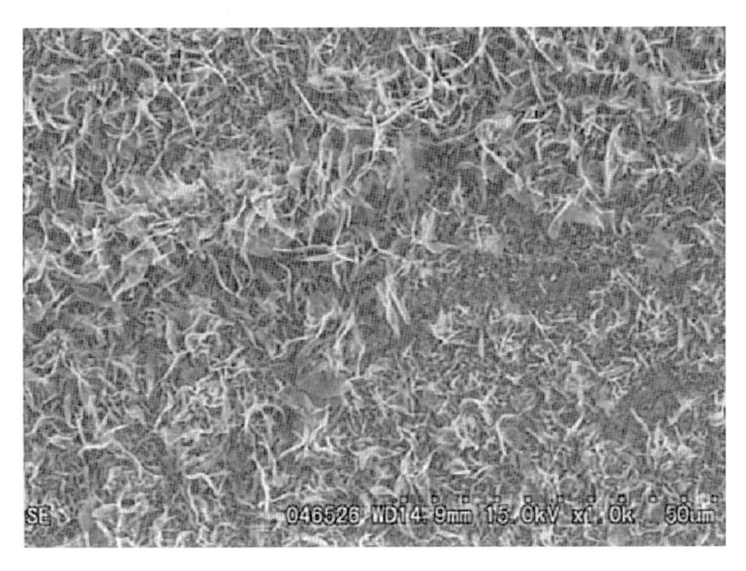

Figure 5.7 SEM image of the ZMG 232 surface annealed at 1023 K for 1000 h in humidified 4% H_2/96% N_2 atmosphere. (Source: Reproduced with permission from Ref. [47].)

microstructure feature in the oxide scale of Ni–Cr alloys oxidized in the anode environment of IT-SOFCs, which has been paid less attention. As stated earlier, the oxygen partial pressure in the anode atmosphere is very low. In the H_2 atmosphere wet by a bubbler at 65 °C, the oxygen partial pressure at 750 °C is 2.5×10^{-21} atm [27], which is much lower than that required for Ni oxidation. Thus, during the oxidation of Haynes 230 alloy at 750 °C in humidified H_2, Ni remains metallic after 1000 h of exposure to the environment, while Cr and Mn are selectively oxidized. Once Cr_2O_3 is nucleated on the original alloy surface, it grows by outward diffusion of Cr ions. The metallic Ni is expelled and enriched at the alloy–Cr_2O_3 interface. Owing to the volume expansion associated with Cr oxidation to Cr_2O_3, the expelled Ni is under compressive stress and subsequently diffuses toward the free surface to form metallic Ni nodules, as shown in Figure 5.8 [26, 27]. It is expected that a complete layer of Ni will form with further oxidation, as seen in the case of superalloys oxidized in the anode environment of molten carbonate fuel cells [97].

The distinctiveness of the anode atmosphere in IT-SOFCs is the low oxygen partial pressure and the presence of water vapor. It has been noted that water vapor enhances oxidation of the alloys, but it can improve the adherence of the oxide scale, especially during thermal cycles [93, 98, 99]. The mechanism for the effect of water vapor on oxidation of alloys is not yet clear. One possibility is the incorporation of hydrogen in oxides, affecting their defect structure and hence changing the diffusion behavior of metal and oxygen ions [100]. In H_2/H_2O gas atmospheres, the following reaction may happen:

$$H_2O\,(g) \rightarrow H^{\bullet} + OH' \tag{5.3}$$

Since the radius of the hydroxide ion ($r = 0.10$ nm) is smaller than that of the oxide ion ($r = 0.14$ nm), the dissociated hydroxide ions may inwardly diffuse with a higher mobility, crossing the formed oxide layer to reach the oxide–metal substrate interface through oxygen vacancies, which are the predominant point defects in the dense Cr_2O_3 grown in the anode atmosphere with a low oxygen activity [101, 102]; that is,

$$OH' + V_O^{\bullet\bullet} \rightarrow (OH)_O^{\bullet} \tag{5.4}$$

Equivalently, the protons may bond to the oxygen ions in the oxide, forming hydroxide species.

$$H^{\bullet} + O_O^{\times} \rightarrow (OH)_O^{\bullet} \tag{5.5}$$

As a result, inward diffusion of oxygen is promoted via the formation of the hydroxide, accelerating the oxidation of the alloy and enhancing the adherence of the oxide scale. On the other hand, as hydroxide ions occupy the lattice positions of oxide ions, a positive charge is produced at each of these positions. Therefore, metal ion vacancies with negative charges are required to be generated for electrical neutrality, which may further increase the outward flux of the cations, again leading to an accelerated oxidation rate.

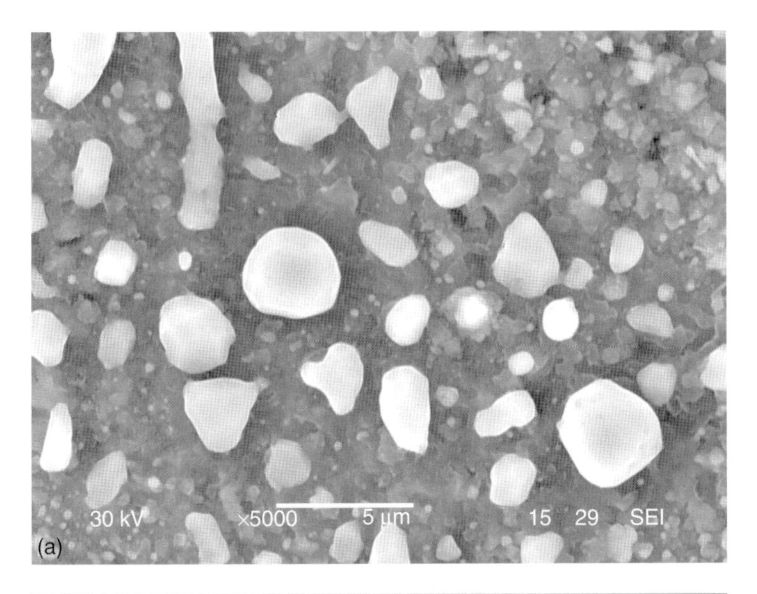

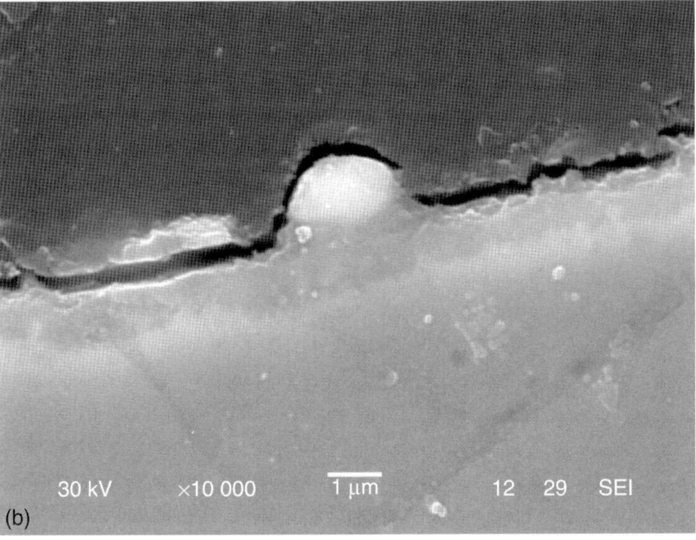

Figure 5.8 Ni nodules on the surface of Haynes 230 alloy: (a) surface morphology and (b) cross-sectional morphology. (Source: Reproduced with permission from Ref. [26].)

The ability to directly use hydrocarbon fuels is one of the advantages of SOFCs, and the stability of potential interconnect materials in carbon-containing atmospheres is important. Horita *et al.* [33, 35, 74, 87, 103–106] studied the oxidation of different ferritic stainless steels, such as ZMG 232, SUS 430, and Fe–Cr–W alloys, in CH_4/H_2O atmosphere. A thick oxide scale consisting of Mn–Fe–Cr spinel and Cr_2O_3 was formed, similar to that observed in H_2/H_2O atmosphere, with a higher oxidation rate than that in H_2/H_2O for ZMG 232, regardless of the content of CH_4 in the mixture, between 3.5 and 12 kPa [33, 87]. The Si concentration in Fe–Cr alloy can change the microstructures of oxide scales, elemental distribution, and oxide scale growth rates [106]. Li *et al.* [107] investigated the oxidation behavior of Crofer 22 and Haynes 230 alloys in a simulated coal syngas ($29.1CO-28.5H_2-11.8CO_2-27.6H_2O-2.1N_2-0.01CH_4$) at 800 °C for 500 h. $(Cr,Fe)_2O_3$, Mn–Cr spinel, and Fe_3O_4 are formed in the oxide scale of Crofer 22, and Cr_2O_3 is the primary phase in that of Haynes 230. The morphology of surface oxide is affected by oxygen partial pressure of the gas, whisker-like and angular oxides are observed on the surface of the scale of Crofer 22 and Haynes 230, respectively. In addition, exposure to the syngas makes the alloys to form porous oxide scale, which increases the electrical resistance and decreases the mechanical stability; however, no carbide formation and metal dusting are noticed in samples of Crofer 22 APU and Haynes 230 exposed to coal syngas at 800 °C for 100 h [108]. The oxide scale on alloy surfaces is found to be protective for alloys against dusting corrosion; the Cr_2O_3-type oxide is more effective than the spinel. High Cr content in alloys is beneficial in resisting metal dusting, and the metal dusting rate of Ni-based alloys is usually lower than that of Fe-based alloys because less of spinel is formed in the oxide scale of Ni-based alloys. High-humidity and low-pressure environments can reduce the susceptibility of the alloys to metal dusting corrosion [34, 109].

5.2.3 Oxidation in Dual Atmospheres

The interconnect is exposed simultaneously to both the cathode and anode atmospheres with an oxygen partial pressure varying from 0.21 atm in air to 10^{-18} to 10^{-8} atm in the fuel gas. The oxidation behavior of candidate metallic interconnect materials in dual atmospheres, that is, one side exposed to air and the other to the fuel, is of great importance. It has been reported that the oxidation behavior of the alloys in this dual atmosphere is different from that exposed to the single atmosphere [31, 53, 110–112]. Specifically, the composition and/or microstructure of the scale grown on the air side in the dual atmosphere is significantly different from that with both sides exposed to air, depending on the Cr content in the Fe–Cr alloys, while the oxidation behavior on the fuel side is comparable to that with both sides exposed to the hydrogen fuel. In the dual atmosphere, the oxide scales of the Fe–Cr alloys formed in the air side contains iron-rich spinel or Fe_2O_3 nodules, as shown in Figure 5.9 for SUS 430 alloy [53], enhancing the growth of the oxide scale. Such anomalous scale growth related to the formation of Fe oxides was also observed in dual-atmosphere oxidation (800 °C) in ZMG 232-M0 [113], SUS 441 [114], and CoMnO-coated SUS 441 [115] alloys.

This anomalous growth of oxide scale on the air side in dual atmospheres is considered to be the consequence of hydrogen transport across the alloy substrate from the fuel side to the air side and the presence of hydrogen at the oxide scale–metal interface and in the scale on the air side [53, 111, 114, 115]. Hydrogen or proton in the oxide scale tends to bond with oxygen to form hydroxide point

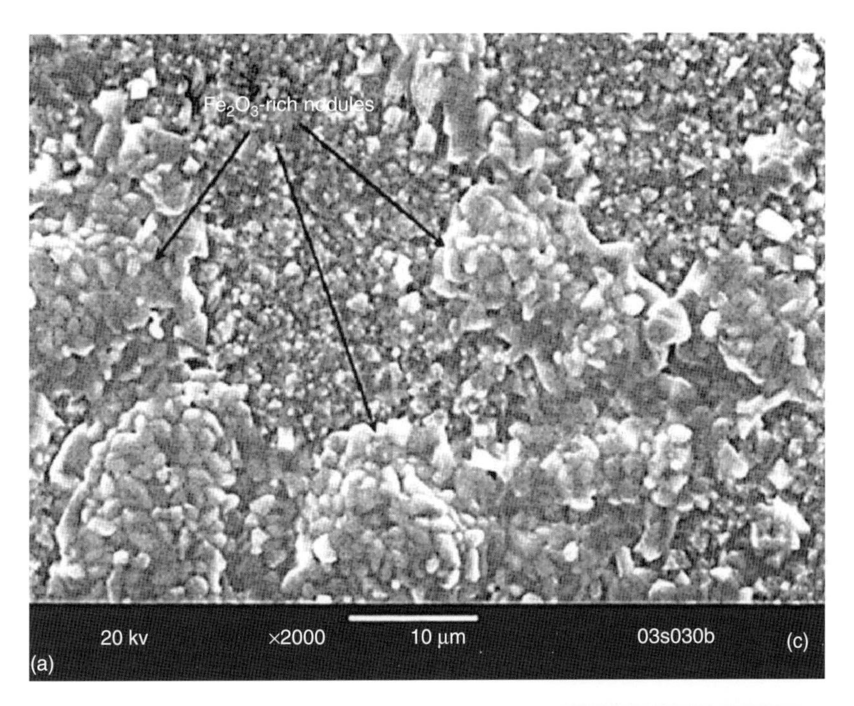

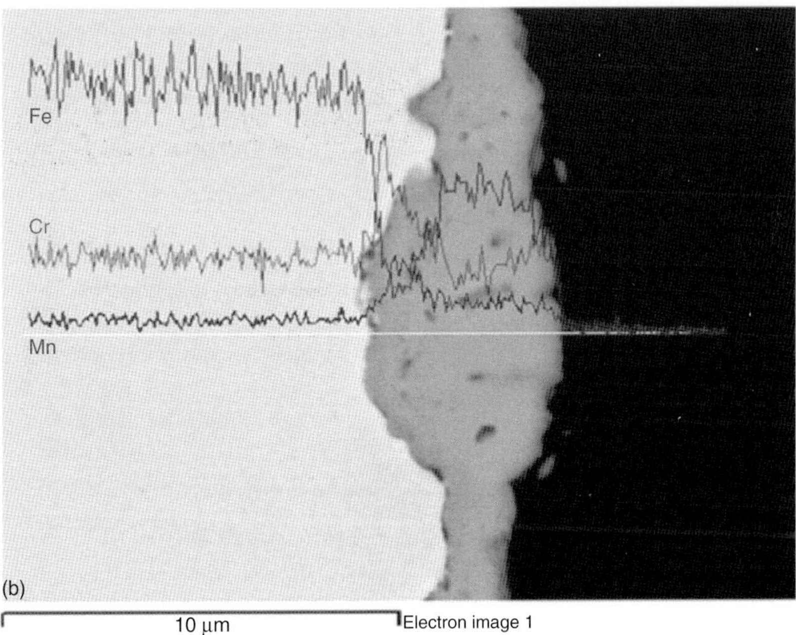

Figure 5.9 SEM observation (a) and EDS analysis (b) of the oxide scale formed on the air side of SUS 430 alloy exposed to dual atmosphere at 800 °C for 300 h. (Source: Reproduced with permission from Ref. [53].)

defects, with charge compensation provided by forming cation vacancies in the oxide. The increased content of cation vacancy inside the oxide scale may result in the enhanced diffusivity of Fe and formation of Fe-rich oxides in the scale. Hydrogen permeation tests on ferritic stainless steels [116] have supported the assumption that hydrogen can diffuse through the alloys, even though the permeation is drastically decreased by the formation of an oxide scale. The detailed mechanism behind the anomalous growth of oxide scale on the cathode side in the dual atmosphere needs to be further understood.

Both alloy composition and water vapor content affect the oxidation behavior of Fe–Cr alloys on the air side in the dual atmosphere. With increase in the Cr content in the alloy, the effect of the dual atmosphere decreases. SUS 430, containing 17 wt% Cr, is subjected to localized attack due to the formation of Fe_2O_3 hematite-rich nodules on the air side of dually exposed samples, whereas the Crofer 22 APU alloy, containing 22 wt% Cr, only forms Fe-rich spinel on top of the scale on the air side. As the Cr content is further increased to 27 wt%, such as in the E-Brite [53, 111] and SUS 446 [117] alloys, no hematite or Fe-rich phases are found in the air side scale, which is only less dense and appears to be more prone to defects than when grown in air only. This difference in the Cr content in alloys can be explained by the fact that the Cr_2O_3 protective layer is more readily established in alloys with higher content of Cr, which slows down the diffusion of Fe ions to the surface of the oxide scale. Increased water vapor partial pressure in the air on the air side can also accelerate anomalous oxidation, resulting in localized nucleation and growth of hematite in the scale. This may be a result of penetration of hydroxide from the moist air into the oxide scale as explained in Section 5.2.2. In contrast, Kurokawa *et al.* [118] exposed SUS 430 alloy to a dual atmosphere (air and $Ar–H_2–H_2O$ mixed fuel) at 800 °C for 300 h, but the morphology and phase of the oxide scale formed on both the air and the fuel sides are almost identical to those formed separately in air and $Ar–H_2–H_2O$ mixed fuel without hydrogen potential gradient, consisting of Cr_2O_3, $MnCr_2O_4$, and trace amounts of $FeCr_2O_4$. The effect of hydrogen potential gradient on the oxidation behavior of the alloy has not been significantly reported.

In addition to the Fe–Cr alloys, the effect of the dual atmosphere on other alloys is also investigated, such as on Al_2O_3-forming alloys and Ni–Cr alloys. In the case of Fe–Cr–Al alloy (Fecralloy®), exposure to the dual atmosphere of moist air and hydrogen at 800 °C results in the formation of a slightly thicker scale and a smoother scale–metal interface on the air side compared to those formed in single air atmosphere, and the alumina scale formed remains protective against dual-atmosphere oxidation [112]. In the case of Ni–Cr-based alloys, exposure to the dual atmosphere leads to the formation of a uniform and adherent scale on the air side; no localized Fe_2O_3 is observed. The dual atmosphere has little effect on changing the oxidation behavior of Haynes 230, Hastelloy S, and Haynes 242 alloys, even though it slightly suppresses the formation of the NiO layer on top of the scale [31]. In general, the Fe–Cr alloys are more susceptible to the dual atmosphere than the Ni–Cr alloys.

5.2.4 Chromium Evaporation from Metallic Interconnects

In addition to oxidation, another major challenge in the use of Cr_2O_3-forming alloys as interconnect materials is the formation of Cr-containing gaseous species from the formed oxide scale, subsequently poisoning the electrode and degrading the electrochemical performance of SOFCs. This phenomenon is recognized as Cr poisoning of the electrode, which is an inherent defect of Fe–Cr and Ni–Cr alloys that form Cr-containing oxide scale in SOFC environment. For example, the Cr_2O_3 scale is thermodynamically unstable at high temperatures and may form volatile Cr species through one of the following reactions:

$$Cr_2O_3\,(s) + 1.5O_2\,(g) \rightleftharpoons 2CrO_3\,(g) \tag{5.6}$$

$$Cr_2O_3\,(s) + 1.5O_2\,(g) + 2H_2O\,(g) \rightleftharpoons 2CrO_2(OH)_2\,(g) \tag{5.7}$$

$$Cr_2O_3\,(s) + O_2\,(g) + H_2O\,(g) \rightleftharpoons 2CrO_2\,(OH)\,(g) \tag{5.8}$$

$$Cr_2O_3\,(s) + H_2O\,(g) \rightleftharpoons 2CrOOH\,(g) \tag{5.9}$$

These volatile Cr species have a tendency to electrochemically or chemically deposit on the surface of the cathode and/or at the cathode–electrolyte interfaces, leading to a significantly reduced cathode active area and a subsequent drastic deterioration in cell performance [119–122]. This degradation can be represented by either a decrease in cell voltage or an increase in cell overvoltage.

From reactions (5.6) to (5.9), it can be seen that $CrO_3(g)$, $CrO_2(OH)_2(g)$, $CrO_2(OH)(g)$, and $CrOOH(g)$ are the possible volatile Cr species and their formation from the Cr-containing oxide scale is strongly dependent on the temperature and the partial pressure of oxygen and water vapor in the atmosphere. The oxygen partial pressure is much higher in the cathode compartment with flowing air than in the anode compartment with flowing fuel gas (the oxygen partial pressure in the fuel gas is in the range between 10^{-18} and 10^{-8} atm). Thus, the effect of Cr vapor species on the performance of anode via reactions (5.6) to (5.8) can be ignored. Furthermore, reaction (5.9) will not occur at temperatures below 1100 °C [123]; therefore, Cr poisoning caused by the metallic interconnects is most likely to occur at the cathode [124]. The evaporation rate of Cr_2O_3 via reaction (5.6) to form the most abundant species in dry air is negligibly small at temperatures lower than 1000 °C; addition of water vapor to air increases the evaporation rate and lowers the evaporation temperature of Cr_2O_3 by forming hydroxides via reactions (5.7) and (5.8) [123–125], resulting in significantly higher Cr deposition and cathode poisoning in humidified air [126]. The vapor pressure of each Cr oxyhydroxide increases with increasing water vapor pressure; the presence of water vapor in air will significantly increase the overall vapor pressure of Cr-containing vapor species through the formation of $CrO_2(OH)_2(g)$. This explains why the dominant Cr-containing vapor species is $CrO_2(OH)_2(g)$ in humid air and its partial pressure increases linearly with water partial pressure. Hilpert *et al.* [124] indicated that CrO_3 is the most abundant Cr-containing gaseous species in the cathode of an

SOFC fed with dry air as the oxidant; the partial pressure of $CrO_2(OH)_2$(g) exceeds that of CrO_3(g) at 1223 K while H_2O partial pressure in air is higher than 90 Pa (equivalent to air, with $pH_2O = 2 \times 10^3$ Pa at 25 °C), showing an increase in Cr volatility by more than one order of magnitude as compared to that in dry air. When a leakage exists between the anode and the cathode of the SOFC, Cr volatility will further increase because of the transport of water vapor from the anode to the cathode.

Cr volatility depends on the composition and microstructure of oxide scales. Transition tests [127, 128] indicate that the transport rate of Cr species from a Fe-based alloy is much lower than that from a Cr-based alloy because of the formation of Fe-rich oxides on the surface of the Fe-based alloy. For instance, SUS 446 alloy has a Cr transport rate that is 1/20th of that of the $Cr5Fe1Y2O_3$ alloy at 850 °C [128]. It is also noticed that the formation of $(Cr,Mn)_3O_4$ spinel on top of the oxide scale significantly reduces the vaporization of volatile Cr-containing species, even though such a scale structure is not adequate to satisfactorily suppress the Cr vaporization from the scale and further improvement is still desired [41, 46, 129, 130]. Stanislowski *et al.* [131] performed a series of systematic experiments to demonstrate the Cr vaporization of Cr-, Fe-, Ni-, and Co-based alloys in air and H_2 atmospheres and the influence of outer scale layers of $(Cr,Mn)_3O_4$, $(Fe,Cr)_3O_4$, Co_3O_4, TiO_2, and Al_2O_3. The results indicated that the Cr vaporization rates of the ferritic stainless steels developed for SOFC interconnect applications, such as Crofer 22 APU, ZMG 232, IT-10, IT-11, and IT-14, are equivalent to each other, as they all form a well-adhered outer $(Cr,Mn)_3O_4$ spinel layer that reduces Cr vaporization rates in humid air at 800 °C by 61–75%, in comparison to alloys such as Ducrolloy and E-Brite that form a pure Cr_2O_3 scale. For different Ni-, Co-, and Fe-based austenitic alloys with varying contents of Mn, Ti, Al, Si, and W, oxidation at 800 °C in humid air results in the formation of different thermally grown outer oxide layers of $(Cr,Fe)_3O_4$, $(Cr,Mn)_3O_4$, Al_2O_3, TiO_2, and Co_3O_4. Al_2O_3 and Co_3O_4 outer scales present the lowest Cr vaporization rate. With a Co_3O_4 outer layer, the Cr vaporization rate is only 10% of that with a pure Cr_2O_3 scale, suggesting that Co_3O_4 coating on metallic interconnects is a promising way to solve Cr vaporization in metallic interconnects.

Cr-containing vapor species formed over the metallic interconnect can be electrochemically or chemically reduced on the electrode surface or electrode–electrolyte interface. As the vapor pressure of volatile Cr-containing species at the anode side of SOFCs is too small to cause any problem, the deterioration mainly occurs at the cathode. The resulting deposition can block the electrode active sites and rapidly degrade cell performance. Therefore, it is important to understand the transport and deposition processes of the Cr-containing species at the cathode. One of the proposed mechanisms for Cr deposition is electrochemical reduction of high-valent vapor species [10, 124, 132–136]. The precipitated Cr_2O_3 solid phase in the region of cathode–electrolyte interface can react with Sr-doped $LaMnO_3$ (LSM) cathode to form $(La,Sr)(Mn,Cr)O_3$ or $(Cr,Mn)_3O_4$ phase at the three phase boundary (TPB), blocking the active sites for oxygen reduction reaction and also the pores in the cathode. Consequently, the TPB area is substantially reduced, inhibiting the reaction of oxygen reduction and increasing the cathodic polarization. The electrochemical deposition mechanism appears to explain the observed correlation between the intensity of Cr-containing species deposited at the electrode–electrolyte interface and the performance degradation for the oxygen reduction reaction occurring on the LSM electrode.

Jiang *et al.* [120, 137–145] provided evidence that the deposition reaction of Cr-containing species on the cathode is not a process of electrochemical reduction of high-valent Cr-containing vapor species. They examined the deposition that occurred under anodic or cathodic polarization and observed that Cr-containing species show no preferential deposition at the TPB of O_2, LSM electrode, and Y_2O_3-stabilized ZrO_2 (YSZ) electrolyte. Furthermore, the deposition of Cr species on the YSZ electrolyte surface, where electrons are not available for an electrochemical reaction, is random at the early stage of the reaction and the Cr deposits are much smaller than those on pure LSM electrode. The results demonstrate that the deposition of Cr-containing vapor species is not an electrochemical reduction reaction to form solid Cr_2O_3 at the TPB region in competition with O_2 reduction; instead, it is essentially a chemical dissociation process initiated by the nucleation reaction between the gaseous Cr species and Mn^{2+} ions. The Mn^{2+} ions generated from LSM electrode under polarization potential are driven to the YSZ electrolyte surface at high temperatures and then react with gaseous Cr species to form Cr–Mn–O nuclei as the nucleation sites for crystallization and growth of Cr_2O_3 and/or $(Cr,Mn)_3O_4$ phases. This mechanism shows that the deposition of Cr-containing vapor species is strongly dependent on the electrode and electrolyte materials, and the mechanism for Cr deposition in other material systems is far from clear.

Applying a protective coating onto metallic interconnects is an effective way to reduce Cr vaporization in Cr-containing oxide scales. The protective coating can serve as a stable dense barrier to minimize both the formation and evaporation of Cr-containing oxide scales on the surface of metallic interconnects. As a consequence, the performance of the cathodes in contact with the coated metallic interconnects is significantly improved [135, 146–154]. Cr-free ceramic coatings prepared from submicrometer powders of mixed conducting oxides, such as LSM, $La_{0.6}Sr_{0.4}Co_{0.8}Fe_{0.2}O_3$ (LSCF), and $MnCo_2O_4$ (MCO), can decrease Cr vaporization by a factor up to ~40 at 800 °C. The density of the ceramic coatings dominates the suppression of the Cr release rate [146]. A uniform and dense Co coating can effectively mitigate Cr migration when applied onto the surface of E-Brite alloy, leading to an improved electrochemical stability of LSCF electrodes [148]. A Mn–Co spinel coating on SUS 430 alloy has demonstrated capability as a diffusion barrier for the Cr-derived species from the steel, lowering Cr evaporation by a factor of 2 [149]. The sputtered perovskite coatings, such as LSM and Sr-doped $LaCoO_3$, are not effective in Cr retention because of the pores in the coatings formed during crystallization of the amorphous sputtered coatings at high temperature. On the other hand, the metallic coatings of Co, Ni, or Cu can reduce the Cr release by more than 99% by forming stable, adherent, and conductive Co_2O_3, NiO, or CuO layers respectively [151], demonstrating their potential as a promising coating material for interconnect alloys. In another study [152], Mn_3O_4 or $(Mn,Co,Fe)_3O_4$ spinel coating on alloys is found to be effective in lowering Cr evaporation by one to two orders of magnitude, compared with the uncoated ones. Generally, applying a coating is beneficial to Cr retention; however, the quality of the coating, such as density and surface uniformity, is of great importance for achieving the goal. In addition to coatings, ion implantation of carbon or nitrogen on the surface of Crofer 22 APU alloy also demonstrates an improvement in Cr evaporation by a factor of 3 due to the suppression of grain boundary oxides and formation of uniform fine surface grains [153].

5.2.5 Compatibility with Cell and Stack Components

The high operating temperature of SOFCs requires metallic interconnects to be thermally and chemically compatible with adjacent cell and stack components, such as materials for electrodes, electrolytes, and seals. The TEC of the cell is typically in the range $10.5{-}12.5 \times 10^{-6}\,K^{-1}$, which conventionally requires the candidate metallic alloys for the interconnect to have a TEC that matches that of the ceramic cell components in order to avoid generating thermal stresses that cause the failure of the cell and contact in a stack. Cr-based alloys, such as $Cr_5Fe_1Y_2O_3$, have a good TEC match with the ceramic components, $11.8 \times 10^{-6}\,K^{-1}$ in the range 20–1000 °C, which is equivalent to that of the cell. Austenitic stainless steels and Ni-based alloys usually have a relatively high TEC, in the range $14.0{-}19.0 \times 10^{-6}\,K^{-1}$ from room temperature to 800 °C, which is considered significantly higher than that of the cell. Thus, austenitic alloys are less attractive as interconnect materials, although Ni-based alloys possess excellent oxidation resistance in SOFC operating environments. In contrast, the ferritic stainless steels, especially Fe–Cr alloys, typically have a TEC in the range $11.0{-}13.0 \times 10^{-6}\,K^{-1}$ from room temperature to 800 °C, which closely matches that of the cell components and makes the Fe–Cr alloys an attractive option for application as interconnects in SOFC stacks. In planar SOFCs, the requirement of TEC match between the metallic interconnect and the cell may not be as stringent as one imagines. For a better compressive contact in the stack, special conductive contact materials, frequently, mixed conducting perovskites in powdery form, are applied between the interconnect and the cell, and their connection is not rigid. In such a stack design, a TEC mismatch between the metallic alloy and the cell to a certain extent, for example, 30%, may be allowed without significantly affecting the performance of the stack during steady operation and thermal cycles. The TEC match between the interconnect and the seal usually needs seal materials to be developed to satisfy the requirement.

The chemical compatibility of metallic interconnects with adjacent components is another key requirement that can affect the applicability of an alloy as interconnect in an SOFC stack. With the contact material in between the interconnect and the cathode, the interaction between the Cr_2O_3-forming alloys and cathode materials is mainly a transport of Cr from the alloys into the electrode via gaseous Cr-containing species, as discussed in Section 5.2.4, and this subsequently degrades the stack performance. As mentioned earlier, the contact materials are usually mixed conducting perovskites, which generally show good chemical compatibility with conventional metallic interconnect materials. Unless Sr is contained in the perovskite, especially LSCF, it reacts with the oxide scale to form $SrCrO_4$, appearing to increase the contact ASR of the stack [155]. The chemical compatibility between the anode and the metallic interconnect is not an issue, as the interconnect is in contact with Ni foams in most of the planar stack design. But the elemental interdiffusion between the interconnect and Ni foam will significantly affect the electrical property and microstructural characteristics of the oxide scale formed on the surface of the interconnect and the Ni foam.

As for the chemical compatibility between the metallic interconnect and the seal, it may be problematic if glass is used in the seal at SOFC operating temperatures. Diffusion, devitrification, and chemical reaction usually occur at the alloy–seal interface, which decreases the stability of the formed oxide scales and degrades the performance of the components [156–160]. The formed Cr_2O_3 scale may be dissolved into the glass, forming a chromate and resulting in bond failure between the seal and the oxide scale due to thermal expansion mismatch [161–163]. With Al_2O_3-forming alloys, no chromate is formed; however, the interface will become porous [164], possibly because of interdiffusion of elements at the operating temperature. It is found that the sealing glass Sr–Ca–Y–B–Si–Zn reacts with Crofer 22 APU alloy after 500 h of testing at 850 °C in air, even with a Mn–Co spinel coating [165]. In the case of bare Crofer 22 APU, the glass reacts with chromium oxide to form a thick layer of $SrCrO_4$, possibly in the presence of SiO_2; in the case of coated Crofer 22 APU, in addition to the formation of the chromate, the originally dense Mn–Co spinel coating is attacked by the glass seal, breaking it into discrete particles or islands. This kind of reaction is detrimental to the effectiveness of the sealing and the coating. Aluminization on the sealing surface is helpful in minimizing or preventing the undesired chromate formation between alkaline earth sealing glass and the oxide scale [166]. Other than chromate formation due to the interaction, anomalous oxidation is frequently observed on the surface of metallic interconnects near the glass seal, as shown in Figure 5.10. Nodular thick oxides are formed both inwardly (50 mm) and outwardly (50 mm) on the original surface. It is considered that the vaporized components from the alkaline earth glass react with previously formed Cr–Mn spinel and Cr_2O_3 scales, resulting in the breakdown of the oxide scale and the formation of a thick oxide consisting of Fe_2O_4 or Fe_2O_3. The proposed mechanism is presented in Figure 5.11 [113, 167]. In a reducing atmosphere, Fe or Mn component in the alloy or oxide scale may react with silicon components in the sealing glass to form $FeSiO_3$ or Mn_2SiO_4, respectively [75, 168]. Minor constituents in the alloys, such as Al and Si, have a detrimental effect on the interaction behavior with the glass-sealing materials [169]. The long-term compatibility between metallic interconnects and glass seals and its effect on performance degradation need to be carefully investigated, especially for anode-supported planar SOFCs.

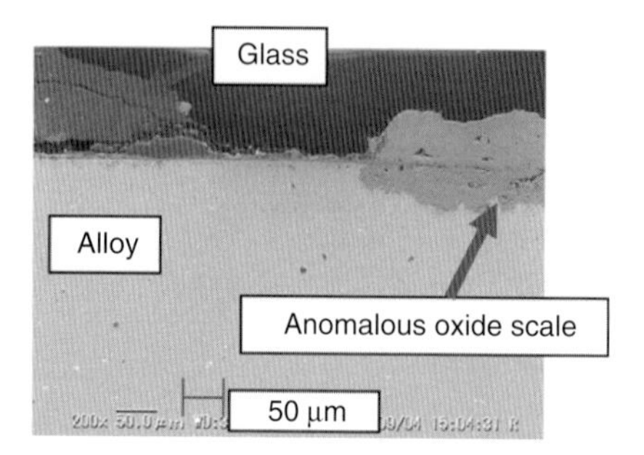

Figure 5.10 Scanning electron microscopic image of the anomalous oxidation near the glass seal–metallic interconnect contact interface. (Source: Reproduced with permission from Ref. [113].)

5.3 Electrical Properties of Oxide Scale

One of the critical requirements for interconnect materials is a relatively low and stable electrical resistance during SOFC operation to minimize the electrical losses. For this reason, metallic alloys are an attractive option for interconnect applications. As described previously, metallic alloys are inevitably oxidized in the environments of SOFC operation, and the electrical resistance contributed by the oxide scale must be taken into account. The ASR (ohms square centimeter), which reflects both the electrical conductivity and thickness of the oxide scale, is usually adopted for the evaluation of electrical properties of the oxidized metallic interconnect alloys. The electrical resistance of the substrate alloy can be negligible as it is very small compared with that of the formed oxide scale, and the main contribution of metallic interconnects to the ASR comes from the oxide scale. Generally, the thermally grown oxide scale is a semiconductor, the electrical resistance of which decreases with increasing temperature. Thus, the ASR of oxidized metallic alloys increases with decreasing temperature and shows an electrical characteristic of semiconductors as described by the Arrhenius equation

$$\frac{\mathrm{ASR}}{T} = A \exp\left(\frac{E_\mathrm{a}}{kT}\right) \tag{5.10}$$

where A is a preexponential constant, T is the absolute temperature, E_a is the activation energy, and k is the Boltzmann constant. And $\log(\mathrm{ASR}/T)$ is linearly proportional to $1/T$. By plotting $\log(\mathrm{ASR}/T)$ against $1/T$, the activation energy is the slope of such a curve. With an oxide scale of thickness X and electrical resistivity ρ, the ASR of the metallic alloy is expressed as

$$\mathrm{ASR} = \rho X \tag{5.11}$$

reflecting the contributions from both the scale thickness and electrical resistivity. The ASR can be related to the oxidation time at a fixed temperature by

$$\mathrm{ASR}^2 = \rho^2 K_\mathrm{p} t \tag{5.12}$$

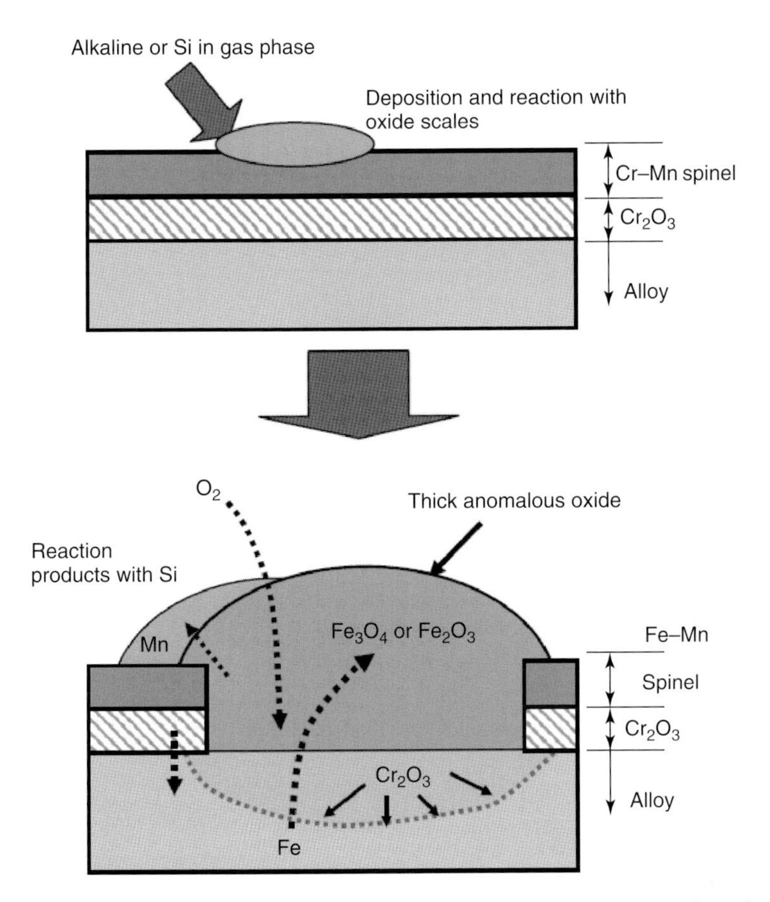

Figure 5.11 Schematic diagram showing the breakdown of protective oxide scale and the resulting anomalous oxidation. (Source: Reproduced with permission from Ref. [113].)

Thus, the ASR should change in a parabolic manner with time if the resistivity of the oxide scale is influenced by the growth kinetics of the oxide scale during long-term exposure to SOFC atmospheres. For an assumed constant oxide scale resistivity during thermal exposure, the ASR after 40 000 h of oxidation (conventional SOFC life expectation) can be approximately extrapolated from a relatively short-term oxidation test. The target ASR value for metallic interconnect materials is below 50 $m\Omega\,cm^2$ [5, 8, 170] (corresponding to 25 mV drop at 500 $mA\,cm^{-2}$), which is used in practice to judge the suitability of metallic alloys as interconnect materials. The ASR is also affected by cracks in the scale and the adherence of oxide scale to the substrate, other than scale thickness and resistivity. In principle, alloys with an oxide scale of slow growing rate, high electronic conductivity, and strong adherence are highly preferred for SOFC interconnect applications.

The ASR of a given oxidized alloy is normally measured by the "four-wire" probe technique as shown in Figure 5.12. Two surfaces of the oxidized coupon are covered with Pt (or Ag) paste, and Pt (or Ag) meshes spark-welded with four Pt

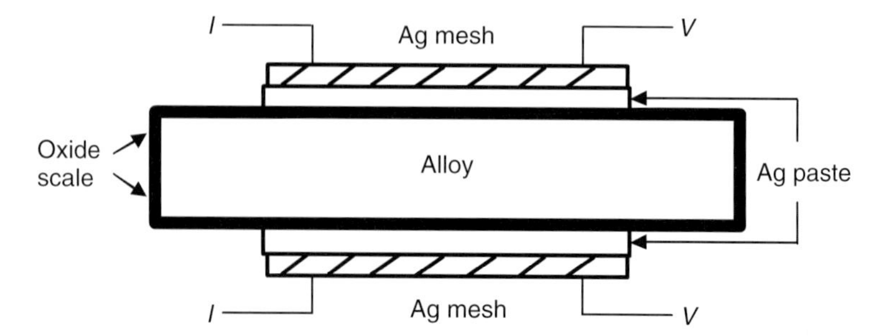

Figure 5.12 Schematic principle of the area-specific resistance measurement.

(or Ag) leads are compressively placed on top of the paste for applying electrical current and measuring the voltage. A constant current (I) is applied through two of the four leads attached on opposite surfaces, while the voltage drop (V) across the sample is taken with the other two leads. The ASR is calculated according to Ohm's law, $\text{ASR} = (V/2I) \times S$, where S is the area covered by the paste and a factor of 1/2 is used to represent the contribution of the oxide scale on one surface to the ASR.

The dominant phase in the oxide scale of Cr_2O_3-forming alloys is Cr_2O_3; its conductivity changes with temperature [171, 172]. In the temperature region above 1000 °C, Cr_2O_3 is an intrinsic electronic conductor, whereas in temperature ranges below 1000 °C, its conductivity exhibits extrinsic behavior, depending primarily on impurities and on oxygen and hydrogen activities to some extent. The conductivity value of the Cr_2O_3 scale is reported in the range 1×10^{-3} to 5×10^{-2} S cm^{-1} at temperatures between 800 and 1000 °C, affected by impurity levels in the scale, oxygen activity in the atmosphere, and equilibrium condition. The conductivity of Cr_2O_3 can be improved by substitution of various oxides into the Cr_2O_3 scale [173].

Among Ni-based alloys, such as Inconel 625, Inconel 718, Hastelloy X, and Haynes 230, Haynes 230 displays the lowest ASR with samples oxidized at 800 °C in both air and reducing environments, respectively, for up to 10 000 h, because of its slow oxidation kinetics and formation of an oxide scale that consists predominantly of Cr_2O_3 and more conductive $(Cr,Mn)_3O_4$ spinel [24, 25]. Figure 5.13 shows the ASR of Haynes 230 alloy as a function of oxidation time at 800 °C in air, in comparison with other Ni-based alloys [24]. For the parabolic oxidation kinetics that is obeyed by the Ni-based alloys, the oxide thickness, X, is proportional to the square root of the oxidation time. The ASR versus square root of oxidation time should be a straight line, as described by Eq. (5.12), assuming the resistivity of the oxide scale to be a constant for a given alloy at a given temperature. The nonlinearity of the ASR against square root of oxidation time suggests that the resistivity of the oxide scale is not constant with time; the scale composition varies while it is thickened by the depletion of minor constituents of thin foil samples. The ASR of Haynes 230 alloy oxidized in H_2/H_2O atmosphere at 800 °C is significantly higher than that in air, which can be attributed to the thicker and less conducting

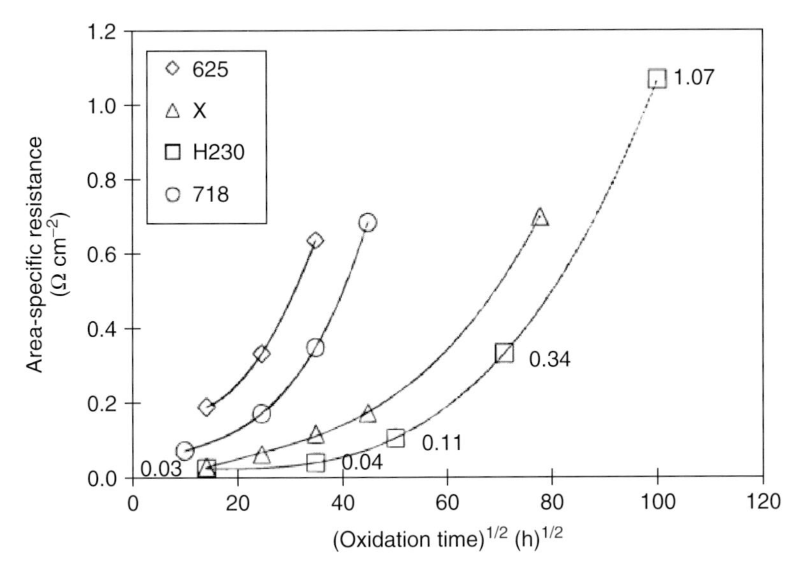

Figure 5.13 Area-specific resistance of Ni-based alloys as a function of oxidation time at 800 °C in air. (Source: Reproduced with permission from Ref. [24].)

oxide scale formed in the reducing atmosphere with low oxygen partial pressure [65, 174].

The electrical resistance of ferritic Fe–Cr alloys is usually higher than that of the Ni-based alloys, as they are less oxidation resistant in SOFC environments. The thermally grown oxide scales of the ferritic stainless steels may overgrow to as thick as tens of micrometers during an SOFC's lifetime, resulting in unacceptable degradation in SOFC stack performance [52, 53]. Increasing the Cr content in the Fe–Cr alloys can enhance the oxidation resistance, leading to reduction of the electrical resistance of the scale. For the commercially available SUS 430 ferritic stainless steel containing ~16 wt% of Cr, oxidized in air at 750 °C for 850 h, the ASR at 750 and 800 °C in air is around 43 and 19 mΩ cm^2, respectively [175]. Crofer 22 APU, containing ~23 wt% of Cr and small amounts of Mn, Ti, and La, is one of the promising Fe–Cr alloys for interconnect applications; however, its ASR is 9.3 mΩ cm^2 after oxidation in air at 800 °C for 500 h and increases to 13 mΩ cm^2 for 1800 h [41]. All these ASR values indicate that both SUS 430 and Crofer 22 APU, with unacceptably high ASR values, will not meet the requirement of 50 mΩ cm^2 for 40 000 h of SOFC operation at intermediate temperatures.

As a matter of fact, most cost-effective commercial alloys cannot meet this standard at an SOFC operating temperature higher than 700 °C in terms of electrical conductivity, and improving the ASR of metallic alloys for the purpose of SOFC interconnects still remains challenging. Coatings and new low-cost metallic alloys are desired to solve the issue of metallic interconnects for intermediate-temperature SOFCs, which is currently a bottleneck.

5.4 Surface Modifications and Coatings

Cr_2O_3-forming ferritic stainless steels are considered the most promising interconnect material; however, as stated earlier, they are facing challenges of insufficient oxidation resistance and Cr evaporation in SOFC environments. Among the various approaches to reduce the rate of oxidation and Cr evaporation, surface modification via protective coatings is expected to be effective, especially on the cathode side where Cr evaporation is considered an issue. A viable coating material should be

1) chemically stable in SOFC environments and compatible with the substrate alloy and neighboring SOFC components;
2) a mass transport barrier to the migration of both outward cations and inward oxygen anions;
3) thermally compatible with the substrate alloy;
4) electrically conductive.

Various kinds of protective coatings have been extensively investigated for the purpose of improving metallic interconnects, such as RE oxides, perovskites, spinels, and metallic elements Mn, Co, Ni, or the others. For the oxide coatings, the phase structure is relatively stable, without any dramatic changes other than element doping; in the case of metallic element coatings, they react with the elements in the alloy to form different ceramic coatings on heating.

5.4.1 RE and Metallic Element Coatings

It is well known that the addition of reactive elements, such as La, Ce, and Y, to the Cr_2O_3-forming alloys can greatly reduce their oxidation rate and increase the scale adhesion at high temperatures [176–182]. Thus, a few attempts have been made to add the REs or their oxides to the alloy as a protective coating for metallic interconnects. Such coatings typically reduce scale thickness and enhance scale adhesion; as a result, a lowered ASR is expected. Y is one of the frequently used elements, and 304 stainless steel with sol–gel Y coating demonstrated a significantly low oxidation rate and remarkably increased scale adherence during isothermal or thermal cycling exposure to air at 1000 °C [183]. With SUS 430 alloy as the substrate, Y coating derived from Y nitrate solution in a reducing environment noticeably enhances scale adhesion, shifting the failure location away from the oxide scale–alloy interface [48]. Surface modification by Y implantation seems to provide more effective protection against oxidation for stainless steels than sol–gel coating, although they both cause a significant reduction in the scale growth rate [184]. The application of an Y_2O_3 coating by metal organic chemical vapor deposition (MOCVD) leads to reduction in the oxidation rate and improvement of the oxidation resistance by one order of magnitude, eliminating cavities and scale spallation [185]. TEM study reveals that the presence of Y at the oxide grain boundaries changes the growth direction of Cr_2O_3 from predominantly outward to predominantly inward, slowing down scale growth; Y prevents S segregation at the oxide scale–alloy interface, enhancing the adhesion of the scale to the alloy [186]. La [187, 188], Ce [46], and other oxides also have beneficial effects on oxidation resistance, electrical conductivity, and scale adhesion of Crofer 22 APU or Inconel 600 alloy. The effects of Y, Y–Co, Ce–Co, Co, and Sm–Co coatings prepared by sol–gel [189, 190], electrodeposition [191], and electron beam vapor deposition [192] or magnetron sputtering [193] on the oxidation behavior and electrical conductivity of Cr_2O_3-forming Fe–Cr ferritic alloys were investigated. Y-containing coatings exhibit the highest oxidation resistance and lowest electrical resistances; Co-containing coatings accelerate the growth of oxide scale by forming spinel oxides with improved electrical resistance. Addition of 5% Sm to Co coating forms multilayered $CoFe_2O_4$, Co_3O_4, and $SmCoO_3$, suppressing the thermally grown oxides Cr_2O_3 and $MnCr_2O_4$. Nitriding of Sm–Co leads to a more compact coating and controlled oxidation to a scale, with a thinner Sm depletion zone and less Fe and improved ASR [193]. Y is either dissolved in Cr_2O_3 without forming a distinct Y-rich phase [189] or in the form of $YCrO_4$ in between Cr_2O_3 and Cr–Mn spinel [192], and Co is primarily presented in the spinel phase. Ce is presented as distinct CeO_2 particles at the Cr_2O_3–spinel interface. Comparison of Y, Ce, and La as the coating shows that Y is the more effective reactive element in reducing the kinetics of oxide scale growth and ASR of Cr_2O_3 scale [194, 195].

It is recognized that Y and light lanthanides (La, Ce, Pr, Nd, Pm, Sm, and Eu) appear to be more effective in improving the high-temperature oxidation behavior of Cr_2O_3-forming alloys than heavy lanthanides (Gd, Tb, Dy, Ho, Er, Th, Yb, and Lu). However, their effectiveness on oxidation resistance and electrical conductivity is alloy dependent. With Fe–30Cr substrate alloy, the effectiveness of oxide coatings on oxidation resistance appears to be in the order $Yb_2O_3 < Nd_2O_3 < Y_2O_3$ [99, 196–198]. In the case of the popular Crofer 22 APU alloy oxidized at 800 °C in air or H_2/10% H_2O atmosphere, La_2O_3 coating is the best choice for reducing oxidation rate and electrical resistance due to the formation of the $LaCrO_3$ phase [99, 198, 199].

There is abundant evidence to demonstrate the success of RE coating in suppressing oxidation of metallic alloys for SOFC interconnect applications; however, the mechanism behind this phenomenon is still not very clear. Two different mechanisms have been proposed to explain the effect of REs on oxide growth: the grain boundary segregation model and the interface poisoning model [137, 200, 201]. The first model suggests that the segregation of REs to oxide grain boundaries greatly retards cation outward transport and that scale growth is dominated by anion inward transport without forming cavities at the scale–alloy interface; consequently, oxidation rate is lowered and scale adherence is improved. This model is supported by TEM analysis on RE-modified Cr_2O_3-forming steels [82, 178, 186, 202, 203]. However, this model fails to explain why the segregation of REs prohibits the transport of cations, while allowing inward transport of anions along the grain boundaries. The second model [204] considers oxide growth not only as diffusion through the scale but also as reaction at the interfaces. It suggests that the segregated large-sized REs poison the scale–alloy interface by eliminating the available sites for cation vacancy annihilation, thus impeding cation outward transport, whereas anion inward transport is not affected by RE blocking at the scale–alloy interface due to vacancy annihilation at the scale–gas interface in this case. In general, the presence of REs in Cr_2O_3-forming alloys can (i) enhance the formation of continuous Cr_2O_3 scale, (ii) reduce growth rate of the scale by altering the transport mechanism in oxide growth, and (iii) increase adherence of the scale to the underlying alloy.

In recent years, heat-resistant coatings such as Cr–Al–N [205], (Ti,Al)N [206], Cr–Al–O–N [207], MCrAlYO (M = Ti, Co, and/or Mn) [208, 209], Cr–Al–Y–O [210], Co–Cr–Al–O–N [211], and CrAlYO–CoMnO [210, 212] are also considered for metallic interconnects. These coatings can function as a barrier to oxidation and Cr evaporation; however, they may not be a favored coating for metallic interconnects because of their high content of Al that may form insulating Al_2O_3. Ni is a special metallic element for SOFCs; it will not be oxidized in the anode atmosphere because of the low oxygen partial pressure. Therefore, Ni can be used as the anode side coating in metallic interconnects to reduce oxidation and promote contact. Prepared by electroplating [213] or atmospheric plasma spraying [214] on Fe–22Cr (Sandvik 0YC44) or SUS 430 alloys, it results in low ASR degradation and oxidation rate on the anode side of the interconnect.

5.4.2
Perovskite Oxide Coatings

Recently, La-containing perovskites have attracted much attention as coating materials for improving oxidation behavior, ASR, and Cr evaporation of Cr_2O_3-forming ferritic Fe–Cr alloys. One important reason for this approach is that these perovskites have proved to be electronically conductive, and thermally matched to and chemically compatible with the alloys. Research efforts have mainly focused on applying $LaCrO_3$ and $(La,Sr)MnO_3$ ceramics in SOFCs, which are conventional ceramic interconnect and cathode materials, respectively.

$LaCrO_3$, either pure or doped, has been widely used as coating material for metallic interconnects because of its low Cr evaporation rate and acceptable electronic conductivity in SOFC operating environments. However, the process for preparing the $LaCrO_3$ coating is of importance for its effectiveness. With the solution method, a dense, crack-free, and well-adhering $LaCrO_4$ coating layer increases isothermal and/or cyclic oxidation resistance and reduces the ASR of ferritic Fe–Cr alloys, such as E-Brite and SUS 444 stainless steels [215, 216]. Other processes, such as reactive formation [217] and magnetron sputtering [218–220], were also used for $LaCrO_3$ coating fabrication on E-Brite, SUS 446, or Crofer 22 APU alloys. Compared with the solution method, voids and pores are observed in the coatings prepared by these two processes because of either the solid reaction of coated La_2O_3 and thermally grown Cr_2O_3 or the crystallization of sputtered amorphous La–Cr–O in air. Such defects may degrade the function of the coating. Crystallization in a reducing environment can lead to a dense and compact coating, which is a good barrier to oxygen diffusion. Using a proprietary coating and heat treating technique, Elangovan *et al.* [221] demonstrated that the $LaCrO_3$ protective coating on commercial stainless steels significantly reduced the oxidation rate of the alloys in both air and dual atmosphere at 750 °C and maintained stable electrical conductivity and low Cr evaporation rate.

For improving the electrical conductivity and TEC, $LaCrO_3$ is often doped at La and/or Cr sites of perovskite structures. Sr-doped $LaCrO_3$ (LSCr, La-site doping) has been intensively investigated as a protective coating material for metallic interconnects, as Sr doping remarkably promotes electrical conductivity. Different approaches, such as reactive formation [222] and RF magnetron sputtering [222–225], were used to prepare the doped protective coatings. The results show that LSCr coating significantly improves the oxidation resistance of commercial ferritic Fe–Cr alloys, such as Crofer 22 APU, E-Brite, AL 453, and SUS 430, at 700–800 °C in air and reduces the ASR of E-Brite, AL453, and SUS 430 alloys, compared with the uncoated ones. However, in the case of Crofer 22 APU, the coated sample has a higher ASR than the uncoated one. As a matter of fact, the ASR depends on the electrical conductivity of the coating and the subsequently grown oxide scale and their thickness. If a more conductive oxide scale is formed without the coating, the coated one may have higher ASR. Other than Sr, Ca is frequently used as the dopant to form $(La,Ca)CrO_3$ coating [226] to promote electrical conductivity, and co-doping with $(La,Sr)(Cr,Zn)O_3$ [227] and $(La,Sr)(Cr,V)O_3$ [228] has proved to be effective in reducing oxidation rate, increasing ASR, and suppressing Cr evaporation.

As known, the ionic conductivity of $LaCrO_3$-based perovskites is low, in the order of 10^{-5} S cm^{-1} at 1000 °C in air, while their electronic conductivity is somewhat high, for example, 24 S cm^{-1} at 800 °C in air for $La_{0.8}Sr_{0.2}CrO_3$. In order to further improve the electrical conductivity of the coating, Ni–$LaCrO_3$ [229, 230] and Co–$LaCrO_3$ [231] composite coatings have been experimented on. With a high concentration of Ni or Co in the coating, the coating itself is oxidized during oxidation at 800 °C in air, enhancing oxidation by forming Ni- or Co-rich oxide and generating significant amounts of pores in the coating and at the scale–alloy interface. Even though the ASR is lowered, compared with the uncoated one, such coatings are not reliable because they are prone to spallation in the long-term run of SOFCs. With less content of pure metal [232], the coating is more promising.

LSM, a typical cathode material of SOFCs, is another type of perovskite oxide favored as coating material for metallic interconnects. The coating can be made by various techniques, such as slurry dipping [233], plasma spraying [234, 235], sputtering [236–239], screen painting [237, 240], physical vapor deposition [241], aerosol deposition [242, 243], thermal spray [244], and slurry spray [245]. Their functions are the same as a barrier to prevent oxide growth, to improve ASR and to suppress Cr evaporation from the scale of the Cr_2O_3-forming alloys. The ASR of the coated alloy depends on the chemical stability of the coating and the phase of the oxide scale formed during the coating process and SOFC operation between the alloy and the coating; the coating process before oxidation is critical to obtain a stable and crack-free coating layer and an interfacial oxide with low resistivity. For example, the LSM-coated SUS 430, prepared by sintering at 1200 °C in inert gas for 2 h and heat treatment at 1000 °C in air for 3 h, holds a low ASR of 0.074 Ω cm^2

for up to 2600 h at 750 °C in air because of the formation of the stable LSM coating [233]. For plasma-spray LSM coatings, post spray, heat treatment at 1000 °C for 2 h heals the microcracks formed during spraying [234], without significantly changing the coating microstructure [235]; but the heat treatment may cause formation of insulating oxides between the coating and the substrate, slightly increasing the ASR of the coated alloys.

For the coating purpose, Ti-doped $LaMnO_3$ (LTM) is reported to be more effective than LSM. Sr doping increases oxygen ionic conductivity of $LaMnO_3$, while Ti doping increases electronic conductivity. Therefore, LTM-coated Haynes 230 alloy presents the lowest ASR, compared with that of the LSM-coated and uncoated ones, as shown in Figure 5.14 [239]. Interdiffusion may occur between the coating and the oxide scale formed under the coating, and this may effectively help further lower the ASR of the LTM-coated samples. However, excessive interdiffusion is not desired [239]. Composite coating of LSM–YSZ [243] may be more effective than LSM coating because of the microstructure stability of YSZ against grain growth. Results on the effectiveness of LSM-type coating on Cr evaporation are controversial [244, 245]; therefore, detailed and long-term studies are needed.

Other than the two types of perovskites mentioned earlier, $LaCoO_3$- and $LaFeO_3$-based materials, such as $(La,Sr)CoO_3$ (LSCo) [135, 225], LSCF [240, 242, 246], and $(La,Sr)FeO_3$ (LSF) [223, 224, 247, 248], are also considered for the cathode side coating of metallic interconnects. They are high-performance cathode materials with mixed conductivity; therefore, these types of materials used as a coating are

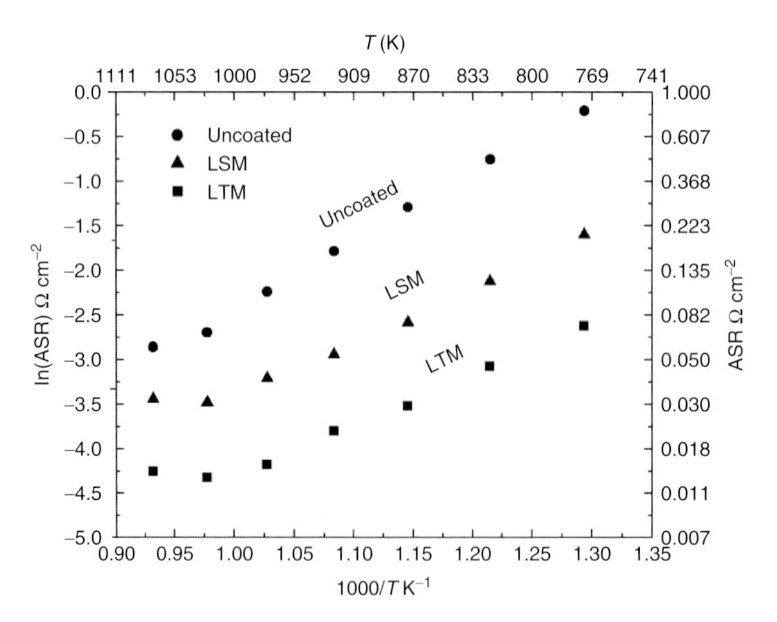

Figure 5.14 ASR as a function of measuring temperature for the LTM- and LSM-coated and uncoated Haynes 230 alloy preoxidized at 800 °C in air for 504 h. (Source: Reproduced with permission from Ref. [239].)

helpful in reducing Cr evaporation and lowering the ASR but may not be perfect as an oxidation barrier because of their high oxygen ion conductivity and reactivity with the thermally grown scale, even though resistance to short-time oxidation is increased [248].

5.4.3 Spinel Oxides

For a higher oxidation resistance and electrical conductivity and lower Cr evaporation than the oxide scale thermally grown on the conventional Cr_2O_3-forming alloys, spinel oxides are often selected as the coating for the metallic interconnect alloys. $(Cr,Mn)_3O_4$ spinel, often formed on Cr_2O_3-forming alloys with a small amount of Mn, is the primary consideration for the coating [65, 228, 249]; its conductivity changes with composition and reaches a maximum with $Cr_{1.5}Mn_{1.5}O_4$. However, Cr vaporization from $(Mn,Cr)_3O_4$ and subsequent cathode poisoning is still a concern for the Cr-bearing spinel. This is why Cr-free spinel oxides with high conductivity are preferred as the coating materials. Among binary spinel oxides composed of metals Mg, Al, Cr, Mn, Fe, Co, Ni, Cu, and Zn, the promising candidates for metallic interconnect coating on ferritic stainless steels are Co_3O_4, $Mn_xCo_{3-x}O_4$, and $Cu_xMn_{3-x}O_4$ ($1 < x < 1.5$); Co_3O_4 has the lowest conductivity and the greatest thermal expansion mismatch with the alloy, and MCO and $Cu_{1.3}Mn_{1.7}O_4$ have the highest conductivity in air among transition metal spinel oxides, that is, 60 $S\,cm^{-1}$ at 800 °C and 225 $S\,cm^{-1}$ at 750 °C, respectively [250]. Electroplated Co on SUS 430 [251] and Crofer 22 APU [252], as shown in Figure 5.15, which diffuses into the alloy and reacts with alloying elements, forms an oxide scale consisting of a Co-doped Cr_2O_3 sublayer underneath the nearly Cr-free Co_3O_4, thus offering oxidation protection, alleviating Cr evaporation, and reducing the ASR of SUS 430 alloy (26 $m\Omega\,cm^2$ at 800 °C after oxidation for 1900 h). Once again, in such coating, voids are formed during Co oxidation, which is a concern for long-term stability of the coating.

$(Mn,Co)_3O_4$ as a coating of metallic interconnects has been prepared by various kinds of processes, such as slurry coating [129, 253–259], atmospheric plasma spraying [260, 261], sol–gel coating [175, 262], magnetron sputtering [263, 264], electroplating [265–269], and spray pyrolysis [270]. All the coatings demonstrate positive effect on oxidation behavior, electrical conductivity, and Cr evaporation for ferritic Cr_2O_3-forming Fe–Cr alloys, with excellent thermal cyclicability during long-term oxidation in air [262, 175, 254]. The coating as a barrier blocks both Cr outward and oxygen inward diffusions, suppressing the growth of oxide scale formed between the coating and alloy substrate and Cr poisoning on the cathode. The life of $(Mn,Co)_3O_4$-coated alloy is significantly longer than the uncoated one, as predicted by an integrated experimental/modeling approach [271]. It is worth noting that the Cr-rich scale formed between the coating and the alloy is important for blocking oxygen inward diffusion, as shown by the ^{18}O isotope distribution profile in Figure 5.16 [272]. Oxygen ions can still diffuse through the MCO-dense coating at 800 °C and are obstructed by the thermally grown thin Cr-rich oxides.

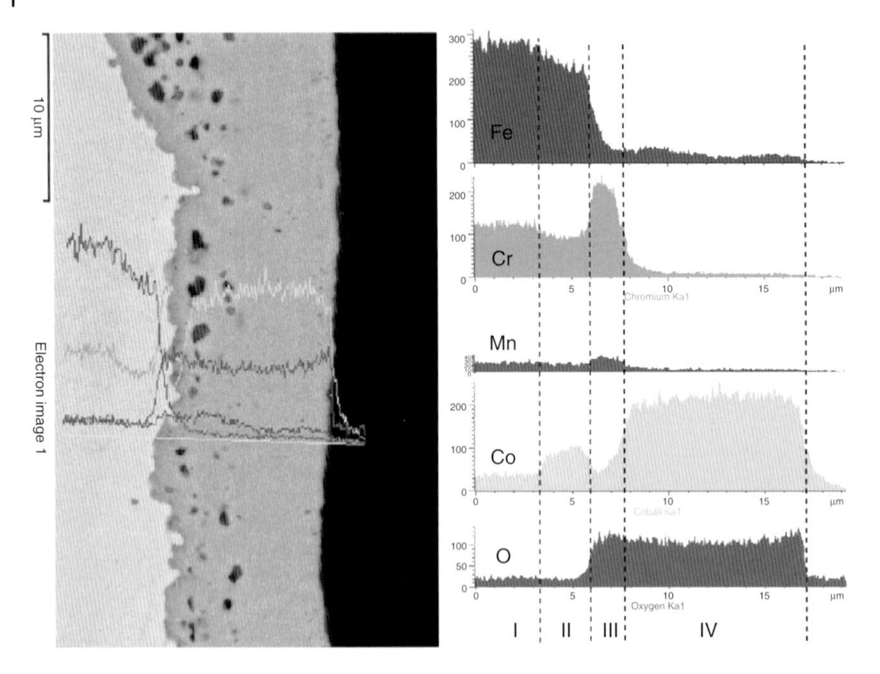

Figure 5.15 Morphology and compositional analysis of oxidized Co on Crofer 22 APU: I, steel substrate; II, Co diffusion layer; III, Fe-, Mn-, and Co-doped Cr_2O_3 layer; and IV, Co_3O_4 top layer. (Source: Reproduced with permission from Ref. [252].)

For the Ni–Cr alloy, such as Haynes 230, the effectiveness of $(Mn,Co)_3O_4$ coating is not obvious [259] and may not be necessary. Even the electrical conductivity of $(Mn,Co)_3O_4$ is three to four and one to two orders of magnitude higher than that of Cr_2O_3 and $MnCr_2O_4$, respectively [129, 250, 253]; thinning the coating, improving coating adherence, and increasing its conductivity by composition modification are still necessary. It is shown that addition of Ce [273], double coating with LSM [274] or La_2O_3 [275], and partial Fe substitution for Co in $(Mn,Co)_3O_4$ [247, 276] improve the long-term stability and electrical performance. Fe substitution for Co particularly increases the electrical conductivity of MCO by enhancing electron hopping between Fe^{2+} and Fe^{3+} ions [64]; however, the level of substitution needs to be carefully controlled to maintain the stability. $MnCo_{1.9}Fe_{0.1}O_4$ seems to be the preferred composition.

Other Cr-free spinels such as $(Cu,Mn)_3O_4$ [265, 147, 277, 278], Cu–Fe spinel [278], $NiCo_2O_4$ [279], $CoFe_2O_4$ [280], and $NiFe_2O_4$ [281, 282] are also evaluated as potential coatings for the metallic interconnects. They all result in improvement in oxidation resistance, electrical conductivity, and Cr evaporation, except for Cu–Fe spinel, which increases the ASR by forming Fe oxides. Evidence suggests that loss of Cu from the spinel occurs at 800 °C in air, depositing on the surface of LSM cathode in contact possibly via volatile Cu species [147]. In this regard, more attention should be paid to the effect of Cu loss from $(Cu,Mn)_3O_4$ on the coating and cathode performances. Sol–gel prepared $NiCo_2O_4$ coating on SUS 430 alloy

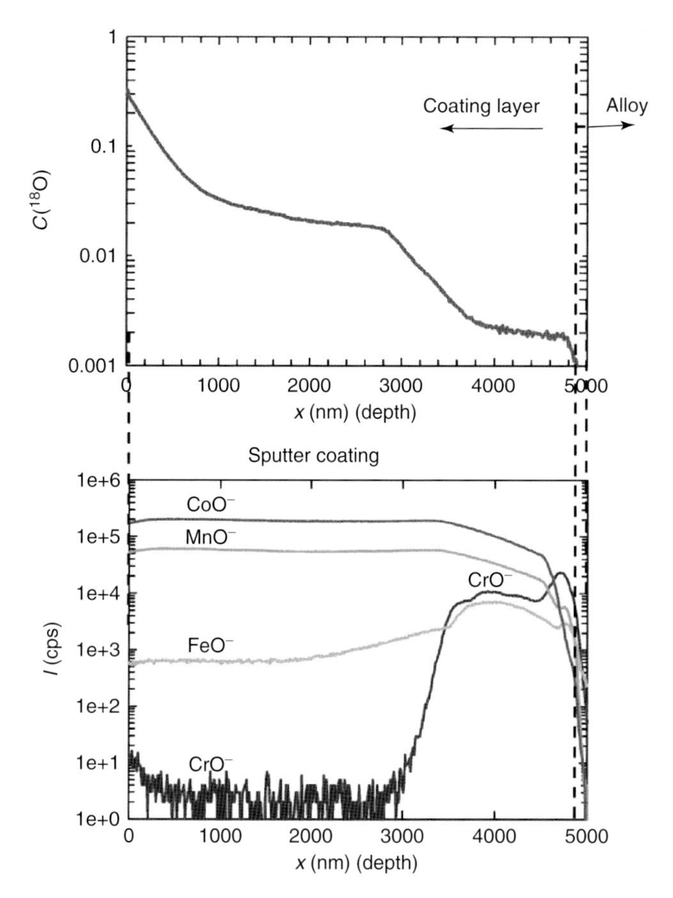

Figure 5.16 Isotope oxygen (^{18}O) diffusion profile and elemental distributions near the $MnCo_2O_4$ coating–alloy interface ($MnCo_2O_4$ coating by RF sputtering, MO^- counts by SIMS (secondary ion mass spectrometry)). (Source: Reproduced with permission from Ref. [272].)

achieved a low parabolic oxidation rate of 8.1×10^{-15} g^2 cm^{-4} s^{-1} and a low ASR of ~3 mΩ cm^2 at 800 °C in air after preoxidation at 800 °C for 200 h, which can be attributed to the thin Cr_2O_3 formed underneath the coating [279]. In order to fully confirm the effectiveness of $NiCo_2O_4$ coating, long-term experiments are required.

5.5
New Alloy Development

As can be seen from the above sections, some Cr_2O_3-forming Fe–Cr or Ni–Cr alloys are promising candidates for applications as interconnects of SOFCs. However, insufficient oxidation resistance and electrical conductivity on long-term oxidation prevent them from being satisfactorily employed. Besides, Cr vaporization from

Cr-containing oxides and the subsequent cathode poisoning is another challenge they face. Applying surface coating seems to be effective in enhancing oxidation resistance, improving the electrical conductivity, and inhibiting Cr evaporation. However, all kinds of coatings are not as effective as expected and lack long-term evaluation. Furthermore, coating process of any kind adds significant cost to that of the interconnect, compromising the advantage of metallic interconnects as a cost-effective substitution for their ceramic counterparts. Based on this situation, an alternative approach is to search or develop other metallic alloys that form a desired oxide scale for oxidation protection, electronic conduction, and Cr retention on the surface during SOFC operation.

Generally, Cr content of alloys is important for oxidation resistance, which usually increases as Cr content increases. Alloys with insufficient Cr content cannot form a continuous Cr_2O_3 protective scale, which usually requires at least 13 wt% Cr in Fe–Cr and Ni–Cr alloys; on the other hand, excessive Cr in the alloy may result in enhanced growth of Cr_2O_3 and unacceptable cathode Cr poisoning, as described in Section 5.2.4. Appropriate Cr content must be able to balance oxidation resistance and Cr retention at SOFC operating temperatures in the oxidizing and reducing environments. Mn and Ti in Fe–Cr and Ni–Cr alloys have a significant effect on oxidation rate and scale ASR. Mn enhances the formation of Mn–Cr spinel on the surface of the scale because of its faster diffusion in Cr_2O_3 than Cr; Ti causes the formation of TiO_2 or Ti-doped Cr_2O_3, increasing the oxidation rate. Thus, excessive Mn and Ti in alloys are not desired. On the other hand, the formation of Mn–Cr spinel is beneficial to suppress the growth of Cr_2O_3 underneath, alleviate Cr evaporation to some extent, and improve scale ASR because of its lower electrical resistance than that of Cr_2O_3. Ti-doped Cr_2O_3 and TiO_2 are more conductive than Cr_2O_3, improving the electrical conductivity of the scale. Moreover, Ti tends to form TiO_2 precipitates at oxide scale–alloy interface, which improves the mechanical property and adherence of the oxide scale. Addition of reactive elements (REs = La, Y, Ce, etc.) to Cr_2O_3-forming alloys enhances the formation of continuous Cr_2O_3 scale, reduces scale growth rate, and increases scale adherence. Addition of Co can enhance conductivity by forming Co-doped spinel and Cr_2O_3. Elements such as Mo and W are added to lower the TEC of the alloy for a better match to other components in contact. Addition of the Laves-type phase forming elements, such as Nb and Mo, to the alloy is effective in controlling the cation diffusivity at grain boundaries by forming the Laves-type phases, which can reduce the oxidation rate and improve electrical property. In addition, high alloy purity is beneficial to oxidation resistance and oxide scale adherence. C and S in the alloys have a detrimental effect on the oxide scale growth rate and mechanical properties. A low level of Al and Si is beneficial for preventing the formation of nonconductive Al_2O_3 and SiO_2, respectively. Thus, in new alloy development, the compositional modification of alloys is critical to have a desired oxide scale property.

Modified Fe–Co–Ni alloys with low thermal expansion and low Cr content as listed in Table 5.1 were selected [39] and evaluated as a new type of metallic interconnect materials. After up to 500 h of oxidation at 800 °C in air, an extremely thick oxide scale (>100 μm) was formed, which consisted of Cr-free and electrically

Table 5.1 Nominal composition of modified Fe–Co–Ni low-thermal-expansion alloys (wt%).

	Fe	Co	Ni	Cr	Nb	Ti	Si	Al	B
Three-Phase	balance	18	42	—	3.0	1.5	—	6.0	—
Exp 4005	balance	31	33	—	3.0	0.6	—	5.3	—
Thermo-Span	balance	29	24.5	5.5	4.8	0.85	0.35	0.45	0.004
HRA 929C	balance	22.5	29.5	2.0	4.0	1.25	0.3	0.55	0.0045

conductive Co_3O_4 and/or $CoFe_2O_4$ surface oxides on top of mixed oxides (Fe, Ni, Co, Al, Ti, and Nb), Al_2O_3, Cr-rich oxide, or internal oxides. Even though the ASR of such oxide scale is lower than that of Crofer 22 APU alloy and the Cr-free top layer can act as a barrier to Cr evaporation, such excessive oxidation, due to the lack of Cr to form continuous Cr_2O_3 protective layer, makes them unacceptable as an interconnect material. As a matter of fact, Cr evaporation of the alloys with such a low Cr content is not an issue. With a modified composition of Fe–30Co–25Ni–6Cr–5Nb–1.5Si–0.1Y [283], the oxidation resistance is improved to a rate similar to that of Crofer 22 APU at 800 °C in air for up to 2000 h after the initial fast oxidation due to the formation of $(Fe,Co,Ni)_3O_4$ spinel. A continuous Cr_2O_3 layer is formed between the spinel and the alloy because of the Cr enrichment in this area and the assistance of Si, Nb, and Y elements. However, the TEC of the Fe–Co–Ni alloys is significantly lower than that of the cell components ($10–13 \times 10^{-6}$), which may still cause thermal expansion mismatch.

In an approach to modify ferritic Fe–Cr alloys, Mo (0.5 wt%) and Nb (0.35 wt%) are added to Fe–20Cr alloy [90]. When oxidized in H_2/H_2O atmosphere at 800 °C for up to 1100 h, the grain boundary Laves phase, such as Fe_2Nb, is formed, in addition to the frequently observed Cr_2O_3 and Cr–Mn spinel in the scale. It is the grain boundary Laves phase that blocks rapid cation outward diffusion along grain boundaries, resulting in an increase in oxidation resistance and decrease in ASR. This subtle modification in composition of Fe–20Cr alloy seems very effective for performance improvement; the same approach should be employed for low-cost commercial ferritic Fe–Cr alloys, such as SUS 430. In order to develop a more conductive oxide scale, Ti-containing alloys Fe–(6–22)Cr–0.5Mn–1Ti were prepared with a purpose of forming conductive outer TiO_2 scale and evaluated in moist air (3% H_2O) at 800 °C [284]. With Cr content lower than 12 wt%, the alloys are not oxidation resistant, forming Fe-rich oxides irrespective of whether the surface is Ce-treated or not. The Ce surface treatment suppresses Fe-rich oxides and promotes TiO_2 formation in alloys containing higher Cr in the range 12–22 wt%, leading to overgrowth of a nonstoichiometric TiO_2 layer with the potential to prevent Cr vaporization. It is once again demonstrated that the Cr content needs to reach around 18 wt% for a performed oxidation resistance.

Recently, Hua *et al.* [59] developed a promising Fe–Cr alloy of Fe–17Cr–1Mn–0.5Ti–2.1Mo with the addition of La, Y, and Zr. It has a TEC of 12.2×10^{-6} between 35 and 800 °C, well matched to that of the cell components. After isothermal or cyclic oxidation at 750 °C in air for up to 1000 h, it forms a multilayered conductive oxide scale consisting of top Mn_2O_3, middle Mn–Cr spinel, and bottom Cr_2O_3, as shown in Figure 5.17 [59], with a low oxidation rate between 5.1×10^{-14} and 7.6×10^{-14} $g^2\,cm^{-4}\,s^{-1}$ and a low ASR of ~10 $m\Omega\,cm^2$. Mn_2O_3 nodules form on the surface of the scale initially and gradually cover the whole area densely with prolonged oxidation time. TiO_2 internal oxide is also formed sporadically in the alloy underlying the scale and along the grain boundaries. The high mobility of Ti and Mn ions promotes their outward diffusion into the formed oxide scale, resulting in doped Cr_2O_3 and spinel. Cyclic oxidation seems to improve the density of the Mn_2O_3 layer, scale adherence, and ASR.

Other than Fe–Cr alloys, a low-Cr-content Ni–Mo–Cr alloy of Ni–20.76Mo–12.24Cr–4.12W–3.02Co–1.18Ti–0.98Mn with minor addition of Y, La, C, and Si [285, 100] has been developed. It has a TEC of 13.92×10^{-6}, close to that of the cell components. After oxidation in air and moist H_2 at 750 °C for up to 1000 h, it forms an oxide scale consisting of top Cr-free $NiMn_2O_4$ and bottom Cr_2O_3 in air and top $MnCr_2O_4$ and bottom Cr_2O_3 in moist H_2, as shown in Figure 5.18 and Figure 5.19, with a parabolic rate constant below 1.56×10^{-14} $g^2\,cm^{-4}\,s^{-1}$ and 4.94×10^{-14} $g^2\,cm^{-4}\,s^{-1}$, respectively. Additionally, a layer of $MoNi_3$ intermetallic compound is formed in the alloy adjacent to the oxide scale, which slows down the outward diffusion of Cr, Ni, or Mn ions for further development of the oxide scale. The ASR of the formed oxide scale at 750 °C is 4.48 and 8.34 $m\Omega\,cm^2$ in air and moist H_2, respectively, with an extrapolation of 22 $m\Omega\,cm^2$ to oxidation of 40 000 h in air. The dense layer of Cr-free $NiMn_2O_4$ prevents Cr evaporation from inner Cr-containing oxides, thus alleviating Cr poisoning on the LSM cathode [286]. The overall results suggest that the newly developed Ni–Mo–Cr alloy is a promising candidate for metallic interconnect applications.

5.6 Summary

Interconnect materials play a crucial role in the fabrication of SOFC stacks. The use of metallic interconnects, instead of their ceramic counterparts, offers many advantages in terms of thermal, electrical, and mechanical properties, as well as material and fabrication costs. The promising candidate materials for metallic interconnects are Cr_2O_3-forming ferritic Fe–Cr alloys. However, challenges of long-term oxidation resistance, oxide scale stability and conductivity, and Cr evaporation from Cr-containing oxides need to be overcome for successful deployment of metallic interconnects. Surface modification of metallic alloys via conductive and protective coatings has been proved to be effective in addressing these issues; in particular, the Cr-free spinels are believed to be the most promising coating materials for this purpose. Moreover, development of new alloys with compatible TEC and desired oxide scale property is an alternative approach to meet all of

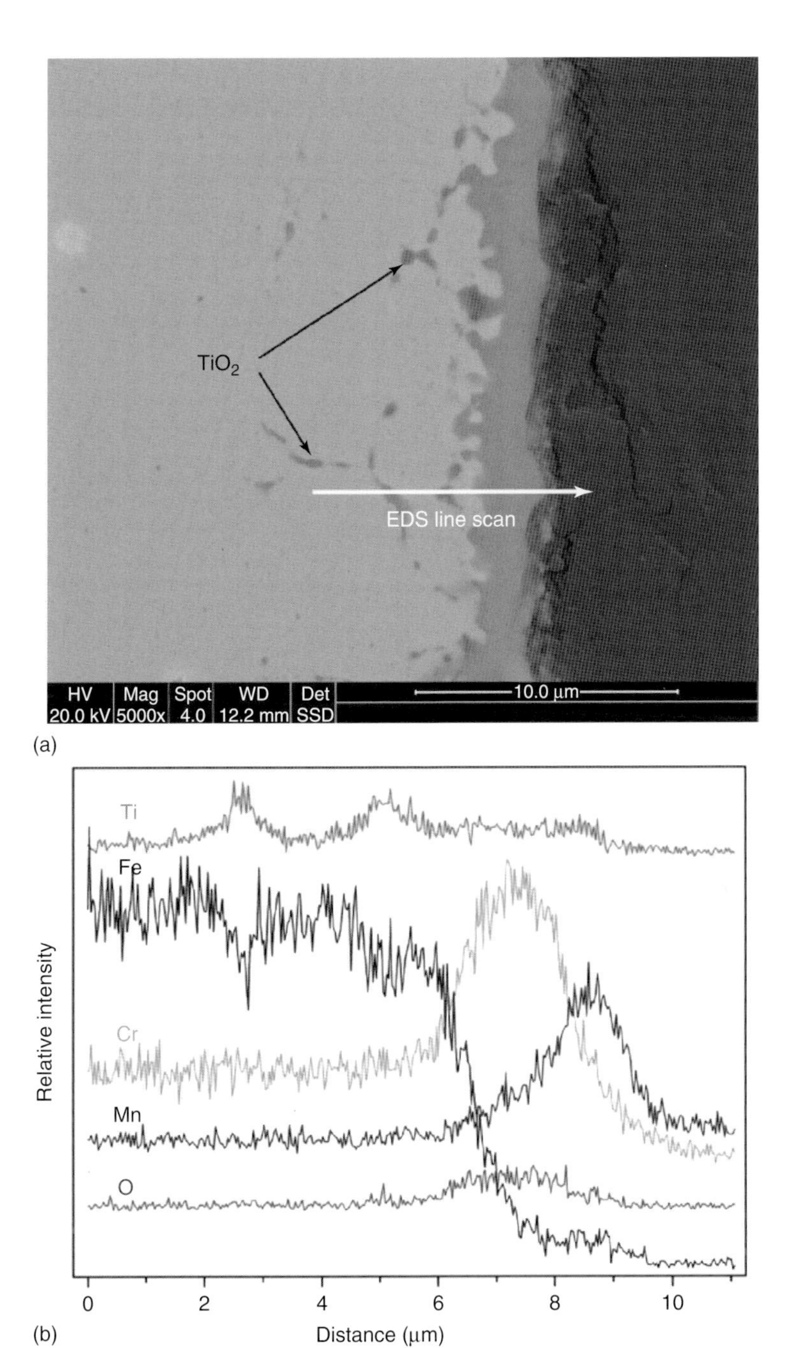

Figure 5.17 SEM backscattering electron morphology (a) and EDS line scan profile (b) of the cross section of Fe–Cr alloy cyclically oxidized at 750 °C in air for 1000 h. (Source: Reproduced with permission from Ref. [59].)

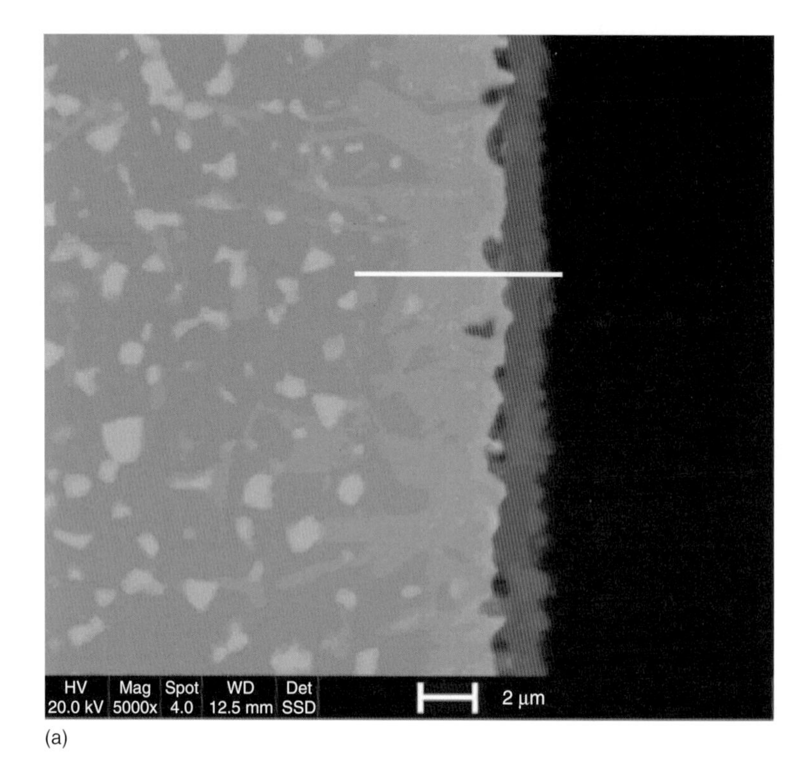

(a)

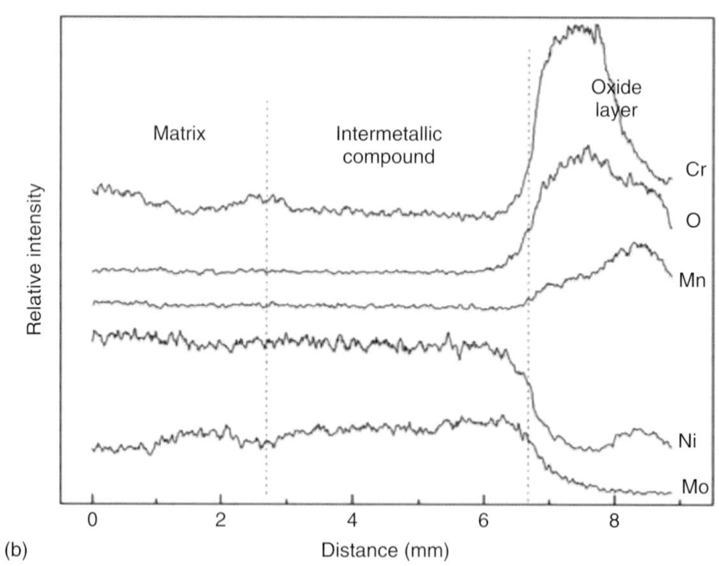

(b)

Figure 5.18 SEM backscattering image (a) and EDS line scan profile (b) of the cross section of Ni–Mo–Cr alloy oxidized at 750 °C for 1000 h in air. (Source: Reproduced with permission from Ref. [285].)

(a)

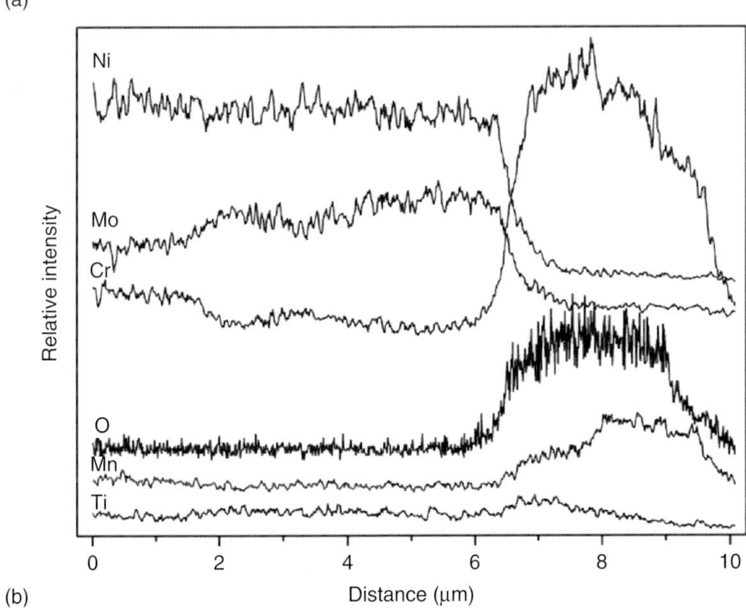

(b)

Figure 5.19 (a) SEM morphology and (b) EDS line scan profile of the cross section of Ni–Mo–Cr alloy oxidized at 750 °C in moist H_2 for 1000 h. (Source: Reproduced with permission from Ref. [100].)

the requirements for metallic interconnect materials. Further work in the area of long-term oxidation behavior, compatibility with components in contact, and cost reduction of the metallic interconnect are still necessary for metallic alloys as the interconnect in SOFC stacks.

References

1. Singh, P. and Minh, N.Q. (2004) Solid oxide fuel cells: technology status. *Int. J. Appl. Ceram. Technol.*, **1** (1), 5–15.
2. Williams, M.C. and Singhal, S.C. (2000) Mass-produced ceramic fuel cells for low-cost power: the solid state energy conversion alliance. *Fuel Cells Bull.*, **3** (24), 8–11.
3. Singhal, S.C. (2002) Solid oxide fuel cells for stationary, mobile, and military applications. *Solid State Ionics*, **152–153**, 405–410.
4. Zhu, W.Z. and Deevi, S.C. (2003) Development of interconnect materials for solid oxide fuel cells. *Mater. Sci. Eng., A*, **A348** (1–2), 227–243.
5. Fergus, J.W. (2005) Metallic interconnects for solid oxide fuel cells. *Mater. Sci. Eng., A*, **397** (1–2), 271–283.
6. Zhu, W.Z. and Deevi, S.C. (2003) Opportunity of metallic interconnects for solid oxide fuel cells: a status on contact resistance. *Mater. Res. Bull.*, **38** (6), 957–972.
7. Yang, Z. (2008) Recent advances in metallic interconnects for solid oxide fuel cells. *Int. Mater. Rev.*, **53** (1), 39–54.
8. Yang, Z., Weil, K.S., Paxton, D.M., and Stevenson, J.W. (2003) Selection and evaluation of heat-resistant alloys for SOFC interconnect applications. *J. Electrochem. Soc.*, **150** (9), A1188–A1201.
9. Yang, Z., Xia, G., Li, X., Maupin, G.D., Coleman, J.E., Nie, Z., Bonnett, J.F., Simner, S.P., Stevenson, J.W., and Singh, P. (2006) Advanced SOFC interconnect development at PNNL. *ECS Trans.*, **5** (1), 347–356.
10. Badwal, S.P.S., Deller, R., Foger, K., Ramprakash, Y., and Zhang, J.P. (1997) Interaction between chromia forming alloy interconnects and air electrode of solid oxide fuel cells. *Solid State Ionics*, **99** (3–4), 297–310.
11. Singhal, S.C. (2000) Science and technology of solid-oxide fuel cells. *MRS Bull.*, **25** (3), 16–21.
12. Minh, N.Q. (1993) Ceramic fuel cells. *J. Am. Ceram. Soc.*, **76** (3), 563–588.
13. Badwal, S.P.S. (2001) Stability of solid oxide fuel cell components. *Solid State Ionics*, **143** (1), 39–46.
14. Huang, K., Hou, P.Y., and Goodenough, J.B. (2000) Characterization of iron-based alloy interconnects for reduced temperature solid oxide fuel cells. *Solid State Ionics*, **129** (1), 237–250.
15. Shaigan, N., Qu, W., Ivey, D.G., and Chen, W. (2010) A review of recent progress in coatings, surface modifications and alloy developments for solid oxide fuel cell ferritic stainless steel interconnects. *J. Power Sources*, **195** (6), 1529–1542.
16. Quadakkers, W.J., Greiner, H., Hänsel, M., Pattanaik, A., Khanna, A.S., and Malléner, W. (1996) Compatibility of perovskites contact layers between cathode and metallic interconnector plates of SOFCs. *Solid State Ionics*, **91** (1–2), 55–67.
17. Hatchwell, C., Sammes, N.M., Brown, I.W.M., and Kendall, K. (1999) Current collectors for a novel tubular design of solid oxide fuel cell. *J. Power Sources*, **77** (1), 64–68.
18. Quadakkers, W.J., Hänsel, M., and Rieck, T. (1998) Carburization of Cr-based ODS alloys in SOFC relevant environments. *Mater. Corros.*, **49** (4), 252–257.
19. Larring, Y. and Norby, T. (2000) Spinel and perovskite functional layers between plansee metallic interconnect (Cr-5 wt% Fe-1 wt% Y_2O_3) and ceramic $(La_{0.85}Sr_{0.15})_{0.91}MnO_3$ Cathode

materials for solid oxide fuel cells. *J. Electrochem. Soc.*, **147** (9), 3251–3256.

20. Konysheva, E., Penkalla, H., Wessel, E., Mertens, J., Seeling, U., Singheiser, L., and Hilpert, K. (2006) Chromium poisoning of perovskites cathodes by the ODS alloy $Cr5Fe1Y_2O_3$ and the high chromium ferritic steel Crofer22APU. *J. Electrochem. Soc.*, **153** (4), A765–A773.

21. Zhu, J.H., Geng, S.J., Lu, Z.G., and Porter, W.D. (2007) Evaluation of binary Fe-Ni alloys as intermediate-temperature SOFC interconnect. *J. Electrochem. Soc.*, **154** (12), B1288–B1294.

22. Brady, M.P., Pint, B.A., Lu, Z.G., Zhu, J.H., Milliken, C.E., Kreidler, E.D., Miller, L., Armstrong, T.R., and Walker, L.R. (2006) Comparison of oxidation behavior and electrical properties of doped NiO- and Cr_2O_3-forming alloys for solid oxide fuel cell metallic interconnects. *Oxid. Met.*, **65** (3–4), 237–261.

23. Church, B.C., Sanders, T.H., Speyer, R.F., and Cochran, J.K. (2007) Thermal expansion matching and oxidation resistance of Fe-Ni-Cr interconnect alloys. *Mater. Sci. Eng., A*, **452–453**, 334–340.

24. England, D.M. and Virkar, A.V. (1999) Oxidation kinetics of some nickel-based superalloy foils and electronic resistance of the oxide scale formed in air part I. *J. Electrochem. Soc.*, **146** (9), 3196–3202.

25. England, D.M. and Virkar, A.V. (2001) Oxidation kinetics of some nickel-based superalloy foils in humidified hydrogen and electronic resistance of the oxide scale formed part II. *J. Electrochem. Soc.*, **148** (4), A330–A338.

26. Li, J., Pu, J., Xiao, J., and Qian, X. (2005) Oxidation of Haynes 230 alloy in reduced temperature solid oxide fuel cell environments. *J. Power Sources*, **139** (1–2), 182–187.

27. Li, J., Pu, J., Xie, G., Wang, S., and Xiao, J. (2006) Heat resistant alloys as interconnect materials of reduced temperature SOFCs. *J. Power Sources*, **157** (1), 368–376.

28. Li, J., Pu, J., Hua, B., and Xie, G. (2006) Oxidation kinetics of Haynes 230 alloy in air at temperatures between 650 and 850 °C. *J. Power Sources*, **159** (1), 641–645.

29. Geng, S.J., Zhu, J.H., and Lu, Z.G. (2006) Evaluation of Haynes 242 alloy as SOFC interconnect material. *Solid State Ionics*, **177** (5–6), 559–568.

30. Geng, S.J., Zhu, J.H., and Lu, Z.G. (2006) Investigation on Haynes 242 alloy as SOFC interconnect in simulated anode environment. *Electrochem. Solid-State Lett.*, **9** (4), A211–A214.

31. Yang, Z., Xia, G., and Stevenson, J.W. (2006) Evaluation of Ni-Cr-base alloys for SOFC interconnect applications. *J. Power Sources*, **160** (2), 1104–1110.

32. Yang, Z., Singh, P., Stevenson, J.W., and Xia, G. (2006) Investigation of modified Ni-Cr-Mn base alloys for SOFC interconnect applications. *J. Electrochem. Soc.*, **153** (10), A1873–A1879.

33. Horita, T., Xiong, Y., Yamaji, K., Sakai, N., and Yokokawa, H. (2002) Characterization of Fe-Cr alloys for reduced operation temperature SOFCs. *Fuel Cells*, **2** (3–4), 189–194.

34. Jian, L., Huezo, J., and Ivey, D.G. (2003) Carburisation of interconnect materials in solid oxide fuel cells. *J. Power Sources*, **123** (2), 151–162.

35. Horita, T., Xiong, Y., Kishimoto, H., Yamaji, K., Sakai, N., and Yokokawa, H. (2004) Application of Fe-Cr alloys to solid oxide fuel cells for cost-reduction oxidation behavior of alloys in methane fuel. *J. Power Sources*, **131** (1–2), 293–298.

36. Brylewski, T., Dabek, J., and Przybylski, K. (2004) Oxidation kinetics study of the iron-based steel for solid oxide fuel cell application. *J. Therm. Anal. Calorim.*, **77** (1), 207–216.

37. Pu, J., Li, J., Hua, B., and Xie, G. (2006) Oxidation kinetics and phase evolution of a Fe-16Cr alloy in simulated SOFC cathode atmosphere. *J. Power Sources*, **158** (1), 354–360.

38. Geng, S.J. and Zhu, J.H. (2006) Promising alloys for intermediate-temperature solid oxide fuel cell interconnect application. *J. Power Sources*, **160** (2), 1009–1016.

39. Zhu, J.H., Geng, S.J., and Ballard, D.A. (2007) Evaluation of several low

thermal expansion Fe-Co-Ni alloys as interconnect for reduced-temperature solid oxide fuel cell. *Int. J. Hydrogen Energy*, **32** (16), 3682–3688.
40. Antepara, I., Villarreal, I., Rodríguez-Martínez, L.M., Lecanda, N., Castro, U., and Laresgoití, A. (2005) Evaluation of ferritic steels for use as interconnects and porous metal supports in IT-SOFCs. *J. Power Sources*, **151** (1–2), 103–107.
41. Yang, Z., Hardy, J.S., Walker, M.S., Xia, G., Simner, S.P., and Stevenson, J.W. (2004) Structure and conductivity of thermally grown scales on ferritic Fe-Cr-Mn steel for SOFC interconnect applications. *J. Electrochem. Soc.*, **151** (11), A1825–A1831.
42. Simner, S.P., Anderson, M.D., Xia, G., Yang, Z., Pederson, L.R., and Stevenson, J.W. (2005) SOFC performance with Fe-Cr-Mn alloy interconnect. *J. Electrochem. Soc.*, **152** (4), A740–A745.
43. Han, M., Peng, S., Wang, Z., Yang, Z., and Chen, X. (2007) Properties of Fe-Cr based alloy alloys as interconnects in a solid oxide fuel cell. *J. Power Sources*, **164** (1), 278–283.
44. Mustala, S., Veivo, J., Auerkari, P., and Kiviaho, J. (2007) Thermal degradation of selected alloys for SOFC interconnectors. *ECS Trans.*, **2** (9), 151–157.
45. Toji, A. and Uehara, T. (2007) Stability of oxidation resistance of ferritic Fe-Cr alloy for SOFC interconnects. *ECS Trans.*, **7** (1), 2117–2124.
46. Alman, D.E. and Jablonski, P.D. (2007) Effect of minor elements and a Ce surface treatment on the oxidation behavior of an Fe-22Cr-0.5Mn (Crofer 22 APU) ferritic stainless steel. *Int. J. Hydrogen Energy*, **32** (16), 3743–3753.
47. Ogasawara, K., Kameda, H., Matsuzaki, Y., Sakurai, T., Uehara, T., Toji, A., Sakai, N., Yamaji, K., Horita, T., and Yokokawa, H. (2007) Chemical stability of ferritic alloy interconnect for SOFCs. *J. Electrochem. Soc.*, **154** (7), B657–B663.
48. Belogolovsky, I., Hou, P.Y., Jacobson, C.P., and Visco, S.J. (2008) Chromia scale adhesion on 430 stainless steel: effect of different surface treatments. *J. Power Sources*, **182** (1), 259–264.
49. Sun, X., Liu, W.N., Stephens, E., and Khaleel, M.A. (2008) Determination of interfacial adhesion strength between oxide scale and substrate for metallic SOFC interconnects. *J. Power Sources*, **176** (1), 167–174.
50. Cooper, L., Benhaddad, S., Wood, A., and Ivey, D.G. (2008) The effect of surface treatment on the oxidation of ferritic stainless steels used for solid oxide fuel cell interconnects. *J. Power Sources*, **184** (1), 220–228.
51. Liu, W.N., Sun, X., and Khaleel, M.A. (2010) Effect of creep of ferritic interconnect on long-term performance of solid oxide fuel cell stacks. *Fuel Cells*. doi: 10.1002/fuce. 200900075
52. Horita, T., Xiong, Y., Yamaji, K., Sakai, N., and Yokokawa, H. (2003) Evaluation of Fe-Cr alloys as interconnects for reduced operation temperature SOFCs. *J. Electrochem. Soc.*, **150** (3), A243–A248.
53. Yang, Z., Walker, M.S., Singh, P., Stevenson, J.W., and Norby, T. (2004) Oxidation behavior of ferritic stainless steels under SOFC interconnect exposure conditions. *J. Electrochem. Soc.*, **151** (12), B669–B678.
54. Shaigan, N., Ivey, D.G., and Chen, W. (2009) Metal-oxide scale interfacial imperfections and performance of stainless steels utilized as interconnects in solid oxide fuel cells. *J. Electrochem. Soc.*, **256** (6), B765–B770.
55. Tortorelli, P.F. and DeVan, J.H. (1992) Behavior of iron aluminides in oxidizing and oxidizing/sulfidizing environments. *Mater. Sci. Eng., A*, **A153** (1–2), 573–577.
56. DeVan, J.H. and Tortorelli, P.F. (1993) The oxidation-sulfidation behavior of iron alloys containing 16–40 AT% aluminum. *Corros. Sci.*, **35** (5–8), 1065–1071.
57. Pint, B.A., Martin, J.R., and Hobbs, L.W. (1993) $_{18}O$/SIMS characterization of the growth mechanism of doped and undoped α-Al_2O_3. *Oxid. Met.*, **39** (3–4), 167–195.

58. Pint, B.A. (1996) Experimental observations in support of the dynamic-segregation theory to explain the reactive-element effect. *Oxid. Met.*, **45** (1–2), 1–37.
59. Hua, B., Pu, J., Lu, F., Zhang, J., Chi, B., and Li, J. (2010) Development of a Fe-Cr alloy for interconnect application in intermediate temperature solid oxide fuel cells. *J. Power Sources*, **195** (9), 2782–2788.
60. Wild, R.K. (1977) High temperature oxidation of austenitic stainless steel in low oxygen pressure. *Corros. Sci.*, **17** (2), 87–104.
61. Lobnig, R.E., Schmidt, H.P., Hennesen, K., and Grabke, H.J. (1992) Diffusion of cations in chromia layers grown on iron-base alloys. *Oxid. Met.*, **37** (1–2), 81–93.
62. Cox, M.G.C., McEnaney, B., and Scott, V.D. (1972) Chemical diffusion model for partitioning of transition elements in oxide scales on alloys. *Philos. Mag.*, **26** (4), 839–851.
63. Hong, S.H., Phaniraj, M.P., Kim, D.-I., Ahn, J.-P., Cho, Y.W., Han, S.H., and Han, H.N. (2010) Improvement in oxidation resistance of ferritic stainless steel by carbon ion implantation. *Electrochem. Solid-State Lett.*, **13** (4), B40–B42.
64. Sakai, N., Horita, T., Xiong, Y., Yamaji, K., Kishimoto, H., Brito, M.E., Yokokawa, H., and Maruyama, T. (2005) Structure and transport property of manganese-chromium-iron oxide as a main compound in oxide scales of alloy interconnects for SOFCs. *Solid State Ionics*, **176** (7–8), 681–686.
65. Lu, Z., Zhu, J., Andrew, P.E., and Paranthaman, M.P. (2005) Electrical conductivity of the manganese chromite spinel solid solution. *J. Am. Ceram. Soc.*, **88** (4), 1050–1053.
66. Geng, S.J., Zhu, J.H., and Lu, Z.G. (2006) Evaluation of several alloys for solid oxide fuel cell interconnect application. *Scr. Mater.*, **55** (3), 239–242.
67. Yasuda, N., Uehara, T., Okamoto, M., Aoki, C., Ohno, T., and Toji, A. (2009) Improvement of oxidation resistance of Fe-Cr ferritic alloy sheets for SOFC interconnects. *ECS Trans.*, **25** (2), 1447–1453.
68. Caplan, D., Beaubien, P.E., and Cohen, M. (1965) *Trans. AIME*, **233**, 766.
69. Marasco, A.L. and Young, D.J. (1991) The oxidation of iron-chromium-manganese alloys at 900 °C. *Oxid. Met.*, **36** (1–2), 157–174.
70. Cox, M.G.C., McEnaney, B., and Scott, V.D. (1975) Kinetics of initial oxide growth on Fe-Cr alloys and the role of vacancies in film breakdown. *Philos. Mag.*, **31** (2), 331–338.
71. Matsunaga, S. and Homma, T. (1976) Influence on the oxidation kinetics of metals by control of the structure of oxide scales. *Oxid. Met.*, **10** (6), 361–376.
72. Atkinson, H.V. (1987) Evolution of grain structure in nickel oxide scales. *Oxid. Met.*, **28** (5–6), 353–389.
73. Kofstad, P. and Lillerud, K.P. (1982) Chromium transport through Cr_2O_3 scales. I. On Lattice diffusion of chromium. *Oxid. Met.*, **17** (3–4), 177–194.
74. Horita, T., Xiong, Y., Yamaji, K., Kishimoto, H., Sakai, N., Brito, M.E., and Yokokawa, H. (2004) Imaging of mass transports around the oxide scale/Fe-Cr alloy interfaces. *Solid State Ionics*, **174** (1–4), 41–48.
75. Sakai, N., Horita, T., Yamaji, K., Xiong, Y., Kishimoto, H., Brito, M.E., and Yokokawa, H. (2006) Material transport and degradation behavior of SOFC interconnects. *Solid State Ionics*, **177** (19–25), 1933–1939.
76. Horita, T., Kishimoto, H., Yamaji, K., Xiong, Y., Sakai, N., Brito, M.E., and Yokokawa, H. (2008) Effect of grain boundaries on the formation of oxide scale in Fe-Cr alloy for SOFCs. *Solid State Ionics*, **179** (27–32), 1320–1324.
77. Froitzheim, J., Meier, G.H., Niewolak, L., Ennis, P.J., Hattendorf, H., Singheiser, L., and Quadakkers, W.J. (2008) Development of high strength ferritic steel for interconnect application in SOFCs. *J. Power Sources*, **178** (1), 163–173.
78. Jablonski, P.D., Cowen, C.J., and Sears, J.S. (2010) Exploration of alloy 441

chemistry for solid oxide fuel cell interconnect application. *J. Power Sources*, **195** (3), 813–820.

79. Tsai, S.C., Huntz, A.M., and Dolin, C. (1996) Growth mechanism of Cr_2O_3 scales: oxygen and chromium diffusion, oxidation kinetics and effect of yttrium. *Mater. Sci. Eng., A*, **A212** (1), 6–13.
80. Moosa, A.A. and Rothman, S.J. (1985) Effect of yttrium additions on the oxidation of nickel. *Oxid. Met.*, **24** (3–4), 133–148.
81. Moosa, A.A., Rothman, S.J., and Nowicki, L.J. (1985) Effect of yttrium addition to nickel on the volume and grain boundary diffusion of Ni in the scale formed on the alloy. *Oxid. Met.*, **24** (3–4), 115–132.
82. Cotell, C.M., Yurek, G.J., Hussey, R.J., Mitchell, D.F., and Graham, M.J. (1990) The influence of grain-boundary segregation of Y in Cr_2O_3 on the oxidation of Cr metal. *Oxid. Met.*, **34** (3–4), 173–200.
83. Kim, K.Y., Kim, S.H., Kwon, K.W., and Kim, L.H. (1994) Effect of yttrium on the stability of aluminide-yttrium composite coatings in a cyclic high-temperature hot-corrosion environment. *Oxid. Met.*, **41** (3–4), 179–201.
84. Brylewski, T., Nanko, M., Maruyama, T., and Przybylski, K. (2001) Application of Fe-16Cr ferritic alloy to interconnector for a solid oxide fuel cell. *Solid State Ionics*, **143** (2), 131–150.
85. Quadakkers, W.J., Malkow, T., Prion-Abbellán, J., Flesch, U., Shemet, V., and Singheiser, L. (2000) in *Proceeding of the 4th Solid Oxide Fuel Cell Forum, Oberrohrdorf, Swilzerland*, European Fuel Cell Forum, Vol. 2 (ed. A.J. McEvoy), Elsevier, Amsterdam, pp. 827–836.
86. Prion-Abbellán, J., Tidtz, F., Shemet, V., Gil, A., Ledwein, T., Singheiser, L., and Quadakkers, W.J. (2002) in (ed. J. Huijsmans) *Proceedings of the 5th Solid Oxide Fuel Cell Forum*, European Fuel Cell Group, Lucerne, Switzerland, pp. 248–256.
87. Horita, T., Yamaji, K., Xiong, Y., Kishimoto, H., Sakai, N., and Yokokawa, H. (2004) Oxide scale formation of Fe-Cr alloys and oxygen diffusion in the scale. *Solid State Ionics*, **175** (1–4), 157–163.
88. Horita, T., Yamaji, K., Yokokawa, H., Toji, A., Uehara, T., Ogasawara, K., Kameda, H., Matsuzaki, Y., and Yamashita, S. (2008) Effects of Si and Al concentrations in Fe-Cr alloy on the formation of oxide scales in H_2-H_2O. *Int. J. Hydrogen Energy*, **33** (21), 6308–6315.
89. Liu, Y. (2008) Performance evaluation of several commercial alloys in a reducing environment. *J. Power Sources*, **179** (1), 286–291.
90. Horita, T., Kishimoto, H., Yamaji, K., Xiong, Y., Sakai, N., Brito, M.E., and Yokokawa, H. (2008) Evaluation of Laves-phase forming Fe-Cr for SOFC interconnects in reducing atmosphere. *J. Power Sources*, **176** (1), 54–61.
91. Asteman, H., Evensson, J.E., and Johansson, L.G. (2002) Oxidation of 310 steel in H_2O/O_2 mixtures at 600 °C: the effect of water-vapour-enhanced chromium evaporation. *Corros. Sci.*, **44** (11), 2635–2649.
92. Mikkelsen, L. and Linderoth, S. (2003) High temperature oxidation of Fe-Cr alloy in O_2-H_2-H_2O atmosphere: microstructure and kinetics. *Mater. Sci. Eng., A*, **A361** (1–2), 198–212.
93. Saunders, S.R.J., Monteiro, M., and Rizzo, F. (2008) The oxidation behaviour of metals and alloys at high temperatures in atmospheres containing water vapour: a review. *Prog. Mater. Sci.*, **53** (5), 775–837.
94. Othman, N.K., Othman, N., Zhang, J., and Young, D.J. (2009) Effects of water vapour on isothermal oxidation of chromia-forming alloys in Ar/O_2 and Ar/H_2 atmospheres. *Corros. Sci.*, **51** (12), 3039–3049.
95. Essuman, E., Meier, G.H., Żurek, J., Hänsel, M., and Quadakkers, W.J. (2008) The effect of water vapor on selective oxidation of Fe-Cr alloys. *Oxid. Met.*, **69** (3–4), 143–162.
96. Sánchez, L., Hierro, M.P., and Pérez, F.J. (2009) Effect of chromium content on the oxidation behaviour of ferritic

steels for applications in steam atmospheres at high temperatures. *Oxid. Met.*, **71** (3–4), 173–186.

97. Jian, L., Yuh, C.Y., and Farooque, M. (2000) Oxidation behavior of superalloys in oxidizing and reducing environments. *Corros. Sci.*, **42** (9), 1573–1585.
98. Hänsel, M., Quadakkers, W.J., and Young, D.J. (2003) Role of water vapor in chromia-scale growth at low oxygen partial pressure. *Oxid. Met.*, **59** (3–4), 285–301.
99. Fontana, S., Chevalier, S., and Caboche, G. (2009) Metallic interconnects for solid oxide fuel cell: effect of water vapour on oxidation resistance of differently coated alloys. *J. Power Sources*, **193** (1), 136–145.
100. Hua, B., Lu, F., Zhang, J., Kong, Y., Pu, J., Chi, B., and Jian, L. (2009) Oxidation behavior and electrical property of a Ni-based alloy in SOFC anode environment. *J. Electrochem. Soc.*, **156** (10), B1261–B1266.
101. Young, E.W.A., Stiphout, P.C.M., and de Wit, J.H.W. (1985) N-type of chromium(III) oxide. *J. Electrochem. Soc.*, **132** (4), 884–886.
102. Maris-Sida, M.C., Meier, G.H., and Pettit, F.S. (2003) Some water vapor effects during the oxidation of alloys that are α-Al2O_3 formers. *Metall. Mater. Trans. A*, **34A** (11), 2609–2619.
103. Horita, T., Xiong, Y., Yamaji, K., Sakai, N., and Yokokawa, H. (2003) Stability of Fe-Cr alloy interconnects under CH_4-H_2O atmosphere for SOFCs. *J. Power Sources*, **118** (1–2), 35–43.
104. Horita, T., Xiong, Y., Kishimoto, H., Yamaji, K., Sakai, N., Brito, M.E., and Yokokawa, H. (2005) Oxidation behavior of Fe-Cr- and Ni-Cr-based alloy interconnects in CH_4-H_2O for solid oxide fuel cells. *J. Electrochem. Soc.*, **152** (11), A2193–A2198.
105. Horita, T., Xiong, Y., Kishimoto, H., Yamaji, K., Sakai, N., Brito, M.E., and Yokokawa, H. (2006) Surface analyses and depth profiles of oxide scales formed on alloy surface. *Surf. Interface Anal.*, **38** (4), 282–286.
106. Horita, T., Kishimoto, H., Yamaji, K., Sakai, N., Xiong, Y., Brito, M.E., and Yokokawa, H. (2006) Effects of silicon concentration in SOFC alloy interconnects on the formation of oxide scales in hydrocarbon fuels. *J. Power Sources*, **157** (2), 681–687.
107. Li, Y., Wu, J., Johnson, C., Gemmen, R., Scott, X.M., and Liu, X. (2009) Oxidation behavior of metallic interconnects for SOFC in coal syngas. *Int. J. Hydrogen Energy*, **34** (3), 1489–1496.
108. Liu, K., Luo, J., Johnson, C., Liu, X., Yang, J., and Mao, S.C. (2008) Conduction oxide formation and mechanical endurance of potential solid-oxide fuel cell interconnects in coal syngas environment. *J. Power Sources*, **183** (1), 247–252.
109. Zeng, Z. and Natesan, K. (2004) Corrosion of metallic interconnects for SOFC in fuel gases. *Solid State Ionics*, **167** (1–2), 9–16.
110. Yang, Z., Walker, M.S., Singh, P., and Stevenson, J.W. (2003) Anomalous corrosion behavior of stainless steels under SOFC interconnect exposure conditions. *Electrochem. Solid-State Lett.*, **6** (10), B35–B37.
111. Yang, Z., Xia, G., Singh, P., and Stevenson, J.W. (2005) Effects of water vapor on oxidation behavior of ferritic stainless steels under solid oxide fuel cell interconnect exposure conditions. *Solid State Ionics*, **176** (17–18), 1495–1503.
112. Yang, Z., Xia, G., Walker, M.S., Wang, C.-M., Stevenson, W.J., and Singh, P. (2007) High temperature oxidation/corrosion behavior of metals and alloys under a hydrogen gradient. *Int. J. Hydrogen Energy*, **32** (16), 3770–3777.
113. Horita, T., Kshimoto, H., Yamaji, K., Sakai, N., Xiong, Y., Brito, M.E., and Yokokawa, H. (2008) Anomalous oxidation of ferritic interconnects in solid oxide fuel cells. *Int. J. Hydrogen Energy*, **33** (14), 3962–3969.
114. Rufner, J., Gannon, P., White, P., Deibert, M., Teintze, S., Smith, R., and Chen, H. (2008) Oxidation behavior of stainless steel 430 and 441 at 800 °C in single (air/air) and dual atmosphere (air/hydrogen) exposures. *Int. J. Hydrogen Energy*, **33** (4), 1392–1398.

115. Gannon, P.E. and White, P.T. (2009) Oxidation of ferritic steels subjected to simulated SOFC interconnect environments. *ECS Trans.*, **16** (44), 53–56.
116. Kurokawa, H., Oyama, Y., Kawamura, K., and Maruyama, T. (2004) Hydrogen permeation through Fe-16Cr alloy interconnect in atmosphere simulating SOFC at 1073 K. *J. Electrochem. Soc.*, **151** (8), A1264–A1268.
117. Zeng, Z., Natesan, K., and Cai, S.B. (2008) Characterization of oxide scale on alloy 446 by X-ray nanobeam analysis. *Electrochem. Solid-State Lett.*, **11** (1), C5–C8.
118. Kurokawa, H., Kawamura, K., and Maruyama, T. (2004) Oxidation behavior of Fe-16Cr alloy interconnect for SOFC under hydrogen potential gradient. *Solid State Ionics*, **168** (1–2), 13–21.
119. Jiang, S.P., Zhang, S., and Zhen, Y.D. (2005) Early interaction between Fe-Cr alloy metallic interconnect and Sr-doped $LaMnO_3$ cathodes of solid oxide fuel cells. *J. Mater. Res.*, **20** (3), 747–758.
120. Jiang, S.P., Zhen, Y.D., and Zhang, S. (2006) Interaction between Fe-Cr metallic interconnect and (La, Sr)MnO_3/YSZ composite cathode of solid oxide fuel cells. *J. Electrochem. Soc.*, **153** (8), A1511–A1517.
121. Chen, X., Zhang, L., and Jiang, S.P. (2008) Chromium deposition and poisoning on $(La_{0.6}Sr_{0.4-x}Ba_x)(Co_{0.2}Fe_{0.8})O_3 (0 \leq x \leq 0.4)$ Cathodes of solid oxide fuel cells. *J. Electrochem. Soc.*, **155** (11), B1093–B1101.
122. Fergus, J.W. (2007) Effect of cathode and electrolyte transport properties on chromium poisoning in solid oxide fuel cells. *Int. J. Hydrogen Energy*, **32** (16), 3664–3671.
123. Yamauchi, A., Kurokawa, K., and Takahashi, H. (2003) Evaporation of Cr_2O_3 in atmospheres containing H_2O. *Oxid. Met.*, **59** (5–6), 517–527.
124. Hilpert, K., Das, D., Miller, M., Peck, D.H., and Weiβ, R. (1996) Chromium vapor species over solid oxide fuel cell interconnect materials and their potential for degradation processes. *J. Electrochem. Soc.*, **143** (11), 3642–3647.
125. Gindorf, C., Singheiser, L., and Hilpert, K. (2005) Vaporisation of chromia in humid air. *J. Phys. Chem. Solids*, **66** (2–4), 384–387.
126. Chen, X., Zhen, Y., Li, J., and Jiang, S.P. (2010) Chromium deposition and poisoning in dry and humidified air at $(La_{0.8}Sr_{0.2})_{0.9}MnO_{3+\delta}$ cathodes of solid oxide fuel cells. *Int. J. Hydrogen Energy*, **35** (6), 2477–2485.
127. Gindorf, C., Singheiser, L., Hilpert, K., Schroeder, M., Martin, M., Greiner, H., and Richter, F. (1999) *Solid Oxide Fuel Cells (SOFC VI). Proceedings of the 6th International Symposium*, The Electrochemical Society, Hawaii, pp. 774–782.
128. Gindorf, C., Hilpert, K., and Singheiser, L. (2001) *Solid Oxide Fuel Cells (SOFC VII). Proceedings of the Seventh International Symposium*, The Electrochemical Society, Pennington, NJ, pp. 793–802.
129. Yang, Z., Xia, G., Li, X., and Stevenson, J.W. (2007) $(Mn,Co)_3O_4$ spinel coatings on ferritic stainless steels for SOFC interconnect applications. *Int. J. Hydrogen Energy*, **32** (16), 3648–3654.
130. Fujita, K., Hashimoto, T., Ogasawara, K., Kameda, H., Matsuzaki, Y., and Sakurai, T. (2004) Relationship between electrochemical properties of SOFC cathode and composition of oxide layer formed on metallic interconnects. *J. Power Sources*, **131** (1–2), 270–277.
131. Stanislowski, M., Wessel, E., Hilpert, K., Markus, T., and Singheiser, L. (2007) Chromium vaporization from high-temperature alloys I. Chromia-forming steels and the influence of outer oxide layers. *J. Electrochem. Soc.*, **154** (4), A295–A306.
132. Taniguchi, S., Kadowaki, M., Kawamura, H., Yasuo, T., Akiyama, Y., Miyake, Y., and Saitoh, T. (1995) Degradation phenomena in the cathode of a solid oxide fuel cell with an alloy separator. *J. Power Sources*, **55** (1), 73–79.
133. Matsuzaki, Y. and Yasuda, I. (2000) Electrochemical properties of a SOFC

cathode in contact with a chromium-containing alloy separator. *Solid State Ionics*, **132** (3), 271–278.

134. Matsuzaki, Y. and Yasuda, I. (2001) Dependence of SOFC cathode degradation by chromium-containing alloy on compositions of electrodes and electrolytes. *J. Electrochem. Soc.*, **148** (2), A126–A131.

135. Fujita, K., Ogasawara, K., Matsuzaki, Y., and Sakurai, T. (2004) Prevention of SOFC cathodes degradation in contact with Cr-containing alloy. *J. Power Sources*, **131** (1–2), 261–269.

136. Paulson, S.C. and Birss, V.I. (2004) Chromium poisoning of LSM-YSZ SOFC cathodes I. Detailed study of the distribution of chromium species at a porous, single-phase cathode. *J. Electrochem. Soc.*, **151** (11), A1961–A1968.

137. Hou, P.Y. and Stringer, J. (1995) The effect of reactive element additions on the selective oxidation, growth and adhesion of chromia scales. *Mater. Sci. Eng., A*, **A202** (1–2), 1–10.

138. Jiang, S.P., Zhang, J.P., Apateanu, L., and Foger, K. (1999) Deposition of chromium species on Sr-doped $LaMnO_3$ cathodes in solid oxide fuel cells. *Electrochem. Commun.*, **1** (9), 394–397.

139. Jiang, S.P., Zhang, J.P., Apateanu, L., and Foger, K. (2000) Deposition of chromium species at Sr-doped $LaMnO_3$ electrodes in solid oxide fuel cells I. Mechanism and kinetics. *J. Electrochem. Soc.*, **147** (11), 4013–4022.

140. Jiang, S.P., Zhang, J.P., and Foger, K. (2000) Deposition of chromium species at Sr-doped $LaMnO_3$ electrodes in solid oxide fuel cells. II. Effect on O_2 reduction reaction. *J. Electrochem. Soc.*, **147** (9), 3195–3205.

141. Jiang, S.P., Zhang, J.P., and Foger, K. (2001) Deposition of chromium species at Sr-doped $LaMnO_3$ electrodes in solid oxide fuel cells III. Effect of air flow. *J. Electrochem. Soc.*, **148** (7), C447–C455.

142. Jiang, S.P., Zhang, J.P., and Zheng, X.G. (2002) A comparative investigation of chromium deposition at air electrodes of solid oxide fuel cells. *J. Eur. Ceram. Soc.*, **22** (3), 361–373.

143. Jiang, S.P., Zhang, S., and Zhen, Y.D. (2006) Deposition of Cr species at $(La,Sr)(Co,Fe)O_3$ cathodes of solid oxide fuel cells. *J. Electrochem. Soc.*, **153** (1), A127–A134.

144. Zhen, Y.D., Li, J., and Jiang, S.P. (2006) Oxygen reduction on strontium-doped $LaMnO_3$ cathodes in the absence and presence of an iron-chromium alloy interconnect. *J. Power Sources*, **162** (2), 1043–1052.

145. Jiang, S.P. and Zhen, Y.D. (2008) Mechanism of Cr deposition and its application in the development of Cr-tolerant cathodes of solid oxide fuel cells. *Solid State Ionics*, **179** (27–32), 1459–1464.

146. Kurokawa, H., Jacobson, C.P., DeJonghe, L.C., and Visco, S.J. (2007) Chromium vaporization of bare and of coated iron-chromium alloys at 1073K. *Solid State Ionics*, **178** (3–4), 287–296.

147. Paulson, S.C., Bateni, M.R., Wei, P., Petric, A., and Birss, V.I. (2007) Improving LSM cathode performance using $(Cu,Mn)_3O_4$ spinel coated UNS430 ferritic stainless steel SOFC interconnects. *ECS Trans.*, **7** (1), 1097–1106.

148. Li, X., Lee, J., and Popov, B.N. (2009) Performance studies of solid oxide fuel cell cathodes in the presence of bare and cobalt coated E-brite alloy interconnects. *J. Power Sources*, **187** (2), 356–362.

149. Collins, C., Lucas, J., Buchanan, T.L., Kopczyk, M., Kayani, A., Gannon, P.E., Deibert, M.C., Smith, R.J., Choi, D.S., and Gorokhovsky, V.I. (2006) Chromium volatility of coated and uncoated steel interconnects for SOFCs. *Surf. Coat. Technol.*, **201** (7), 4467–4470.

150. Nielsen, K.A., Persson, A., Beeaff, D., Høgh, J., Mikkelsen, L., and Hendriksen, P.V. (2007) Initiation and performance of a coating for countering chromium poisoning in a SOFC stack. *ECS Trans.*, **7** (1), 2145–2154.

151. Stanislowski, M., Froitzheim, J., Niewolak, L., Quadakkers, W.J., Hilpert, K., Markus, T., and Singheiser, L. (2007) Reduction of chromium vaporization from SOFC interconnectors

by highly effective coatings. *J. Power Sources*, **164** (2), 578–589.

152. Trebbels, R., Markus, T., and Singheiser, L. (2009) Reduction of chromium evaporation with manganese-based coatings. *ECS Trans.*, **25** (2), 1417–1422.
153. Hong, S.H., Madakashira, P., Kim, D.-I., Cho, Y.W., Han, S.H., and Han, H.N. (2009) Oxidation behavior and chromium vaporization of ion implanted Crofer22APU. *ECS Trans.*, **25** (2), 1437–1446.
154. Trebbels, R., Markus, T., and Singheiser, L. (2010) Investigation of chromium vaporization form interconnector steels with spinel coatings. *J. Electrochem. Soc.*, **157** (4), B490–B495.
155. Yang, Z., Xia, G., Singh, P., and Stevenson, J.W. (2006) Electrical contacts between cathodes and metallic interconnects in solid oxide fuel cells. *J. Power Sources*, **155** (2), 246–252.
156. Mahapatra, M.K. and Lu, K. (2009) Interfacial study of crofer 22 APU interconnect-SABS-0 seal glass for solid oxide fuel/electrolyzer cells. *J. Mater. Sci.*, **44** (20), 5569–5578.
157. Widgeon, S.J., Corral, E.L., Spilde, M.N., and Loehman, R.E. (2009) Glass-to-metal seal interfacial analysis using electron probe microscopy for reliable solid oxide fuel cells. *J. Am. Ceram. Soc.*, **92** (4), 781–786.
158. Horita, T., Kishimoto, H., Yamaji, K., Brito, M.E., Xiong, Y., and Yokokawa, H. (2009) Anomalous oxide scale formation under exposure of sodium containing gases for solid oxide fuel cell alloy interconnects. *J. Power Sources*, **193** (1), 180–184.
159. Goel, A., Tulyaganov, D.U., Kharton, V.V., Yaremchenko, A.A., and Ferreira, J.M.F. (2010) Electrical behavior of aluminosilicate glass-ceramic sealants and their interaction with metallic solid oxide fuel cell interconnects. *J. Power Sources*, **195** (2), 522–526.
160. Jin, T. and Lu, K. (2010) Compatibility between AISI441 alloy interconnect and representative seal glasses in solid oxide fuel/electrolyzer cells. *J. Power Sources*, **195** (15), 4853–4864.
161. Yang, Z., Meinhardt, K.D., and Stevenson, J.W. (2003) Chemical compatibility of barium-calcium-aluminosilicate-based sealing glasses with the ferritic stainless steel interconnect in SOFCs. *J. Electrochem. Soc.*, **150** (8), A1095–A1101.
162. Menzler, N.H., Batfalsky, P., Blum, L., Bram, M., Groß, S.M., Haanappel, V.A.C., Malzbender, J., Shemet, V., Steinbrech, R.W., and Vinke, I. (2007) Studies of material interaction after long-term stack operation. *Fuel Cells*, **7** (5), 356–363.
163. Chou, Y., Stevenson, J.W., and Singh, P. (2007) Novel refractory alkaline earth silicate sealing glasses for planar solid oxide fuel cells. *J. Electrochem. Soc.*, **154** (7), B644–B651.
164. Yang, Z., Stevenson, J.W., and Meinhardt, K.D. (2003) Chemical interactions of barium-calcium-aluminosilicate-based sealing glasses with oxidation resistant alloys. *Solid State Ionics*, **160** (3–4), 213–225.
165. Chou, Y.-S., Stevenson, J.W., Xia, G.-G., and Yang, Z.-G. (2010) Electrical stability of a novel sealing glass with (Mn,Co)-spinel coated Crofer22APU in a simulated SOFC dual environment. *J. Power Sources*, **195** (17), 5666–5673.
166. Chou, Y., Stevenson, J.W., and Singh, P. (2008) Effect of aluminizing of Cr-containing ferritic alloys on the seal strength of a novel high-temperature solid oxide fuel cell sealing glass. *J. Power Sources*, **185** (2), 1001–1008.
167. Horita, T., Kishimoto, H., Yamaji, K., Xiong, Y., Sakai, N., Brito, M.E., and Yokokawa, H. (2006) Oxide scale formation and stability of Fe-Cr alloy interconnects under dual atmospheres and current flow conditions for SOFCs. *J. Electrochem. Soc.*, **153** (11), A2007–A2012.
168. Wiener, F., Bram, M., Buchkremer, H.P., and Sebold, D. (2007) Chemical interaction between crofer 22 APU and mica-based gaskets under simulated SOFC conditions. *J. Mater. Sci.*, **42** (8), 2643–2651.
169. Menzler, N.H., Sebold, D., Zahid, M., Gross, S.M., and Koppitz, T. (2005) Interaction of metallic SOFC interconnect

materials with glass-ceramic sealant in various atmospheres. *J. Power Sources*, **152** (1–2), 156–167.

170. Quadakkers, W.J., Prion-Abbellán, J., Shemet, V., and Singheiser, L. (2003) Metallic interconnectors for solid oxide fuel cells-a review. *Mater. High Temp.*, **20** (2), 115–127.

171. Holt, A. and Kofstad, P. (1994) Electrical conductivity and defect structure of Cr_2O_3. I. High temperatures (>~1000°C). *Solid State Ionics*, **69** (2), 127–136.

172. Holt, A. and Kofstad, P. (1994) Electrical conductivity and defect structure of Cr_2O_3. II. Reduced temperatures (<~1000°C). *Solid State Ionics*, **69** (2), 137–143.

173. Huang, K., Hou, P.Y., and Goodenough, J.B. (2001) Reduced area specific resistance for iron-based metallic interconnects by surface oxide coatings. *Mater. Res. Bull.*, **36** (1–2), 81–95.

174. Park, J.H. and Natesan, K. (1990) Electronic transport in thermally grown Cr_2O_3. *Oxid. Met.*, **33** (1–2), 31–54.

175. Hua, B. *et al.* (2010) The electrical property of $MnCo_2O_4$ and its application for SUS 430 metallic interconnect. *Chin. Sci. Bull.*, **55**, 3831–3837.

176. Kvernes, I.A. (1973) The role of yttrium in high-temperature oxidation behavior of Ni-Cr-Al alloys. *Oxid. Met.*, **6** (1), 45–64.

177. Golightly, F.A., Stott, F.H., and Wood, G.C. (1976) The influence of yttrium additions on the oxide-scale adhesion to an iron-chromium aluminum alloy. *Oxid. Met.*, **10** (3), 163–187.

178. Ramanarayanan, T.A., Ayer, R., Petkovic-Luton, R., and Leta, D.P. (1988) The influence of yttrium on oxide scale growth and adherence. *Oxid. Met.*, **29** (5–6), 445–472.

179. Moon, D.P. (1989) The reactive element effect on the growth rate of nickel oxide scales at high temperature. *Oxid. Met.*, **32** (1–2), 47–66.

180. Biegun, T., Danielewski, M., and Skrzypek, Z. (1992) The reactive-element effect in the high-temperature oxidation of Fe-23Cr-5Al commercial alloys. *Oxid. Met.*, **38** (3–4), 207–215.

181. Roure, S., Czerwinski, F., and Petric, A. (1994) Influence of CeO2-coating on the high-temperature oxidation of chromium. *Oxid. Met.*, **42** (1–2), 75–102.

182. Molin, S., Kusz, B., Gazda, M., and Jasinski, P. (2009) Protective coatings for stainless steel for SOFC applications. *J. Solid State Electrochem.*, **13** (11), 1695–1700.

183. Riffard, F., Buscail, H., Caudron, E., Cueff, R., Issartel, C., and Perrier, S. (2003) Yttrium sol–gel coating effects on the cyclic oxidation behaviour of 304 stainless steel. *Corros. Sci.*, **45** (12), 2867–2880.

184. Riffard, F., Buscail, H., Caudron, E., Cueff, R., Issartel, C., and Perrier, S. (2002) Effect of yttrium addition by sol–gel coating and ion implantation on the high temperature oxidation behaviour of the 304 steel. *Appl. Surf. Sci.*, **199** (1–4), 107–122.

185. Cabouro, G., Caboche, G., Chevalier, S., and Piccardo, P. (2006) Opportunity of metallic interconnects for ITSOFC: reactivity and electrical property. *J. Power Sources*, **156** (1), 39–44.

186. Hamid, A.U. (2002) TEM study of the effect of Y on the scale microstructures of Cr_2O_3. and Al_2O_3-forming alloys. *Oxid. Met.*, **58** (1–2), 23–40.

187. Oishi, N., Namikawa, T., and Yamazaki, Y. (2000) Oxidation behavior of an La-coated chromia-forming alloy and the electrical property of oxide scales. *Surf. Coat. Technol.*, **132** (1), 58–64.

188. Piccardo, P., Chevalier, S., Molins, R., Viviani, M., Caboche, G., Barbucci, A., Sennour, M., and Amendola, R. (2006) Metallic interconnects for SOFC: Characterization of their corrosion resistance in hydrogen/water atmosphere and at the operating temperatures of differently coated metallic alloys. *Surf. Coat. Technol.*, **201** (7), 4471–4475.

189. Qu, W., Li, J., and Ivey, D.G. (2004) Sol–gel coatings to reduce oxide growth in interconnects used for solid oxide fuel cells. *J. Power Sources*, **138** (1–2), 162–173.

190. Qu, W., Li, J., Ivey, D.G., and Hill, J.M. (2006) Yttrium, cobalt and yttrium/cobalt oxide coatings on ferritic stainless steels for SOFC interconnects. *J. Power Sources*, **157** (1), 335–350.

191. Tondo, E., Boniardi, M., Cannoletta, D., Riccardis, M.F.D., and Bozzini, B. (2010) Electrodeposition of yttria/cobalt oxide and yttria/gold coatings onto ferritic stainless steel for SOFC interconnects. *J. Power Sources*, **195** (15), 4772–4778.

192. Kim, S.-H., Huh, J.-Y., Jun, J.-H., and Favergeon, J. (2010) Thin elemental coatings of yttrium, cobalt, and yttrium/cobalt on ferritic stainless steel for SOFC interconnect applications. *Curr. Appl. Phys.*, **10** (10), S86–S90.

193. Wu, J., Li, C., Johnson, C., and Liu, X. (2008) Evaluation of SmCo and SmCoN magnetron sputtering coatings for SOFC interconnect applications. *J. Power Sources*, **175** (2), 833–840.

194. Jun, J. and Kim, D. (2007) Effects of REM coatings on electrical conductivity of ferritic stainless steels for SOFC interconnect applications. *ECS Trans.*, **7** (1), 2385–2390.

195. Seo, H., Jin, G., Jun, J., Kim, D., and Kim, K. (2008) Effect of reactive elements on oxidation behaviour of Fe-22Cr-0.5Mn Ferritic stainless steel for a solid oxide fuel cell interconnect. *J. Power Sources*, **178** (1), 1–8.

196. Chevalier, S. and Larpin, J.P. (2003) Influence of reactive element oxide coatings on the high temperature cyclic oxidation of chromia-forming steels. *Mater. Sci. Eng., A*, **A363** (1–2), 116–125.

197. Chevalier, S., Valot, C., Bonnet, G., Colson, J.C., and Larpin, J.P. (2003) The reactive element effect on thermally grown chromia scale residual stress. *Mater. Sci. Eng., A*, **A343** (1–2), 257–264.

198. Fontana, S., Amendola, R., Chevalier, S., Piccardo, P., Caboche, G., Viviani, M., Molins, R., and Sennour, M. (2007) Metallic interconnects for SOFC: characterisation of corrosion resistance and conductivity evaluation at operating temperature of differently coated alloys. *J. Power Sources*, **171** (2), 652–662.

199. Piccardo, P., Amendola, R., Fontana, S., Chevalier, S., Caboches, G., and Gannon, P. (2009) Interconnect materials for next-generation solid oxide fuel cells. *J. Appl. Electrochem.*, **39** (4), 545–551.

200. Simkovich, G. (1995) The change in growth mechanism of scales due to reactive elements. *Oxid. Met.*, **44** (5–6), 501–504.

201. Stott, F.H., Wood, G.C., and Stringer, J. (1995) The influence of alloying elements on the development and maintenance of protective scales. *Oxid. Met.*, **44** (1–2), 113–145.

202. Cotell, C.M., Yurek, G.J., Hussey, R.J., Mitchell, D.F., and Graham, M.J. (1990) The influence of grain-boundary segregation of Y in Cr_2O_3 on the oxidation of Cr metal. II. Effects of temperature and dopant concentration. *Oxid. Met.*, **34** (3–4), 201–216.

203. Chevalier, S., Bonnet, G., Dufour, P., and Larpin, J.P. (1998) The REE: a way to improve the high-temperature behavior of stainless steels. *Surf. Coat. Technol.*, **100–101** (1–3), 208–213.

204. Pieraggi, B. and Rapp, R.A. (1993) Chromia scale growth in alloy oxidation and the reactive element effect. *J. Electrochem. Soc.*, **140** (10), 2844–2850.

205. Kayani, A., Buchanan, T.L., Kopczyk, M., Collins, C., Lucas, J., Lund, K., Hutchison, R., Gannon, P.E., Deiber, M.C., Smith, R.J., Choi, D.S., and Gorokhovsky, V.I. (2006) Oxidation resistance of magnetron-sputtered CrAlN coatings on 430 steel at 800 °C. *Surf. Coat. Technol.*, **201** (7), 4460–4466.

206. Liu, X., Johnson, C., Li, C., Xu, J., and Cross, C. (2008) Developing TiAlN coatings for intermediate temperature solid oxide fuel cell interconnect applications. *Int. J. Hydrogen Energy*, **33** (1), 189–196.

207. Gannon, P.E., Kayani, A., Ramana, V., Deibert, M.C., Smith, R.J., and Gorokhovsky, V.I. (2008) Simulated SOFC interconnect performance of crofer 22 APU with and without filtered arc CrAlON coatings. *Electrochem. Solid-State Lett.*, **11** (4), B54–B58.

208. Gannon, P., Deibert, M., White, P., Smith, R., Chen, H., Priyantha, W.,

Lucas, J., and Gorokhovsky, V. (2008) Advanced PVD protective coatings for SOFC interconnects. *Int. J. Hydrogen Energy*, **33** (14), 3991–4000.

209. Chen, H., Lucas, J.A., Priyantha, W., Kopczyk, M., Smith, R.J., Lund, K., Key, C., Finsterbusch, M., Gannon, P.E., Deibert, M., Gorokhovsky, V.I., Shutthanandan, V., and Nachimuthu, P. (2008) Thermal stability and oxidation resistance of TiCrAlYO coatings on SS430 for solid oxide fuel cell interconnect applications. *Surf. Coat. Technol.*, **202** (19), 4820–4824.

210. Gannon, P.E., Gorokhovsky, V.I., Deibert, M.C., Smith, R.J., Kayani, A., White, P.T., Sofie, S., Yang, Z., McCready, D., Visco, S., Jacobson, C., and Kurokawa, H. (2007) Enabling inexpensive metallic alloy as SOFC interconnects: an investigation into hybrid coating technologies to deposit nanocomposite functional coatings on ferritic stainless steels. *Int. J. Hydrogen Energy*, **32** (16), 3672–3681.

211. Gorokhovsky, V.I., Gannon, P.E., Deibert, M.C., Smith, R.J., Kayani, A., Kopczyk, M., VanVorous, D., Yang, Z., Stevenson, J.W., Visco, S., Jacobson, C., Kurokawa, H., and Sofie, S.W. (2006) Deposition and evaluation of protective PVD coatings on ferritic stainless steel SOFC interconnects. *J. Electrochem. Soc.*, **153** (10), A1886–A1893.

212. Piccardo, P., Gannon, P., Chevalier, S., Viviani, M., Barbucci, A., Caboche, G., Amendola, R., and Fontana, S. (2007) ASR evaluation of different kinds of coatings on a ferritic stainless steel as SOFC interconnects. *Surf. Coat. Technol.*, **202** (4–7), 1221–1225.

213. Nielsen, K.A., Dinesen, A.R., Korcakova, L., Mikkelsen, L., Hendriksen, P.V., and Poulsen, F.W. (2006) Testing of Ni-plated ferritic steel interconnect in SOFC stacks. *Fuel Cells*, **6** (2), 100–106.

214. Fu, C., Sun, K., Chen, X., Zhang, N., and Zhou, D. (2008) Effects of the nickel-coated ferritic stainless steel for solid oxide fuel cells interconnects. *Corros. Sci.*, **50** (7), 1926–1931.

215. Lu, Z., Zhu, J., Pan, Y., Wu, N., and Ignatiev, A. (2008) Improved oxidation resistance of a nanocrystalline chromite-coated ferritic stainless steel. *J. Power Sources*, **178** (1), 282–290.

216. Yoon, J.S., Lee, J., Hwang, H.J., Whang, C.M., Moon, J., and Kim, D. (2008) Lanthanum oxide-coated stainless steel for bipolar plates in solid oxide fuel cells (SOFCs). *J. Power Sources*, **181** (2), 281–286.

217. Zhu, J.H., Zhang, Y., Basu, A., Lu, Z.G., Paranthaman, M., Lee, D.F., and Payzant, E.A. (2004) $LaCrO_3$-Based coatings on ferritic stainless steel for solid oxide fuel cell interconnect applications. *Surf. Coat. Technol.*, **177–178**, 65–72.

218. Orlovskaya, N., Coratolo, A., Johnson, C., and Gemmen, R. (2004) Structural characterization of lanthanum chromite perovskite coating deposited by magnetron sputtering on an iron-based chromium-containing alloy as a promising interconnect material for SOFCs. *J. Am. Ceram. Soc.*, **87** (10), 1981–1987.

219. Johnson, C., Gemmen, R., and Orlovskaya, N. (2004) Nano-structured self-assembled $LaCrO_3$ thin film deposited by RF-magnetron sputtering on a stainless steel interconnect material. *Composites Part B*, **35** (2), 167–172.

220. Johnson, C., Orlovskaya, N., Coratolo, A., Cross, C., Wu, J., Gemmen, R., and Liu, X. (2009) The effect of coating crystallization and substrate impurities on magnetron sputtered doped $LaCrO_3$ coatings for metallic solid oxide fuel cell interconnects. *Int. J. Hydrogen Energy*, **34** (5), 2408–2415.

221. Elangovan, S., Balagopal, S., Hartvigsen, J., Bay, I., Larsen, D., Timper, M., and Pendleton, J. (2006) Selection and surface treatment of alloys in solid oxide fuel cell systems. *J. Mater. Eng. Perform.*, **15** (4), 445–452.

222. Linderoth, S. (1996) Controlled reactions between chromia and coating on alloy surface. *Surf. Coat. Technol.*, **80** (1–2), 185–189.

223. Yang, Z., Xia, G., Maupin, G.D., and Stevenson, J.W. (2006) Conductive protection layers on oxidation resistant alloys for SOFC interconnect applications. *Surf. Coat. Technol.*, **201** (7), 4476–4483.

224. Yang, Z., Xia, G., Maupin, G.D., and Stevenson, J.W. (2006) Evaluation of perovskite overlay coatings on ferritic stainless steels for SOFC interconnect applications. *J. Electrochem. Soc.*, **153** (10), A1852–A1858.

225. Lee, C. and Bae, J. (2008) Oxidation-resistant thin film coating on ferritic stainless steel by sputtering for solid oxide fuel cells. *Thin Solid Films*, **516** (18), 6432–6437.

226. Brylewski, T., Przybylski, K., and Morgiel, J. (2003) Microstructure of Fe-25Cr/(La,Ca)CrO_3 composite interconnector in solid oxide fuel cell operating conditions. *Mater. Chem. Phys.*, **81** (2–3), 434–437.

227. Belogolovsky, I., Zhou, X., Kurokawa, H., Hou, P.Y., Visco, S., and Anderson, H.U. (2007) Effects of surface-deposited nanocrystalline chromite thin films on the performance of a ferritic interconnect alloy. *J. Electrochem. Soc.*, **154** (9), B976–B980.

228. Mikkelsen, L., Chen, M., Hendriksen, P.V., Persson, Å., Pryds, N., and Rodrigo, K. (2007) Deposition of $La_{0.8}Sr_{0.2}Cr_{0.97}V_{0.03}$ and $MnCr_2O_4$ thin films on ferritic alloy for solid oxide fuel cell application. *Surf. Coat. Technol.*, **202** (4–7), 1262–1266.

229. Shaigan, N., Ivey, D.G., and Chen, W. (2008) Oxidation and electrical behavior of nickel/lanthanum chromite-coated stainless steel interconnects. *J. Power Sources*, **183** (2), 651–659.

230. Shaigan, N., Ivey, D.G., and Chen, W. (2008) Electrodeposition of Ni/$LaCrO_3$ composite coatings for solid oxide fuel cell stainless steel interconnect applications. *J. Electrochem. Soc.*, **155** (4), D278–D284.

231. Shaigan, N., Ivey, D.G., and Chen, W. (2008) Co/$LaCrO_3$ composite coatings for AISI 430 stainless steel solid oxide fuel cell interconnects. *J. Power Sources*, **185** (1), 331–337.

232. Feng, Z.J. and Zeng, C.L. (2010) $LaCrO_3$-Based coatings deposited by high-energy micro-arc alloying process on a ferritic stainless steel interconnect material. *J. Power Sources*, **195** (13), 4242–4246.

233. Kim, J., Song, R., and Hyun, S. (2004) Effect of slurry-coated $LaSrMnO_3$ on the electrical property of Fe-Cr alloy for metallic interconnect of SOFC. *Solid State Ionics*, **174** (1–4), 185–191.

234. Lim, D.P., Lim, D.S., Oh, J.S., and Lyo, L.W. (2005) Influence of post-treatments on the contact resistance of plasma-sprayed $La_{0.8}Sr_{0.2}MnO_3$ Coating on SOFC metallic interconnector. *Surf. Coat. Technol.*, **200** (5–6), 1248–1251.

235. Nie, H.W., Wen, T.L., and Tu, H.Y. (2003) Protection coatings for planar solid oxide fuel cell interconnect prepared by plasma spraying. *Mater. Res. Bull.*, **38** (9–10), 1531–1536.

236. Jan, D., Lin, C., and Ai, C. (2008) Structural characterization of $La_{0.67}Sr_{0.33}MnO_3$ Protective coatings for solid oxide fuel cell interconnect deposited by pulsed magnetron sputtering. *Thin Solid Films*, **516** (18), 6300–6304.

237. Chu, C., Lee, J., Lee, T., and Cheng, Y. (2009) Oxidation behavior of metallic interconnect coated with La-Sr-Mn film by screen painting and plasma sputtering. *Int. J. Hydrogen Energy*, **34** (1), 422–434.

238. Chu, C., Wang, J., and Lee, S. (2008) Effects of $La_{0.67}Sr_{0.33}MnO_3$ Protective coating on SOFC interconnect by plasma-sputtering. *Int. J. Hydrogen Energy*, **33** (10), 2536–2546.

239. Pattarkine, G.V., Dasgupta, N., and Virkar, A.V. (2008) Oxygen transport resistant and electrically conductive perovskite coatings for solid oxide fuel cell interconnects. *J. Electrochem. Soc.*, **155** (10), B1036–B1046.

240. Lee, S., Chu, C.-L., Tsai, M.-J., and Lee, J. (2010) High temperature oxidation behavior of interconnect coated with LSCF and LSM for solid oxide fuel cell by screen printing. *Appl. Surf. Sci.*, **256** (6), 1817–1824.

241. Kunschert, G., Kailer, K.H., Schlichtherle, S., and Strauss, G.N. (2007) Ceramic PVD coatings as dense/thin barrier layers on interconnect components for SOFC applications. *ECS Trans.*, **7** (1), 2407–2416.

242. Choi, J., Lee, J., Park, D., Hahn, B., Yoon, W., and Lin, H. (2007) Oxidation resistance coating of LSM and LSCF on SOFC metallic interconnects by the aerosol deposition process. *J. Am. Ceram. Soc.*, **90** (6), 1926–1929.

243. Choi, J.-J., Ryu, J., Hahn, B.-D., Yoon, W.-H., Lee, B.-K., Choi, J.-H., and Park, D.-S. (2010) Oxidation behavior of ferritic steel alloy coated with LSM-YSZ composite ceramics by aerosol deposition. *J. Alloys Compd.*, **492** (1–2), 488–495.

244. Hwang, H. and Choi, G.M. (2009) The effects of LSM coating on 444 stainless steel as SOFC interconnect. *J. Electroceram.*, **22** (1–3), 67–72.

245. Pyo, S.-S., Lee, S.-B., Lim, T.-H., Song, R.-H., Shin, D.-R., Hyun, S.-H., and Yoo, Y.-S. (2011) Characteristic of $(La_{0.8}Sr_{0.2})_{0.98}MnO_3$ Coating on Crofer22APU used as metallic interconnects for solid oxide fuel cell. *Int. J. Hydrogen Energy*, **36** (2), 1868–1881.

246. Tsai, M.-J., Chu, C.-L., and Lee, S. (2010) $La_{0.6}Sr_{0.4}Co_{0.2}Fe_{0.8}O_3$ Protective coatings for solid oxide fuel cell interconnect deposited by screen printing. *J. Alloys Compd.*, **489** (2), 576–581.

247. Montero, X., Jordán, N., Pirán-Abellán, J., Tietz, F., Stöver, D., Cassir, M., and Villarreal, I. (2009) Spinel and perovskite protection layers between Crofer22APU and $La0.8Sr0.2FeO_3$ Cathode materials for SOFC interconnects. *J. Electrochem. Soc.*, **156** (1), B188–B196.

248. Fu, C.J., Sun, K.N., Zhang, N.Q., Chen, X.B., and Zhou, D.R. (2008) Evaluation of lanthanum ferrite coated interconnect for intermediate temperature solid oxide fuel cells. *Thin Solid Films*, **516** (8), 1857–1863.

249. Qu, W., Jian, L., Hill, J.M., and Ivey, D.G. (2006) Electrical and microstructural characterization of spinel phases as potential coatings for SOFC metallic interconnects. *J. Power Sources*, **153** (1), 114–124.

250. Petric, A. and Ling, H. (2007) Electrical conductivity and thermal expansion spinels at elevated temperatures. *J. Am. Ceram. Soc.*, **90** (5), 1515–1520.

251. Deng, X., Wei, P., Bateni, M.R., and Petric, A. (2006) Cobalt plating of high temperature stainless steel interconnects. *J. Power Sources*, **160** (2), 1225–1229.

252. Fu, Q.-X., Sebold, D., Tietz, F., and Buchkremer, H.-P. (2010) Electrodeposited cobalt coating on Crofer22APU steels for interconnect applications in solid oxide fuel cells. *Solid State Ionics*. doi: 10.1016/j.ssi.2010.03.010

253. Chen, X., Hou, P.Y., Jacobson, C.P., Visco, S.J., and De Jonghe, L.C. (2005) Protective coating on stainless steel interconnect for SOFCs: oxidation kinetics and electrical properties. *Solid State Ionics*, **176** (5–6), 425–433.

254. Yang, Z., Xia, G., Simner, S.P., and Stevenson, J.W. (2005) Thermal growth and performance of manganese cobaltite spinel protection layers on ferritic stainless steel SOFC interconnects. *J. Electrochem. Soc.*, **152** (9), A1896–A1901.

255. Simner, S.P., Anderson, M.D., Xia, G., Yang, Z., and Stevenson, J.W. (2005) Long-term SOFC stability with coated ferritic stainless steel interconnect. *Ceram. Eng. Sci. Proc.*, **26** (4), 83–90.

256. Xia, G., Yang, Z., and Stevenson, J.W. (2006) Manganese-cobalt spinel oxides as surface modifiers for stainless steel interconnects of solid oxide fuel cells. *ECS Trans.*, **1** (7), 325–332.

257. Yang, Z., Xia, G., Wang, C., Nie, Z., Templeton, J., Stevenson, J.W., and Singh, P. (2008) Investigation of iron-chromium-niobium-titanium ferritic stainless steel for solid oxide fuel cell interconnect applications. *J. Power Sources*, **183** (2), 660–667.

258. Alvarez, E., Meier, A., Weil, K.S., and Yang, Z. (2009) Oxidation kinetics of manganese cobaltite spinel protection layers on sanergy HT for solid oxide fuel cell interconnect applications. *Int. J. Appl. Ceram. Technol.* doi: 10.1111/j.1744-7402.2009.02421.x

259. Chen, L., Sun, E.Y., Yamanis, J., and Magdefrau, N. (2010) Oxidation kinetics of $Mn_{1.5}Co_{1.5}O_4$-Coated Haynes 230 and crofer 22 APU for solid oxide fuel cell interconnects. *J. Electrochem. Soc.*, **157** (6), B931–B942.

260. Garcia-Vargas, M.J., Zahid, M., Tietz, F., and Aslanides, A. (2007) Use of SOFC metallic interconnect coated with spinel protective layers using the APS technology. *ECS Trans.*, **7** (1), 2399–2405.

261. Saoutieff, E., Bertrand, G., Zahid, M., and Gautier, L. (2009) APS deposition of $MnCo_2O_4$ on commercial alloys K41X used as solid oxide fuel cell interconnect: the importance of post heat-treatment for densification of the protective layer. *ECS Trans.*, **25** (2), 1397–1402.

262. Hua, B., Pu, J., Gong, W., Zhang, J., Lu, F., and Jian, L. (2008) Cyclic oxidation of Mn-Co spinel coated SUS 430 alloy in the Cathodic atmosphere of solid oxide fuel cells. *J. Power Sources*, **185** (1), 419–422.

263. Mardare, C.C., Asteman, H., Spiegel, M., Savan, A., and Ludwig, A. (2008) Investigation of thermally oxidised Mn-Co thin films for application in SOFC metallic interconnects. *Appl. Surf. Sci.*, **255** (5), 1850–1859.

264. Mardare, C.C., Spiegel, M., Savan, A., and Ludwig, A. (2009) Thermally oxidized Mn-Co thin films as protective coatings for SOFC interconnects. *J. Electrochem. Soc.*, **156** (12), B1431–B1439.

265. Bateni, M.R., Wei, P., Deng, X., and Petric, A. (2007) Spinel coatings for UNS 430 stainless steel interconnects. *Surf. Coat. Technol.*, **201** (8), 4677–4684.

266. Wei, W., Chen, W., and Ivey, D.G. (2009) Oxidation resistance and electrical properties of anodically electrodeposited Mn-Co oxide coatings for solid oxide fuel cell interconnect applications. *J. Power Sources*, **186** (2), 428–434.

267. Wu, J., Jiang, Y., Johnson, C., and Liu, X. (2008) DC electrodeposition of Mn-Co alloys on stainless steels for SOFC interconnect application. *J. Power Sources*, **177** (2), 376–385.

268. Wu, J., Johnson, C.D., Jiang, Y., Gemmen, R.S., and Liu, X. (2008) Pulse plating of Mn-Co alloys for SOFC interconnect applications. *Electrochim. Acta*, **54** (2), 793–800.

269. Wu, J., Johnson, C.D., Gemmen, R.S., and Liu, X. (2009) The performance of solid oxide fuel cells with Mn-Co electroplated interconnect as cathode current collector. *J. Power Sources*, **189** (2), 1106–1113.

270. Xie, Y., Qu, W., Yao, B., Shaigan, N., and Rose, L. (2010) Dense protective coatings for SOFC interconnect deposited by spray pyrolysis. *ECS Trans.*, **26** (1), 357–362.

271. Liu, W.N., Sun, X., Stephens, E., and Khaleel, M.A. (2009) Life prediction of coated and uncoated metallic interconnect for solid oxide fuel cell applications. *J. Power Sources*, **189** (2), 1044–1050.

272. Horita, T., Kishimoto, H., Yamaji, K., Xiong, Y., Brito, M.E., Yokokawa, H., Baba, Y., Ogasawara, K., Kameda, H., Matsuzaki, Y., Yamashita, S., Yasuda, N., and Uehara, T. (2008) Diffusion of oxygen in the scales of Fe-Cr alloy interconnects and oxide coating layer for solid oxide fuel cells. *Solid State Ionics*, **179** (38), 2216–2221.

273. Yang, Z., Xia, G., Nie, Z., Templeton, J., and Stevenson, J.W. (2008) Ce-modified $(Mn,Co)_3O_4$ spinel coatings on ferritic stainless steels for SOFC interconnect applications. *Electrochem. Solid-State Lett.*, **11** (8), B140–B143.

274. Yoo, J., Woo, S., Yu, J.H., Lee, S., and Park, G.W. (2009) $La_{0.8}Sr_{0.2}MnO_3$ And $(Mn_{1.5}Co_{1.5})O_4$ Double layer coated by electrophoretic deposition on Crofer22 APU for SOFC interconnect applications. *Int. J. Hydrogen Energy*, **34** (3), 1542–1547.

275. Balland, A., Gannon, P., Deibert, M., Chevalier, S., Caboche, G., and Fontana, S. (2009) Investigation of La_2O_3 and/or $(Co,Mn)_3O_4$ deposits on Crofer22APU for the SOFC interconnect application. *Surf. Coat. Technol.*, **203** (20–21), 3291–3296.

276. Montero, X., Tietz, F., Sebold, D., Buchkremer, H.P., Ringuede, A., Cassir, M., Laresgoiti, A., and Villarreal, I. (2008) $MnCo_{1.9}Fe_{0.1}O_4$ Spinel protection layer on commercial

ferritic steels for interconnect applications in solid oxide fuel cells. *J. Power Sources*, **184** (1), 172–179.

277. Huang, W., Gopalan, S., Pal, U.B., and Basu, S.N. (2008) Evaluation of electrophoretically deposited $CuMn_{1.8}O_4$ Spinel coatings on crofer 22 APU for solid oxide fuel cell interconnects. *J. Electrochem. Soc.*, **155** (11), B1161–B1167.

278. Wei, P., Deng, X., Bateni, M.R., and Petric, A. (2007) Oxidation behavior and conductivity of UNS 430 stainless steel and crofer 22 APU with spinel coatings. *ECS Trans.*, **7** (1), 2135–2143.

279. Hua, B., Zhang, W., Wu, J., Pu, J., Chi, B., and Jian, L. (2010) A promising $NiCo_2O_4$ protective coating for metallic interconnects of solid oxide fuel cells. *J. Power Sources*, **195** (21), 7375–7379.

280. Bi, Z.H., Zhu, J.H., and Batey, J.L. (2010) $CoFe_2O_4$ Spinel protection coating thermally converted from the electroplated Co-Fe alloy for solid oxide fuel cell interconnect application. *J. Power Sources*, **195** (11), 3605–3611.

281. Liu, Y. and Chen, D.Y. (2009) Protective coatings for Cr_2O_3-forming interconnects of solid oxide fuel cells. *Int. J. Hydrogen Energy*, **34** (22), 9220–9226.

282. Geng, S., Li, Y., Ma, Z., Wang, L., Li, L., and Wang, F. (2010) Evaluation of electrodeposited Fe-Ni alloy on ferritic stainless steel solid oxide fuel cell interconnect. *J. Power Sources*, **195** (10), 3256–3260.

283. Geng, S., Zhu, J., Brady, M.P., Anderson, H.U., Zhou, Z., and Yang, Z. (2007) A low-Cr metallic interconnect for intermediate-temperature solid oxide fuel cells. *J. Power Sources*, **172** (2), 775–781.

284. Jablonski, P.D. and Alman, D.E. (2008) Oxidation resistance of novel ferritic stainless steels alloyed with titanium for SOFC interconnect applications. *J. Power Sources*, **180** (1), 433–439.

285. Hua, B., Pu, J., Zhang, J., Lu, F., Chi, B., and Jian, L. (2009) Ni-Mo-Cr alloy for interconnect applications in intermediate temperature solid oxide fuel cells. *J. Electrochem. Soc.*, **156** (1), B93–B98.

286. Chen, X., Hua, B., Pu, J., Li, J., Zhang, L., and Jiang, S.P. (2009) Interaction between $(La,Sr)MnO_3$ cathode and Ni-Mo-Cr metallic interconnect with suppressed chromium vaporization for solid oxide fuel cells. *Int. J. Hydrogen Energy*, **34** (14), 5737–5748.

6
Sealants for Planar Solid Oxide Fuel Cells

Qingshan Zhu, Lian Peng, and Tao Zhang

6.1
Introduction

In planar solid oxide fuel cell (pSOFC) design, sealing is required to keep the fuel gas separated from air, which would otherwise directly combust to cause reduced power generation efficiency and local overheating. As illustrated in Figure 6.1 [1], the positions need to be sealed in pSOFCs include metal interconnect and metal interconnect, metal interconnect and positive electrode–electrolyte–negative electrode (PEN), as well as metal interconnect and frame. pSOFC seals need to work at high temperatures (700–850 °C) under both oxidizing and wet reducing atmospheres, with a lifetime greater than 40 000 h together with hundreds of thermal cycles for stationary applications or a lifetime of at least 5000 h and not less than 3000 thermal cycles for transportation applications [2, 3]. To achieve these goals, the seals must be chemically stable as well as chemically compatible with other solid oxide fuel cell (SOFC) components during the long-term operation at high temperatures under both the atmospheres. Moreover, the seal should be mechanically robust enough to accommodate the thermal stresses caused by thermal cycling (start-up and shutdown) and by temperature gradients induced by gas flowing and electrochemical reactions. All these impose a great challenge for seal material development. Accordingly, sealing has been identified as one of the toughest technical challenges to the development of pSOFCs [4, 5].

Until now, four types of sealants have been mainly investigated for pSOFCs: glass (including glass–ceramic), mica, metal braze, and composite seal. Glass can wet the surface of adjacent components by viscous flow at high temperatures (sealing temperature) and then solidify at low temperatures (operating temperature), which is similar to glue. Under external compressive force, mica can deform itself to fill in the space between two adjacent components, which is similar to elastic loop. Similar to glass, metal braze also seals adjacent components by wetting the corresponding surfaces. Composite sealants, which mostly consist of two phases (glass–ceramic, ceramic–ceramic, glass–metal, etc.), were also used for pSOFCs. Composite sealants reported in the literature have two types: rigid sealant such as glass or braze and compressive sealant such as mica. For rigid composite sealants,

Materials for High-Temperature Fuel Cells, First Edition. Edited by San Ping Jiang and Yushan Yan.

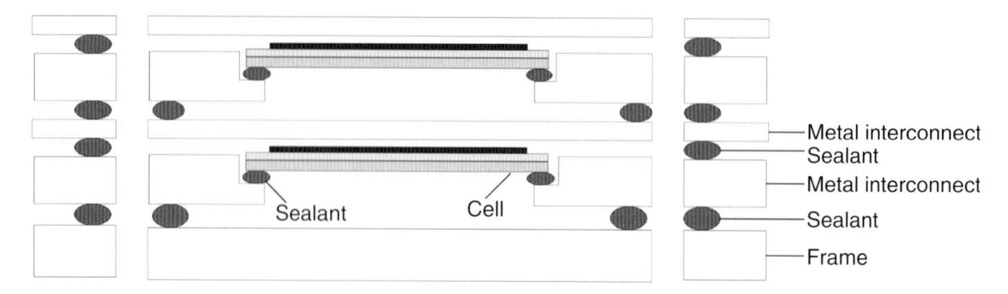

Figure 6.1 A scheme of pSOFC sealing.

glass is used to wet the surface of adjacent components, and ceramic or metal is used to tailor the properties of composite sealants, such as the thermal expansion coefficient (TEC), the viscosity, and the mechanical properties. On the other hand, for compressive composite sealants, the ceramic phase is used to provide the skeleton and the glass–ceramic phase is used to fill in the pores of the skeleton. In this chapter, the development of various types of sealants is discussed, with emphasis on long-term stability of various sealing materials.

6.2 Glass and Glass–Ceramic Sealants

Glass and glass–ceramics have been widely investigated as pSOFC sealing materials. The main advantages of glass-based sealants are (i) the property can be tailored in a wide range by manipulating glass composition and heat-treatment schemes; (ii) glass and glass–ceramics are insulating, which simplifies the design of pSOFCs; and (iii) glass and glass–ceramics are cheap and easy to be processed. Properties concerning pSOFC sealants are TEC, viscosity, chemical stability, thermal stability, and chemical compatibility with other SOFC components. Since the properties of glass and glass–ceramic may significantly change with service time at high temperatures, some properties such as TEC and viscosity are only meaningful for short-term service of pSOFCs. For long-term applications, other properties such as chemical stability, thermal stability, and chemical compatibility with other SOFC components are much more important. In addition, glass and glass–ceramic are brittle and vulnerable to cracking under tensile stress, which has been identified as one of the main failure modes of glass-based sealants. Fracture is not only influenced by material properties but also determined by the sealing structure; accordingly, stress optimization through sealing geometry design is also important for improving the reliability of pSOFC sealants.

6.2.1 Properties Related to Short-Term Performance

Viscosity is an important glass property since it determines the sealing and operating temperature of glass seals. The viscosity of glass decreases continuously with increasing temperature and can be estimated by the so-called

Vogel–Fulcher–Tammann (VFT) equation:

$$\log \eta = \frac{-A + B}{(T - T_0)} \tag{6.1}$$

where A and B are temperature-independent constants and T_0 is a temperature-dependent constant. These constants can be estimated by regression using measured viscosity data for a given glass system. Practically, the viscosity–temperature relationship of a glass is normally characterized by several characteristic temperatures such as the glass-transition temperature T_g, the softening temperature T_s, and the working temperature T_w, which could be determined by techniques such as differential scanning calorimetry (DSC) and dilatometry. T_g is the temperature corresponding to the glass viscosity of $\sim 10^{12}$ Pa·s at which glass is transformed from the solid state to the liquid state. T_s corresponds to a viscosity of $10^{6.6}$ Pa·s and is the low temperature limit of shape forming of a glass. T_w corresponds to a viscosity of 10^3 Pa·s and is the upper temperature limit where glass can be formed into shapes. Accordingly, the sealing temperature should be between the T_w and the T_s, and a viscosity of 10^5 Pa·s is normally sufficient to obtain good sealing ability. The operating temperature of a glass seal is best above T_g, because at this temperature, stress in a glass can be released through plastic deformation. The viscosity at the operating temperature should not be too low, for example, $<10^9$ Pa·s, in order to have enough rigidity to maintain the integrity of pSOFCs [5].

Instead of measuring T_g and T_s, shape change experiments via hot-stage microscopy (HSM) were sometimes employed to determine the sealing and operating temperatures of glass sealants [6–14]. As illustrated in Figure 6.2 [6], the shape of a glass cube changes with increasing temperature. At temperature $T1$, the edge of the glass becomes round and this characteristic temperature is named as the *round edge temperature*. The ball temperature $T2$ corresponds to the temperature at which the cube becomes a ball. Similarly, $T3$ is called the *hemisphere temperature*. Sealing temperature is generally near the ball temperature, while the operating temperature is generally below the round edge temperature. It must be noted that the three characteristic temperatures are dependent on the external compressive force and the annealing time.

Sealing is best performed below 850 °C to prevent excess oxidation of the interconnect during the sealing process. Therefore, one of the primary goals of the sealing glass design is to achieve a sufficiently low viscosity to facilitate effective sealing below 850 °C. The viscosity of a glass can be adjusted through composition, for example, increasing the concentration of low-melting-point oxides such as B_2O_3, P_2O_5, and Bi_2O_3 decreases the viscosity of glass, whereas the viscosity

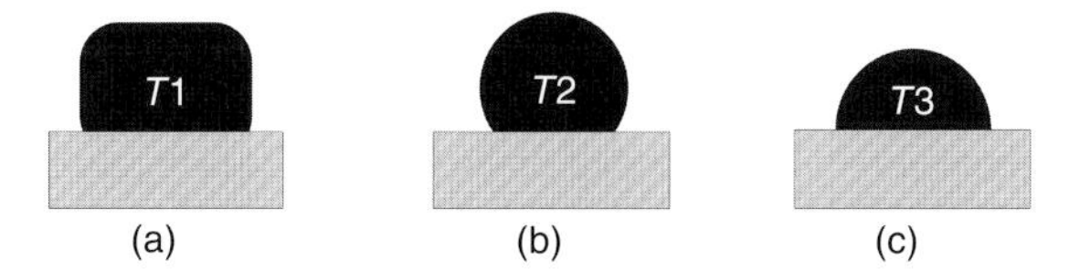

Figure 6.2 (a–c) Shape change behavior of glass with increasing temperature.

increases with increasing the concentration of high-melting-point oxides such as SiO_2 and Al_2O_3. For glass–ceramic sealants, viscosity is dependent on both the resultant glass phase and the amount of crystalline phase and is generally higher than that of the parent glass. Since after crystallization glass–ceramics may be difficult to sinter to form a dense seal below 850 °C, the crystallization process is normally performed together with the sintering process, where the sealing process consists of two steps: parent glass powder is first sintered to dense to seal adjacent components, followed by crystallization of the dense glass seal to form a glass–ceramic seal. To achieve this, the glass sintering temperature should be lower than the crystallization temperature. If, however, the crystallization temperature is lower than the sintering temperature, glass crystallizes to form a glass–ceramic before densification, resulting in a porous glass–ceramic [15].

The viscosity–temperature relationship also determines the self-healing ability of glass-based sealants [16–18]. If at operating temperature the viscosity of a glass sealant is low enough to facilitate simultaneous sintering, cracks formed during thermal cycling can be healed through viscous flow at the operating temperature, which is extremely helpful for improving the thermal cyclicability of the glass seal. The leak rate of a glass seal may be high at low temperatures due to the cracks formed during cooling down. However, if crack healing occurs, the leak rate will decrease at high temperatures [16]. Singh [17] observed that crack healing occurred in three stages. In the first stage, the crack tip became round and cracks became cylindrical; in the second stage, cylindrical cracks became spherical; and in the third stage, spherical cracks became small with annealing time and vanished finally. Owing to the self-healing ability, impressive results have been obtained by Singh *et al.*, where the seal still functioned well after 300 thermal cycles between room temperature (RT) and 800 °C. The low viscosity may however cause problems, such as the glass seal may flow out of the sealing area, which has adverse effects on the rigidity of a stack [19]. To solve the problem, Liu *et al.* proposed to use stoppers to avoid excess compression that squeezed out the low-viscosity seal. At low viscosity, the reactivity of a glass is also higher, which may cause problems for long-term interfacial stability. Alternatively, healing of cracks can be achieved by raising the temperature after certain numbers of thermal cycles even if the viscosity is not sufficiently low to heal cracks at the operating temperature. Figure 6.3 shows such tests as an example, where the thermal cycle was performed from 150 to 700 °C. Before the leak rate approached the up limit after ~20 thermal cycles, the temperature was raised to 810 °C (sealing temperature) for 30 min for crack healing, after which the leak rate decreased over one order of magnitude. Such an operation is also very helpful for improving thermal cycle performance of a glass seal.

TEC is the most extensively studied property for glass-based sealants, as TEC mismatch with other SOFC materials is the origin of thermal stress that causes fracture of glass sealants. Consequently, another goal of sealing glass design is to minimize the TEC difference with the anode and the interconnect to be sealed. The TEC of Ni–YSZ anode varies from 11.0×10^{-6} to $14.0 \times 10^{-6}\ K^{-1}$ depending on nickel content and measured temperature range. The TEC of interconnect would

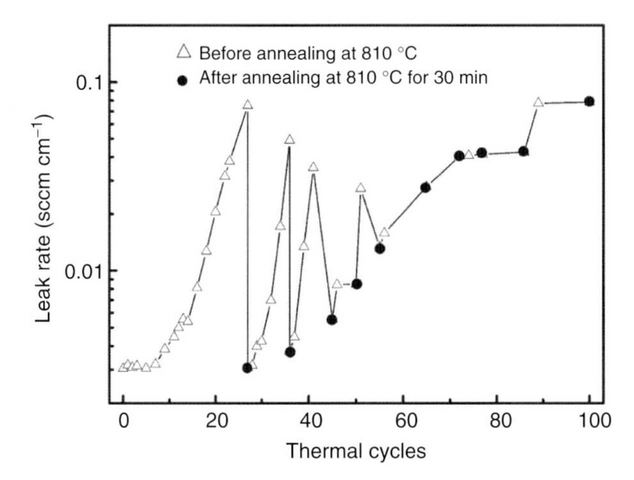

Figure 6.3 The effect of self-healing ability on the leak rate of glass.

be dependent on the type of alloys used, for example, $11.5 \times 10^{-6}\,K^{-1}$ for Crofer 22 APU and $12.2 \times 10^{-6}\,K^{-1}$ for SS410 [3, 20]. Therefore, the TEC values of glass sealants are best in the range $11.0–13.5 \times 10^{-6}\,K^{-1}$ (RT–T_g). The TEC of glasses can be approximately predicted by the Appen model expressed as follows:

$$\alpha = \sum_i \alpha_i c_i \tag{6.2}$$

where α is the TEC of the glass ($10^{-6}\,K^{-1}$), α_i is the TEC contribution factor of the ith component ($10^{-6}\,K^{-1}$), and c_i is the mole fraction of the ith component in a glass. Some TEC contribution values of common oxides are listed in Table 6.1. Silicate glasses would normally have much lower TEC values than those needed by pSOFC sealing applications. It is often necessary to increase the TEC of pSOFC sealing application. Table 6.1 shows that alkali metal oxides have high TEC contributions, so the TEC of silicate glasses normally increases with increasing alkali metal oxide contents. However, alkali metal oxides easily react with other SOFC components at high temperatures, which causes cell performance degradation. Consequently,

Table 6.1 TEC contribution values for various metal oxides [3].

Substance	α_i ($10^{-6}\,K^{-1}$)	Substance	α_i ($10^{-6}\,K^{-1}$)
SiO_2	0.005–0.038	B_2O_3	−0.050 to 0.00
Li_2O	0.270	SrO	0.160
Na_2O	0.395	BaO	0.200
K_2O	0.465	PbO	0.130–0.190
MgO	0.060	Al_2O_3	−0.030
CaO	0.130	ZrO_2	−0.060
ZnO	0.050	P_2O_5	0.140
MnO	0.105	Fe_2O_3	0.055

alkali metal oxide addition is not recommended in most cases [21]. To achieve high TEC values, alkaline earth metal oxides are often used [21], among which BaO addition is most attractive due to its relatively large contribution to TEC increase among all alkaline earth metal oxides. Even with the addition of BaO, it is not so easy to obtain a glass with a TEC greater than 11.5×10^{-6} K^{-1}, since excess addition of alkaline earth metal oxides will make the glass network weak, which results in low viscosity and poor thermal stability.

Another way to increase the TEC of a glass is through controlled crystallization to form a glass–ceramic. Forming glass–ceramics will provide additional benefits since glass–ceramic is normally more stable and mechanically more robust than the corresponding glass. The control of crystallization can be achieved by controlling both thermodynamic and kinetic parameters. Thermodynamically, crystalline phases formed during crystallization are determined by the composition of a glass, where by manipulating the glass composition crystalline phases of high TEC may be formed. Table 6.2 shows the TEC values of various crystalline phases [21–26] related to the pSOFC sealing applications. It is clear from Table 6.2 that aluminates such as $Mg_2Al_4Si_5O_{18}$, $CaAl_2SiO_8$, $SrAl_2Si_2O_8$, and $BaAl_2Si_2O_8$ have low TEC values and should be prevented from being formed. From a thermodynamic point

Table 6.2 The TEC of crystalline phases containing alkaline earth metal oxides.

Crystalline phase	TEC (10^{-6} K^{-1})
$MgSiO_3$ (enstatite)	9.0–12.0
$MgSiO_3$ (clinoenstatite)	7.8–13.5
$MgSiO_3$ (protoenstatite)	9.8
Mg_2SiO_4 (forsterite)	9.4
$CaSiO_3$ (wollastonite)	9.4
Ca_2SiO_4 (calcium orthosilicate)	10.8–14.4
$Ba_3CaSi_2O_8$ (barium calcium orthosilicate)	12.2–13.8
$BaSi_2O_5$ (barium silicate)	14.1
$Ba_2Si_3O_8$ (barium silicate)	12.6
$BaSiO_3$ (barium silicate)	9.4–12.5
BaB_2O_4 (barium borate)	$\alpha_a = 4$, $\alpha_c = 16$
$BaZrO_3$ (barium zirconate)	7.9
$CaZrO_3$ (calcium zirconate)	10.4
$BaCrO_4$ (barium chromate)	21–23
$SrCrO_4$ (strontium chromate)	21–23
$Mg_2Al_4Si_5O_{18}$ (cordierite)	2
$CaAl_2SiO_8$ (anorthite)	4.5
$SrAl_2Si_2O_8$ (hexacelsian)	7.5–11.1
$SrAl_2Si_2O_8$ (monocelsian)	2.7
$SrAl_2Si_2O_8$ (orthorhombic celsian)	5.4–7.6
$BaAl_2Si_2O_8$ (hexacelsian)	6.6–8.0
$BaAl_2Si_2O_8$ (monocelsian)	2.3
$BaAl_2Si_2O_8$ (orthorhombic celsian)	4.5–7.1

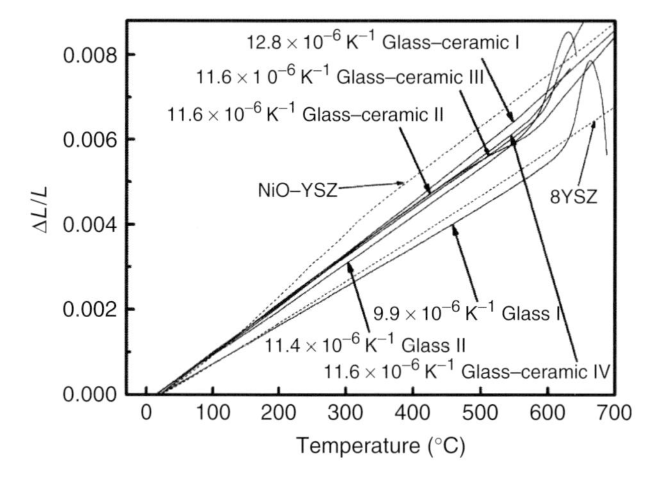

Figure 6.4 Thermal expansion curves of various glasses and glass–ceramics.

of view, glass–ceramics crystallized from parent glasses with low alumina content or even without alumina could avoid the formation of the low TEC aluminates and thus may achieve glass–ceramics of high TEC values [27–37]. With this strategy, the researchers have developed several glass–ceramic seals with a TEC ranging from 11.6×10^{-6} to $12.8 \times 10^{-6}\ K^{-1}$ in the $BaO–B_2O_3–SiO_2$ system, as illustrated in Figure 6.4. Similarly, Chou *et al.* [38] reported a series of glass–ceramics with a TEC of $11.2–11.7 \times 10^{-6}\ K^{-1}$ in the parent glass of the $SrO–CaO–Y_2O_3–B_2O_3–SiO_2$ system. Reis *et al.* [15] developed a glass–ceramic with a TEC of $11.5 \times 10^{-6}\ K^{-1}$ in the $SrO–CaO–ZnO–B_2O_3–Al_2O_3–SiO_2–TiO_2$ system with a low Al_2O_3 content of 1.96 mol%. It must be pointed out that the above-mentioned alumina effect is just a simple example illustrating how composition can manipulate the TEC of glass–ceramics. In reality, the influence of parent glass composition on the crystallized phases is quite complex, for example, adding one component inhibits one particular phase formation while promoting the formation of other phases, as summarized by Fergus [21]. Until now, no general rules can be applied to predict crystalline phase formation and the glass–ceramic development is much dependent on trial-and-error approaches. Crystallization can also be used to transform chemically less stable components in the glass to more stable phases, for example, transformation of amorphous boron oxide and phosphoric oxide in the glass network to crystalline borates and phosphates can improve the chemical stability of the glass.

Although particular crystalline phases formed are crucial to controlling the properties of the crystals, crystallization kinetics also plays an important role in the crystallization process. Kinetically, the crystallization process can be controlled by adjusting crystallization temperature and time, nucleation agents, nucleating temperature, and so on. Crystallization is normally performed at temperatures around the crystallization temperature, which can be determined through DSC. Crystal growth rate may be characterized by measuring the grain size with crystallization time. By measuring crystallization speed at different temperatures, activation energy for crystallization can be estimated. The crystallization kinetics can be controlled by the addition of nucleating agents such as ZrO_2, TiO_2, Cr_2O_3, and P_2O_5. Nucleating agents have several important functions: (i) facilitating homogenous nucleation of crystalline phases, (ii) promoting the formation of particular phases, and (iii) suppressing the formation of unwanted phases. The interaction of the nucleating agents with glass network has been discussed in the literature [21] and is not further discussed here. It should be noted that glass and glass–ceramic for pSOFC applications are usually used in a powder form, so the crystallization is mainly controlled by the surface crystallization mechanism, whose particle size has significant influence on the crystallization kinetics. For this reason, the crystallization speed should be lower than the sintering speed to obtain a dense seal, as discussed before.

6.2.2 Properties Related to Long-Term Performance

pSOFC seals need to function under both oxidizing and wet reducing atmospheres at high temperatures for a long period, during which seals undergo continuous changes, including the loss of seal materials due to evaporation, TEC change due to crystallization, and new phase formation due to interfacial reaction. These changes will accumulate with time and may finally reach the critical level that causes failure of the seal. It is therefore of great importance to test the properties related to long-term performance before applying a new glass-based seal.

The chemical stability of glass seals containing B_2O_3 and P_2O_5 are of great concern since B_2O_3 and P_2O_5 may evaporate significantly at high temperatures, especially under a wet reducing atmosphere. B_2O_3 is used to reduce the viscosity in sealing glasses. Investigations showed that sealing glasses containing B_2O_3 are stable in air at high temperatures; however, under a humidified fuel atmosphere, the boron oxide in glasses reacts with wet fuel to form highly volatile species such as $B_2(OH)_2$ and $B_2(OH)_3$, which can cause significant seal loss, for example, the weight loss can reach 20% for glasses using B_2O_3 as the sole network former. Reis and Brow [39] investigated the effect of the content of B_2O_3 on the chemical stability of glass in wet reducing gas. Four glasses with different contents of B_2O_3 were annealed at 750 °C for 10 days in reducing gas (10% H_2 + 90% N_2). The results showed that the mass loss of glass increased with increasing the content of B_2O_3: 2.0×10^{-5} g cm^{-2} for the glass containing 2 mol% B_2O_3 and 1.0×10^{-4} g cm^{-2} for the glass containing 7 mol% B_2O_3. Accordingly, from the long-term chemical stability point of view, the amount of B_2O_3 in sealing glasses should be kept as small as possible. Another way to improve the chemical stability is to transform amorphous B_2O_3 in glass network into crystalline borate as demonstrated by Reis and Brow [39]. Similar to borate glasses, phosphate glasses also evaporate at high temperatures; in air, P_2O_5 can only evaporate in the form of P_2O_5 gas, whereas in wet reducing atmosphere, P_2O_5 can evaporate in the form of both P_2O_5 gas and

P_2O_3 gas [40]. The volatile nature together with severe reactivity with other SOFC components makes phosphate glasses unsuitable for pSOFC sealing applications.

The operating temperature (700–850 °C) of pSOFCs coincides with the crystallization temperature range of most glasses. So, glass-based seals may crystallize during operating under pSOFC environments. Uncontrolled crystallization of glass can cause significant TEC changes, as demonstrated by Sohn *et al.* [27], where TEC values of glasses based on the $BaO–Al_2O_3–B_2O_3–SiO_2$ system decrease about 17.8–39.6% after keeping at 800 °C for only 1000 h because of the formation of the low TEC celsian phase. The large change in TEC is extremely harmful since failure of sealing becomes inevitable with significant thermal stress build-up. The thermal stability of glass is generally characterized by the variations in TEC with annealing time. Achieving long-term thermal stability is one of the main issues in developing a suitable sealing glass and glass–ceramic for advanced planar SOFC stacks. Although glasses are thermodynamically unstable phases susceptible to devitrification, our work showed that glass with superior thermal stability can be attainable through glass design. As shown in Figure 6.5, the initial TEC of the glass is $9.8 \times 10^{-6}\ K^{-1}$ (RT–T_g) and after keeping at 700 °C in air for 5000 h, the change in the TEC is marginal. Transformation of glass into glass–ceramic is a common practice to improve the thermal stability. Superior thermal stability can only be achieved through proper composition design. One of the primary goals of glass design is to achieve a composition that could avoid the formation of low TEC phases while promoting the formation of particularly high TEC phases, as discussed before. Another goal for the parent glass design is to achieve a composition that undergoes fast and homogenous crystallization at crystallization temperature while crystallization is controlled to a minimal speed at operating temperatures. To achieve this, the operating temperature is best to be controlled at least 100 °C lower than the crystallization temperature. If the operating temperature is too close to the crystallization temperature, the seal would undergo continuous crystallization

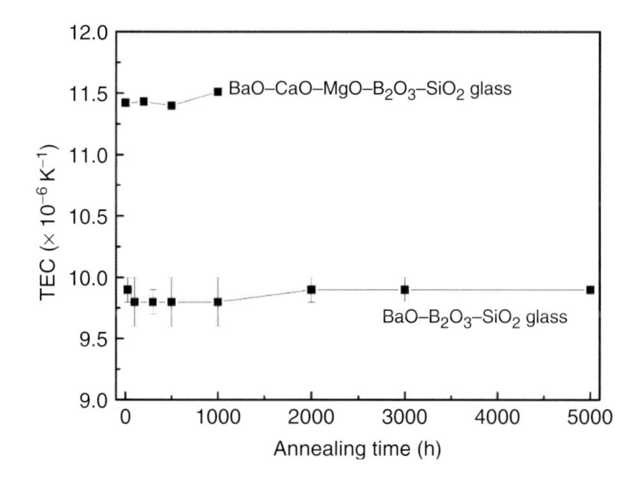

Figure 6.5 Thermal stability of the glasses in Figure 6.4 at 700 °C.

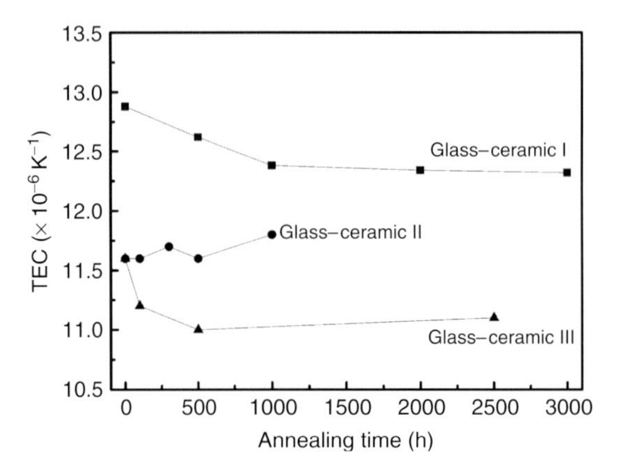

Figure 6.6 Thermal stability of the glass–ceramics in Figure 6.4 at 700 °C.

during operation; consequently, microstructure changes may accumulate to the state that affects the integrity of the seal after long-term operation. Over the past decade, significant progress has been achieved in this area. We found that the low TEC $BaAl_2Si_2O_8$ phase can be eliminated by reducing the alumina content in glasses in the SiO_2–B_2O_3–BaO–Al_2O_3 system, and therefore, the seal development was focused on the RO–B_2O_3–SiO_2 (R = Mg, Ca, Sr, Ba) system. Through glass composition design, three glass–ceramics with a TEC of $11.5–13.0 \times 10^{-6}\ K^{-1}$ were successfully developed. As shown in Figure 6.6, the three glass–ceramics have good thermal stability, where after keeping at 700 °C for 3000, 1000, and 2500 h, the TEC changes are 4.3, 1.7, and 4.3%, respectively. Chou *et al.* [38] developed several thermally stable glass–ceramics in the SrO–CaO–Y_2O_3–B_2O_3–SiO_2 system. As listed in Table 6.3, after annealing under 30% H_2O/2.7% H_2/67.3% Ar at 900 °C for 1000 h, the TEC changes were marginal. Reis *et al.* [15] recently reported a glass–ceramic with the initial TEC of about $11.7 \times 10^{-6}\ K^{-1}$ in the SrO–CaO–ZnO–B_2O_3–SiO_2–Al_2O_3 system, where after annealing at 800 °C for 2880 h, the TEC changes were less than 5%, exhibiting superior thermal stability.

Table 6.3 The TEC comparison between parent glass and glass–ceramic.

	Parent glass ($\times 10^{-6}\ K^{-1}$)	Glass–ceramic ($\times 10^{-6}\ K^{-1}$)		
		As received	After 1000 h in wet reducing gas	After 2000 h in air
YSO-1	12.1 (RT–695 °C)	11.7 (RT–870 °C)	11.5 (RT–945 °C)	11.6 (RT–970 °C)
YSO-4	11.7 (RT–713 °C)	11.5 (RT–938 °C)	11.9 (RT–940 °C)	—
YSO-5	11.6 (RT–735 °C)	11.3 (RT–910 °C)	11.6 (RT–970 °C)	—
YSO-7	11.4 (RT–685 °C)	11.4 (RT–877 °C)	11.6 (RT–975 °C)	—
YSO-8	11.5 (RT–673 °C)	11.2 (RT–896 °C)	11.6 (RT–980 °C)	—

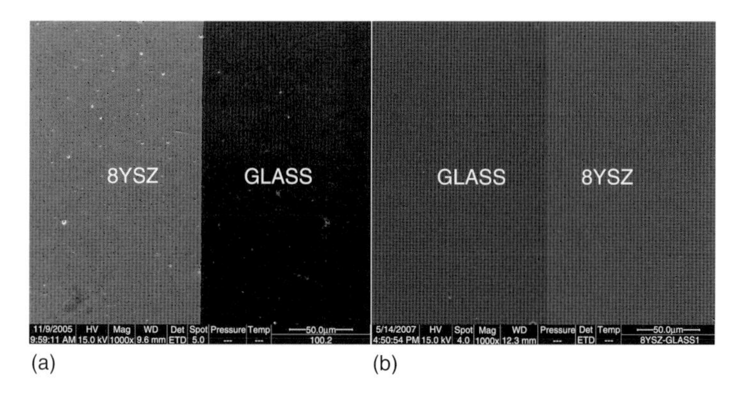

Figure 6.7 SEM cross-sectional views of glass–8YSZ interface heated at 700 °C for (a) 100 h and (b) 5000 h.

For anode-supported pSOFCs, the chemical compatibility of glass with electrolyte and metal interconnect is of main concern in the literature [41–45]. Interfacial reactions are generally considered in two stages. In the first stage, the elements of two adjacent components diffuse within each other to form a diffusion zone, which strengthens the bonding between the two adjacent components. In the second stage, the contents of the elements in the diffusion zone reach a critical level and new phases are formed. The formation of new phases generally weakens the bonding of adjacent components, so interfacial reactions are best to be restricted to the first stage. Glass and glass–ceramic are generally compatible with 8 mol% Y_2O_3–ZrO_2 (8YSZ) at the operating temperature of pSOFCs, as illustrated in Figure 6.7; even after annealing at 700 °C for 5000 h, new phases were not found at the glass–8YSZ interface. Moreover, the elemental distribution at the glass–8YSZ interface was found to be insignificant. As illustrated in Figure 6.8, after annealing at 700 °C for 5000 h, the diffusion zone is ~5 μm. The only potential problem may be caused by the diffusion of Y_2O_3 into glass, which results in the transformation of cubic ZrO_2 into monoclinic ZrO_2 [46]. The diffusion force of Y_2O_3 can be weakened by adding Y_2O_3 to glasses, and consequently, a glass containing Y_2O_3 was developed in our earlier research [3].

Compared with the 8YSZ electrolyte, reactions between glass-based seals and metal interconnects are more complex and problematic. Until now, investigation on seal–metal interaction is scarce [47–50]. Yang *et al.* [47] studied interfacial reactions of a barium silicate glass (G18) with a ferritic stainless steel (AISI 446), a Ni-based superalloy (Nicrofer6025), and a Fecralloy. Since silicate glasses with alkaline earth metal oxide addition are most promising for pSOFC applications, the use of a barium silicate glass to study glass–metal interaction is a good justification. Yang *et al.* found that the extent of reaction and the nature of products are much dependent on the oxide scale on the metal surface. The 446 ferritic stainless steel and the Nicrofer6025 have a chromia-rich scale on the surface. In such cases, $BaCrO_4$ is the main product as formed through the following reactions:

$$Cr_2O_3\,(s) + 2BaO + 1.5O_2 \rightarrow 2BaCrO_4\,(s) \qquad (6.3)$$

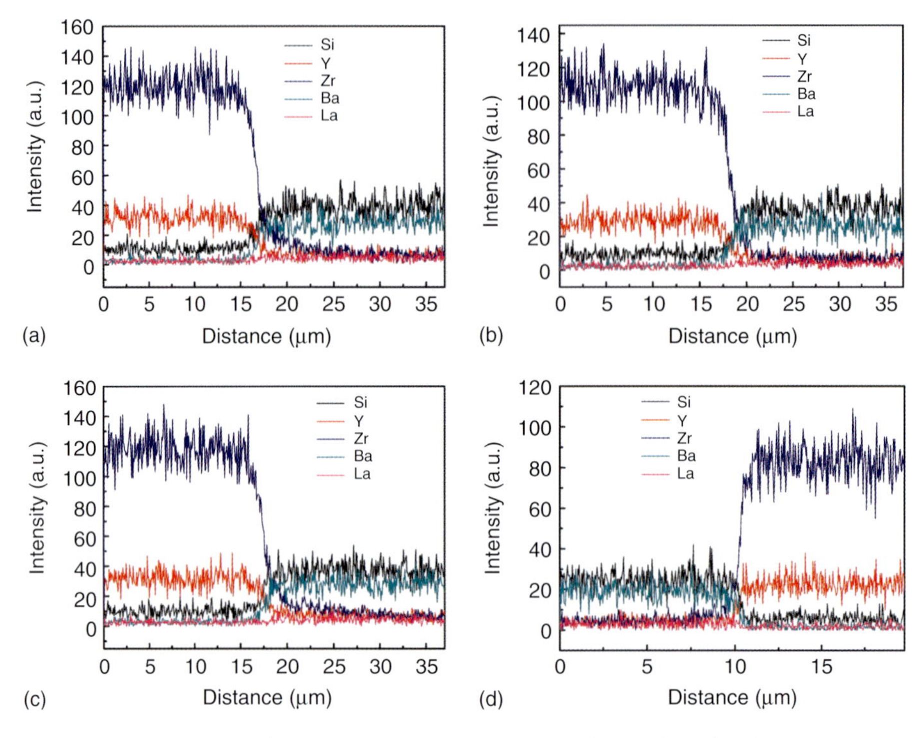

Figure 6.8 Energy-dispersive X-ray spectrometry (EDX) line analyses for glass–8YSZ interface heat treated at 700 °C for (a) 0 h, (b) 1 h, (c) 100 h, and (d) 5000 h.

$$CrO_2(OH)_2\,(g) + BaO \rightarrow BaCrO_4\,(s) + H_2O\,(g) \tag{6.4}$$

If the barium in glasses is crystallized to $BaSiO_3$, $BaCrO_4$ can also be formed through the following reaction:

$$2BaSiO_3\,(s) + Cr_2O_3\,(s) + 1.5O_2\,(g) \rightarrow 2BaCrO_4\,(s) + 2SiO_2\,(s) \tag{6.5}$$

Since $BaCrO_4$ has a very high TEC of $21–23 \times 10^{-6}\,K^{-1}$ (20–1000 °C) [38], extensive formation of $BaCrO_4$ not only leads to depletion of barium in the glass–ceramic but also causes the separation of the glass–ceramic from the alloy matrix because of the thermal expansion mismatch [47]. For the Nicrofer6025, the surface scale consists also of NiO, Al_2O_3, and Fe_2O_3, apart from Cr_2O_3, and the formation of $BaCrO_4$ was hindered to some extent as compared with AISI 446. It should be noted that reactions (6.3) and (6.4) need oxygen to form $BaCrO_4$, so in the interior region of the interface where the access to oxygen from air was blocked, instead of forming $BaCrO_4$, chromia dissolved into the barium–calcium–aluminosilicate (BCAS) sealing glass and formed a chromium-rich solid solution. Another characteristic feature shown by Yang *et al.* is the formation of large pores and voids along the glass–alloy interface after joining, which is a quite short period, for example, exposure of the metal–glass–metal

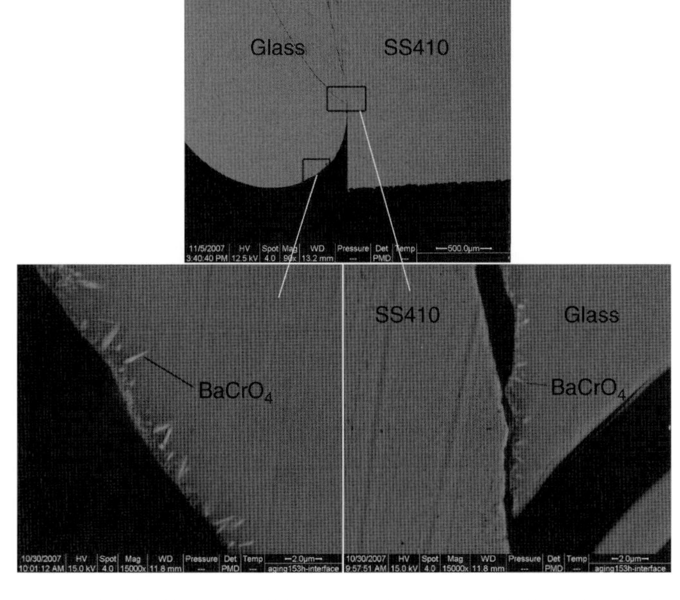

Figure 6.9 The formation of $BaCrO_4$ at the glass–SS410 interface and on the free surface of glass after annealing at 700 °C for 150 h.

coupon in air at 800 °C for 1 h and 750 °C for 4 h. The voids were proposed by Yang *et al.* to be caused by vapor phase formation via the interaction of alloy elements, specifically the chromium with the dissolved water in the glass. The surface of the Fecralloy is mainly composed of alumina, so no $BaCrO_4$ was observed at the interface. Again, large pores with the diameter equals the width of the seal zone appeared, which was proposed to be due to reactions of alloy elements, such as Al, Y, and Cr, with the dissolved water and sodium oxide residue in the sealing glass. The pores formed along the interface are extremely detrimental to the performance of the seal since reaction-induced pores will greatly reduce the interfacial bonding strength.

We have also studied the reaction between a barium silicate glass developed for pSOFC applications and SS410 [51]. Similar to the results of Yang *et al.*, the formation of $BaCrO_4$ was observed at the metal–glass interface accessible by air as well as on the free surface of glass through reactions (6.3) and (6.4), as illustrated in Figure 6.9. In our case, reaction-induced pores were not observed at the interface after annealing at 700 °C for 150 h.

While such static tests give information about the extent of reaction and the nature of the product, thermal cyclic tests are more relevant to real pSOFC applications, as thermal stress induced by thermal cycles also plays a very important role in the seal–metal interaction. We have studied the influence of the interfacial reaction on the thermal cyclic behavior and found that the formation of $BaCrO_4$ at the interface

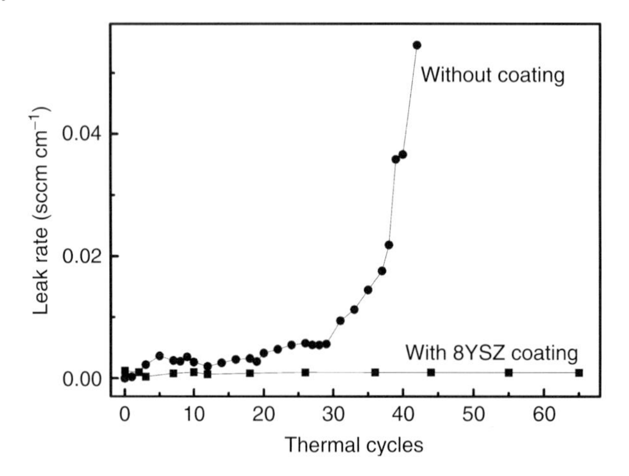

Figure 6.10 Comparison between the leak rates of glass with and without 8YSZ coating.

caused seal cracking and cracking promotes $BaCrO_4$ formation at crack tip since the cracks formed acted as the oxygen diffusion channels [51]. Hence, crack extends with increasing the number of thermal cycles, resulting in increased leak rate with increasing thermal cycles. Figure 6.10 shows the comparison between the leak rates of glass with and without the 8YSZ coating. The increase in the leak rate appeared in two stages. In the first stage, the leak rates increased slowly with thermal cycles, which might be attributed to the incomplete self-healing behavior of the glass seal. In the second stage, the leak rates increased rapidly with thermal cycles, which was caused by the formation of a more or less continuous $BaCrO_4$ layer at the interface that hindered crack healing.

Adding a barrier coating to the metal interconnect can prevent the formation of the high-TEC $BaCrO_4$ phase, thus improving the thermal cycle stability of glass. It must be pointed out that the barrier coating needs to be chemical compatible with the glass sealant and the interconnect at high temperatures. Otherwise, the coating itself may cause detrimental reactions. Mahapatra and Lu [52] investigated the interfacial compatibility of $SrO-La_2O_3-Al_2O_3-SiO_2$ glass with AISI 441 alloy and $(Mn,Co)_3O_4$ coating. Although $(Mn,Co)_3O_4$ coating hinders iron and chromium diffusion into glass to a certain extent, it accelerates the AISI 441–glass interface degradation because of its instability and reactivity with the glass seal. Similar results were also observed for $(Mn,Co)_3O_4$-coated Crofer 22 APU alloy [53]. We have also used 8YSZ as the coating material for the SS410 interconnect to improve the thermal cycle stability of a $BaO-B_2O_3-SiO_2$ glass [24]. As shown in Figure 6.10, the leak rate for the interconnect without the 8YSZ coating increased rapidly after 30 thermal cycles, while it remained unchanged even after 65 thermal cycles for the interconnect with the 8YSZ coating, demonstrating that the thermal cyclicability of glass-based seal can be improved through reduction of adverse interfacial reaction using a barrier coating.

6.2.3
Sealing Structure Optimization

Owing to the brittle nature of glasses and glass–ceramics, glass-based seals are susceptible to fracture under tensile thermal stress. Consequently, fracture has been reported to be one of the main failure modes for glass-based seals [17, 54–57]. Until now, much effort has been devoted to material development aiming at achieving a good TEC match between the seal and the adjacent components [41, 58–61]. However, residual thermal stress cannot be totally eliminated because the TECs of the anode, cathode, and interconnect are different. On the other hand, thermal stress is determined by material properties and also influenced largely by seal geometry. It is therefore equally important to reduce the thermal stress through sealing structure optimization, which has not been paid sufficient attention by the SOFC community.

In practice, there exist two typical failure phenomena: the interfacial debonding of adhesive joints [62] and the cracking of the seal layer [54, 55]. The failure mode of the sealing is dependent on the nature of the glass–metal bonding, which may be weak (the interfacial fracture energy is lower than the fracture energy of the seal) [63] or strong (the interfacial fracture energy is higher than the fracture energy of the seal) [55, 59]. The first case has been addressed by Muller *et al.* [62], in which finite element modeling was employed to analyze the residual stress distribution in a typical seal-and-crack propagation along the interface. They found that a crack would grow if and only if both the released energy and the local stresses exceed the critical values, and the resistance of the seal to debonding would increase with increasing width and decreasing thickness of the seal.

We have developed a model based on the classical beam bending theory and the fracture theory of ceramic materials for predicting cracking of the SOFC seals. Through a series of equation derivation, which can be found in the literature [63], a criterion was obtained for predicting whether a crack will extend in a seal as

$$t \leq t_c = \frac{\Gamma E'}{0.42\sigma_s^2} \tag{6.6}$$

$$E' = \frac{E}{1-\nu} \tag{6.7}$$

where t is the seal thickness, t_c is the critical seal thickness, σ_s is biaxial compressive stresses, Γ is the critical strain energy release rate of the sealing material, E is Young's modulus, and ν is Poisson's ratio. Eq. (6.6) indicates that cracking (crack extension) can be prevented if the seal thickness is kept less than the critical thickness for a given seal system. Consequently, the so-called cracking diagram can be obtained for a given seal system at specified operational conditions, as shown in Figure 6.11. The lines in Figure 6.11 represent the boundaries between the "cracking" and the "no-cracking" area, which show that the "no-cracking" area increases with decrease in TEC mismatch, meaning that a larger TEC mismatch would require a thinner seal to avoid cracking; for example, when the TEC mismatch

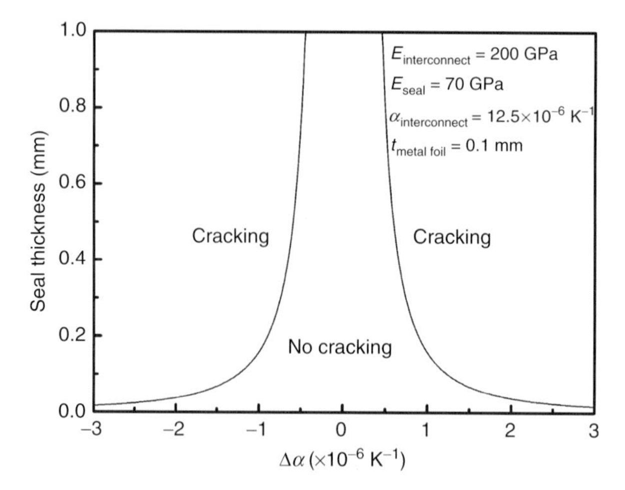

Figure 6.11 Cracking diagram for a glass–SS410 sealing system.

is greater than 1.2×10^{-6} K^{-1}, seals less than 0.1 mm should be employed to avoid cracking, as illustrated in Figure 6.11. On the other hand, when the TEC mismatch is less than 0.6×10^{-6} K^{-1}, t_c increases rapidly with decreasing TEC mismatch and the dependence of t_c on $\Delta\alpha$ becomes not so critical. For example, a TEC mismatch of 0.5×10^{-6} K^{-1} would allow a maximum seal thickness of 0.8 mm, which already exceeds the seal thickness of 0.1–0.5 mm, as commonly reported in the literature [58, 60, 64–67]. The cracking diagram has been verified by experiments [63] and can be used as the guide to sealing structure design.

6.3 Mica

Mica is a compressive sealant that needs external force to seal the surface of adjacent components. The mica used for pSOFCs has two forms: muscovite ($KAl_2(AlSi_3O_{10})(F,OH)_2$) and phlogopite ($KMg_3(AlSi_3O_{10})(OH)_2$). Mica has a layered structure with adjacent layers connected by K^+ cations. The space between adjacent layers can be compressed, so mica can deform itself to seal adjacent components. The bond between adjacent layers is very weak, so mica can easily separate itself along cleavage planes. This special structure makes mica possess the following features for pSOFC applications.

1. Because of the weak bond between adjacent layers, mica can release thermal stress by sliding between layers, which is helpful to achieve long-term thermal cycle stability.
2. Unlike glass, mica does not adhere with adjacent components, so mica can tolerate large TEC mismatch, which makes it possible to use Ni-based alloys as metal interconnect for pSOFCs. In addition, mica can easily detach from adjacent components, which enhances the flexibility of pSOFCs.

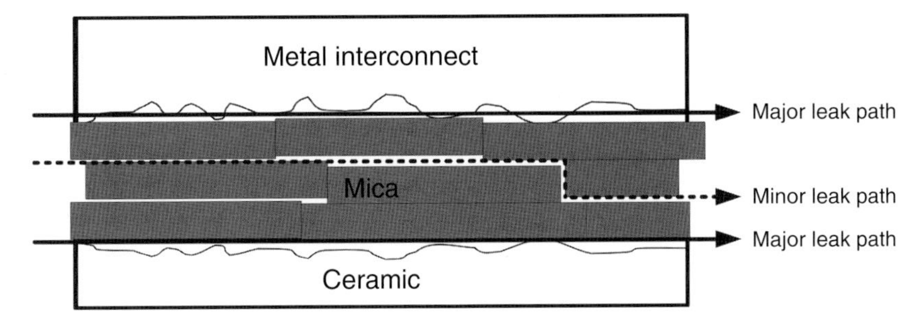

Figure 6.12 Schematic drawing of the major and minor leak paths in mica sealant.

Mica has been investigated for use in pSOFCs by the PNNL (Pacific Northwest National Laboratory) in detail, which includes the leakage mechanism of mica, effect of compressive stress and differential pressure on the leak rate of mica, effect of long-term aging on the leak rate of mica, effect of thermal cycles on the leak rate of mica, and combined effect of aging and thermal cycles on the leak rate of mica [66, 68–79]. In addition, a leak rate model has been established for mica sealants [80].

6.3.1
The Leakage Mechanism of Mica

Mica flakes generally overlap each other under external compressive force in pSOFCs. Unlike glass sealants, which are dense and wet the surface of adjacent components, mica sealants are not compact and do not adhere to adjacent components. Therefore, there are two main leak paths for mica sealants, as illustrated in Figure 6.12. One is formed at the interface between the mica flake and the adjacent components due to the roughness of contact surfaces, which was identified as the main leak path. The other is formed in between the mica flakes, which is the minor leak path [69]. According to Chou and Stevenson [69], the leak rate of mica sealants decreased if the major and minor leak paths can be diminished, where a glass or a Ag interlayer between the mica flake and the adjacent components has been proposed as the "hybrid mica sealant" for improving the gastightness of mica sealants, as shown in Figure 6.13. In addition, Chou *et al.* [77] further developed composite mica sealants that not only used glass as interlayer to diminish major leak paths but also used glass-infiltrating mica to diminish minor leak paths. As a result, composite mica exhibits the lowest leak rate among all mica sealants tested.

6.3.2
The Effect of Compressive Stress and Differential Pressure on the Leak Rate of Mica

The leak rate of mica decreased with increasing compressive stress and decreasing differential pressure. Chou *et al.* [73, 75] investigated the leak rates of muscovite single crystal, muscovite paper, and phlogopite paper under different compressive stresses with a gas pressure gradient of 13.79 KPa. Mica was investigated according

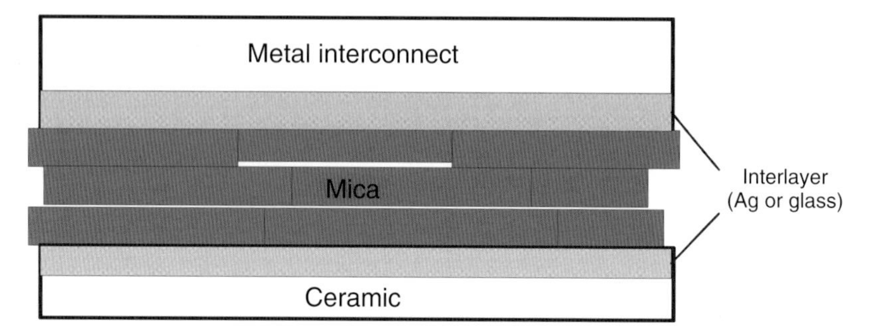

Figure 6.13 Schematic drawing of a hybrid mica compressive sealant.

to three different forms: plain mica, mica with glass interlayer, and mica with Ag interlayer. Plain muscovite single crystal has much lower leak rates than plain muscovite paper and plain phlogopite paper under identical compressive stress. Moreover, the leak rate of mica with glass interlayer was lower than that of mica with Ag interlayer. According to Chou *et al.*, the leak rate of plain muscovite single crystal is 0.65 $sccm\,cm^{-1}$ at 689.5 KPa with a differential pressure of 13.79 KPa. It must be noted that this compressive stress is too high for pSOFCs to endure. Compared with plain mica, hybrid mica has a much lower leak rate and needs a much lower compressive stress, which is caused by diminishing major leak paths. For muscovite single crystal with glass interlayer, the leak rate is 3.59×10^{-4} $sccm\,cm^{-1}$ under a 172.4 KPa compressive stress. Figure 6.14 shows the dependence of leak rate on compressive stress under the differential pressure of 13.79 KPa. It can be noted from Figure 6.14 that the leak rate of hybrid mica with glass interlayer is on the order of 10^{-3} magnitude, the leak rate of hybrid mica with Ag interlayer is on

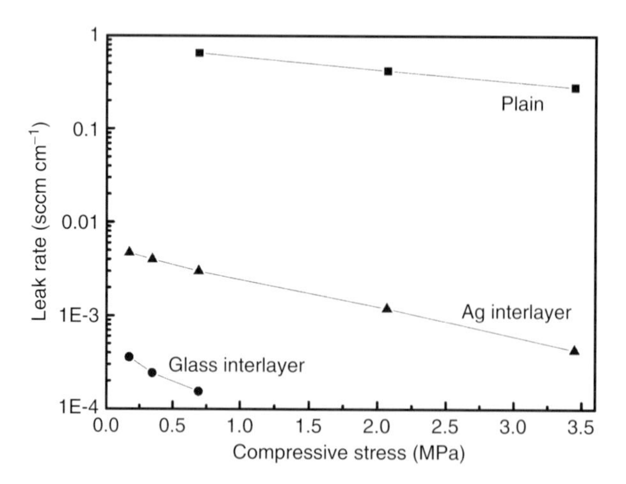

Figure 6.14 The dependence of leak rate on compressive stress for plain mica, hybrid mica with glass interlayer, and hybrid mica with Ag interlayer.

the order of 10^{-2} magnitude and the leak rate of plain mica is on the order of 10^{-1} magnitude. Although leak rate and compressive stress are decreased for the hybrid mica seal with glass interlayer, the glass interlayer can cause some problems, such as TEC mismatch and poor thermal stability and chemical compatibility. In addition, the merits of mica, such as tolerance of large TEC mismatch and convenience of disassembly, can never be possessed by hybrid mica because the presence of glass or Ag interlayer will more or less transform the seal to a rigid bonded type. In addition, precompression on mica is helpful to decrease its leak rate, where the leak rate three times lower than the mica without precompression was reported by Simner and Stevenson [70].

6.3.3 The Effect of Long-Term Aging on the Leak Rate of Mica

Mica is thermally stable at operating temperatures of pSOFCs. The problems faced by glass sealants, such as crystallization and TEC variation, are not faced by mica. Many concerns are currently focused on the thermal stability of hybrid mica because of the presence of an interlayer. Chou *et al.* [66] investigated the thermal stability of hybrid micas with three different glass interlayers at 800 °C. They found that chemical reactions of glass layer with mica had an obvious effect on the thermal stability of the hybrid mica seal, where one glass crystallized and reacted easily with mica, for which the leak rate of the corresponding hybrid mica dropped to about $1–2 \times 10^{-3}$ sccm cm^{-1} after annealing at 800 °C for 350 h because the crystalline phase and the reaction product blocked partially the leak paths. However, after annealing at 800 °C for about 850 h, the leak rate increased rapidly to about 0.04 sccm cm^{-1}, indicating that excessive reaction created large leak paths. The other two hybrid seals showed better thermal stability due to limited reactivity of glasses with mica, where the leak rates were about 0.02 sccm cm^{-1} at 800 °C for 1036 h and about 0.001–0.004 sccm cm^{-1} at 800 °C for 508 h. It can obviously be concluded that the long-term thermal stability of hybrid mica is absolutely determined by the thermal stability and chemical compatibility of the glass interlayer with mica. In addition, glass interlayer can continuously infiltrate into mica during isothermal aging, which results in the decrease in the leak rate of hybrid mica [76]. Chou and Stevenson also investigated the thermal stability of hybrid mica with Ag interlayer at 800 °C in a wet reducing atmosphere (2.64% H_2/Ar + 3% H_2O). The leak rate of the hybrid mica with Ag interlayer was about 0.01–0.02 sccm cm^{-1} during the first 3000 h of annealing time and then decreased to about 0.01 sccm cm^{-1} during the period of 3000–30 000 h, which showed the superior thermal stability of the hybrid mica with Ag interlayer [78]. In addition, the chemical stability of mica in wet reducing gas is a potential problem for its long-term thermal stability. According to Chou and Stevenson, two reactions may occur for mica in wet reducing gas:

$$F_2\,(g) + H_2\,(g) \rightarrow 2HF\,(g) \tag{6.8}$$

$$SiO_2\,(s) + 4HF\,(g) \rightarrow SiF_4\,(g) + 2H_2O\,(g) \tag{6.9}$$

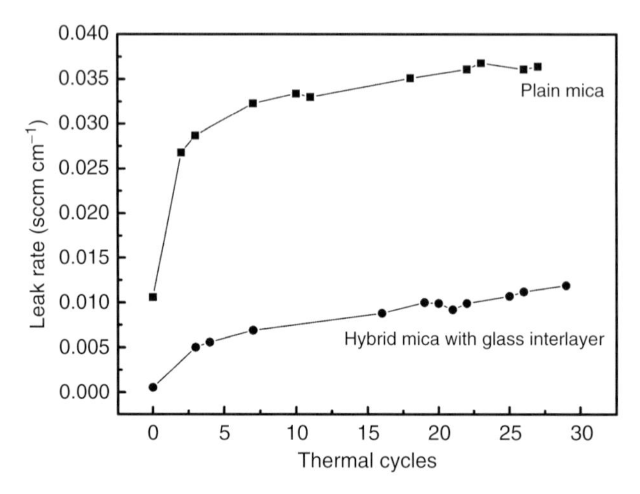

Figure 6.15 The effect of thermal cycles on the leak rates of plain/hybrid mica.

According to reaction (6.8), fluorine in mica may be released as HF in wet reducing atmospheres. Because mica is a silicate mineral, the formed HF may be corrosive to the mica itself according to reaction (6.9).

6.3.4
The Effect of Thermal Cycles on the Leak Rate of Mica

For most mica sealants, the dependence of leak rates on thermal cycles is similar, that is, leak rate increases rapidly during the first couple of thermal cycles and then increases slowly. Figure 6.15 shows the typical effect of thermal cycles on the leak rates of plain/hybrid mica. The thermal cycle stability of mica has been investigated from many aspects, including the thickness of mica, volume fraction of glass, different interlayer, and substrate with different TEC and compressive stress. Both thick (about 0.5 mm) and thin (about 0.1 or 0.2 mm) hybrid mica have good thermal stability over 50 thermal cycles, and the leak rate of thick hybrid mica is generally higher than that of thin hybrid mica [76]. The volume fraction of glass seems to have no obvious influence on the thermal cycle stability of hybrid mica seals [77]. Chou and Stevenson [72] compared the thermal cycle stability of plain mica, hybrid mica with glass interlayer, and hybrid mica with Ag interlayer. They found that the Ag interlayer was suitable for the substrate metal with a high TEC rather than a low TEC due to the high TEC of Ag (19×10^{-6} K^{-1}). For the substrate metal with a low TEC (e.g., SS430), the leak rate of the hybrid mica with Ag interlayer was even higher than that of the plain mica after 10 thermal cycles due to its TEC mismatch with SS430. The leak rate of the hybrid mica with a glass interlayer was less than that of the hybrid mica with an Ag interlayer regardless of the substrate metal because glass can release thermal stress above T_g. In addition, the leak rate of the hybrid mica with a glass interlayer decreases with increasing compressive stress [77]. The leak rate of a glass-infiltrated mica even decreases with thermal cycles under a high compressive stress because of the continuous infiltration of glass [77, 79].

6.3.5 The Combined Effect of Aging and Thermal Cycles on the Leak Rate of Mica

The leak rate of the hybrid mica with a G18 glass interlayer increased rapidly with thermal cycles with prior isothermal aging at 800 °C for 1036 h [66]. However, the leak rate of this hybrid mica remained almost constant with thermal cycles without prior isothermal aging at 800 °C. This difference was caused by the reaction of the glass interlayer with mica. On the other hand, the hybrid mica with a G6 glass interlayer had good thermal cycle stability with prior isothermal aging at 800 °C for 508 h because the G6 glass interlayer had better chemical compatibility with mica. Therefore, the thermal stability and chemical compatibility of the glass interlayer is very important for hybrid mica. As a result, the problems faced by glass sealants are also faced by the hybrid mica with a glass interlayer.

6.4 Metal Braze

Like glass sealants, metal braze seals adhere to adjacent components also by wetting the corresponding surfaces. The advantage of applying metal brazes as the pSOFC sealing materials is that they have lower stiffness as compared to ceramics and can undergo plastic deformation more easily, both of which allow for accommodation of thermal and mechanical stresses, which would be beneficial to reduce the failure probability of seals [81–87]. On the other hand, several obstacles have to be overcome before metal brazes can be applied for pSOFC sealing applications; for example, (i) metal brazes are electrically conductive, so proper insulation should be achieved to avoid current short-circuit; (ii) most metal brazes have insufficient long-term oxidation resistance at high temperatures under SOFC atmospheres, so only noble metals such as Au, Ag, Pt, and Pd may be used, which increases the capital cost of pSOFCs [55, 81–87]; and (iii) the TEC of most metal brazes is much greater than that of the PEN and the interconnect, which would generate high thermal stress during thermal cycling. Owing to these problems, metal braze seals are less popularly studied. Among the various metal brazes, most investigations focused on silver, possibly because of its low cost as compared to other noble metals. PNNL developed a novel metal braze consisting of 4% CuO and 96% Ag (mol%) [55, 81–83]. The reason for the addition of CuO is to improve the wetting ability of Ag to the ceramic and metal components of pSOFCs. The Ag–CuO metal braze exhibited superior thermal stability and thermal cycle stability because of the superior mechanical properties of Ag. Weil *et al.* [82] found that the leak rate of Ag–CuO remained at a very low level under both air and wet reducing wetting atmospheres after being kept at 800 °C for 1000 h. Moreover, they also investigated the thermal cycle stability of the Ag–CuO metal braze between 70 and

750 °C with a high heating/cooling rate of 75 °C min^{-1} and found that the leak rate of the Ag–CuO seal remained at a low level under both air and wet reducing wetting atmospheres after 50 thermal cycles, which showed the superior thermal cycle stability of the Ag–CuO seal. Although the Ag–CuO braze possesses more superior mechanical property than glass and mica sealants, the chemical stability of Ag at operating temperatures is a potential problem for pSOFCs. Weil *et al.* found that bubbles of submicrometer size were formed in the bulk of Ag–CuO braze after aging at 750 °C for 200 h under a wet reducing atmosphere. After annealing at 750 °C for 800 h, the bubble size increased to ~0.5 μm [82]. These pores were formed due to the diffusion of H_2 into Ag, causing the reduction of CuO and the formation of water vapor, as shown in the following reaction:

$$CuO + 2H\,(\text{dissolved}) \rightarrow Cu + H_2O\,(\text{vapor}) \tag{6.10}$$

The reaction between Ag–4% CuO and Fe–Cr alloy is limited, and this reaction strengthens the bonding between braze and Fe–Cr alloy [82]. However, the reaction between Ag–8% CuO and Fe–Cr alloy resulted in the formation of a partially porous Cu–Fe–Cr–Mn-mixed oxide layer at the side of air, which weakened the bonding of braze with Fe–Cr alloy. As a result, the mechanical property of Ag–8% CuO braze is even worse than that of CuO-free braze. Therefore, Ag–4% CuO seems to be the most optimal braze composition, considering both rupture strength and wetting ability [55].

6.5
Composite Sealants

A composite sealant generally consists of two different phases: glass–ceramic and glass–metal or ceramic–ceramic [57, 58, 60, 65, 88–105]. The properties of composite sealants are determined by both phases. Therefore, the property of composite sealants may be tailored in a wide range. Composite sealants can be classified into two types according to the sealing procedure. One is like the glass sealant that seals adjacent components by wetting the corresponding surfaces. The other is like the mica sealant that seals adjacent components by external compressive stress.

Most composite sealants consist of glass phases and ceramic phases and seal adjacent components by wetting the corresponding surfaces. For glass–ceramic composite sealants, glass is used to wet the surface of adjacent components, while ceramic phases are used to tailor properties such as TEC, viscosity, self-healing ability, flexure strength, and fracture toughness [57, 58, 60, 65, 88–98]. Adding the ceramic phase can obviously increase the rigidity of seals at high temperature. Brochu *et al.* [95, 96] investigated the effect of the content of YSZ on the contact angle of the glass–YSZ composite seal in detail and found that the maximum content of YSZ was 5 vol%, otherwise the glass–YSZ composite cannot wet the surface of adjacent components, showing that the addition of ceramic phases increased the viscosity of the glass seal extensively. Therefore, the content of ceramic phase should be low, considering the suitable sealing temperature of the composite. As a result, the effect of YSZ on the TEC of glass–YSZ composite is very limited due to the low content and medium TEC of YSZ. To increase the TEC of the composite with low content of ceramic phase, the ceramic phases with high TEC, such as NiO, MgO, $KAlSi_2O_6$, and $KAlSiO_4$, were employed [57, 58, 90, 91]. In addition, the chemical compatibility of the ceramic phase with the glass phase may also be a problem. Brochu *et al.* [95] found that a new phase of $BaZrO_3$ was formed for the glass–YSZ composite at 850 °C, which caused a decrease in the TEC, since $BaZrO_3$ had low TEC of $7.9 \times 10^{-6}\ K^{-1}$. Coillot *et al.* [88] reported a glass–VB (vanadium boride, CAS number: 12045-27-1) composite with self-healing ability, where VB can be oxidized into B_2O_3 and V_2O_5, which have low viscosity that facilitates crack filling. Choi *et al.* [93] used a boron nitride nanotube to reinforce a glass seal and showed that the strength and fracture toughness can be increased up to 90 and 35%, respectively, as compared with those of the unreinforced glass. Similar to glass–ceramic composites, glass–metal composites also seal adjacent components by wetting the corresponding surfaces [99, 100]. For glass–metal composites, the metal phase is mainly used to improve the mechanical properties.

Like mica sealants, the ceramic–ceramic composites generally seal adjacent components by external compressive stress. Le *et al.* [101, 102] developed an $Al_2O_3–SiO_2$ ceramic fiber sealant infiltrated by fumed silica. When the $Al_2O_3–SiO_2$ composite sealant was applied for an SS430-SS430 coupon, the leak rate was 0.05 sccm cm^{-1} under 50 KPa compressive stress with a differential pressure of 1.4 KPa. Sang *et al.* [103] developed an Al_2O_3–Al composite sealant, where metal Al was added to fill in the pores in Al_2O_3 tape. Moreover, Al can be oxidized into Al_2O_3 at operating temperatures to further fill in the pores in Al_2O_3 bulk due to the increase in volume. They found that the leak rate of Al_2O_3–Al composite was about 0.03 sccm cm^{-1}. Sang *et al.* [104] also established the leak rate model of Al_2O_3 compressive sealants. Recently, Zhang *et al.* [105] proposed a new concept for sealants. They developed a crystalline material that can be self-healed at a temperature between 800 and 840 °C. Unlike glass sealants, the crystalline sealant is thermodynamically stable at operating temperatures. Therefore, the crystalline sealant will exhibit super thermal stability. However, the sealing procedure of the crystalline sealant needs to be further investigated.

The properties of composite sealants can be tailored in a wider range than those of traditional sealants because of the mixture of two different phases. However, the chemical compatibility of one phase with the other is a critical problem for glass-based composite sealants. The leak rate of ceramic–ceramic composites seems to be higher than that of mica sealants.

6.6
Conclusion

Glass and glass–ceramics are the most promising sealants and have been extensively investigated for pSOFCs. Over the past decade, significant progress has been

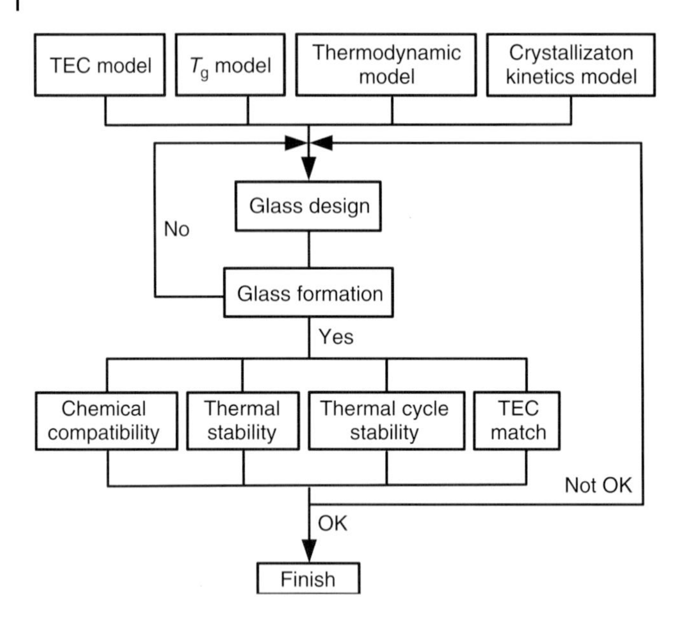

Figure 6.16 A schematic procedure for sealing glass design.

achieved, where several glass and glass–ceramic sealants with good TEC match, superior thermal stability, and thermal cyclicability have been developed. These glass-based seals were developed mainly through "trial-and-error" approach, which would be low in efficiency to find suitable glasses for pSOFC sealing applications, since pSOFC seals need to meet many requirements including matched TEC, suitable viscosity, chemical compatibility with other materials, chemical stability under fuel cell atmospheres, and long-term thermal stability. It is absolutely necessary to develop a quantitative design approach to speed up the sealing glass development. We have proposed a sealing glass development strategy, as shown in Figure 6.16, which includes several steps: (i) establish TEC model, T_g model, thermodynamic model, and crystallization kinetics model to design glass composition, which simultaneously fulfills the TEC, T_g, chemical compatibility, and sealing requirements; (ii) prepared a glass based on the designed composition to check whether it can be formed or not; and (iii) if the glass can be formed, the chemical compatibility, thermal stability, thermal cycle stability, and TEC match of the glass will be experimentally checked. If the properties are all fulfilled, the design process is finished. Otherwise, the glass design would start from the beginning again according to the experimental results. To realize the approach, a reliable TEC and viscosity model needs to be established for glass compositions related to pSOFC sealing applications. Reliable thermodynamic model and crystallization model are also needed to be established before the approach can really work, which is even more difficult than establishing the TEC and viscosity model. Self-healing ability can effectively overcome the brittleness of glass sealants, which is helpful to achieve long-term thermal cycle stability.

Plain mica sealants can tolerate much larger TEC mismatch than glass sealants because of their nonadhesive nature, which is also helpful in disassembly. However, the leak rate of plain mica is too high even under a high compressive stress, which limits its application in pSOFCs. Moreover, too high compressive stresses (e.g., 689.5 KPa) make it difficult for PENs and stack design. Consequently, hybrid micas with glass or Ag interlayer were proposed and developed. The thermal cycle stability of the hybrid mica with an Ag interlayer is inferior to that of the hybrid mica with a glass interlayer because of the too high TEC of Ag. Currently, the hybrid mica with a glass interlayer was mainly developed. However, problems faced by glass sealants will also be faced by mica seals with glass interlayer. Although the combination of a glass or Ag interlayer with mica reduces the leak rate, the interlayer transforms mica seals from compressive type to rigid bonded type, which makes the hybrid mica seal less tolerable to a large TEC mismatch.

Brazes are not suitable for traditional pSOFCs due to its conductivity, although they exhibit superior long-term thermal cycle stability. Composite sealants are also promising sealing materials for pSOFCs and deserved to be further investigated.

Acknowledgment

The financial support from the National Natural Science Foundation of China under the Contract Numbers 50730002, 21006111, and 20876159 are highly appreciated.

References

1. Zhu, Q.S., Peng, L., Huang, W.L., and Xie, Z.H. (2008) Sealing glass and a method of sealing for intermediate temperature solid oxide fuel cell. Chinese Patent 100376046C.
2. Chou, Y.S., Stevenson, J.W., and Choi, J.P. (2010) Alkali effect on the electrical stability of a solid oxide fuel cell sealing glass. *J. Electrochem. Soc.*, **157** (3), B348–B353.
3. Zhu, Q.S., Peng, L., and Zhang, T. (2007) in *Fuel Cell Electronics Packing*, 1st edn (eds K. Kuang and K. Easler), Springer Science+Business Media, LLC, New York, pp. 33–60.
4. Singh, P. and Misra, A. (2004) NASA, pacific northwest team on SOFC sealing. *Fuel Cells Bull.*, **2**, 6.
5. Eichler, K., Solow, G., Otschik, P., and Schaffrath, W. (1999) BAS ($BaO{\cdot}Al_2O_3{\cdot}SiO_2$)-glasses for high temperature applications. *J. Eur. Ceram. Soc.*, **19**, 1101–1104.
6. Schwickert, T., Sievering, R., Geasee, P., and Conradt, R. (2002) Glass-ceramic materials as sealants for SOFC applications. *Materialwiss. Werkstofftech.*, **33**, 363–366.
7. Flügel, A., Dolan, M.D., Varshneya, A.K., Zheng, Y., Coleman, N., Hall, M., Earl, D., and Misture, S.T. (2007) Development of an improved devitrifiable fuel cell sealing glass. *J. Electrochem. Soc.*, **154** (6), B601–B608.
8. Sun, T., Xiao, H.N., Guo, W.M., and Hong, X.C. (2010) Effect of Al_2O_3 content on BaO-Al_2O_3-B_2O_3-SiO_2 glass sealant for solid oxide fuel cell. *Ceram. Int.*, **36**, 821–826.
9. Goel, A., Tulyaganov, D.U., Ferrari, A.M., Shaaban, E.R., Prange, A., Bondioli, F., and Ferreira, J.M.F. (2010) Structure, structure, sintering, and crystallization kinetics of alkaline-earth aluminosilicate glass-ceramic sealants

for solid oxide fuel cells. *J. Am. Ceram. Soc.*, **93** (3), 830–837.

10. Meinhardt, K.D., Kim, D.S., Chou, Y.S., and Weil, K.S. (2008) Synthesis and properties of a barium aluminosilicate solid oxide fuel cell glass-ceramic sealant. *J. Power Sources*, **182**, 188–196.
11. Goel, A., Pascual, M.J., and Ferreira, J.M.F. (2010) Stable glass-ceramic sealants for solid oxide fuel cells: influence of Bi_2O_3 doping. *Int. J. Hydrogen Energy*, **35**, 6911–6923.
12. Wang, R.F., Lü, Z., Liu, C.Q., Zhu, R.B., Huang, X.Q., Wei, B., Ai, N., and Su, W.H. (2007) Characteristics of a SiO_2-B_2O_3-Al_2O_3-$BaCO_3$-PbO_2-ZnO glass-ceramic sealant for SOFCs. *J. Alloys Compd.*, **432**, 189–193.
13. Zheng, R., Wang, S.R., Nie, H.W., and Wen, T.L. (2004) SiO_2-CaO-B_2O_3-Al_2O_3 ceramic glaze as sealant for planar ITSOFC. *J. Power Sources*, **128**, 165–172.
14. Smeacetto, F., Salvo, M., Bytner, F.D.D., Leone, P., and Ferraris, M. (2010) New glass and glass-ceramic sealants for planar solid oxide fuel cells. *J. Eur. Ceram. Soc.*, **30**, 933–940.
15. Reis, S.T., Pascual, M.J., Brow, R.K., Ray, C.S., and Zhang, T. (2010) Crystallization and processing of SOFC sealing glasses. *J. Non-Cryst. Solids*, **356**, 3009–3012.
16. Huang, X.Y. (2005) Low-cost Integrated Composite Seal for SOFC: Materials and Design Methodologies. 2005 Office of Fossil Energy Fuel Cell Program Annual Report. SECA 121.
17. Singh, R.N. (2007) Sealing technology for solid oxide fuel cells (SOFC). *Int. J. Appl. Ceram. Technol.*, **4** (2), 134–144.
18. Liu, W.N., Sun, X., Koeppel, B., and Khaleel, M. (2010) Experimental study of the aging and self-healing of the glass/ceramic sealant used in SOFCs. *Int. J. Appl. Ceram. Technol.*, **7** (1), 22–29.
19. Liu, W.N., Sun, X., and Khaleel, M.A. (2011) Study of geometric stability and structural integrity of self-healing glass seal system used in solid oxide fuel cells. *J. Power Sources*, **196**, 1750–1761.
20. Mahapatra, M.K. and Lu, K. (2010) Thermochemical compatibility of a seal glass with different solid oxide cell components. *Int. J. Appl. Ceram. Technol.*, **7** (1), 10–21.
21. Fergus, J.W. (2005) Sealants for solid oxide fuel cells. *J. Power Sources*, **147**, 46–57.
22. Mahapatra, M.K. and Lu, K. (2010) Glass-based seals for solid oxide fuel and electrolyzer cells - a review. *Mater. Sci. Eng., R*, **67**, 65–85.
23. Weil, K.S., Deibler, J.E., Hardy, J.S., Kim, D.S., Xia, G.G., Chick, L.A., and Coyle, C.A. (2004) Rupture testing as a tool for developing planar solid oxide fuel cell seals. *J. Mater. Eng. Perform.*, **13** (3), 316–326.
24. Peng, L. (2008) Sealing glass design and characterization in the BaO-B_2O_3-SiO_2 system for intermediate-temperature planar SOFC applications. Doctoral thesis, Institute of Process Engineering, Chinese Academy of Sciences.
25. Peng, L. and Zhu, Q.S. (2008) The development of thermally stable sealing glass in the BaO-B_2O_3-SiO_2 system for planar SOFC applications. *J. Fuel Cell Sci. Technol.*, **5** (3), 031210.
26. Namwong, P., Laorodphan, N., Thiemsorn, W., Jaimasith, M., Wannakon, A., and Chairuangsri, T. (2010) A barium-calcium silicate glass for use as seals in planar SOFCs. *Chiang Mai J. Sci.*, **37** (2), 231–242.
27. Sohn, S.B., Choi, S.Y., Kim, G.H., Song, H.S., and Kim, G.D. (2004) Suitable glass-ceramic sealant for planar solid-oxide fuel cells. *J. Am. Ceram. Soc.*, **87** (2), 254–260.
28. Ghosh, S., Sharma, A.D., Kundu, P., and Basu, R.N. (2008) Glass-ceramic sealants for planar IT-SOFC: a bilayered approach for joining electrolyte and metallic interconnect. *J. Electrochem. Soc.*, **155** (5), B473–478.
29. Goel, A., Tulyaganov, D.U., Kharton, V.V., Yaremchenko, A.A., and Ferreira, J.M.F. (2010) Electrical behavior of aluminosilicate glass-ceramic sealants and their interaction with metallic solid oxide fuel cell interconnects. *J. Power Sources*, **195**, 522–526.
30. Chang, H.T., Lin, C.K., Liu, C.K., and Wu, S.H. (2011) High-temperature

mechanical properties of a solid oxide fuel cell glass sealant in sintered forms. *J. Power Sources*, **196**, 3583–3591.

31. Zhang, T., Brow, R.K., Reis, S.T., and Ray, C.S. (2008) Isothermal crystallization of a solid oxide fuel cell sealing glass by differential thermal analysis. *J. Am. Ceram. Soc.*, **91** (10), 3235–3239.
32. Bansal, N.P. and Gamble, E.A. (2005) Crystallization kinetics of a solid oxide fuel cell seal glass by differential thermal analysis. *J. Power Sources*, **147** (1–2), 107–115.
33. Pascual, M.J., Lara, C., and Durán, A. (2006) Non-isothermal crystallisation kinetics of devitrifying RO-BaO-SiO_2 (R=Mg, Zn) glasses. *Phys. Chem. Glasses B*, **47** (5), 572–581.
34. Lara, C., Pascual, M.J., and Durán, A. (2004) Glass-forming ability, sinterability and thermal properties in the systems RO-BaO-SiO_2 (R=Mg, Zn). *J. Non-Cryst. Solids*, **348**, 149–155.
35. Zhu, Q., Peng, L., Huang, W., and Xie, Z. (2007) Ultra stable sealing glass for intermediate temperature solid oxide fuel cells. *Key Eng. Mater.*, **336–338**, 481–485.
36. Peng, L., Zhu, Q.S., Xie, Z.H., and Huang, W.L. (2006) Thermal stability investigation of a newly developed sealing glass as IT-SOFC sealant. *J. Inorg. Mater.*, **21** (4), 867–872.
37. Chou, Y.S., Stevenson, J.W., and Gow, R.N. (2007) Novel alkaline earth silicate sealing glass for SOFC part I. The effect of nickel oxide on the thermal and mechanical properties. *J. Power Sources*, **168**, 426–433.
38. Chou, Y.S., Stevenson, J.W., and Singh, P. (2007) Novel refractory alkaline earth silicate sealing glasses for planar solid oxide fuel cells. *J. Electrochem. Soc.*, **154** (7), B644–B651.
39. Reis, S.T. and Brow, R.K. (2006) Designing sealing glasses for solid oxide fuel cells. *J. Mater. Eng. Perform.*, **15**, 410–413.
40. Larsen, P.H. and James, P.F. (1998) Chemical stability of $MgO/CaO/Cr_2O_3$-Al_2O_3-B_2O_3-phosphate glasses in solid oxide fuel cell environment. *J. Mater. Sci.*, **33**, 2499–2507.
41. Jin, T. and Lu, K. (2010) Compatibility between AISI441 alloy interconnect and representative seal glasses in solid oxide fuel/electrolyzer cells. *J. Power Sources*, **195**, 4853–4864.
42. Ghosh, S., Sharma, A.D., Mukhopadhyay, A.K., Kundu, P., and Basu, R.N. (2010) Effect of BaO addition on magnesium lanthanum alumino borosilicate-based glass-ceramic sealant for anode-supported solid oxide fuel cell. *Int. J. Hydrogen Energy*, **35**, 272–283.
43. Ghosh, S., Kundu, P., Sharma, A.D., Basu, R.N., and Maiti, H.S. (2008) Microstructure and property evaluation of barium aluminosilicate glass-ceramic sealant for anode-supported solid oxide fuel cell. *J. Eur. Ceram. Soc.*, **28**, 69–76.
44. Chou, Y.S., Stevenson, J.W., Xia, G.G., and Yang, Z.G. (2010) Electrical stability of a novel sealing glass with (Mn, Co)-spinel coated Crofer22APU in a simulated SOFC dual environment. *J. Power Sources*, **195**, 5666–5673.
45. Donald, I.W., Metcalfe, B.L., and Gerrard, L.A. (2008) Interfacial reactions in glass-ceramic-to-metal seals. *J. Am. Ceram. Soc.*, **91** (3), 715–720.
46. Horita, T., Sakai, N., Kawada, T., Yokokawa, H., and Dokiya, M. (1993) Reaction of SOFC components with sealing materials. *Denki Kagaku*, **61** (7), 760–762.
47. Yang, Z.G., Stevenson, J.W., and Meinhardt, K.D. (2003) Chemical interactions of barium-calcium-aluminosilicate-based sealing glasses with oxidation resistant alloys. *Solid State Ionics*, **160**, 213–225.
48. Lahl, N., Bahadur, D., Singh, K., Singheiser, L., and Hilpert, K. (2002) Chemical interactions between aluminosilicate base sealants and the components on the anode side of solid oxide fuel cells. *J. Electrochem. Soc.*, **149** (5), A607–A614.
49. Yang, Z.G., Meinhardt, K.D., and Stevenson, J.W. (2003) Chemical compatibility of barium-calcium-aluminosilicate-based sealing glasses with the ferritic stainless steel interconnect in SOFCs. *J. Electrochem. Soc.*, **150** (8), A1095–A1101.

50. Kumar, V., Pandey, O.P., and Singh, K. (2010) Effect of A_2O_3 (A = La, Y, Cr, Al) on thermal and crystallization kinetics of borosilicate glass sealants for solid oxide fuel cells. *Ceram. Int.*, **36**, 1621–1628.
51. Peng, L. and Zhu, Q.S. (2009) Thermal cycle stability of BaO-B_2O_3-SiO_2 sealing glass. *J. Power Sources*, **194**, 880–885.
52. Mahapatra, M.K. and Lu, K. (2010) Seal glass compatibility with bare and $(Mn,Co)_3O_4$ coated AISI 441 alloy in solid oxide fuel/electrolyzer cell atmospheres. *Int. J. Hydrogen Energy*, **35**, 11908–11917.
53. Mahapatra, M.K. and Lu, K. (2011) Seal glass compatibility with bare and $(Mn,Co)_3O_4$ coated Crofer 22 APU alloy in different atmospheres. *J. Power Sources*, **196**, 700–708.
54. Malzbender, J. and Steinbrech, R.W. (2007) Advanced measurement techniques to characterize thermo-mechanical aspects of solid oxide fuel cells. *J. Power Sources*, **173** (1), 60–67.
55. Weil, K.S., Coyle, C.A., Hardy, J.S., Kim, J.Y., and Xia, G.G. (2004) Alternative planar SOFC sealing concepts. *Fuel Cells Bull.*, **5**, 11–16.
56. Singh, R.N. (2006) High-temperature seals for solid oxide fuel cells (SOFC). *J. Mater. Eng. Perform.*, **15** (4), 422–426.
57. Chou, Y.S., Stevenson, J.W., and Gow, R.N. (2007) Novel alkaline earth silicate sealing glass for SOFC part II. Sealing and interfacial microstructure. *J. Power Sources*, **170**, 395–400.
58. Nielsen, K.A., Solvang, M., Nielsen, S.B.L., Dinesen, A.R., Beeaff, D., and Larsen, P.H. (2007) Glass composite seals for SOFC application. *J. Eur. Ceram. Soc.*, **27** (2–3), 1817–1822.
59. Pascual, M.J., Guillet, A., and Duran, A. (2007) Optimization of glass-ceramic sealant compositions in the system MgO-BaO-SiO_2 for solid oxide fuel cells (SOFC). *J. Power Sources*, **169** (1), 40–46.
60. Smeacetto, F., Salvo, M., Ferraris, M., Casalegno, V., and Asinari, P. (2008) Glass and composite seals for the joining of YSZ to metallic interconnect in solid oxide fuel cells. *J. Eur. Ceram. Soc.*, **28**, 611–616.
61. Smeacetto, F., Salvo, M., Ferraris, M., Cho, J., and Boccaccini, A.R. (2008) Glass-ceramic seal to join Crofer 22 APU alloy to YSZ ceramic in planar SOFCs. *J. Eur. Ceram. Soc.*, **28** (1), 61–68.
62. Muller, A., Becker, W., Stolten, D., and Hohe, J. (2006) A hybrid method to assess interface debonding by finite fracture mechanics. *Eng. Fract. Mech.*, **73** (8), 994–1008.
63. Zhang, T., Zhu, Q.S., and Xie, Z.H. (2009) Modeling of cracking of the glass-based seals for solid oxide fuel cell. *J. Power Sources*, **188** (1), 177–183.
64. Weil, K.S. and Koeppel, B.J. (2008) Thermal stress analysis of the planar SOFC bonded compliant seal design. *Int. J. Hydrogen Energy*, **33** (14), 3976–3990.
65. Taniguchi, S., Kadowaki, M., Yasuo, T., Akiyama, Y., Miyake, Y., and Nishio, K. (2000) Improvement of thermal cycle characteristics of a planar-type solid oxide fuel cell by using ceramic fiber as sealing material. *J. Power Sources*, **90**, 163–169.
66. Chou, Y.S., Stevenson, J.W., Hardy, J., and Singh, P. (2006) Material degradation during isothermal ageing and thermal cycling of hybrid mica seals under solid oxide fuel cell exposure conditions. *J. Power Sources*, **157**, 260–270.
67. Gross, S.M., Koppitz, T., Remmel, J., and Reisgen, J.B.B.U. (2006) Joining properties of a composite glass-ceramic sealant. *Fuel Cells Bull.*, **9**, 12–15.
68. Chou, Y.S. and Stevenson, J.W. (2005) Long-term thermal cycling of phlogopite mica-based compressive seals for solid oxide fuel cells. *J. Power Sources*, **140**, 340–345.
69. Chou, Y.S. and Stevenson, J.W. (2002) Thermal cycling and degradation mechanisms of compressive mica-based seals for solid oxide fuel cells. *J. Power Sources*, **112**, 376–383.
70. Simner, S.P. and Stevenson, J.W. (2001) Compressive mica seals for SOFC applications. *J. Power Sources*, **102**, 310–316.

71. Chou, Y.S., Stevenson, J.W., and Singh, P. (2005) Combined ageing and thermal cycling of compressive mica seals for solid oxide fuel cells. *Ceram. Eng. Sci. Proc.*, **26** (4), 265–272.
72. Chou, Y.S. and Stevenson, J.W. (2003) Novel silver/mica multilayer compressive seals for solid-oxide fuel cells: the effect of thermal cycling and material degradation on leak behavior. *J. Mater. Res.*, **18** (9), 2243–2250.
73. Chou, Y.S., Stevenson, J.W., and Chick, L.A. (2003) Novel compressive mica seals with metallic interlayers for solid oxide fuel cell applications. *J. Am. Ceram. Soc.*, **86** (6), 1003–1007.
74. Chou, Y.S. and Stevenson, J.W. (2003) Mid-term stability of novel mica-based compressive seals for solid oxide fuel cells. *J. Power Sources*, **115**, 274–278.
75. Chou, Y.S., Stevenson, J.W., and Chick, L.A. (2002) Ultra-low leak rate of hybrid compressive mica seals for solid oxide fuel cells. *J. Power Sources*, **112**, 130–136.
76. Chou, Y.S. and Stevenson, J.W. (2003) Phlogopite mica-based compressive seals for solid oxide fuel cells: effect of mica thickness. *J. Power Sources*, **124**, 473–478.
77. Chou, Y.S., Stevenson, J.W., and Singh, P. (2005) Thermal cycle stability of a novel glass-mica composite seal for solid oxide fuel cells: effect of glass volume fraction and stresses. *J. Power Sources*, **152**, 168–174.
78. Chou, Y.S. and Stevenson, J.W. (2009) Long-term ageing and materials degradation of hybrid mica compressive seals for solid oxide fuel cells. *J. Power Sources*, **191**, 384–389.
79. Chou, Y.S. and Stevenson, J.W. (2005) *Development in Solid Oxide Fuel Cells and Lithium Ion Batteries*, Ceramic Transaction Series, Vol. **161**, American Ceramic Society, Westerville, OH, pp. 89–98.
80. Sang, S.B., Pu, J., Jiang, S.P., and Li, J. (2008) Prediction of H_2 leak rate in mica-based seals of planar solid oxide fuel cells. *J. Power Sources*, **182**, 141–144.
81. Weil, K.S. (2006) The state-of-the-art in sealing technology for solid oxide fuel cells. *JOM*, **58** (8), 37–44.
82. Weil, K.S., Coyle, C.A., Darsell, J.T., Xia, G.G., and Hardy, J.S. (2005) Effects of thermal cycling and thermal aging on the hermeticity and strength of silver-copper oxide air-brazed seals. *J. Power Sources*, **152**, 97–104.
83. Weil, K.S., Kim, J.Y., and Hardy, J.S. (2005) Reactive air brazing: a novel method of sealing SOFCs and other solid-state electrochemical devices. *Electrochem. Solid-State Lett.*, **8** (2), A133–A136.
84. Kuhn, B., Wessel, E., Malzbender, J., Steinbrech, R.W., and Singheiser, L. (2010) Effect of isothermal aging on the mechanical performance of brazed ceramic/metal joints for planar SOFC-stacks. *Int. J. Hydrogen Energy*, **35**, 9158–9165.
85. Kuhn, B., Wetzel, F.J., Malzbender, J., Steinbrech, R.W., and Singheiser, L. (2009) Mechanical performance of reactive-air-brazed (RAB) ceramic/metal joints for solid oxide fuel cells at ambient temperature. *J. Power Sources*, **193**, 199–202.
86. Le, S.R., Shen, Z.M., Zhu, X.D., Zhou, X.L., Yan, Y., Sun, K.N., Zhang, N.Q., Yuan, Y.X., and Mao, Y.C. (2010) Effective Ag-CuO sealant for planar solid oxide fuel cells. *J. Alloys Compd.*, **496**, 96–99.
87. Jiang, W.C., Tu, S.T., Li, G.C., and Gong, J.M. (2010) Residual stress and plastic strain analysis in the brazed joint of bonded compliant seal design in planar solid oxide fuel cell. *J. Power Sources*, **195**, 3513–3522.
88. Coillot, D., Podor, R., Méar, F.O., and Montagne, L. (2010) Characterization of self-healing glassy composites by high-temperature environmental scanning electron microscopy (HT-ESEM). *J. Electron Microsc.*, **59** (5), 359–366.
89. Suda, S., Matsumiya, M., Kawahara, K., and Jono, K. (2010) Thermal cycle reliability of glass/ceramic composite gas sealing materials. *Int. J. Appl. Ceram. Technol.*, **7** (1), 49–54.
90. Wang, S.F., Wang, Y.R., Hsu, Y.F., and Chuang, C.C. (2009) Effect of additives

on the thermal properties and sealing characteristic of $BaO-Al_2O_3-B_2O_3-SiO_2$ glass-ceramic for solid oxide fuel cell application. *Int. J. Hydrogen Energy*, **34**, 8235–8244.

91. Sakuragi, S., Funahashi, Y., Suzuki, T., Fujishiro, Y., and Awano, M. (2008) Non-alkaline glass-MgO composites for SOFC sealant. *J. Power Sources*, **185**, 1311–1314.
92. Caron, N., Bianchi, L., and Méthout, S. (2008) Development of a functional sealing layer for SOFC applications. *J. Therm. Spray Technol.*, **175** (5–6), 598–602.
93. Choi, S.R., Bansal, N.P., and Garg, A. (2007) Mechanical and microstructural characterization of boron nitride nanotubes-reinforced SOFC seal glass composite. *Mater. Sci. Eng., A*, **460–461**, 509–515.
94. Lee, J.C., Kwon, H.C., Kwon, Y.P., Lee, J.H., and Park, S. (2007) Porous ceramic fiber glass matrix composites for solid oxide fuel cell seals. *Colloids Surf. A Physicochem. Eng. Asp.*, **300**, 150–153.
95. Brochu, M., Gauntt, B.D., Shah, R., Miyake, G., and Loehman, R.E. (2006) Comparison between barium and strontium-glass composites for sealing SOFCs. *J. Eur. Ceram. Soc.*, **26** (5), 3307–3313.
96. Brochu, M., Gauntt, B.D., Shah, R., and Loehman, R.E. (2006) Comparison between micrometer- and nano-scale glass composites for sealing solid oxide fuel cells. *J. Am. Ceram. Soc.*, **89** (3), 810–816.
97. Hong, S.J. and Kim, D.J. (2007) Polymer derived ceramic seals for application in SOFC. *Mater. Sci. Forum*, **534–536**, 1061–1064.
98. Lee, J.C., Kwon, H.C., Kwon, Y.P., Lee, J.H., and Park, S. (2007) Sealing properties of ceramic fiber composites for SOFC application. *Solid State Phenom.*, **124–126**, 803–806.
99. Deng, X.H., Duquette, J., and Petric, A. (2007) Silver-glass composite for high temperature sealing. *Int. J. Appl. Ceram. Technol.*, **4** (2), 145–151.
100. Story, C., Lu, K., Reynolds, W.T. Jr., and Brown, D. (2008) Shape memory alloy/glass composite seal for solid oxide electrolyzer and fuel cells. *Int. J. Appl. Ceram. Technol.*, **33**, 3970–3975.
101. Le, S.R., Sun, K.N., Zhang, N.Q., An, M.Z., Zhou, D.R., Zhang, J.D., and Li, D.G. (2006) Novel compressive seals for solid oxide fuel cells. *J. Power Sources*, **161**, 901–906.
102. Le, S.R., Sun, K.N., Zhang, N.Q., Shao, Y.B., An, M.Z., Fu, Q., and Zhu, X.D. (2007) Comparison of infiltrated ceramic fiber paper and mica base compressive seals for planar solid oxide fuel cells. *J. Power Sources*, **168**, 447–452.
103. Sang, S.B., Li, W., Pu, J., and Li, J. (2008) Al_2O_3-based compressive seals for planar intermediated temperature solid oxide fuel cells. *J. Inorg. Mater.*, **23** (4), 841–846.
104. Sang, S.B., Pu, J., Chi, B., and Li, J. (2009) Model-oriented cast ceramic tape seals for planar solid oxide fuel cells. *J. Power Sources*, **193**, 723–729.
105. Zhang, T., Tang, D., and Yang, H.W. (2011) Can crystalline phase be self-healing sealants for solid oxide fuel cells? *J. Power Sources*, **196**, 1321–1323.

7
Degradation and Durability of Electrodes of Solid Oxide Fuel Cells

Kongfa Chen and San Ping Jiang

7.1
Introduction

To improve the stability and reliability and to reduce the production cost of solid oxide fuel cells (SOFCs), it is necessary to reduce the operating temperature from the conventional 1000 °C to the intermediate-temperature ranges 600–800 °C [1, 2]. One significant advantage of intermediate-temperature solid oxide fuel cells (IT-SOFCs) lies in the use of metallic interconnect materials with higher electronic conductivity, better mechanical properties, higher thermal conductivity, better workability, and lower cost, as compared to lanthanum–chromite–perovskite-based ceramic interconnect materials [3, 4]. The most widely developed and studied metallic interconnect materials are the chromia-forming alloys because of their good thermal properties, high corrosion resistance at high temperature, and conductive chromium oxide scales. However, on the air side of fuel cells, chromium oxide scales can further react with oxygen and moisture to form gaseous hexavalent Cr species such as CrO_3 and $CrO_2(OH)_2$ under the operating temperatures of SOFCs [5]. The gaseous chromium species from the interconnect based on chromia-forming alloy will migrate through the cathode and deposit in the form of solid chromium(III) oxides either at the electrode–electrolyte interface, inside the bulk of the electrode, and/or on the electrode surface. The accumulation of Cr deposits eventually leads to a deactivation of the cathode and degradation of SOFC performance [6–9].

Sulfur dioxide in the airstream is another potential contaminant of SOFC cathode. The main source of sulfur dioxide in the air is the industrial activities that process materials or fuels containing sulfur, for example, the generation of electricity from coal, oil, or gas that contains sulfur. Although the concentration of SO_x may be low (0.02–0.20 ppm, depending on the location and time of the day), the accumulation of sulfur is shown to be detrimental to the cathode performance of SOFCs [10]. However, little is known about the deposition and interaction between SO_x and the cathode, and the source of sulfur on the air side is not well recognized in SOFCs, but the presence of sulfur in natural gas is well known.

Materials for High-Temperature Fuel Cells, First Edition. Edited by San Ping Jiang and Yushan Yan.

Other impurities such as Na, Al, and Si have also been found on the cathode during the long-term operation of flattened tube type SOFC stacks, and their concentration increases with operation time [11, 12]. The exact source of these elements is not clear but they may come from thermal insulating materials and gas supply tubes and condense on the cathode surface and cathode–electrolyte interface during the long-term operation of SOFC stacks. Chemical reaction at the cathode–electrolyte interface, microstructural changes, and grain growth during operation at high temperature are also detrimental to the electrocatalytic activities of the electrodes toward oxygen reduction reaction.

One of the distinct advantages of SOFCs operating at high temperatures is the fuel flexibility including syngas derived from coal and biomass gasification. However, direct use of hydrocarbon fuels as feed over Ni-based anodes results in significant deactivation from carbon deposition, blocking active sites on the anode and reducing electrical efficiency and durability of the fuel cells. In particular, tars are one of the primary impurities in biomass and coal gasification fuel gas and can degrade anodes by carbon deposition [13]. Ni-based anodes are very active for H_2 oxidation reaction, but are highly vulnerable to deactivation and poisoning by sulfur [14, 15]. Sulfur is a common contaminant in readily available hydrocarbon fuels. Also, the poisoning by impurities such as silica in the raw materials could not be neglected. In the case of Ni-based cermet anodes, Ni sintering and redox cycling are primarily responsible for the microstructure degradation [16, 17]. Degradation of cell performance can also be caused by aging of the Y_2O_3–ZrO_2 (YSZ) electrolyte [18–20].

In this chapter, we begin by examining the effects of various degradation, deposition, and poisoning processes of common contaminants on the electrochemical activity and performance durability of anodes and cathodes of SOFCs, reviewing the prospects of developing contaminant-tolerant electrodes either by modifying the present electrode system or by developing alternative electrode materials. This is concluded by a brief review of the recent progress in the understanding of the degradation behavior of high-temperature solid oxide electrolysis cells (SOECs).

7.2 Anodes

Electrochemical reactions at the anode in SOFCs predominantly occur close to the triple-phase boundaries (TPBs) among the ionic conductor, electronic conductor, and gas phase. Therefore, changes in the microstructure and surface properties of anode materials can have a significant impact on the performance and durability of SOFCs.

7.2.1 Sintering and Agglomeration of Ni Particles

Sintering and grain growth of Ni phases in Ni-based cermet anodes is a major concern for long-term operation of SOFCs. In the Ni–YSZ cermet anode, the Ni phase has poor wetting properties with the YSZ electrolyte phases because of the high interfacial energy [21]. Thus, Ni particles have a tendency to grow and agglomerate during reduction and operation at high temperatures. Ni agglomeration can cause a loss of connectivity between Ni particles and decrease the active surface area, leading to degradation of cell performance. Iwata [22] observed that during the operation at 1008 °C for 1015 h, the Ni–YSZ anode degrades at a rate of 14 μV h^{-1} at a current density of 0.3 A cm^{-2}. The initial diameter of Ni particles in the Ni–YSZ anode is 0.1 μm, which increases to 1 μm after operating at 927 °C for 211 h. Simwonis *et al.* [23] exposed the Ni–YSZ anode in 93% Ar/4% H_2/3% H_2O at 1000 °C for over 4000 h and observed the agglomeration of Ni particles. The electrical conductivity decreases by up to 33.3% in addition to the Ni agglomeration. Ni agglomeration can also cause a loss in percolation of the anodes.

Under steady-state operating conditions, temperature and water vapor are important parameters affecting the agglomeration of Ni phase in Ni-based cermet anodes. Sehested *et al.* [24, 25] studied the growth of Ni particles as a function of operating temperature in a Ni steam reforming reactor and observed an Arrhenius-type behavior for the change in specific surface area of Ni. Reduction temperature also has an impact on Ni agglomeration. Iwanschitz *et al.* [26] studied Ni agglomeration as a function of reducing temperatures between 555 and 1140 °C and observed significant Ni-particle growth at reduction temperatures above 850 °C for Ni–GDC (gadolinium-doped ceria) anodes. The growth in Ni particles is generally accelerated at high steam concentration, at high fuel utilization, and under high currents [27–30]. Matsui *et al.* [30] studied the performance stability of Ni–YSZ anode at high fuel humidity between 20 and 40%. In the case of 30% steam, the anode degrades gradually up to 69 h of operation, followed by a sudden drop in current density. The microstructure of the Ni particles changed significantly under current load in 40% of steam. FIB-SEM examination shows a drastic decrease in the TPB areas after the polarization test. This may be due to the formation of volatile $Ni(OH)_2$ species. On the other hand, stable operations have been reported for cells operating at a high steam concentration [31, 32]. de Haart *et al.* [32] operated an anode-supported SOFC in mixtures of methane/water (1 : 2) (internal reforming) and hydrogen/water (2 : 1) at 800 °C. The influence of the water content in hydrogen fuel has been found to be mainly on the change in the open cell voltage.

It is well known that adding ceramic phases such as YSZ and doped ceria can significantly slow down Ni agglomeration through the formation of a rigid network. Thus, for Ni-based cermet anodes, Ni agglomeration depends on both the continuous particle size distribution of ceramic phase [33] and the ratio between Ni and ceramic phases [27]. Jiang [17] studied the sintering behavior of Ni–YSZ cermets as a function of weight ratio of Ni/YSZ at 1000 °C for up to 2000 h. Figure 7.1 shows the SEM images of Ni–YSZ cermet after sintering in moist 10% H_2/90% N_2 for different periods at 1000 °C [17]. For pure Ni anodes, the change in the morphology of electrode coating is dramatic, characterized by the formation of cracked structure and isolated islands. For Ni–YSZ cermet anodes, the change in microstructure is much slower and depends on the Ni and YSZ content in the cermet. For Ni (70 vol%)–YSZ (30 vol%) cermet electrode coating,

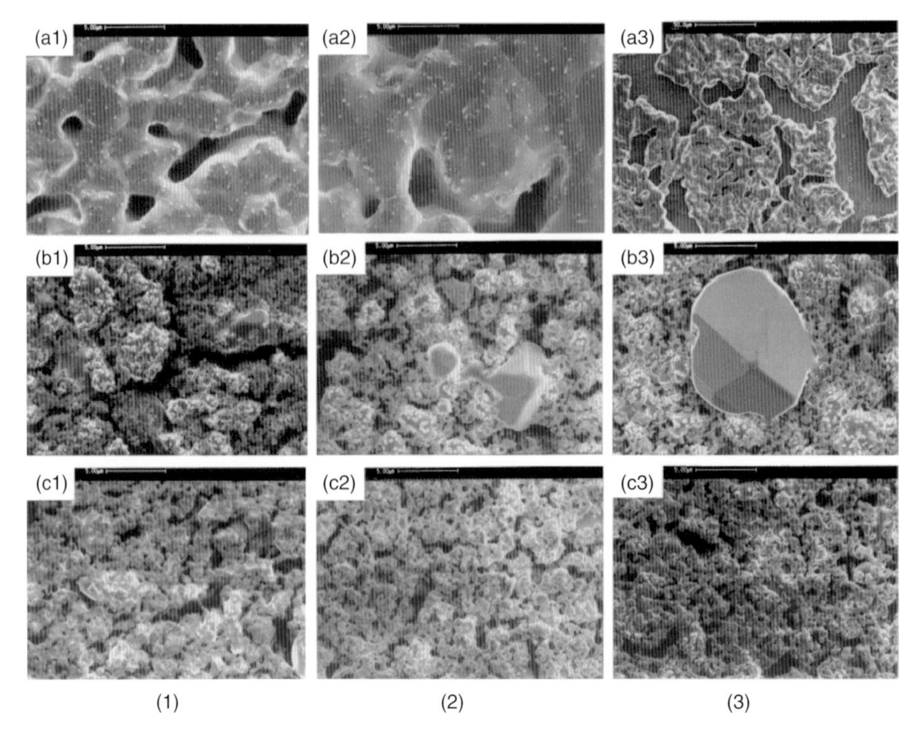

Figure 7.1 SEM pictures of (a) Ni, (b) Ni (70 vol%)–YSZ (30 vol%), and (c) Ni (50 vol%)–YSZ (50 vol%) after sintering under moist 10% H_2/90% N_2 at 1000 °C for 250 (1), 750 (2), and 2000 h (3). (Source: After Ref. [17]. Reproduced with permission.)

bright Ni particles grow outward after sintering at 1000 °C for 250 h and they became bigger after sintering for 750 h (Figure 7.1b1,b2). On the other hand, for Ni (50 vol%)–YSZ (50 vol%) cermet coating, there is no clear observation of outward growth of Ni particles (Figure 7.1c1,c2). The results indicate that the sintering behavior of Ni–YSZ cermet electrodes is dominated by the agglomeration and grain growth of Ni particles in the cermets and is strongly dependent on the Ni and YSZ content in the cermet.

The sintering and agglomeration of Ni particles can be significantly reduced by optimizing the fabrication process. Itoh *et al.* [34, 35] fabricated NiO–YSZ anode with 0.6 μm fine YSZ powder as the starting electrolyte components and observed that both the electrical conductivity and the electrochemical activity of the anode decreased rapidly after 20 h of operation at 1000 °C. However, after substitution of a part of the 0.6 μm fine YSZ particles with 27.0 μm coarse particles, the stability of the Ni–YSZ anode is significantly enhanced. The consolidated framework made from two distinct sized particles has been suggested to promote a more stable YSZ and Ni network.

Cermet anodes prepared by a co-synthesis process are also effective in inhibiting Ni sintering and enhancing the cell stability [36–38]. The co-synthesis involves the mixing of YSZ and NiO particles in nanocrystalline scale, thus creating adequate

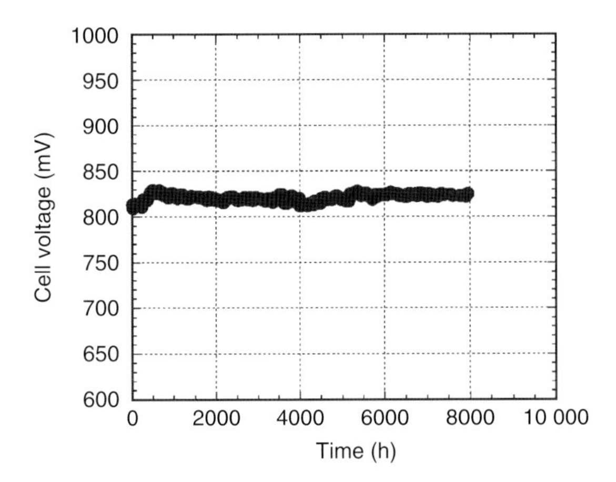

Figure 7.2 Cell voltage as a function of operation time of a single cell with a Ni–YSZ anode prepared by spray pyrolysis and an LSM–YSZ cathode. The cell was operated at 1000 °C and a current density of 300 mA cm^{-2} using H_2/3% H_2O. (Source: After Ref. [39]. Reproduced with permission.)

TPBs for the fuel oxidation reaction. Moreover, as the surface of Ni network is uniformly covered by fine electrolyte particles, the sintering and agglomeration of Ni is significantly inhibited. Fukui *et al.* [39] prepared a Ni–YSZ composite by spray pyrolysis and the single cell based on the Ni–YSZ composite anode is stable over 8000 h at 1000 °C in H_2 (Figure 7.2). The anode still maintained its initial morphology after the stability test. Addition of samaria-doped ceria (SDC) or YSZ nanoparticles into the Ni–YSZ anodes by wet impregnation has also been demonstrated to enhance the electrocatalytic activity and inhibit the sintering and agglomeration of nickel during the high-temperature sintering and reducing stage [40].

7.2.2 **Redox Cycling**

In the commercial lifetime of 5 years, an SOFC system would require to survive 25–100 redox cycles [41]. A particular issue with the redox cycle is the volume change of the Ni phase during the reoxidation of the Ni phase in cermet anodes when the fuel supply is accidentally interrupted or if the fuel utilization is too high [42, 43]. The volume changes in Ni cermet anodes due to redox cycle can cause crack formation in the electrode and delamination at the electrolyte–anode interface.

Microstructural changes have been observed on the anodes during redox cycling. Klemenso *et al.* [44] studied the microstructure change of an anode-supported cell using environmental SEM. Figure 7.3 shows a fixed sample after the first redox cycle [44]. The reduction leads to the appearance of the outline of the grains due to the shrinkage of the Ni particles in size (Figure 7.3b). On the other hand, when

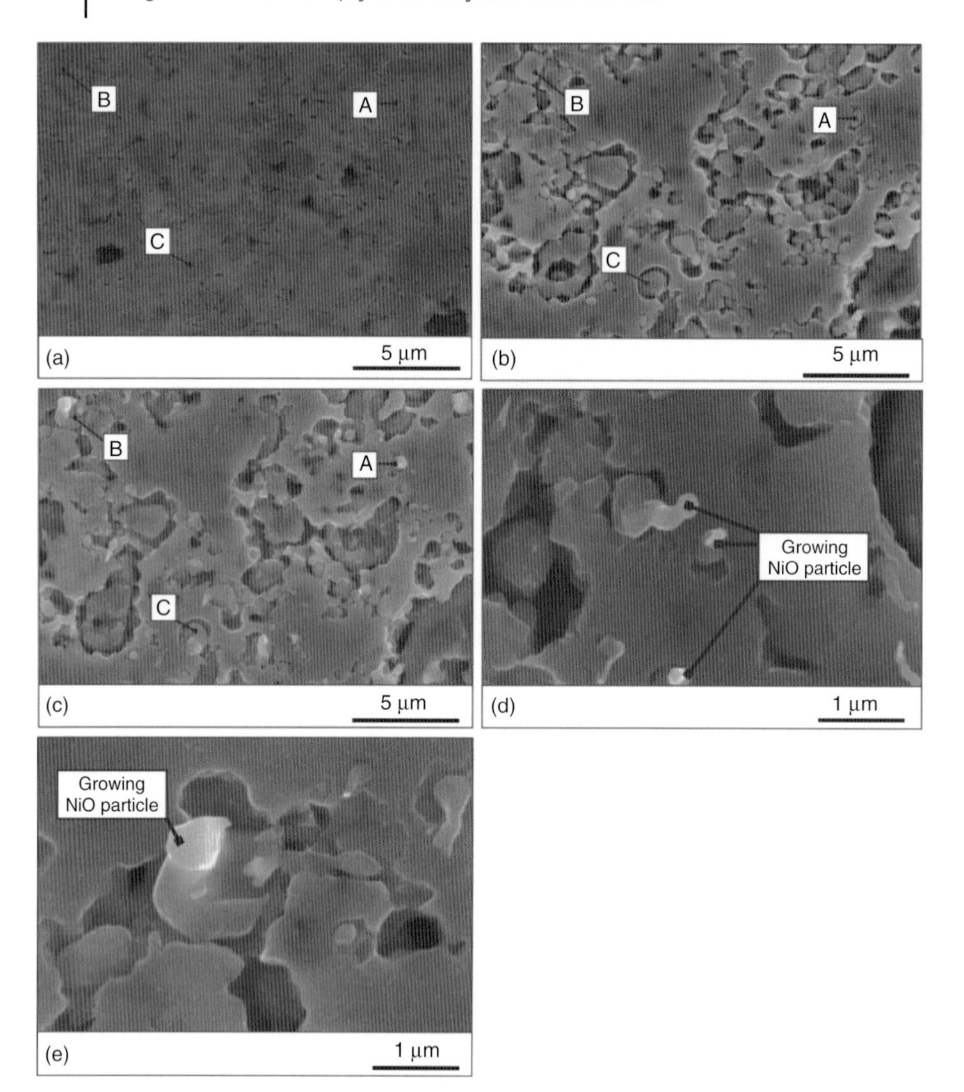

Figure 7.3 ESEM (environmental scanning electron microscopic) images from the first redox cycle. The images (a–c) are of the same area. Three Ni particles are indicated as A, B, and C. (a) As-sintered, (b) reducing conditions, 27 min at 850 °C, (c) reoxidized for 60 min at 850 °C, and (d,e) close-up of growing nickel oxide particles after 60 min reoxidation at 850 °C. (Source: After Ref. [44]. Reproduced with permission.)

reoxidized for the first time at 850 °C, the Ni particles increase in size but do not revert to the as-sintered shape (Figure 7.3c). Instead, the particles are separated into two to four grains and grow out of the polished plane (Figure 7.3d,e). However, the grain growth of Ni phase on reoxidation depends on the oxidation kinetics. Ni phase after reoxidation has been observed to have a higher porosity and requires a larger volume than the initial NiO–YSZ anode before reduction [45]. The change in the microstructure is generally accompanied by the volume expansion of the anode [46]. The expansion of the anode causes the development of tensile stresses in the electrolyte, which may cause fracture of thin electrolyte film and lead to failure of the cell [45, 47–49]. In addition, the volume expansion and shrinkage may also cause crack formation and fracture of the anode itself [41].

The performance of the Ni–YSZ cermet anode is known to degrade when subjected to redox cycling [50, 51]. Waldbillig *et al.* [50] observed that the cell performance decreased after each redox cycle, especially for redox times greater than 1 h. Sumi *et al.* [52] reported that both ohmic loss and polarization resistance of the Ni–YSZ anode were increased after redox cycling. The TPB lengths of the anode are decreased from the initial 2.49 to 2.39 and 2.11 $\mu m\,\mu m^{-3}$ after the first and fourth redox cycles, respectively.

The effect of redox is significantly dependent on the initial microstructure of the anodes. Anodes with coarse structure [53] or a higher porosity are more dimensionally stable during redox cycling. Decreasing the amount of Ni in the anode or reducing the oxidation temperature also lowers the volume expansion [53]. Klemenso *et al.* [41] studied the bulk expansion of Ni–YSZ anodes by dilatometry and the results indicate that the key parameters for achieving redox-stable anodes are the ceramic network strength and the degree of restriction of Ni-particle relocation and Ni-particle coarsening. Waldbillig *et al.* [54] presented a method to improve the redox tolerance of anodes. The tolerance of Ni-based cermet anodes has been enhanced by fabrication of a two-layer graded anode, with the inner layer having less Ni content. Then, a 20 μm thick NiO–YSZ oxidation barrier layer is fabricated on the outer layer of the anode. The barrier layer is porous under reducing conditions but densified under oxidizing conditions, restricting oxygen penetration into the anode. However, as the redox kinetics is a function of sample depth [55], the oxidation barrier could be less effective in the case of longer redox period. A highly dense layer may also adversely affect the electrochemical activity of the anode.

Use of appropriate operational parameters and procedures can help minimize the detrimental effect of redox cycling. Vedasri *et al.* [49] observed that controlling the cooling rate could avoid mechanical damage during redox cycling. When the Ni–YSZ-anode-supported cells are cooled down by air exposure at a rate higher than 3 $^{\circ}C\,min^{-1}$ from 800 °C to temperatures below 600 °C, the Ni oxidation rate could be significantly slowed down and the volume expansion on Ni oxidation is reduced. At a higher cooling rate, less Ni particles, especially those in the inner layers, will be oxidized, which could reduce the volume expansion of the Ni phase in the anode and thus protect the thin electrolyte layer from cracking.

Nanostructured Ni-based cermet produced by infiltrating or impregnating Ni into a structurally robust electrolyte skeleton can minimize the microstructure change and performance degradation during redox cycling. Busawon *et al.* [42] prepared the Ni–YSZ anode by infiltration of 12–16 wt% Ni into a porous YSZ skeleton. The impregnated anode underwent no bulk dimensional changes during reoxidation at 800 °C. However, the electrical conductivity of the anode decreased by 20% after one redox cycle because of agglomeration and growth of fine Ni particles. Hanifi *et al.* [56] incorporated Ni–SDC into the porous YSZ skeleton as the anode and LSM ((La,Sr)MnO_3) into the YSZ skeleton as the cathode by wet impregnation. The OCV (open circuit voltage) of the tubular cell is quite stable for 100 thermal cycles within 400–800 °C and 10 redox cycling tests. Tucker *et al.* [48] fabricated tubular metal-supported SOFCs, with YSZ electrolyte and electrodes comprising the porous YSZ skeleton, and infiltrated Ni and LSM phases. The cell survived five complete anode redox cycles and five rapid thermal cycles. Precoarsening the infiltrated Ni anode at 800 °C before operation also improved the stability. Kim *et al.* [57] reported that anode precursor powder prepared by coating NiO nanoparticles on YSZ powder by the Pechini method also enhanced the thermal and redox tolerance of Ni–YSZ cermet.

Ni-free oxide materials have been investigated as alternative redox-tolerant anodes of SOFCs, such as $La_{0.75}Sr_{0.25}Cr_{0.5}Mn_{0.5}O_3$ (LSCM) [58] and (La,Sr)TiO_3 (LST) [59]. One main concern regarding Ni-free oxide anodes is the generally low electrical conductivity of the conductive metal oxides. For example, at 800 °C in fuel oxidation atmosphere, the electrical conductivity of LSCM is ~1 S cm^{-1} [60, 61] and for LST, it is ~10 to 100 S cm^{-1} [59, 62], which is significantly lower than ~1000 S cm^{-1} measured on Ni-based cermet anodes [63].

7.2.3 Carbon Deposition

Hydrogen is an ideal fuel for SOFCs; however, its generation, storage, and transportation on a large scale remain as an important technological challenge [64]. Thus, direct utilization of hydrocarbon fuels without reforming and cleaning procedures will have significant advantages over conventional hydrogen-based fuel cells [65–67].

On the other hand, in hydrocarbon fuels, Ni tends to catalyze carbon cracking, forming and depositing carbon particles or fibers on the surface of Ni-based anodes (Figure 7.4 [68]), which can hinder fuel transport and block active reaction

Figure 7.4 SEM images of carbon deposited on (a,b) Ni–YSZ cermet and (c) Ni catalyst supported on ZrO_2 during CH_4 decomposition at 1000 °C. (Source: After Ref. [68]. Reproduced with permission.)

sites, and as a consequence degrade the cell performance [68–72]. Metal dusting corrosion of nickel has also been observed due to the dissolution of carbon into the bulk of nickel [73]. In the case of methane, formation of carbon proceeds either via dissociation (Eq. (7.1)) or according to the Boudouard reaction (Eq. (7.2)) [66, 74]:

$$CH_4 \rightleftharpoons C + 2H_2 \tag{7.1}$$

$$2CO \rightleftharpoons C + CO_2 \tag{7.2}$$

The formation of solid carbon via the Boudouard reaction is expected to occur at temperatures below 700 °C, whereas decomposition of methane occurs thermodynamically at temperatures above 500 °C [75].

In the case of heavy hydrocarbons, carbon deposition occurs more rapidly than that in light hydrocarbons, because heavy hydrocarbons are more susceptible to noncatalytic thermal cracking [76–79]. Saunders *et al.* [80] observed that the cell voltage dropped to zero in about 30 min of operation in iso-octane, and about 90% of the molecular carbon was deposited on the nickel cermet.

Carbon deposited on the Ni anode could be removed by the oxygen anions that are electrochemically driven through the electrolyte [81]. The applied current density or overpotential has a significant effect on coke formation [82, 83]. Lin *et al.* [82] observed that the Ni–YSZ anode degraded quickly in methane fuel under open circuit conditions at temperatures over 700 °C, but there is no degradation and no coke formation on the anode under current loads. As carbon pyrolysis occurs readily at a high temperature, relatively large currents are required to avoid carbon formation and cell failure at higher temperatures [82].

Carbon deposition and formation can be avoided by increasing the steam to carbon (S/C) ratio [84–87], CO_2, or air [88, 89] to enhance internal reforming. Kishimoto *et al.* [90] found that at low S/C ratios, that is, S/C = 0.05–0.1, immediate carbon deposition occurs on the Ni–scandium-stabilized zirconia (ScSZ) anode in *n*-dodecane fuel. However, when the S/C is increased to 2, the cell potential is stable at a current density of $\sim$38 mA cm^{-2} over 120 h. A high concentration of steam produced because of high fuel utilization suppresses carbon deposition [91]. On the other hand, the activities of steam reforming are closely related to the operating temperature. At low operating temperatures, steam reforming may not be complete, and the presence of high concentration of steam tends to cause high concentration polarization [85]. However, there are several intrinsic disadvantages of internal steam reforming in SOFCs. A high concentration of steam dilutes the fuel and lowers the electrical efficiency of the cell [92, 93], can oxidize Ni, and can develop excessively large temperature gradients across the cell and interconnect plate, because of the endothermic cooling especially at the fuel inlet zone [93, 94].

Carbon deposition is also affected by the composition and reduction temperature of the cermet anode. Wang *et al.* [95] prepared an anode precursor powder by impregnating SDC powder with Ni nitrate. The Ni–SDC anode with 60 wt% Ni exhibits low polarization, high performance, and good stability in methane. The researchers suggest that the high stability is due to the optimized distribution and connection between Ni–Ni, SDC–SDC, and Ni–SDC particles. Mallon and Kendall [96] found that reduction at different temperatures causes different Ni distribution within the anode and the Ni–YSZ anode reduced at a lower temperature is more stable in methane fuel.

The nature of the electrolyte component in cermet anodes affects the carbon deposition of Ni-based cermet anodes. Sumi *et al.* [97] compared a Ni–ScSZ anode and a Ni–YSZ anode in 3% H_2O/97% CH_4 fuel at 1000 °C. The power density of the Ni–YSZ anode decreases from an initial 0.8 to 0.6 W cm^{-2} after 250 h of operation. However, the cell with the Ni–ScSZ anode shows no degradation in performance over the same period. The Ni–ScSZ anode is also stable in *n*-dodecane without coking over 50 h of operation [98]. Iida *et al.* [71] found that both Ni–$(Sm,Ce)O_2$ (SDC) and Ni–ScSZ anodes in propane are more stable than the Ni–YSZ anode. The superior carbon tolerance property of the Ni–ScSZ and the Ni-doped CeO_2 anodes over the Ni–YSZ anodes is believed to be due to the strong interaction between Ni and ScSZ and doped ceria and/or the higher ionic conductivity of the ScSZ and doped ceria, as compared to that of YSZ [99–102].

Addition of less active metals such as Cu [103–106] and Au [107, 108] that have little catalytic activity for carbon cracking is also effective in enhancing the coking tolerance of Ni anodes. Boder and Dittmeyer [109] found that the addition of Cu to Ni–YSZ anode by impregnation has a little effect on the electrochemical activity of the anode, but the catalytic activity toward steam reforming reaction decreases significantly with the increasing Cu content. Calculation shows that the addition of 25 wt% Cu decreases the reforming rate constant by a factor of 3 at 950 °C and a factor of 10 at 750 °C. Addition of Fe [110, 111], Mo [87, 112, 113], and Sn [114–116] also lowers the driving force for carbon deposition on the surface of Ni-based cermet.

Addition of precious metals into the Ni cermet can promote the electrocatalytic activity of the anode and suppress carbon formation [117, 118]. Takeguchi *et al.* [117] added Pd, Pt, Rh, and Ru into Ni–YSZ anodes and found that the electrochemical activity of the anodes with Ru and Pt is significantly enhanced. The most likely reason is probably the hydrogen spillover from the precious metal surface to the carbon species, which suppresses carbon cracking. Babaei *et al.* [119] revealed that Pd impregnation mainly promoted the adsorption and diffusion processes of the oxidation reaction of H_2, methane, and ethanol on the surface of Ni–GDC anodes because of the significantly promoted hydrogen and oxygen spillover processes over the nanosized Pd–PdO redox couples on the Ni surface.

Oxide additives also show improvement in the carbon tolerance of Ni-based cermet anodes [68, 120, 121]. Takeguchi *et al.* [68] found that an appropriate amount of CaO, SrO, and CeO_2 effectively suppresses carbon deposition, whereas MgO additive promotes carbon deposition. On the other hand, Zhong *et al.* [111] showed that the addition of 5 wt% MgO to the Ni–Fe alloy effectively enhances the cell stability in CH_4. Addition of electrolyte components such as YSZ, doped ceria, and the proton conductor $SrZr_{0.95}Y_{0.05}O_{3-\alpha}$ (SZY) also improves the carbon tolerance of Ni–YSZ anode [40, 122–124]. Wang *et al.* [125] observed that the GDC-impregnated Ni anode is very stable when exposed to 3% H_2O humidified methane either under open circuit conditions or at a constant current density of 20 mA cm^{-2}. SDC-impregnated Ni anode has been reported to be relatively stable in iso-octane/air mixture for over 260 h at 600 °C [126].

Adding a catalyst layer has been shown to be effective in suppressing the carbon deposition on Ni cermet anodes. Murray and Barnett [66] deposited a 0.5 μm thick $(Y_2O_3)_{0.15}(CeO_2)_{0.85}$ (YDC) layer between the Ni–YSZ anode and the YSZ electrolyte, and the cell shows a highly stable performance in methane for 100 h at temperatures below 700 °C. By depositing a thin Ru-CeO_2 catalyst layer on the surface of Ni–YSZ anode, Zhan and Barnett [89] demonstrated a stable performance at 0.6 W cm^{-2} and 770 °C for 50 h in iso-octane fuel. The combination of Ru-CeO_2/PSZ/Ru-CeO_2 catalyst layer also leads to a stable cell performance with the Ni anode in CH_4 for 350 h [127]. Other catalyst layers such as Ni–Al_2O_3 and Cu–CeO_2 [128] have also been demonstrated to be effective in enhancing the stability of the anodes in methane and hydrocarbon fuels. The catalyst layer, however, reduces the gas diffusion rate to the anode and reduces the electrical current collecting ability and thereby may decrease the electrical efficiency and cell power density [89].

Nickel-free anodes such as Cu–CeO_2 anodes have been demonstrated to be stable in hydrocarbon fuels [67, 129, 130]. In this case, Cu serves as the current collector, as it does not have catalytic activity toward carbon cracking. Impregnated ceria serve as electrocatalytic reaction sites for the oxidation reaction of hydrogen and hydrocarbon fuels. The Cu–CeO_2 or Cu–SDC anode is stable in methane and heavy hydrocarbon fuels (Figure 7.5) [131, 132]. Adding lanthanum and/or precious metals enhances the electrochemical activity of Cu–CeO_2 anodes for the oxidation of hydrocarbons [130, 133]. Although the Cu–CeO_2 anode exhibits excellent tolerance in hydrocarbon fuels at 700 °C, there is a concern regarding the thermal stability of Cu at higher temperatures [134, 135]. The continuous Cu

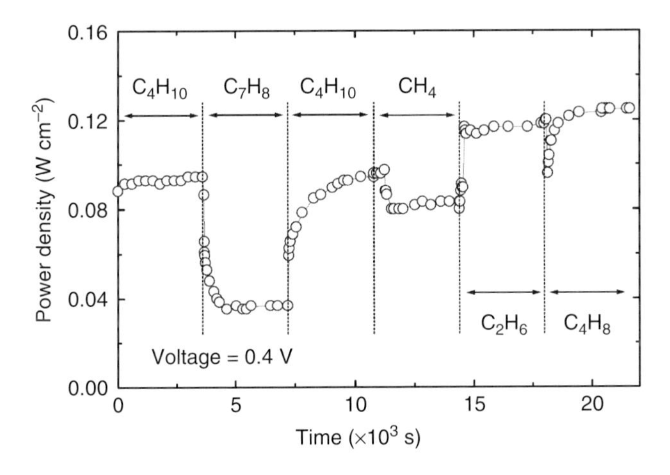

Figure 7.5 Effect of switching fuel type on the cell with the Cu-samaria doped ceria (SDC) composite anode at 700 °C. The power density is shown as a function of time. (Source: After Ref. [132]. Reproduced with permission.)

network formed at low temperatures has been observed to break up into individual Cu particles at 800 and 900 °C because of the agglomeration of the Cu layer [134].

Conductive ceramic oxide materials are promising alternatives to Ni cermet anodes due to their high tolerance toward carbon deposition when operating in hydrocarbon fuels [136–138]. CeO_2 is well known for its ability to store, transport, and release oxygen. Marina and Mogensen [139] reported that $Ce_{0.6}Gd_{0.4}O_3$ (GDC) as anode exhibits a low electrocatalytic activity for methane oxidation and reforming reaction and that no carbon deposition is observed on the GDC anode after operating at 1000 °C for 350 h. The low performance of the ceria anode is most likely due to the low electrical conductivity. On the other hand, its performance could be improved by either doping or adding other components. CeO_2 doped with Mn and Fe to form $Ce_{0.6}Mn_{0.3}Fe_{0.1}O_2$ increases its anodic performance [140]. Addition of a small amount of Ni [141] or precious metals such as Rh, Ru, and Pd [142–144] also enhances the electrocatalytic activity. The cell based on a 1 wt% Pd and 50 wt% CeO_2-impregnated YSZ anode has been reported to exhibit power densities as high as 1.1 W cm^{-2} in H_2/H_2O at 800 °C [145].

Perovskite-based oxides such as $La_{0.75}Sr_{0.25}Cr_{0.5}Mn_{0.5}O_{3-\delta}$ (LSCM) [58, 146, 147], doped $SrTiO_3$ [59, 148–150], $La_4Sr_8Ti_{11}Mn_{0.5}Ga_{0.5}O_{37.5}$ [151], and $Sr_2MgMoO_{6-\delta}$ [152, 153] also exhibit good carbon tolerance. However, the power densities of the cells with the oxide anodes are not as high as those with the Ni-based anodes, because of the relatively low electrical conductivity of the perovskite-based oxides under fuel-reducing conditions [154, 155] and/or the low electrochemical activity of the ceramic oxides for the oxidation reaction of H_2 and CH_4 [155–157]. McIntosh *et al.* [158] studied the electrocatalytic activity of dense LSCM film as anode on YSZ by ultrasonic spray pyrolysis and the results show the low activation activity of bare LSCM film toward methane oxidation. On the other hand, the anode performance can be enhanced by adding electrolyte components, precious metals, and Ni that improve the ionic conductivity, electronic conductivity, and electrocatalytic activity of the oxide anodes. For example, the electrocatalytic activity toward fuel oxidation of LSCM could be improved by addition of YSZ [146], GDC [159], and Pd [160, 161]. However, there is an issue regarding the long-term stability of nanoparticles at relatively high SOFC operating temperatures. Considerable work is needed to address or improve the phase stability of the nanostructured anodes.

7.2.4 Sulfur Poisoning

Sulfur species, which are typically in the form of hydrogen sulfide, commonly exist in coal syngas and fossil fuels such as natural gases [162, 163]. It has been reported that H_2S even in the concentrations of parts per million range can poison the Ni-based cermet anode [164]. The sulfur in the fuel gases can be removed by electrochemical membrane separation [165, 166], a sorbent such as cerium and lanthanum oxide surfaces [167], or the water-scrubber-cleanup technology [168]. However, development of sulfur-tolerant anodes is still required to reduce the overall system cost and ensure cell integrity in case of desulfurization system fault.

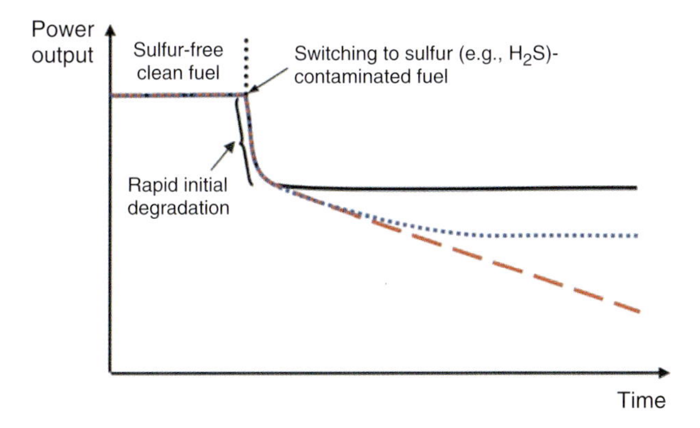

Figure 7.6 Schematic for power output versus time for SOFCs with Ni–YSZ cermet anode showing sulfur poisoning with rapid initial degradation followed by different second-stage behavior: solid line indicates no second-stage slow degradation; dotted line, a slow second-stage voltage versus time degradation followed by a steady state; and dashed line, a slow second-stage degradation showing no steady state after a very long time. (Source: After Ref. [175]. Reproduced with permission.)

The poisoning of H_2S on Ni cermet anodes is affected by a number of operational conditions, such as operating temperature [15, 169], H_2S concentration [169, 170], current density [171], fuel utilization [172], H_2/CO ratio [173], and water content [174]. Sulfur poisoning of Ni cermet anodes generally consists of two stages: an initial rapid drop within a short time in power output on exposure to H_2S in the parts per million level, followed by different second-stage behaviors after exposure to a sulfur-containing fuel for a long period, as shown in Figure 7.6 [175]. On removal of H_2S from the fuel, the electrocatalytic activity of the anodes can be completely [15, 173, 176] or partially recovered [162, 177]. Ishikura *et al.* [178] found that the second stage of poisoning cannot be completely recovered and the irreversible poisoning might be caused by the nickel sulfides at the anode–electrolyte interface. The recovery rate is affected by the operating temperature and the current density [15, 169]. Liu's group at the Georgia Institute of Technology pioneered the *in situ* Raman spectroscopy technique in the study of sulfur poisoning and extensively investigated the poisoning mechanism by a combination of experimental and computational modeling [179–183]. An excellent overview on the topics of materials and electrochemical aspects of sulfur poisoning is given by Liu *et al.* [175] recently.

Sulfur poisoning reduces both the oxidation reaction of hydrogen and the activity of internal reforming of methane [176, 184, 185]. Lohsoontorn *et al.* [170] showed that in the presence of sulfur, charge transfer resistance of the Ni–GDC anode is significantly increased. In the case of high H_2S concentration or low partial pressure of H_2, both the high- and low-frequency arcs are affected. Sulfur poisoning, however, could inhibit carbon formation in Ni cermet anodes by blocking the Ni surface with sulfur [174, 186].

Similar to the case of carbon deposition mentioned in Section 7.2.3, the sulfur tolerance of Ni cermet anodes is also affected by the electrolyte component in the anode. Sasaki *et al.* [173] observed that the Ni–ScSZ anode has higher sulfur tolerance than the Ni–YSZ anode, as ScSZ has a higher ionic conductivity than YSZ, leading to a more active reaction site. A comparative study of sulfur tolerance on the Ni–YSZ and Ni–GDC anodes also showed that the magnitude in the increase of polarization resistance is significantly smaller for the H_2 oxidation reaction on Ni–GDC anodes than that on Ni–YSZ anodes under identical conditions [177], indicating higher sulfur tolerance of Ni–GDC cermets. Similar results have also been reported by Trembly *et al.* [162]. Yang *et al.* [187] replaced the oxygen ion conductor YSZ in a Ni–YSZ cermet anode by a mixed ion conductor such as $BaZr_{0.1}Ce_{0.7}Y_{0.2-x}Yb_xO_3$ (BZCYYb), which allows transport of both protons and oxygen ions. The Ni–BZCYYb cermet anode shows superior coking and sulfur tolerance at 750 °C for up to ~20 ppm H_2S using a cell based on a BZCYYb electrolyte and ~50 ppm H_2S using a cell based on SDC electrolyte. The Ni–BZCYYb anodes for both cells show no observable change in power output as the fuel is switched from clean hydrogen to hydrogen contaminated with 10, 20, or 30 ppm H_2S (Figure 7.7). Its ability to resist deactivation by sulfur and coking is considered to result from the enhanced catalytic activity of BZCYYb for sulfur oxidation as well as enhanced water adsorption capability to facilitate the oxidation of H_2S or elemental sulfur to SO_2 at or near the active sites.

The sulfur tolerance of the Ni–YSZ anode can be improved by introduction of oxide and metal components. Sasaki *et al.* [173] showed that the impregnation of various elements such as Ce, Y, La, Mg, Nb, Sc, Zr, and Ti in the form of oxide or Ru metal under reducing environment is effective in suppressing the degradation in the presence of 20 ppm H_2S. Kurokawa *et al.* [188] showed that

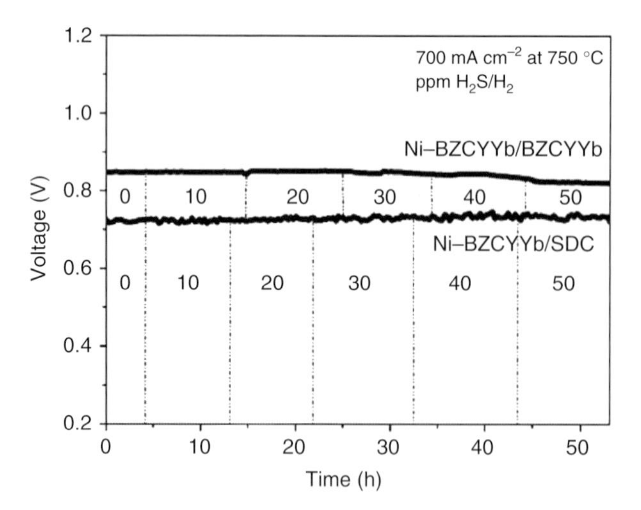

Figure 7.7 Terminal voltages measured at 750 °C as a function of time for two cells with the configuration Ni–BZCYYb|BZCYYb|BZCY–LSCF and Ni–BZCYYb|SDC|LSCF operated at a constant current density of 700 mA cm^{-2} as the fuel was switched from clean H_2 to H_2 contaminated with different concentrations of H_2S. (Source: After Ref. [187]. Reproduced with permission.)

after the impregnation of ceria nanoparticles, the Ni–YSZ anode is stable in H_2 containing 40 ppm H_2S over 500 h at 700 °C. The ceria nanoparticles, which act as a sulfur sorbent to react with H_2S to form $Ce_2O_2S_2$, can avoid the formation of Ni_2S_3 at a H_2S concentration as high as 200 ppm. Choi *et al.* [189] modified a Ni–YSZ anode with niobium oxide (Nb_2O_5). The NbO_x-coated Ni–YSZ cermet anode shows an enhanced sulfur tolerance when exposed to 50 ppm H_2S at 700 °C. Different phases of niobium sulfides (NbS_x) are formed on the surfaces of niobium oxides, and they are electrically conductive and catalytically active for hydrogen oxidation. Marina *et al.* [190] exposed the Ni–YSZ anodes to antimony or tin vapor to form Sb–Ni or Sn–Ni alloys, respectively. The enrichment of Sn or Sb on the alloy surface efficiently weakens the sulfur adsorption on the anode. A recent study by Zheng *et al.* [191] showed that adding palladium nanoparticles slows down the performance degradation of Ni–GDC anodes in sulfur-containing hydrogen fuel.

Ni-free Cu–CeO_2-based anodes have been shown to be highly sulfur tolerant [192, 193]. The anode remains stable when operating in 5 mol% *n*-decane containing 100 ppm H_2S. On the other hand, when operating in 50 mol% *n*-decane containing H_2S as high as 5000 ppm, the current density drops dramatically on the introduction of sulfur. The degradation, however, could be completely recovered by introducing 50 mol% steam in N_2. The high sulfur tolerance of the Cu–CeO_2 anodes is due to the fact that Cu is stable in H_2S, as the formation of Cu_2S is not favorable under SOFC operating conditions. In addition, CeO_2 reacts with sulfur to form Ce_2O_2S at a high H_2S concentration, but heating in steam could remove sulfur and restore CeO_2. H_2S levels up to 450 ppm in H_2 at 800 °C had no effect on the Cu–CeO_2 anode performance [193]. On the other hand, a study by Jiang *et al.* [177] showed that the interaction between CeO_2 and sulfur will significantly change the microstructure of the ceria phase of Ni–GDC cermets, implying the long-term detrimental effects of sulfur on the durability of ceria-based electrodes.

Some metal oxide materials have been demonstrated to be sulfur tolerant. Aguilar *et al.* [194] and Cooper *et al.* [195] reported stable performance of $La_xSr_{1-x}VO_{3-\delta}$ (LSV)-based anodes in H_2S-containing fuel. Huang *et al.* [196] showed that double perovskites, such as Sr_2MgMoO_6, exhibit good sulfur tolerance and no sulfur is detected after operating in 5 ppm H_2S/H_2 for 2 days. Zha *et al.* [197] showed that a pyrochlore material $Gd_2Ti_{1.4}Mo_{0.6}O_7$ shows remarkable tolerance to sulfur deposition. No observable degradation has been detected during the 6 days of operation. $SrTiO_3$-based oxide is also stable in sulfur [195, 198, 199]. Mukundan *et al.* [198] reported that the $La_{0.4}Sr_{0.6}TiO_3$ (LST)–YSZ anode shows no degradation in the presence of up to 5000 ppm of H_2S in hydrogen. Ni cermet in combination with LST oxide also exhibits enhanced sulfur tolerance. Pillai *et al.* [200] fabricated an SOFC with anodes consisting of LST support, Ni–SDC adhesion layer, and Ni–YSZ active layer. In 100 ppm H_2S/H_2, the cell power density initially decreased, but was stable for 80 h of testing. The cell was recovered to the initial performance level by removing H_2S. The enhanced sulfur tolerance of metal-oxide-based anodes is considered to be due to reduced sulfur adsorption [175].

As discussed, one of the main drawbacks of conductive metal oxide anodes is that the electrical conductivity for majority of these anodes is significantly lower than that of Ni-based cermet anodes. Another concern is the difficulties in adopting the ceramic oxide anode materials in the state-of-the-art SOFC fabrication processes based on YSZ electrolyte, especially for an anode-supported cell structure. The difficulties may arise due to the limited thermal, physical, and chemical compatibility between YSZ electrolyte and metal oxides.

7.2.5 Poisoning by Impurities in Coal Gasification Syngas

Coal is one of the most abundant fossil fuels. Coal syngas (H_2, CO, CO_2, N_2, and H_2O) produced by coal gasification is a promising fuel for SOFCs [201–203]. On the other hand, multiple minor and trace impurities are present in coal syngas even after gas cleanup process and could affect the SOFC performance and durability [204, 205]. Thermodynamic analysis predicts that impure elements such as phosphorus (P), arsenic (As), and antimony (Sb) present in coal syngas could react with Ni cermet anodes [206]. Phosphorus is a trace element but potentially is the most detrimental.

Phosphorus readily reacts with Ni to form nickel phosphides (Ni_mP_n) under SOFC operating conditions. Thermodynamic calculations indicate that the formation of nickel phosphide is possible at phosphorus concentrations as low as 1 ppb [207]. Phosphorus has been reported to react with zirconia to form zirconium phosphate [208]. Gong *et al.* [209] studied the performance and stability of $LaSr_2Fe_2CrO_9$–GDC anodes in H_2 and in coal-gas-containing phosphine impurity. Introducing 5–20 ppm PH_3 into the fuel causes significant performance degradation because of the formation of FeP_x and $LaPO_4$ compounds, and the loss in performance cannot be recovered after removing PH_3 from the fuel. Steam has an effect on PH_3 poisoning – a high steam content reduces phosphorus poisoning [206].

The presence of phosphorus significantly degrades the performance of Ni anode during the long-term test [204, 207, 210]. Mariana *et al.* observed that phosphorus is completely captured by the anode and reacts with Ni to form a series of bulk nickel phosphide phases. A sharp boundary between converted and unconverted anode portions can be clearly seen, as shown in Figure 7.8 [211]. The electrode ohmic resistance increases significantly when the entire anode is converted to nickel phosphide. Haga *et al.* [207] also reported the formation of Ni_xP_y, Ni agglomeration, and densification of the outside anode layer in the presence of phosphine. The growth of Ni particles reduces the active TPB sites and the densification of the outside anode layer hinders the fuel diffusion in the porous anodes, while the formation of Ni_xP_y phases leads to the decreased electrical conductivity of the anode. The presence of 20 ppm PH_3 in coal syngas increases both the charge transfer and the diffusion resistances, and the increase in the latter is faster than the former [208]. Liu *et al.* [212] reported crack formation on the Ni–YSZ anode after operation in the presence of 5 ppm PH_3 over 470 h. The phase conversion from Ni to nickel phosphide results in a large localized volume

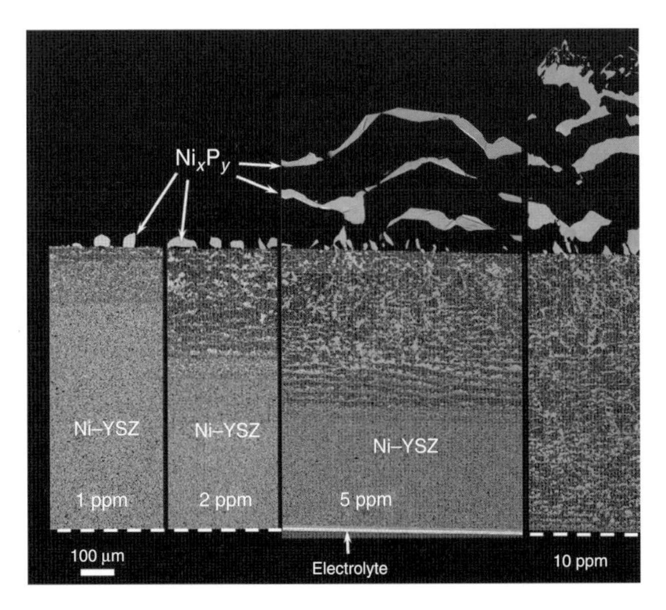

Figure 7.8 Cross-sectional SEM images of Ni–YSZ anode support after 990 h of cell test at 700 °C in simulated coal gas with phosphine. Electrolyte position is marked at the bottom by a dashed line and is seen for a 5 ppm cell. Ni and Ni_xP_y are light gray, YSZ is dark gray, and the pores are black. Not converted Ni–YSZ is seen in images of cells in 1–5 ppm PH_3. All Ni was converted to Ni_xP_y in 10 ppm PH_3. (Source: After Ref. [211]. Reproduced with permission.)

increase and creates a high level of localized stress, leading to structural failure of the anode.

Chlorine is one of the impurities in coal syngas that exists in the form of HCl. Cell degradation has been observed in the presence of HCl [168, 213, 214], but the effect of chloride on performance degradation is much smaller than that of phosphorus. The cell performance degradation due to the presence of chloride has been reported to be recoverable [168]. On the other hand, there are reports that the degradation can only be partially recovered due to microstructural change. For example, Haga *et al.* [213] observed nickel nanoparticle formation on zirconia grains after 150 h in Cl_2- and HCl-containing fuel probably due to sublimation of gaseous $NiCl_2$. Xu *et al.* [214] showed that in the presence of 100 ppm HCl, the surface of the Ni particles becomes scabrous and the Ni particles at the anode surface are rougher than those at the anode–electrolyte interface. Marina *et al.* [215] studied the performance of anode-supported SOFCs in coal-derived syngas containing HCl in the temperature range 650–850 °C. The results show that performance losses increase with the concentration of HCl to 100 ppm, above which losses are insensitive to HCl concentration and the extent of HCl poisoning is not related to cell potential and current density. They have found no evidence for long-term degradation that can be attributed to HCl exposure, suggesting that the effect of HCl poisoning on Ni–YSZ cermet anodes is reversible.

Other impurities in coal gas could also cause degradation of the anode. Marina *et al.* [204, 216, 217] observed that the presence of 0.5–5 ppm arsenic (As), antimony (Sb), and selenide (Se) leads to irreversible SOFC degradation because of the formation of secondary phases in Ni cermet anodes. Bao *et al.* [218] found that 10 ppm As and 5 ppm Cd cause a significant power output reduction. On the other hand, the presence of 9 ppm Zn, 7 ppm Hg, and 8 ppm Sb degrades the cell power density by less than 1% during the 100 h test.

The presence of various kinds of impurities could have synergistic effects on the anode performance and durability. Bao *et al.* [219] showed that the synergistic effect of impurities can be destructive or constructive and is not a simple addition of individual impurities. For example, adding H_2S to As and P accelerates performance degradation, whereas the presence of HCl could mitigate or prevent performance loss of cells with Ni–YSZ cermet anodes.

7.2.6 Silicon Contamination

The presence of Si impurities in the raw materials of Ni-based cermet anodes can lead to the formation of silicate film at the anode–electrolyte interface during long-term operation [220–222]. Deposition of silicate blocks the active TPB sites and degrades anode performance. Liu *et al.* [223] studied the long-term properties of the Ni–YSZ anode using two commercial NiO powders containing different levels of impurities. Anodes containing SiO_2 and Na_2O at a concentration level of hundreds of parts per million degrade faster than those with a few tens of parts per million. Sodium silicate glass has been found to segregate and accumulate at the anode–electrolyte interface, causing serious damage to the YSZ electrolyte in the vicinity of the interface.

Primdahl and Mogensen [224] evaluated the durability of the Ni–YSZ anode over 1300–2000 h at an anodic current of 300 mA cm^{-2} in wet hydrogen. The anode is stable at 1050 °C, but degradation is observed at lower temperatures and the degradation rate increases with the decrease in operating temperature. At 850 °C, the electrode polarization resistance is almost doubled after polarized for 400 h. Liu and Jiao [225] examined the microstructure of the Ni–YSZ anode/YSZ interface after a long-term operation and found that amorphous silicate phase with a composition of ~90 mol% SiO_2 is distributed as nanoscale films along the anode–electrolyte interface and Ni–YSZ grain boundaries (Figure 7.9). It has been found that the impurities come from the raw materials. Gentile and Sofie [226] found that the aluminosilicate for SOFC refractory applications can also cause preferential deposition of silica on YSZ, potentially isolating YSZ grain boundaries with glassy phase. In addition, the presence of aluminosilicate powder in the hot zone of a fuel line impedes peak performance and accelerates degradation. The presence of siloxane impurity in coal syngas could also poison the anode by segregation of silica in the anodes [227].

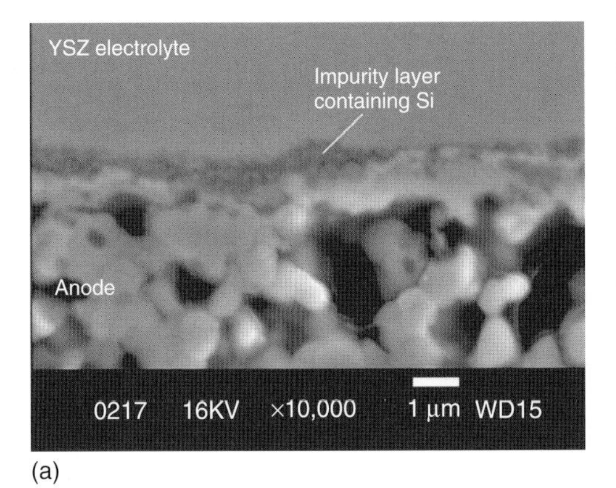

(a)

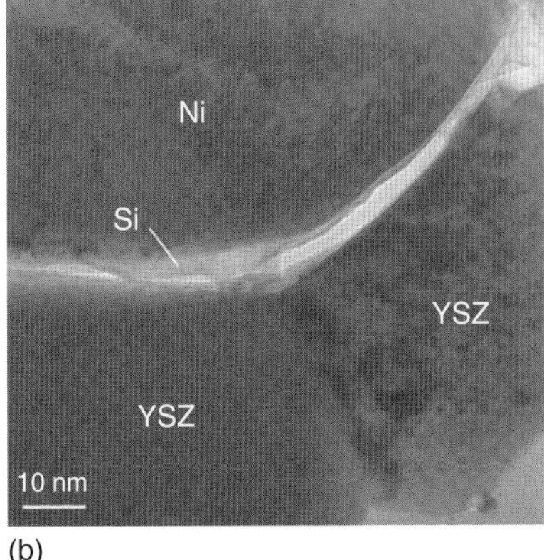

(b)

Figure 7.9 (a) An SEM image showing segregation of impurities at the anode–electrolyte interface and (b) a TEM image showing the presence of the silicate glass film at the Ni–YSZ grain boundary in the anode bulk after the 1800 h of testing. (Source: After Ref. [225]. Reproduced with permission.)

7.3 Cathodes

7.3.1 Degradation due to Interfacial Chemical Reactions

One of the main causes of performance degradation of cathodes of SOFCs is the chemical incompatibility between cathode materials and YSZ-based electrolyte at high temperatures during the sintering process and/or the operation. For example, chemical reaction between LSM cathode and YSZ electrolyte occurs during sintering at high temperatures [228]. Mn has high solubility in YSZ, and the diffusion of Mn from the B site of LSM into the YSZ surface results in excess La and Sr in the A site, which would react with YSZ to form the resistive pyrochlore phases $La_2Zr_2O_7$ and $SrZrO_3$. A high Sr content leads to the preferential formation of $SrZrO_3$. $La_2Zr_2O_7$ has an electrical conductivity of 3.8×10^{-5} S cm^{-1} at 1000 °C [229], which is much lower than that of LSM and YSZ. In addition, the electrochemical activity of $La_2Zr_2O_7$ for O_2 reduction is also negligible, and therefore, the growth of the interfacial product layer can lead to performance degradation of the cathode. Chemical reaction between LSM and YSZ can also take place during cell operation. Lee and Oh [230] observed the formation of $La_2Zr_2O_7$ at the $La_{0.9}Sr_{0.1}MnO_3$–YSZ interface at 900 °C. The electrode polarization resistance increases by up to 40% after being tested for 500 h.

The risk of the solid solution between LSM and YSZ can be minimized using an LSM composition with A-site-deficient nonstoichiometry, such as $(La_{0.8}Sr_{0.2})_{0.9}MnO_3$ [231–233]. A-site-nonstoichiometric LSM reduces the activity between LSM and YSZ and increases its phase stability [232]. The chemical stability between LSM and YSZ is also affected by the LSM/YSZ ratio. Yang *et al.*

[234] found that the formation of secondary phase such as $La_2Zr_2O_7$ and $SrZrO_3$ occurs in 20 vol% $La_{0.65}Sr_{0.3}MnO_3$ (LSM)–YSZ composite, but not in cases where the LSM content is higher than 20 vol%. The effect of secondary phases on electrical conductivity of the 50 vol% LSM–YSZ composite is insignificant even after sintering at 1400 °C for 24 h.

Compared to the electronically conducting LSM cathode, mixed ionic and electronic conducting (MIEC) materials such as $(La,Sr)(Co,Fe)O_3$ (LSCF) and $Ba_{0.5}Sr_{0.5}Co_{0.8}Fe_{0.2}O_{3-\delta}$ (BSCF) exhibit a much better electrocatalytic activity toward oxygen reduction reaction [235–238]. However, lanthanum–strontium–cobalt-based cathode materials are generally much more reactive with YSZ, as compared to that between LSM and YSZ [239]. $SrZrO_3$ has been observed within LSCF and YSZ mixture after heating at a temperature as low as 800 °C [240]. Zhu *et al.* [241] observed no reaction between BSCF and YSZ and GDC powders for the powders sintered at 800 °C, suggesting BSCF is chemical compatible with YSZ and GDC up to 800 °C; above 800 °C, severe reactions are observed. A ceria interlayer is generally used as the barrier layer to prevent the chemical reaction between MIEC cathode and YSZ electrolyte. A recent study [242] found that under cathodic polarization conditions, interaction between BSCF and GDC or SDC occurs at temperatures as low as 700 °C, forming Ba-containing particles/phases at the electrode–electrolyte interface and thus indicating the significant promoting effect of cathodic polarization on the chemical activity between BSCF and ceria-based electrolyte. The accelerated reaction between BSCF and ceria is most likely related to the Ba segregation under the influence of polarization and/or temperature. Although barium cerates, such as $Ba_xCe_{1-y}Gd_yO_3$, are proton and p-type electronic conductors, their presence and growth at the interface is a concern on the long-term performance stability of SOFCs based on the BSCF cathodes–ceria electrolyte.

However, cells with LSCF cathode generally show relatively high degradation during long-term operation. Tietz *et al.* [243] reported a degradation rate of 0.5–2% per 1000 h for single cells with LSCF cathode, which is significantly higher than that of cells with LSM cathodes. Mai *et al.* [240] studied the degradation of LSCF cathodes on anode-supported cells and the cells show performance losses of 2–6% per 1000 h at 700–800 °C. $SrZrO_3$ has been observed at the GDC interlayer–YSZ electrolyte interface after sintering the cathode at 1080 °C. The solid phase continues to form during the long-term operation. Simner *et al.* [244] observed that anode-supported YSZ electrolyte cells with LSCF cathode degrade by more than 30% in power output over 500 h at 750 °C. There are no indications of obvious microstructural changes or chemical reactions, but the degradation could be caused by the observed enrichment of Sr at the cathode–electrolyte and cathode–current collector interfaces.

The mismatch in TEC between the lanthanum–strontium–cobalt-based cathode and the electrolyte can also contribute to the performance and the microstructure degradation. For example, the TEC of LSCF with the composition of $La_{0.6}Sr_{0.4}Fe_{0.8}Co_{0.2}O_3$ is $\sim 20 \times 10^{-6}$ K^{-1} [245] and a similar TEC is also reported for BSCF [246]. In comparison, the TEC of YSZ and GDC electrolytes is 10.3×10^{-6} and 12.5×10^{-6} K^{-1}, respectively [247]. A residual stress between the cathode and the electrolyte layer due to the mismatched TECs can result in crack formation or delamination at the electrode–electrolyte interface.

Surface segregation in Sr-doped lanthanum manganite and cobalt ferrite perovskites is a well-known phenomenon and is, in general, detrimental to the electrocatalytic activities of the SOFC cathodes. Using angle-resolved X-ray photoelectron spectroscopy (XPS), electron energy loss spectroscopy, and the Auger electron spectroscopy, several independent studies have reported evidence for Sr segregation on the surface of LSM perovskites [248–251]. Strontium segregation has also been reported for lanthanum cobatite perovskites. Vovk *et al.* [252] studied the $La_{0.5}Sr_{0.5}CoO_3$ perovskite oxide surfaces under electrochemical polarization using the *in situ* XPS technique. Under cathodic polarization, the Sr/(La + Co) ratio at the oxide surface increases irreversibly by 5%, whereas the La/Co ratio remains constant, indicating the surface enrichment of strontium. Most recently, Norman and Leach [253] reported an *in situ* high-temperature XPS study of BSCF. Although XPS is not able to give the exact surface composition of the BSCF, the analysis shows that the B-site Co/Fe ratio remains fairly constant throughout the heat treatment, while Ba and Sr segregate at the surface as a result of heat treatment at 800 °C. Our recent study also indicates that Ba and Sr segregation plays an important role in chromium deposition and poisoning on BSCF cathodes [254]. The surface segregation of barium may be related to the phase instability of BSCF perovskite [255].

7.3.2
Microstructure Degradation

Microstructural changes have been observed at the LSM electrode–electrolyte interface under SOFC operating conditions. Jorgensen *et al.* [256] observed that after LSM–YSZ cathodes are operated at 300 mA cm^{-2} at 1000 °C for up to 2000 h, the increase in electrode overpotential exceeds 100% of the initial value. However, the cathodes without current load showed little degradation over 2000 h at the same temperature. An increase in the porosity at the electrode–electrolyte interface has been observed in the case of current load, but no such structural changes were observed under open circuit conditions. The observed pore formation at the interface has been considered to decrease the length of the active TPB and thereby degrade electrode performance.

Sintering and particle growth of LSM has been observed during operation at high temperatures. Jiang and Wang [257] studied the sintering behavior of $(La_{0.8}Sr_{0.2})_xMnO_3$ electrodes with $x = 1.0$, 0.9, and 0.8. The results show that the LSM particles grow in size during sintering at 1000 °C under open circuit conditions for 1600 h. LSM with A-site nonstoichiometry shows a much higher sinterability than those with stoichiometric compositions. On the other hand, under a cathodic current density of 500 mA cm^{-2} at 1000 °C, the grain growth of LSM electrodes is significantly smaller than that under open circuit conditions (Figure 7.10). The grains of LSM electrodes, when sintered at 1000 °C, grow less under current load as compared to the ones sintered under open circuit conditions. The hindrance of

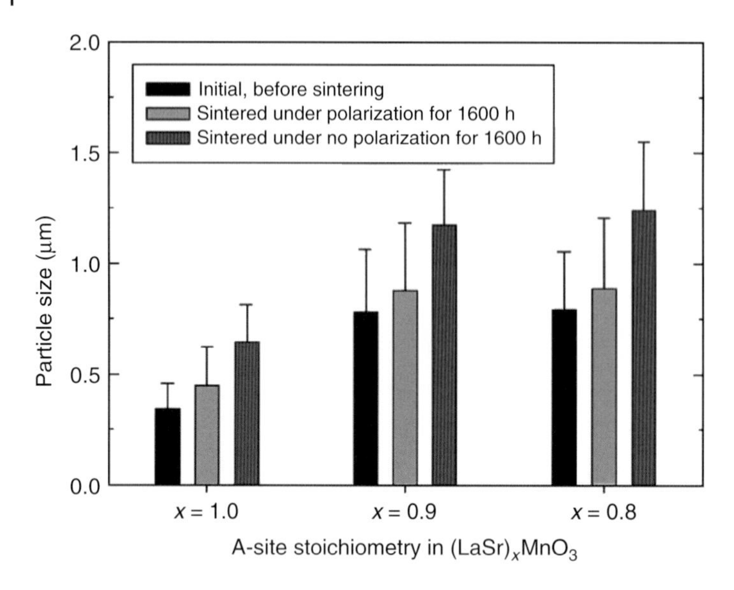

Figure 7.10 Plots of the particle size of LSM electrode coating sintered at 1000 °C in air with and without a current load of 500 mA cm^{-2} as a function of A-site-stoichiometry compositions. (Source: After Ref. [257]. Reproduced with permission.)

sintering during cathodic polarization is attributed to the elimination of the cation vacancies at the A site, thus increasing the resistance to the sintering process.

Nanostructured electrodes of SOFC have been extensively investigated in recent years and wet impregnation/infiltration is the most common method to introduce the nanosized catalytic particles into the porous electrode structure [131, 258, 259]. However, there is an issue on the stability of nanostructured electrodes regarding phase stability and chemical compatibility. Chemical reaction between the impregnated phases and the skeleton has been observed during the operation. Huang *et al.* [260] incorporated $La_{0.6}Sr_{0.4}CoO_3$ (LSC) into the porous YSZ skeleton by impregnation. A 35 wt% LSC impregnated YSZ composite cathode exhibits an area-specific resistance (ASR) as low as 0.03 Ω cm^2 at 700 °C and 0.01 Ω cm^2 at 800 °C. On the other hand, both the ohmic resistance and the polarization resistance of the cathode increase with time over 250 h at 700 °C. The cathode degrades more rapidly at 800 °C. The performance losses are likely due to the formation of insulating phases such as $SrZrO_3$, as the chemical reaction between LSC and YSZ can take place at temperatures as low as 650–700 °C [261]. Ai *et al.* [262] incorporated BSCF nanoparticles into porous LSM skeleton by impregnation. The electrode polarization resistance has been observed to increase from the initial 0.47 to 2.4 Ω cm^2 after operation at 700 °C over 90 h. Degradation is most likely due to the observed grain growth of the impregnated BSCF nanoparticles. On the other hand, Kungas *et al.* [263] found that the incorporation of a dense protective SDC layer within the YSZ skeleton by infiltration could avoid the chemical reactions between LSC and YSZ even after heating at 1100 °C.

Sintering and growth of nanoparticles occur readily during operation because of the high surface energy of the nanoparticles, leading to the decrease in the active surface area and TPBs. Shah *et al.* [264] fabricated LSM-impregnated YSZ cathodes and observed that the electrode polarization resistance increases from an initial value of 0.2 to 0.5 $\Omega\,cm^2$ after aging over 300 h at 800 °C. The LSM particles increase in size from 59 to 105 nm after aging for 300 h. Yoon *et al.* [265] incorporated $Sm_{0.2}Ce_{0.8}O_2$ (SDC) phase into porous LSM skeleton by impregnation, followed by heat treatment at 600 °C. The electrode polarization resistance increased from 0.19 to 0.67 $\Omega\,cm^2$ after testing at 700 °C for 500 h. Wang *et al.* [266] observed that the electrode polarization resistance for the $La_{0.8}Sr_{0.2}FeO_3$ (LSF)-impregnated YSZ cathode increases from 0.13 to 0.55 $\Omega\,cm^2$ after sintering at 700 °C over 2500 h. On the other hand, the stability of the electrodes is significantly improved when the heat-treatment temperature is increased from 850 to 1100 °C. However, the high sintering temperature coarsens the nanostructure and reduces the initial performance of cathodes.

Lowering the operating temperature leads to low grain growth and thereby less degradation. Chen *et al.* [267] showed that an SOFC based on an SDC-impregnated LSM cathode at 700 °C is more stable than that at 800 °C. No degradation has been observed on LSM-impregnated ScSZ cathode under 150 mA cm^{-2} at 650 °C for 500 h [268]. Shah *et al.* [264, 269] investigated the changes of electrode polarization resistance of 12 vol% LSCF-impregnated GDC cathodes at 650–850 °C and observed accelerated cathode degradation at a higher operating temperature (Figure 7.11). The microstructure changes significantly at a high operating temperature. The LSCF particle size grows from the initial value of ~38 to ~60 nm after aging at

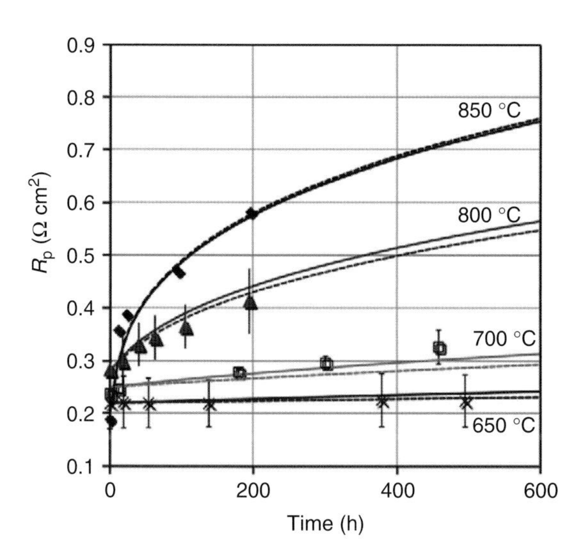

Figure 7.11 Experimental polarization resistance (discrete points) versus time and fitted lines for LSCF-infiltrated SDC cathode aging at various temperatures. (Source: After Ref. [269]. Reproduced with permission.)

850 °C for 289 h. On the other hand, the cathode shows no obvious changes in microstructure after aging at 600 °C over 168 h.

Addition of a second phase is effective in inhibiting the particle growth of nanoparticles. Sholklapper *et al.* [270] reported that the polarization performance of Ag-impregnated ScSZ cathode is not stable under 400 mA cm^{-2} at 700 °C. However, the stability of the cathode is significantly improved after LSM particles are incorporated to prevent Ag agglomeration. Pd has been demonstrated to exhibit high electrocatalytic activity toward the oxygen reduction reaction [236, 271]. However, Pd nanoparticles are not stable at SOFC operating temperatures, and significant agglomeration of Pd nanoparticles has been observed after operation at 750 °C for 30 h [272]. It was reported that addition of 5 mol% Mn can significantly inhibit Pd coarsening [273]. The addition of a small amount of Ag and Co is also effective in enhancing the resistance of Pd against sintering and agglomeration [274]. Kim *et al.* [275] showed recently that highly active and stable Pd catalysts can be obtained by infiltrating Pd-CeO_2 core–shell nanoparticles into YSZ scaffold. The core–shell nanostructure inhibits the growth of Pd nanoparticles.

Zhao *et al.* [14] showed that excellent thermal cycle durability can be achieved by nanostructured LSC–SDC cathodes (Figure 7.12). LSC and SDC are chemically compatible and there is no detectable chemical reaction between them during heat treatment and long-term operation. As shown in Figure 7.12, after 20 times of 500–800 °C and 10 times of room temperature to 800 °C thermal cycles, no performance degradation is observed for the impregnated electrodes. Improved performance and significantly better thermal stability is achieved with the impregnated electrode when compared with the conventional composite electrode. The

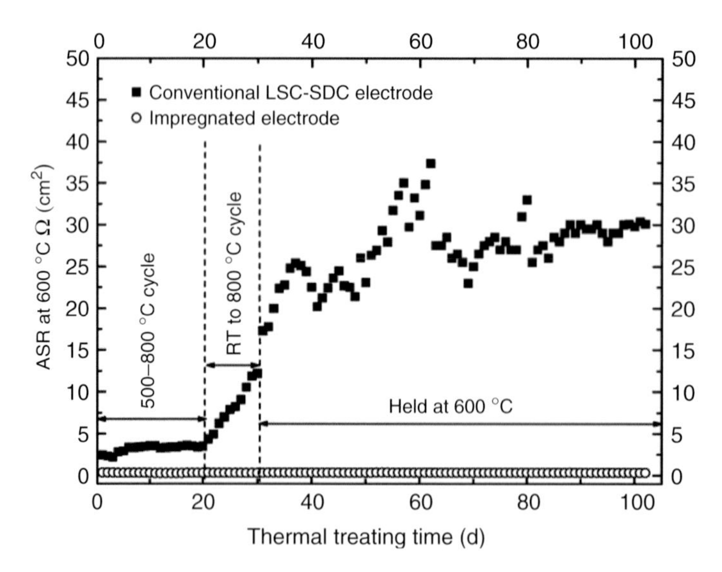

Figure 7.12 Comparison of area-specific resistance (ASR) on thermal cycles and thermal treatment at 600 °C for LSC–SDC composite cathodes prepared by impregnation and screen printing. (Source: After Ref. [14]. Reproduced with permission.)

high stability of the infiltrated LSC–SDC cathodes is most likely due to the formation of continuous and stable LSC nanoparticles anchored on the SDC scaffold. The application of nanocrystalline (Sm,Sr)CoO_3 (SSC)-coated SDC core powder has also been demonstrated to be resistive to thermal cycles [276]. The modeling study suggests that the size of the nanoparticles is probably the most important factor to stabilize the nanostructure during the thermal cycle [277]. Nevertheless, the long-term stability of nanostructured electrodes on large and practical cells has not been demonstrated yet.

7.3.3
Chromium Poisoning

The application of SOFCs in the intermediate-temperature range (650–850 °C) significantly increases the feasibility of application of low cost metallic interconnect materials [278, 279]. However, Cr deposition and poisoning in the presence of the Fe–Cr interconnect has been reported to be one of the most important degradation mechanisms of SOFC cathodes [7]. The gaseous chromium species from the chromia scale (Cr_2O_3) of chromia-forming metallic interconnect poisons the cathodes, leading to significant performance degradation of the SOFCs [9, 280, 281]. Figure 7.13 shows the typical polarization behavior for the O_2 reduction reaction on an LSM cathode in the presence of the Cr–Fe interconnect. The electrode polarization generally shows a sharp increase with the application of a current, followed by a much slow increase in the polarization potential [8]. The

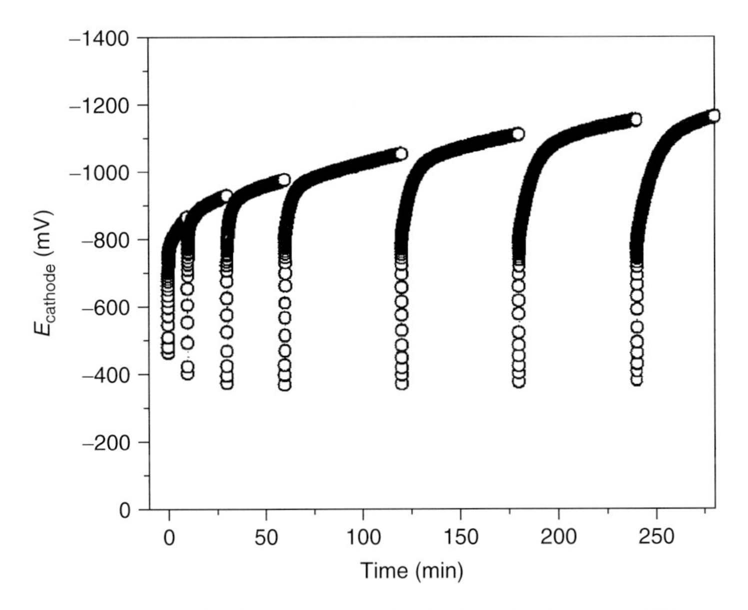

Figure 7.13 Typical polarization curves for the O_2 reduction on an LSM cathode in the presence of the Fe–Cr alloy interconnect as a function of cathodic current passage.

significant increase in the polarization potential is a clear indication of chromium poisoning on the activity of the cathode.

Chromium poisoning is characterized by Cr_2O_3 deposition and formation of Cr-based solid solution in cathodes. It is affected by a number of factors, such as type of cathode materials [278, 282, 283], composition of electrolyte [284], polarization [9], airflow [285, 286], and humidity [287]. In the case of LSM cathodes, Cr_2O_3 and $(Cr,Mn)_3O_4$ spinel have been observed and Cr is mainly deposited at the cathode–electrolyte interface [9, 288]. Cr deposition is generally observed during polarization, whereas under open circuit conditions, little Cr deposition has been observed on the surface of the LSM cathode. In the case of LSCF cathodes, preferential deposition and formation of Cr_2O_3 and $SrCrO_4$ phases on the electrode surface have been observed [289, 290]. Cr deposition occurs on LSCF surface while heating to 900 °C under open circuit conditions [289].

Two main reaction mechanisms proposed in the literature for the deposition and poisoning of chromium on SOFC cathodes. One is the electrochemical deposition mechanism in which the deposition of chromium is considered to be controlled by electrochemical reduction of the gaseous Cr species CrO_3 and $Cr(OH)_2O_2$ to solid-phase Cr_2O_3, followed by the chemical reaction of Cr_2O_3 with LSM to form $(Cr,Mn)_3O_4$ [5, 6, 9, 280, 288, 290]. The electrochemical reduction of high-valence chromium species is in direct competition with the O_2 reduction reaction and the process can be written as follows [9]:

$$2CrO_3\,(g) + 6e' \rightarrow Cr_2O_3\,(s) + 3O^{2-} \tag{7.3}$$

$$2Cr(OH)_2O_2\,(g) + 6e' \rightarrow Cr_2O_3\,(s) + 2H_2O\,(g) + 3O^{2-} \tag{7.4}$$

$$\begin{aligned} 1.5Cr_2O_3\,(s) + 3\,(La, Sr)\,MnO_3 &\rightarrow 3\,(La, Sr)\left(Mr_yCr_{1-y}\right)O_3 \\ &+ \left(Cr_yMr_{1-y}\right)O_4 + 0.25O_2 \end{aligned} \tag{7.5}$$

$$\begin{aligned} 3CrO_3\,(g) + 3\,(La, Sr)\,MnO_3 &\rightarrow 3\,(La, Sr)\left(Mr_yCr_{1-y}\right)O_3 \\ &+ \left(Cr_yMr_{1-y}\right)O_4 + 2.5O_2 \end{aligned} \tag{7.6}$$

We studied extensively the chromium deposition and poisoning processes on a wide range of cathodes including LSM, LSM–YSZ, LSCF, BSCF and $(LaBrSr)(CoFe)O_3$, and $(LaSr)(CoMn)O_3$ [8, 254, 291–296]. The results show that there is no intrinsic relationship between the Cr deposition and the oxygen activity or the O_2 reduction at the electrode–electrolyte interface. Chromium deposition is essentially a chemical process in nature, which is initiated by the nucleation reaction between the gaseous Cr species and nucleation agents such as Mn^{2+} in the case of LSM electrode and YSZ electrolyte cells. Mn^{2+} species can be generated under cathodic polarization. The deposition process for Cr on LSM could be written as follows [8, 293]:

$$Cr_2O_3\,(s) + \frac{3}{2}O_2 \rightarrow 2CrO_3\,(g) \tag{7.7}$$

$$Mn^{2+} + CrO_3 \rightarrow Cr - Mn - O_x\,(nuclei) \tag{7.8}$$

$$Cr - Mn - O_x + CrO_3 \rightarrow Cr_2O_3 \quad (7.9)$$

$$Cr - Mn - O_x + CrO_3 + Mn^{2+} \rightarrow (Cr, Mn)_3O_4 \quad (7.10)$$

In the case of Cr deposition on LSCF, the deposition is initiated by the nucleation reaction between the gaseous Cr and SrO segregated on the electrode surface. The mechanism of Cr deposition on LSCF could be written as follows [289]:

$$SrO + CrO_3 \rightarrow Cr - Sr - O_x \text{ (nuclei)} \quad (7.11)$$

$$Cr - Sr - O_x + CrO_3 \rightarrow Cr_2O_3 \quad (7.12)$$

$$Cr - Mn - O_x + CrO_3 + SrO \rightarrow SrCrO_4 \quad (7.13)$$

Chromium deposition and poisoning has a significant effect on the electrocatalytic activity and stability of the cathodes of SOFCs. The degradation of LSM due to Cr poisoning occurs very fast in the presence of Cr-containing alloy under polarization [284]. Taniguchi *et al.* [280] showed that the electrode polarization of LSM–YSZ cathode increases in correlation with the intensity of chromium at the cathode–electrolyte interface. Jiang *et al.* [291] found that in the presence of chromia-forming alloy, the dissociative adsorption and diffusion of oxygen on the LSM electrode surface are inhibited by the gaseous Cr species (e.g., CrO_3) and the migration processes of oxygen ions into zirconia electrolyte are inhibited by the solid Cr species, for example, Cr_2O_3–$(Cr,Mn)_3O_4$ deposited on the electrolyte surface. The activation energy of the migration process of the oxygen ions into YSZ electrolyte has been found to increase significantly [297]. A recent study on $(La_{0.8}Sr_{0.2})_{0.95}(Mn_{1-x}Co_x)O_3$ (LSMC) cathodes shows that chromium deposition on LSMC cathodes is a strong function of Co substitution at the B site [295]. Figure 7.14 summarizes the chromium deposition on the LSMC electrode surface and GDC electrolyte surface and chromium poisoning on the electrochemical activity of the LSMC cathodes as a function of the Co content at the B site [295]. The results show that as the B-site Mn is substituted by Co, chromium deposition on the electrolyte surface in contact with the LSMC electrode decreases, whereas on the electrode surface it increases. On the other hand, the chromium poisoning effects as measured by the increase in the overpotential and electrode polarization resistance are most pronounced for the LSMC cathode with $x = 0.4$. The occurrence of maximum chromium poisoning is most likely due to combined poisoning effect of the Cr deposits at the electrode and electrolyte surface on the kinetics of the O_2 reduction reactions.

The degradation of LSCF cathodes by Cr poisoning is generally slower as compared to that on LSM [298]. Cr_2O_3 and Sr_2CrO_4 deposition on the electrode surface blocks the reaction site and deteriorates the electrode electrocatalytic activity. In addition, Wachsman *et al.* [299] found that as Sr diffuses from the lattice to form the $SrCrO_4$ phase, LSCF gradually becomes Sr deficient and a phase transition from rhombohedral to cubic perovskite occurs in the near-surface region. The Sr deficiency significantly deteriorates the catalytic activity of LSCF due to reduction of oxygen vacancies and surface Co–Fe concentration.

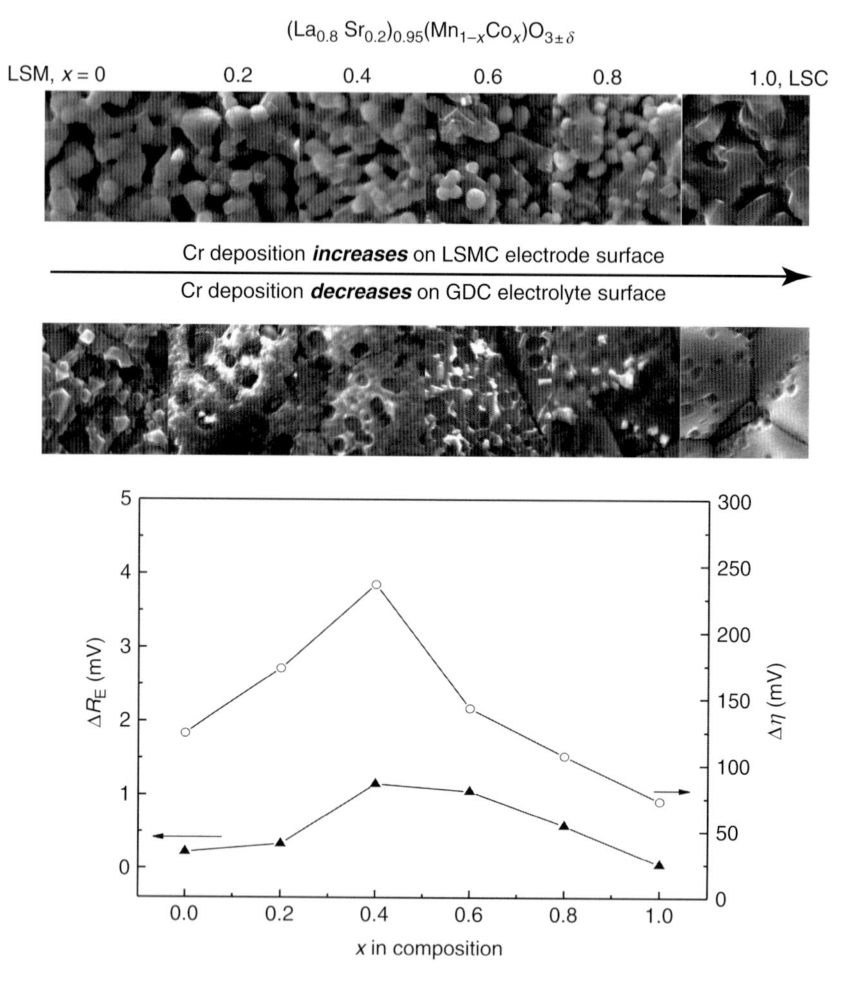

Figure 7.14 Chromium deposition on the LSMC electrode surface and GDC electrolyte surface and chromium poisoning on the electrochemical activity of the LSMC cathodes as a function of the Co content at the B site after current passage at 200 mA cm^{-2} and 900 °C for 1200 min. (Source: After Ref. [295]. Reproduced with permission.)

Some cathode materials show certain degree of chromium tolerance under SOFC operating conditions. We observed that chromium deposition on LSM–YSZ composite cathodes is significantly smaller as compared to that on pure LSM [300, 301]. Different to that on LSM, there is no preferential deposition of Cr species at the electrode–electrolyte interface for the reaction on the LSM–YSZ composite electrode with high YSZ content (30 wt%) and on a GDC-impregnated LSM composite electrode. The reason for the significantly reduced Cr deposition in LSM composite cathodes is most likely the decrease in Mn^{2+} ion generation under cathodic polarization as a result of the addition of the ion-conducting electrolyte phase. This led to significant reduction in the nucleation and grain growth reaction for the Cr deposition.

$La(Ni,Fe)O_3$ (LNF) electrode appears to have a high tolerance against chromium deposition and poisoning [302–304]. Zhen *et al.* [302] showed that LNF is significantly more stable than LSM in the presence of Fe–Cr interconnect. In the case of LNF, there is no deposition of chromium species either on the electrode surface or at the electrode–electrolyte interface after polarization at $200\,mA\,cm^{-2}$ and 900 °C for 20 h (Figure 7.15a–c). By contrast, a significant amount of chromium species is preferentially deposited at the interface regions of the LSM electrode (Figure 7.15d). The high chromium tolerance of LNF is most likely due to the lack of nucleation agents such as Mn^{2+} in the LSM–YSZ electrolyte system for Cr deposition [302]. However, Stodolny *et al.* [305] reported that LNF is chemically unstable at 800 °C when it is in direct contact with Cr_2O_3. Cr can be gradually incorporated into the LNF lattice structure, and the reaction results in a decrease in the electronic conductivity of LNF. Komatsu *et al.* [304] studied the effect of operation on the stability of LNF-cathode-based cell in the presence of an alloy interconnect. The cell is stable at a low current density of $0.3\,A\,cm^{-2}$ over 700 h, but starts to degrade at higher current densities and the degradation is accelerated as the current density increases. The cathode particles are coalesced at the high current loading of 1.5 and $2.3\,A\,cm^{-2}$, and a Cr species layer is formed in the vicinity of the cathode–electrolyte interface at a high current density of $2.3\,A\,cm^{-2}$.

Reducing the strontium content at the A site of LSCF and BSCF can improve the chromium tolerance of the cathodes. Zhen and Jiang [306] substituted Sr in the A

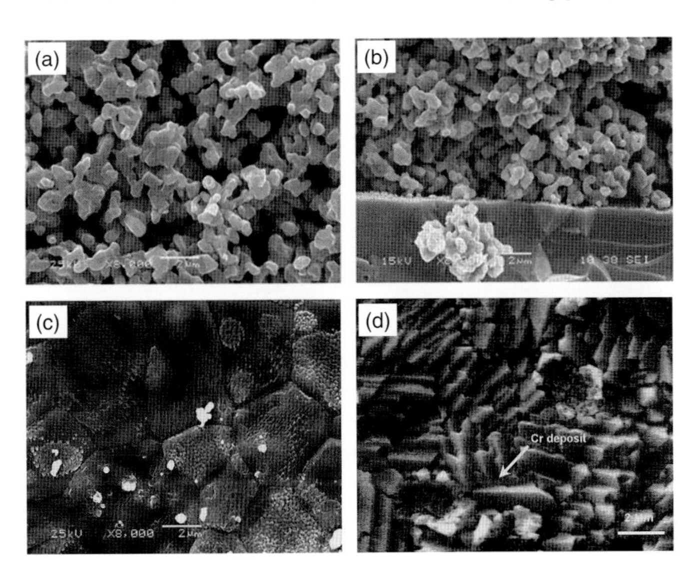

Figure 7.15 SEM images of (a) surface and (b) cross section of LNF and YSZ electrolyte surface in contact with (c) LNF and (d) LSM electrodes after cathodic polarization for 20 h. In (c) and (d), the electrode coating is removed by HCl acid treatment. (Source: After Ref. [302]. Reproduced with permission.)

site of the LSCF with Ba to form (La,Ba)(Co,Fe)O_3 (LBCF) electrode. Cr deposition on the LBCF cathode is very less, and the poisoning effect of chromium for O_2 reduction on the electrode is substantially reduced. Chen *et al.* [307] studied in detail Cr deposition on (La,$Sr_{0.4-x}Ba_x$)(Co,Fe)O_3 (LSBCF) as a function of Ba addition. At $x = 0$, that is, LSCF, Cr deposition is most significant, completely covering the surface of the electrode. As the Sr content in LSBCF is decreased, the Cr deposition on the electrode surface is reduced significantly. In the case of BSCF, chromium poisoning and deposition on BSCF cathodes is decreased significantly as the Sr content of BSCF electrodes is decreased [254, 296].

Besides the development of chromium-tolerant cathodes, chromium poisoning can be reduced by alloying the conventional Fe–Cr interconnect materials with other metals. Yang *et al.* [308] added 0.45% Mn to form a Fe–Cr–Mn alloy. The presence of Mn leads to the formation of an electrically conductive scale of $(Mn,Cr)_3O_4$ spinel on the surface of the alloy. Simner *et al.* [309] studied the performance degradation of anode-supported SOFCs using Mn-containing ferric stainless steel. The cells with LSM, LSF, and LSCF cathodes degrade rapidly in the presence of the interconnect. However, preoxidation of the steel at 800 °C for 500 h results in the formation of a protective $(Mn,Cr)_3O_4$ spinel coating, which slows down the cell degradation rate.

Chromium poisoning can also be reduced by using new interconnect materials. Stanislowski *et al.* [310] systematically studied the Cr vaporization of Cr-, Fe-, Ni-, and Co-based alloys at high temperatures. The results show that Cr vaporization of chromia-forming steels can be reduced by more than 90% by alloying. Chen *et al.* [311] studied the Ni–Mo–Cr interconnect materials containing 20.76% Mo and 3.02% Cr. Chromium deposition on the LSM cathodes in the presence of Ni–Mo–Cr interconnect is remarkably reduced as compared to the Fe–Cr metallic interconnect. The improvement is attributed to the low Cr content and the formation of the Cr-free $NiMn_2O_4$ protective layer on the interconnect.

Coating of a dense and conductive ceramic oxide layer on metallic interconnect materials is effective in reducing chromium deposition and poisoning. Fujita *et al.* [312] studied various oxide coatings such as YSZ, Y_2O_3, La_2O_3, $LaAlO_3$, and (La,Sr)CoO_3 (LSC) on the Fe–Cr interconnect. In these oxides, the LSC coating is found to be most effective in controlling the growth of the Cr_2O_3 layer on the interconnect, leading to enhanced stability of a single cell with an LSCF–SDC cathode. Kurokawa *et al.* [313] studied the effect of LSM, LSCF, and $MnCo_2O_4$ ceramic layers on chromium vaporization. The coating effectively decreases chromium vaporization, but the effect is dependent on the starting powders for coating. Coating a Co–$LaCrO_3$ composite by electrodeposition has been shown to be effective in acting as a barrier against chromium migration into the outer oxide layer [314]. Qu *et al.* [315] fabricated Y, Co, and Y–Co oxide coatings on the oxidation behavior of a ferritic stainless steel with 16–18 wt% Cr. Cr–Mn spinel and Cr_2O_3 are the two main surface oxides, but the morphology of the spinel phase depends on the type of coating. The lowest resistances have been obtained for the Y–Co-coated samples, the ASR values of which are seven times lower than that obtained on the uncoated ferric steels.

A spinel barrier layer has been demonstrated to be effective in reducing Cr poisoning. Yang *et al.* [316] fabricated a $Mn_{1.5}Co_{1.5}O_4$ spinel barrier layer on ferritic stainless steel by screen printing. Thermal and electrical tests conducted over 1000–3000 h confirm the effectiveness of the spinel protection layer in the reduction of chromium migration and decrease of oxidation, while promoting electrical contact and minimizing the cathode–interconnect interfacial resistance. Montero *et al.* [317] found that a reduced Cr migration to the contact layer and the LSF cathode is achieved by application of a $MnCo_{1.9}Fe_{0.1}O_4$ spinel layer on the Fe–Cr interconnect. Bi *et al.* [318] deposited a Co–Fe alloy on the ferritic steel and a $CoFe_2O_4$ spinel layer was attained after thermal oxidation in air. The presence of the dense spinel layer has been shown to be effective in blocking Cr migration/transport and thus contribute to the stability of the cell with LSM cathode over 300 h. $NiCo_2O_4$ spinel coating is also effective in enhancing the oxidation resistance and improving the electrical conductivity of the ferrite steel interconnect [319].

7.3.4
Contaminants from Glass Sealant

Glass–ceramics are promising sealing materials for SOFCs due to their tailorable structure and properties. However, the volatility and reactivity of some components of sealing glass can be significant at the operating temperatures of SOFCs. Batfalsky *et al.* [320] found that the chemical interaction between the glass–ceramic sealant and the Fe–Cr alloy interconnect causes severe corrosion of the interconnect along sealing rims, leading to short-circuiting and stack failure. Silicon from the sealants has also been observed at the electrolyte–electrode interface of Ni-based fuel electrodes [321]. We studied the effect of the borosilicate glass on the microstructure and electrocatalytic activity of LSM cathodes at 1000 °C [286] and the results show that the presence of Na and K, especially Na, in the glass enhances the grain growth of LSM particles. The growth in the particle size causes the reduction in TPB at the interface, leading to the increase in polarization losses.

Boron oxide (B_2O_3) is a key component to tailor the softening temperature and viscosity of the sealing glass for SOFCs. The melting point of boron oxide is low (450–510 °C), and volatile boron species, such as BO_2, are formed under dry conditions and $B_3H_3O_6$ is formed under wet, reducing conditions [322]. Zhou *et al.* [323] studied the effect of boron poisoning on LSM–YSZ and LSCF cathodes. A reduction in power density has been observed in cells with LSM–YSZ cathodes after the introduction of boron sources to the airstream. Boron deposition occurs on the cathode, which probably blocks the active site and causes the degradation. The electrochemical performance of the cell is partially recovered on removal of the boron source. On the other hand, boron deposition has little effect on the electrochemical performance of LSCF cathodes.

Recently, we studied the impact of volatile boron species on the microstructure and performance of nanostructured (Gd,Ce)O_2-infiltrated (La,Sr)MnO_3 (GDC–LSM) cathodes of SOFCs [324]. The results indicate that after the heat treatment of the cathodes at 800 °C in air for 30 days in the presence of borosilicate glass, significant grain growth and agglomeration of infiltrated GDC nanoparticles were observed. Figure 7.16 is the SEM image of the infiltrated GDC–LSM cathodes. As-prepared GDC–LSM electrode is characterized by uniformly distributed GDC nanoparticles on the surface of LSM particles and at the electrode–electrolyte interface, with average particle size of 57 ± 14 nm (Figure 7.16a,b). After sintering at 800 °C in the absence of glass, GDC nanoparticles grow to 83 ± 11 nm in diameter (Figure 7.16c). After sintering at 800 °C in the presence of glass, dense agglomerates and large irregularly shaped particles are formed on the electrode surface and at the interface region (Figure 7.16d,e). The dense agglomerates are as large as 200–600 nm, indicating the significant detrimental effect of volatile boron species on the microstructure of GDC–LSM electrodes. The electrode polarization resistance of the GDC–LSM cathode after the heat treatment in the presence of glass is 3.15 $\Omega\,cm^2$ at 800 °C, which is substantially higher than 0.17 $\Omega\,cm^2$ of the cathode heat treated in the absence of glass under identical conditions. ICP-OES (inductively coupled plasma-optical emission spectroscopic) analysis shows the deposition of boron species on the cathodes. This study indicates the significant poisoning effect of volatile boron species on the microstructure and activity of nanostructured GDC–LSM cathodes.

Silica in the glass sealant could also lead to cathode degradation. Horita *et al.* [11] studied the durability of the performance of flattened tube type SOFC stacks. The ohmic resistance increases with the operation time but the polarization resistance remains more or less constant. The concentrations of several elements such as Na, Al, Si, and Cr have been observed to increase with the operation time on the $LaFeO_3$-based cathode. The concentration of Si in the ceria-based interlayer is also increased from 165 ppm after 24 h to about 910 ppm after 8000 h. Bae and Steele [325] pointed out that the SiO_2 in GDC could affect both the conductivity of the GDC electrolyte and the electrocatalytic activity of the LSCF cathode. The presence of SiO_2 at the interface will restrict the flux of oxygen ions into the electrolyte, and the distribution of SiO_2 over the internal pore surface of the electrode will poison the active catalytic area and inhibit the oxygen exchange reaction.

Komatsu *et al.* [326] examined the long-term operation of an anode-supported SOFC and found that the polarization resistance and the ohmic resistance of the $LaNi_{0.6}Fe_{0.4}O_3$ (LNF) cathode increase with time. The voltage degradation rate is 0.86% per 1000 h over 5200 h and 1.40% per 1000 h over 5670 h and has been found to be closely related to the Si and B impurities from the glass sealant.

7.3.5
Poisoning by Impurities in Ambient Air

In atmospheric air, there are traces of SO_2 that may have an effect on the activity and stability of cathode. Liu *et al.* [327] observed that the LSM cathode is stable at 800 °C when the SO_2 concentration is ~1 ppm, but the poisoning is accelerated and the performance degradation increases with the increase in SO_2 concentration. LSCF shows less tolerance toward SO_2 poisoning, and the cell with the LSCF

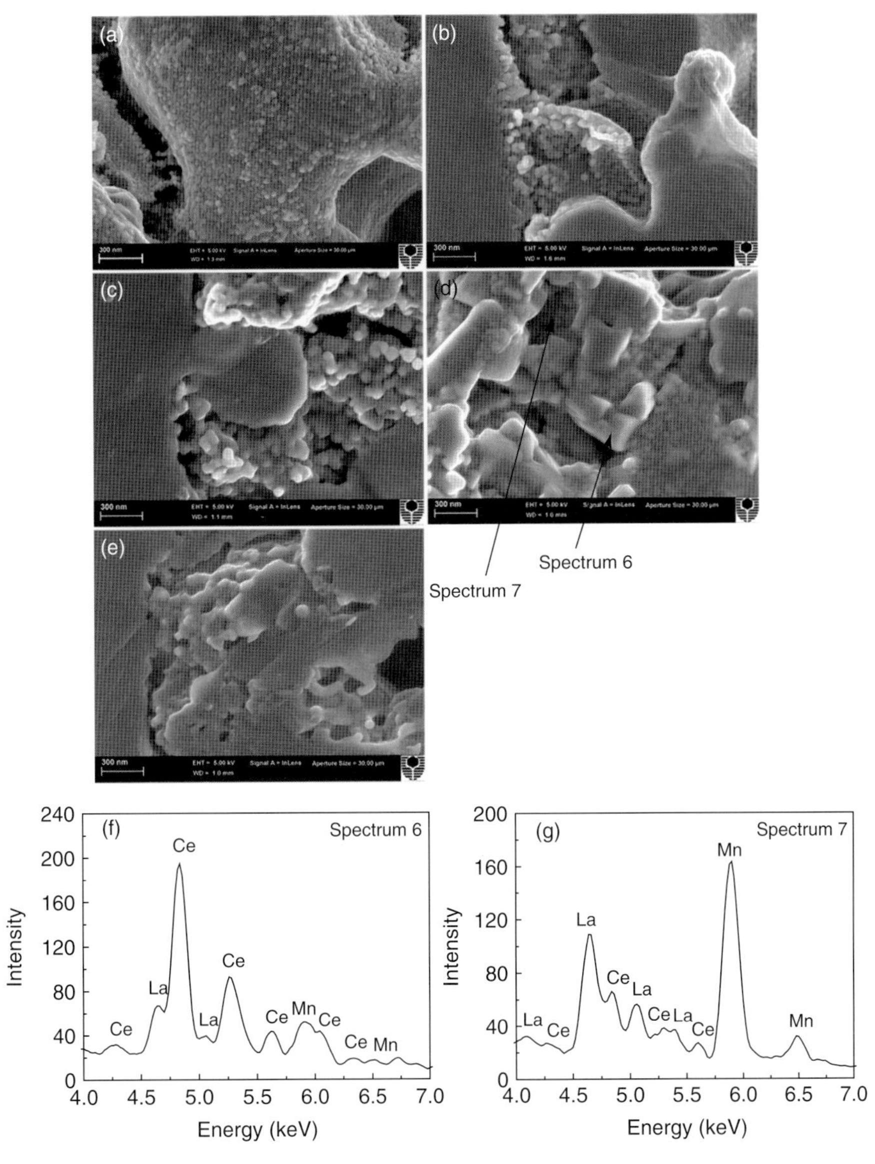

Figure 7.16 SEM images of (a,b) freshly prepared GDC–LSM cathode, (c) GDC–LSM cathode sintered at 800 °C for 30 days in the absence of glass, and (d,e) GDC–LSM cathode sintered at 800 °C for 30 days in the presence of glass. EDS (energy-dispersive spectrometric) patterns of small and irregularly shaped larger particles in (d) are shown in (f) and (g), respectively. (Source: After Ref. [324]. Reproduced with permission.)

cathode is not stable even with 1 ppm SO_2. $SrSO_4$ compounds have been observed in both LSM and LSCF cathodes after exposure to 20 ppm SO_2 over 1000 h.

Schuler *et al.* [328] fabricated a double-layer cathode comprising an LSM–YSZ functional layer and an LSC current collector. When exposed to SO_2-containing air at 800 °C over 1900 h, the LSM–YSZ composite is free of sulfur poisoning, but $SrSO_4$ is found in the LSC layer. In the presence of Cr, $Sr(Cr_{0.85}S_{0.15})O_4$ is formed on the LSC surface.

Yamaji *et al.* [329] reported that the introduction of a small concentration of SO_2 (e.g., 5 ppb and 1 ppm) could lead to the degradation of SSC cathode. $SrSO_4$ has been observed due to the reaction between SO_2 and the SSC cathode. On the other hand, at a high concentration of 100 ppm of SO_2, the degradation of SSC becomes much more rapid and a significant change in the morphology is observed (Figure 7.17) [10]. Sr and Sm species from SSC cathode react with SO_2 to form $SrSO_4$- and $Sm_2O_2SO_4$-related phases, respectively. Cobalt oxides such as Co_3O_4 and CoO are formed as a result of the partial decomposition of SSC.

There is a certain degree of humidity in the ambient air that is dependent on the location and climate. Sakai *et al.* [330] showed that the presence of a small amount of water vapor significantly enhances the surface exchange rate of oxygen on the surface of YSZ. On the other hand, a small amount of water vapor of up to 3 vol% in the air has little effect on the performance of the LSM and LSCF cathodes during the 1000 h tests [331]. However, a large amount of vapor (e.g., 20 vol% H_2O) leads to the segregation of La_2O_3 or Mn_3O_4 on the surface of LSM cathodes and therefore deteriorates the cathode performance [331, 332].

Hagen *et al.* showed that the introduction of 4% H_2O in air decreases the cell voltage, and this effect is partly reversible on switching to dry air. In the presence of the steam, the YSZ surface in contact with the LSM–YSZ cathode forms ripple-like structure. On the other hand, as shown by Chen *et al.* [287], chromium deposition and poisoning on a $(La_{0.8}Sr_{0.2})_{0.9}MnO_3$ (LSM) cathode of SOFCs are more serious in humidified (3% H_2O) air, as compared to that in dry air in the presence of a Fe–Cr metallic interconnect. In the absence of interconnect, the LSM cathode

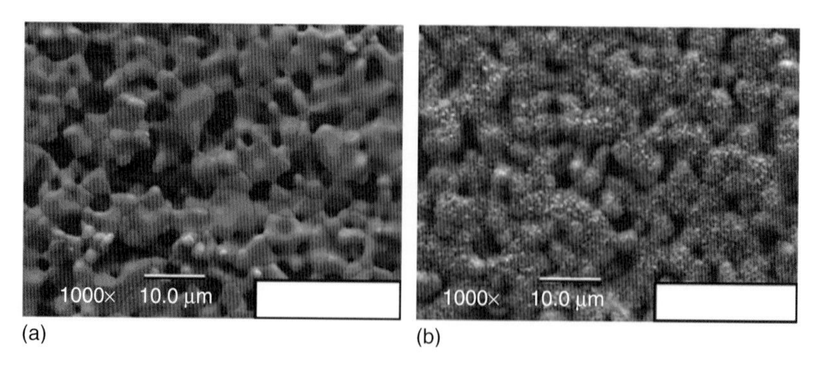

Figure 7.17 Backscattered electron images of the (Sm,Sr)CoO_3 surface at the central area after cell tests (a) without and (b) with 98 ppm of SO_2 in air. (Source: After Ref. [10]. Reproduced with permission.)

shows similar electrochemical activities for the O_2 reduction reaction in both dry and humidified air. However, in the presence of the Fe–Cr metallic interconnect, the electrochemical activity of the LSM cathode in humidified air is much lower than that in dry air. Cr deposition in humidified air is accelerated as compared to that in dry air. The significantly high Cr poisoning and deposition for the reaction in humidified air on the LSM cathode are most likely due to the high partial pressure of the gaseous Cr oxyhydroxide species formed. Thus, in general, for SOFCs based on the Fe–Cr metallic interconnect, the humidity in air on the cathode side should be kept as low as possible.

There is 0.039% CO_2 in the ambient air. The poisoning effect of CO_2 has been observed on alkali-containing perovskite materials typically $(Ba,Sr)(Co,Fe)O_3$ (BSCF) cathodes. Ba in the A site of BSCF tends to react with CO_2 to form $BaCO_3$, which could cause a substantial decrease in oxygen reduction. Yan *et al.* [333, 334] observed that the presence of 0.28–3.07% CO_2 rapidly deteriorates the BSCF performance. The degradation is more significant at a lower temperature, higher barium content in BSCF, or higher concentration of CO_2. Bucher *et al.* [335] observed that at 300–400 °C, the surface of BSCF is passivated with respect to oxygen exchange at $4 \times 10^{-4} \leq pCO_2$ (bar) $\leq 5 \times 10^{-2}$. A pronounced mass increase at 600–800 °C has been observed in CO_2-rich atmospheres (20 vol% O_2 + 5 vol% CO_2 + Ar) due to the significant carbonate formation of BSCF powders. On the other hand, the cathode could be regenerated by removing the CO_2 [333] and by decomposition of the carbonate at temperatures higher than 800 °C [335]. This implies that there is long-term degradation of the performance of BSCF-based cathodes particularly at intermediate-temperature ranges 600–700 °C.

7.4 Degradation of Solid Oxide Electrolysis Cells

High-temperature SOECs have a great potential for efficient and economic production of hydrogen fuel, as it involves less electrical energy consumption compared to conventional low-temperature water electrolysis [336–338]. SOECs are the reversely operated SOFCs and could be used to store renewable energies into H_2 by high-temperature electrolysis of steam. During operation, steam is fed to the fuel electrode side where it is reduced to hydrogen. At the anode side, the oxygen ions immigrated from the electrolyte are oxidized, evolving pure oxygen gas. The high operating temperature that is necessary for efficient electrolysis requires the use of materials that are stable at those operating temperatures. The materials and fabrication technologies that are used for SOFCs are now directly applicable for SOECs, and high electrolysis performance has been demonstrated by the SOECs [339–341]. However, there are issues regarding the stability of state-of-the-art SOFC materials operating under electrolysis conditions. Hauch *et al.* [342] showed that the long-term degradation of SOECs is up to approximately five times higher compared to SOFCs. Thus, understanding of the degradation mechanism in SOECs is important for the development of high-performance SOECs.

7.4.1
Fuel Electrodes

The presence of impurities in industrial-grade inlet gases can poison the fuel electrodes of SOECs. Ebbesen *et al.* [343] studied the stability of CO_2 electrolysis using Ni–YSZ-anode-supported SOECs at current density of 0.25–0.50 A cm^{-2}. The SOECs degrade at a rate between 0.22 and 0.44 mV h^{-1} during the first 500 h and the degradation is decreased to a moderate rate of 0.05–0.09 mV h^{-1} after operation for another 400–500 h. The degradation is independent of current density and it continues when the cell is held under open circuit conditions. The degradation is also irreversible when the steam is introduced, but is partly reversible when hydrogen is introduced. It was suggested that the degradation may be a consequence of impurities in the gas stream, most likely sulfur, adsorbing onto some specific nickel sites in the Ni–YSZ electrode. SOECs could be operated without degradation when the impurities are removed from the supply gases [344–346]. Figure 7.18 shows the enhanced stability of SOECs for CO_2 electrolysis when the inlet gases are cleaned. The poisoning of the fuel electrodes in SOECs by impurities such as sulfur could be similar to those in SOFCs, as discussed in Section 7.2.4.

Silicon impurity can also degrade the fuel electrodes of SOECs. Hauch *et al.* [347] found that during electrolysis, cell voltages increase significantly for the first few days of testing. The degradation is closely related to the segregation of silica-containing impurities at the hydrogen electrode–electrolyte interface. Silica in the glass sealant can cause the degradation of the fuel electrode of SOECs. Using a glass sealant with the composition $NaAlSi_3O_8$ on the fuel electrode side has been

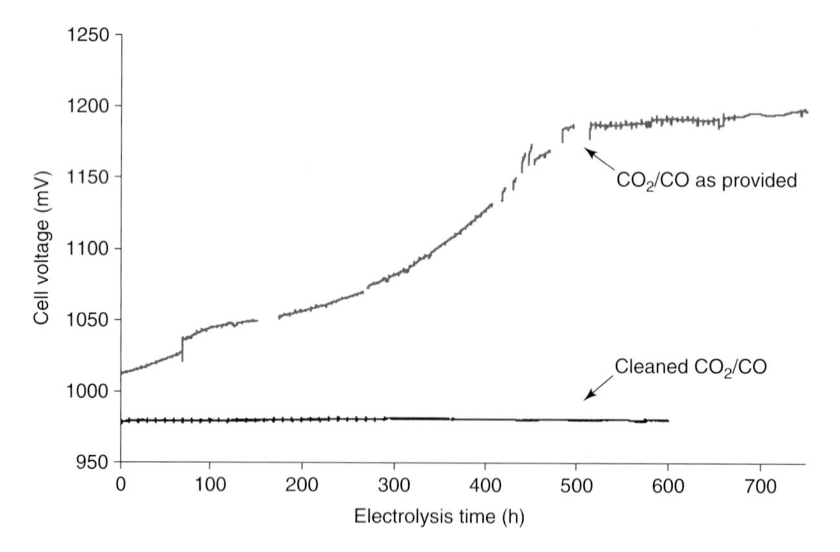

Figure 7.18 Cell voltage measured during CO_2 electrolysis (850 °C, 0.25 A cm^{-2}, 70% CO_2–30% CO) in the Ni–YSZ-based SOEC when applying the inlet gases to the fuel electrode as received and when applying cleaned gases. (Source: After Ref. [345]. Reproduced with permission.)

reported to result in the degradation of cell performance in the first few hundred hours of steam electrolysis, and the degradation has been found to be mainly caused by the increase in polarization resistance of the hydrogen electrode [321]. Silica has been observed in the fuel electrode of the tested SOECs. However, the degradation behavior has not been observed when using gold as the sealant. Knibbe *et al.* [348] studied the degradation behavior of SOECs at high current densities of 1, 1.5, and 2.0 A cm^{-2} with 50/50 H_2O/H_2 gas supplied to the Ni–YSZ electrode. The electrode polarization resistance has been observed to increase; however, the increase is not directly related to the applied current density but rather a consequence of adsorbed SiO_2 in the Ni–YSZ electrode. There is indication of small agglomeration of Ni phases, but no appreciable signs of reduced Ni percolation. SiO_2 has been found in the inlet region of the Ni–YSZ anode but not in the outlet region.

Kim-Lohsoontorn *et al.* [349] found that the Ni–YSZ-based SOECs degraded at a high rate of 2.5 mV h^{-1} over the 200 h electrolysis operation at 0.1 A cm^{-2}. On the other hand, impregnation of GDC into the Ni–YSZ fuel electrode not only enhances the electrocatalytic activity of the Ni–YSZ fuel electrode but also improves the durability of the SOECs. The SOECs with GDC-impregnated Ni–YSZ electrode are stable over 200 h (Figure 7.19).

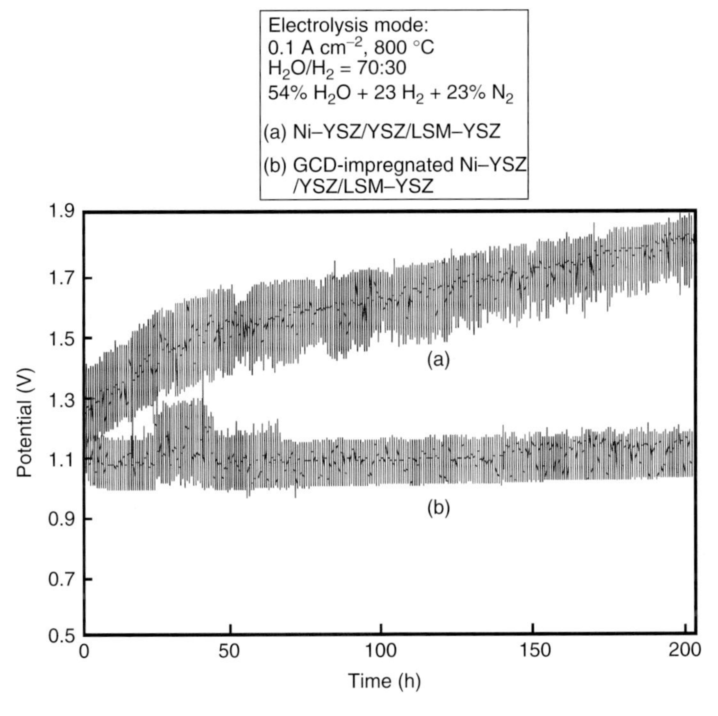

Figure 7.19 Polarization potential of hydrogen-electrode-supported SOECs during steam electrolysis (0.1 A cm^{-2}, 800 °C, $H_2O/H_2 = 70/30$) for two cells with and without GDC impregnation to the Ni–YSZ fuel electrodes. (Source: After Ref. [349]. Reproduced with permission.)

As discussed in Section 7.2.1, Ni tends to undergo oxidation and grain growth at a high steam concentration, and this issue is also critical for Ni-based cermet electrodes operating under severe SOEC operating conditions. Hauch *et al.* [321] observed a significant microstructural change in the Ni–YSZ electrode after operating the SOECs at a high current density and a high partial pressure of steam ($2\ A\ cm^{-2}$, $pH_2O = 0.9$ atm) at 950 °C for over 68 h. A dense Ni–YSZ layer with a thickness of 2–4 μm is formed at the Ni–YSZ electrode–electrolyte interface after the electrolysis operation. The researchers suggested that gaseous $Ni(OH)_2$ formed in the high steam environment would be reduced to Ni especially at the interface, leading to the relocation and redistribution of the Ni and a dense Ni–YSZ layer closest to the YSZ electrolyte. Schiller *et al.* [350] also observed the formation of oxide layer on the porous metallic substrate and coarsening of the Ni particles in their metal-supported cells after hydrogen electrolysis over 2000 h with a steam content of 43% at 800 °C.

Considering the instability of the Ni–YSZ electrode under extremely high steam concentration and high current density, alternative materials have also been investigated. Yang and Irvine [351] prepared 50 wt% $(La_{0.75}Sr_{0.25})_{0.95}Mn_{0.5}Cr_{0.5}O_3$ (LSCM)-impregnated YSZ as the fuel electrode of SOECs and compared it with the Ni–YSZ fuel electrode. In the case of the Ni–YSZ electrode, when the inlet gases are changed from reducing 3% steam/Ar/4% H_2 to 3% steam/Ar inert atmospheres, after a short period of operation, the impedance of the cell increases dramatically, to several thousand ohms. Ni is oxidized to NiO after the electrolysis operation at 1 V. On the other hand, cells with LSCM fuel electrodes operate well in both reducing and inert atmospheres.

7.4.2 Oxygen Electrodes

Degradation and electrode delamination of the oxygen electrodes have been observed by many research groups [352–357]. Very different from the enhanced effect of polarization on the electrochemical activity of LSM-based cathodes for O_2 reduction reaction, the so-called activation under SOFC operating conditions [235, 358–361], the anodic current polarization causes the deactivation of LSM and LSM–YSZ composite electrodes, leading to the increase in electrode polarization resistance for the oxygen oxidation reaction [340, 362–364]. Figure 7.20 shows the typical polarization curves of an LSM oxygen electrode under SOEC operating conditions [365]. The change of electrode anodic polarization potential (E_{anodic}) with polarization time can be characterized by three distinct regions. Region I is characterized by a continuous decrease in polarization potential with the polarization time, showing the characteristics of the activation of the LSM oxygen electrodes under anodic polarization conditions. E_{anodic} becomes more or less stable after the initial activation period and there is basically no change in it with polarization time (i.e., region II). In both regions I and II, the ohmic resistance of LSM oxygen electrodes does not change. However, the anodic potential, ohmic resistance, and polarization resistance were all observed to increase significantly in the final stage

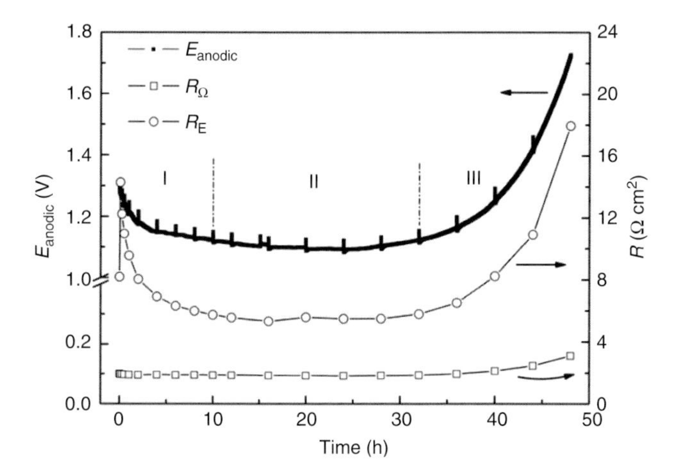

Figure 7.20 Polarization curves on a freshly prepared LSM oxygen electrode as a function of anodic current passage time at 500 mA cm^{-2} and 800 °C in air. (Source: After Ref. [365]. Reproduced with permission.)

of operation, indicating delamination of LSM electrode from the electrolyte. The most common mode of failure has been reported to be delamination at the oxygen electrode–electrolyte interface [352, 353, 366].

The degradation and delamination of oxygen electrode is affected by the level of applied current density. Chen *et al.* [354] found that the LSM oxygen electrode completely delaminates from the YSZ electrolyte after anodic current passage at 1 A cm^{-2} and 800 °C for 22 h. However, after polarized at 200 mA cm^{-2} over the same period, the LSM electrode still adheres well to the YSZ electrolyte. Graves *et al.* [367] found that in the case of low current density such as 0.25 A cm^{-2}, cell degradation is mainly caused by the Ni–YSZ fuel electrode, whereas at higher current densities of 0.5 and 1.0 A cm^{-2}, the Ni–YSZ electrode continues to degrade but the ohmic resistance and the degradation at the LSM–YSZ electrode also play a major role in cell degradation.

There are various proposed mechanisms for the electrode delamination including the processes due to the penetration of evolved oxygen into the closed pores/defects at the electrode–electrolyte interface and the high oxygen partial pressure building up at the electrode–electrolyte interface [352, 368, 369]. Virkar [370] presented an electrochemical model for degradation of SOECs that predicted that the occurrence of the delamination of the oxygen electrode is due to the formation of high internal oxygen pressure within the electrolyte close to the oxygen electrode–electrolyte interface. This model appears consistent with some experimental observations. Knibbe *et al.* [348] studied the degradation behavior of SOECs under high current densities of 1–2 A cm^{-2} and observed hole/pore formation along the grain boundaries of YSZ electrolyte close to the LSM–YSZ oxygen electrode. The researchers suggested that degradation in cell performance is related to the nucleation and formation of oxygen in the YSZ grain boundaries near the oxygen electrode. Laguna-Bercero *et al.* [357] also observed an irreversible damage on the YSZ electrolyte near the oxygen electrode–YSZ electrolyte interface after operation under a high voltage above 1.8 V, suggesting that the damage of YSZ may be due to the electrolyte reduction at high voltages.

Recently, we studied in detail the degradation of LSM oxygen electrode under SOEC operating conditions [365]. The LSM electrode has been observed to undergo delamination after electrolysis at 500 mA cm^{-2} over 48 h. Postmodern examination of the cells indicates the formation of nanoparticle clusters within LSM contact rings on the electrolyte surface (Figure 7.21). The clusters are consist of several layers of nanoparticles in the range 50–100 nm (Figure 7.21b). These nanoparticles within the clusters can be removed by HCl acid treatment, leaving ring-shaped marks on the YSZ surface. These marks are identical to the convex contact rings observed on the YSZ electrolyte surface that is in contact with LSM electrode after sintering at 1100 °C [371]. This indicates that nanoparticles in the clusters formed on the YSZ electrolyte surface have originated from LSM particles in contact with YSZ electrolyte and are most likely due to the disintegration of the large LSM grains at the oxygen electrode–electrolyte interface under the influence of the anodic current passage. The formation of nanoparticles appears to be constrained within the convex contact rings (Figure 7.21b). The results suggest that the shrinkage of the LSM lattice under SOEC operating conditions creates local tensile strains, resulting in the microcrack and subsequent formation of nanoparticles within LSM particles at the electrode–electrolyte interface. The formation of the nanoparticle clusters together with the high internal partial pressure of oxygen is responsible for the delamination and failure of the LSM-based oxygen electrode. Addition of YSZ electrolyte phase to form an LSM–YSZ composite oxygen electrode substantially enhances the electrochemical performance. However, the composite oxygen electrode also degrades significantly under SOEC mode and the formation of LSM nanoparticles has been found at the electrode–electrolyte interface after testing at 500 mA cm^{-2} and 800 °C for 100 h [372]. Severe degradation has also been reported for the LSM–YSZ oxygen electrode/10% Sc_2O_3–1% CeO_2–ZrO_2 electrolyte electrolysis cells, and the possible phase reaction has been speculated for the degradation [373].

Under SOEC operating conditions, MIEC-based oxygen electrodes exhibit much higher electrocatalytic activities than the pure electron-conducting electrodes such as LSM-based oxygen electrode materials. Minh's group [355, 374] studied several kinds of oxygen electrodes of SOECs and found that the performance of oxygen electrodes is in the order LSCF $>$ LSF $>$ LSM–YSZ in both the SOFC and the SOEC modes. In addition, LSCF and LSF exhibit better performance stability in the SOEC mode than the LSM–YSZ composite electrode.

On the other hand, degradation due to delamination has also been observed in MIEC electrodes. Hino *et al.* [356] applied $LaCoO_3$ oxygen electrode on YSZ electrolyte for SOECs. The $LaCoO_3$ electrode undergoes delamination from the electrolyte after one thermal cycle, and the delamination is primarily due to a large mismatch in the TEC between the electrode and the electrolyte. This mismatch also leads to the delamination of BSCF electrode from the electrolyte [366].

Cr poisoning of the oxygen electrodes under SOEC operation mode has also been reported. Mawdsley *et al.* [353] evaluated two SOEC stacks with the oxygen electrodes

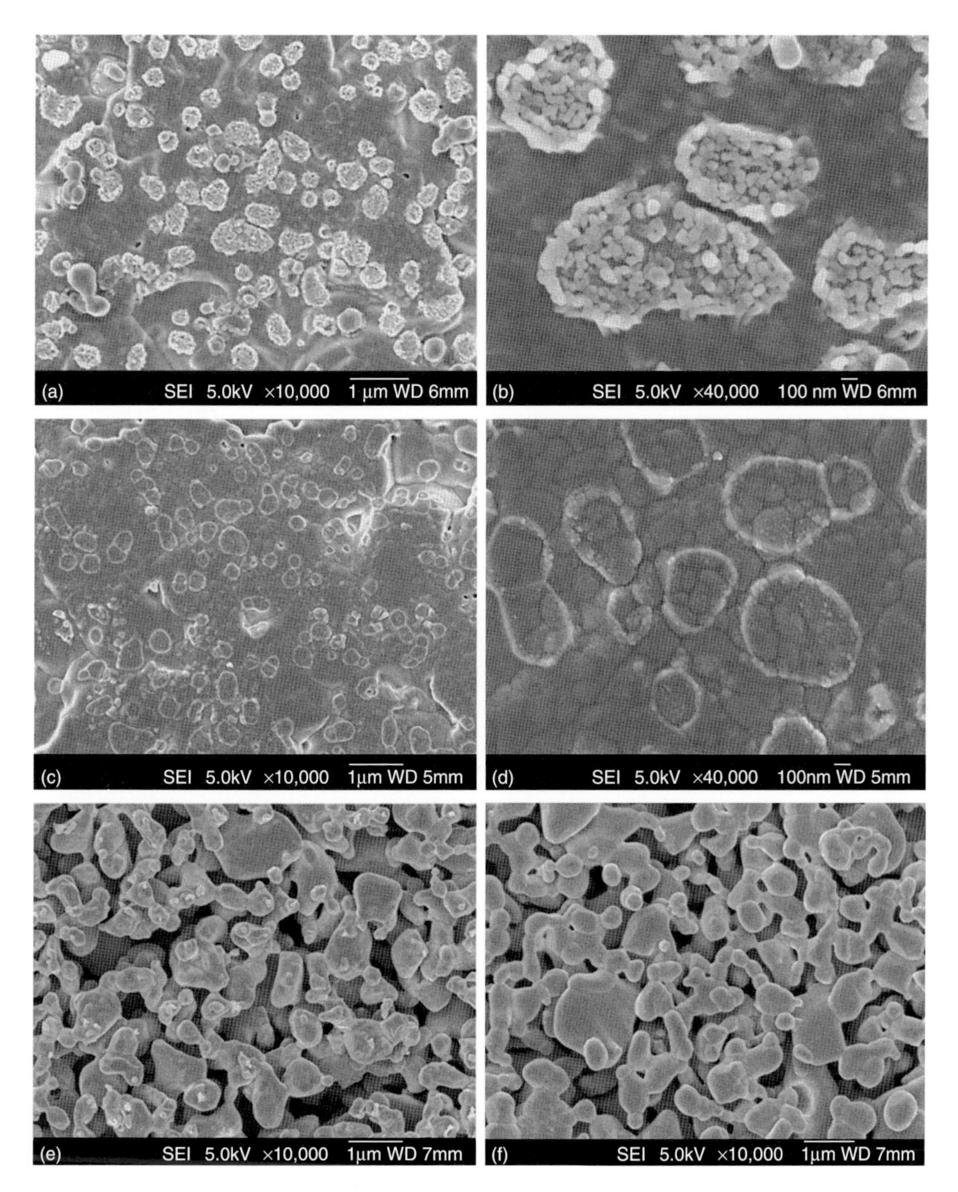

Figure 7.21 SEM images of the YSZ electrolyte surface in contact with the LSM oxygen electrode after anodic current passage at 500 mA cm^{-2} and 800 °C for 48 h: (a,b) before and (c,d) after HCl acid treatment, (e) inner surface of the LSM oxygen electrode in contact with YSZ electrolyte, and (f) outmost surface of the LSM oxygen electrode exposed to the air side. (Source: After Ref. [365]. Reproduced with permission.)

containing manganite–zirconia composite and $(La,Sr)CoO_3$ (LSC) as the current collector. The stacks experience a performance degradation of ~46% over the 2000 h test, and the degradation is closely related to the delamination of the oxygen electrode and Cr contamination of the LSC layer. Secondary phases of Cr_2O_3, $LaCrO_3$, La_2CrO_6, and Co_3O_4 are identified in the LSC layer. The dissociation of LSC is probably driven by the La–Cr–O related thermodynamics under the electrolytic potential and oxygen pressure at the oxygen electrode [375]. Minh [355] observed Sr and Cr migration and interactions at LSCF oxygen electrode–stainless steel interconnect interface after long-term operation. Elangovan *et al.* [376] used a cobalt–ferrite oxygen electrode and protective spinel layer on the air side of the steel interconnects. The stack showed a performance degradation rate as low as $<0.5\%$ per 1000 h.

Nanostructured oxygen electrodes have been demonstrated to significantly enhance the electrochemical activity and stability of oxygen electrodes of SOECs. Chen *et al.* [354] modified the LSM electrode by impregnation of 20–30 nm GDC nanoparticles. The GDC impregnation not only enhances significantly the electrocatalytic activity of the oxygen oxidation reaction but also inhibits electrode delamination at the oxygen electrode–electrolyte interface. Wang *et al.* [363] fabricated nanostructured $(La,Sr)CoO_3$ (LSC)-, $(La,Sr)FeO_3$ (LSF)-, and LSM-impregnated YSZ oxygen electrodes and conventional LSM–YSZ composite electrodes. The LSF–YSZ and LSC–YSZ electrodes are stable in both the SOEC and the SOFC modes for several hours. However, the LSM–YSZ oxygen electrode is not stable during electrolysis over 24 h at 700 °C. On the other hand, Yang *et al.* fabricated LSM-impregnated YSZ oxygen electrodes and showed that the electrode polarization resistance for the oxygen oxidation reaction is reduced as a result of the formation of uniformly distributed LSM nanoparticles on porous YSZ structure [377]. SOECs with LSM-impregnated YSZ oxygen electrode have been demonstrated to be stable during electrolysis at $0.33\ A\ cm^{-2}$ and 800 °C over 50 h. Recent study on nanostructured LSM-infiltrated YSZ oxygen electrodes showed that the microstructural stability of infiltrated LSM nanoparticles is governed by two opposite effects: one is the grain growth by the thermal sintering effect and the other is the LSM lattice shrinkage under the anodic polarization [378]. Which process is reponsible for the microstructural behavior of infiltrated LSM nanoparticles appears to be related to the initial particle size of the infiltrated LSM.

Alternative oxide materials have been investigated as potential SOEC oxygen electrodes. Liu *et al.* [379] applied the perovskite oxide $Sr_2Fe_{1.5}Mo_{0.5}O_6$ (SFM) as both fuel electrode and oxygen electrode in SOECs. The cell polarization resistance is as low as $0.26\ \Omega\ cm^2$ at 900 °C at 60 vol% of humidity at the fuel side, and a relatively stable cell performance over 100 h of electrolysis test at 850 °C has been demonstrated.

7.5 Summary and Conclusions

Performance degradation of the electrodes either due to the microstructure deterioration and/or impurity contamination/poisoning is one of the most important factors governing the long-term stability and durability of SOFC systems. This is particularly the case for the SOFCs operating on coal- and biomass-derived synthesis gas for the near future power generation, giving the abundant coal and biomass resources in the world. However, the coal and biomass syngas contains various impurities such as tar, phosphorus, arsenic, sulfur, zinc, and selenium, which affect the performance and durability of SOFCs. It is difficult to completely remove these impurities from the gas streams and the cleanup process can also be expensive. In addition, impurities such as chromium, boron, and silica are generated from the stack components and may not be completely avoidable. Therefore, it is imperative that fuel cells be tolerant to these impurities and contaminants up to a certain level.

However, as discussed in this chapter, studies on the degradation process and mechanism remain largely at the experimental stage and fundamental knowledge for the understanding of the underlying mechanisms through which the cell performance and microstructure deteriorate lacks far behind the materials development for SOFCs and SOECs. Such knowledge is also essential for developing accurate models to predict contaminant-induced performance degradation. Nevertheless, carefully calibrated empirical models can be very effective tools for design and prediction of the service life of SOFCs in the absence of thorough understanding of the degradation mechanisms. Cayan *et al.* [380] developed a one-dimensional model for the performance behavior in coal-derived syngas against degradation rates reported in the literature for arsine, phosphine, hydrogen sulfide, and hydrogen selenide and reasonable agreement is obtained. Further progress in terms of materials microstructural behavior, material surface chemistry, and diffusion properties under influence of impurities and interaction between materials and impurities under fuel cell operating conditions will be essential to develop and establish more practical and accurate performance models of anodes and cathodes of SOFCs.

In addition to the high electrocatalytic activity for oxygen reduction at the cathode side and fuel oxidation at the anode side, a successful electrode must also have high tolerance toward contaminants under SOFC or SOEC operating conditions. This is generally a less studied area, but the development of electrode materials with high electrocatalytic activity and high tolerance toward contaminants opens new opportunities and the challenges are substantial for both research scientists and engineers working in SOFC and SOEC fields.

References

1. Brett, D.J.L., Atkinson, A., Brandon, N.P., and Skinner, S.J. (2008) Intermediate temperature solid oxide fuel cells. *Chem. Soc. Rev.*, **37**, 1568–1578.
2. Wachsman, E.D. and Lee, K.T. (2011) Lowering the temperature of solid oxide fuel cells. *Science*, **334**, 935–939.
3. Fergus, J.W. (2005) Metallic interconnects for solid oxide fuel cells. *Mater. Sci. Eng., A*, **397**, 271–283.
4. Zhu, W.Z. and Deevi, S.C. (2003) Development of interconnect materials for solid oxide fuel cells. *Mater. Sci. Eng., A*, **348**, 227–243.
5. Hilpert, K., Das, D., Miller, M., Peck, D.H., and Weiss, R. (1996) Chromium vapor species over solid oxide fuel cell interconnect materials and their potential for degradation processes. *J. Electrochem. Soc.*, **143**, 3642–3647.

6. Paulson, S.C. and Birss, V.I. (2004) Chromium poisoning of LSM-YSZ SOFC cathodes – I. Detailed study of the distribution of chromium species at a porous, single-phase cathode. *J. Electrochem. Soc.*, **151**, A1961–A1968.
7. Fergus, J.W. (2007) Effect of cathode and electrolyte transport properties on chromium poisoning in solid oxide fuel cells. *Int. J. Hydrogen Energy*, **32**, 3664–3671.
8. Jiang, S.P., Zhang, J.P., Apateanu, L., and Foger, K. (2000) Deposition of chromium species at Sr-doped $LaMnO_3$ electrodes in solid oxide fuel cells I. Mechanism and kinetics. *J. Electrochem. Soc.*, **147**, 4013–4022.
9. Badwal, S.P.S., Deller, R., Foger, K., Ramprakash, Y., and Zhang, J.P. (1997) Interaction between chromia forming alloy interconnects and air electrode of solid oxide fuel cells. *Solid State Ionics*, **99**, 297–310.
10. Xiong, Y.P., Yamaji, K., Horita, T., Yokokawa, H., Akikusa, J., Eto, H., and Inagaki, T. (2009) Sulfur poisoning of SOFC cathodes. *J. Electrochem. Soc.*, **156**, B588–B592.
11. Horita, T., Kishimoto, H., Yamaji, K., Brito, M.E., Xiong, Y.P., Yokokawa, H., Hori, Y., and Miyachi, I. (2009) Effects of impurities on the degradation and long-term stability for solid oxide fuel cells. *J. Power Sources*, **193**, 194–198.
12. Yokokawa, H., Yamaji, K., Brito, M.E., Kishimoto, H., and Horita, T. (2011) General considerations on degradation of solid oxide fuel cell anodes and cathodes due to impurities in gases. *J. Power Sources*, **196**, 7070–7075.
13. Lorente, E., Millan, M., and Brandon, N.P. (2012) Use of gasification syngas in SOFC: impact of real tar on anode materials. *Int. J. Hydrogen Energy*, **37**, 7271–7278.
14. Zhao, F., Peng, R.R., and Xia, C.R. (2008) A $La_{0.6}Sr_{0.4}CoO_{3-\delta}$-based electrode wit high durability for intermediate temperature solid oxide fuel cells. *Mater. Res. Bull.*, **43**, 370–376.
15. Zha, S.W., Cheng, Z., and Liu, M.L. (2007) Sulfur poisoning and regeneration of Ni-based anodes in solid oxide fuel cells. *J. Electrochem. Soc.*, **154**, B201–B206.
16. Tikekar, N.M., Armstrong, T.J., and Virkar, A.V. (2006) Reduction and reoxidation kinetics of nickel-based SOFC anodes. *J. Electrochem. Soc.*, **153**, A654–A663.
17. Jiang, S.P. (2003) Sintering behavior of Ni/Y_2O_3-ZrO_2 cermet electrodes of solid oxide fuel cells. *J. Mater. Sci.*, **38**, 3775–3782.
18. Fergus, J.W. (2006) Electrolytes for solid oxide fuel cells. *J. Power Sources*, **162**, 30–40.
19. Nomura, K., Mizutani, Y., Kawai, M., Nakamura, Y., and Yamamoto, O. (2000) Aging and raman scattering study of scandia and yttria doped zirconia. *Solid State Ionics*, **132**, 235–239.
20. Yamamoto, O., Arati, Y., Takeda, Y., Imanishi, N., Mizutani, Y., Kawai, M., and Nakamura, Y. (1995) Electrical conductivity of stabilized zirconia with ytterbia and scandia. *Solid State Ionics*, **79**, 137–142.
21. Tsoga, A., Naoumidis, A., and Nikolopoulos, P. (1996) Wettability and interfacial reactions in the systems Ni/YSZ and Ni/Ti-TiO_2/YSZ. *Acta Mater.*, **44**, 3679–3692.
22. Iwata, T. (1996) Characterization of Ni-YSZ anode degradation for substrate-type solid oxide fuel cells. *J. Electrochem. Soc.*, **143**, 1521–1525.
23. Simwonis, D., Tietz, F., and Stöver, D. (2000) Nickel coarsening in annealed Ni/8YSZ anode substrates for solid oxide fuel cells. *Solid State Ionics*, **132**, 241–251.
24. Sehested, J., Gelten, J.A.P., and Helveg, S. (2006) Sintering of nickel catalysts: effects of time, atmosphere, temperature, nickel-carrier interactions, and dopants. *Appl. Catal., A*, **309**, 237–246.
25. Sehested, J. (2003) Sintering of nickel steam-reforming catalysts. *J. Catal.*, **217**, 417–426.
26. Iwanschitz, B., Holzer, L., Mai, A., and Schutze, M. (2012) Nickel agglomeration in solid oxide fuel cells: the influence of temperature. *Solid State Ionics*, **211**, 69–73.

27. Jiang, S.P., Callus, P.J., and Badwal, S.P.S. (2000) Fabrication and performance of Ni/3 mol% Y_2O_3–ZrO_2 cermet anodes for solid oxide fuel cells. *Solid State Ionics*, **132**, 1–14.
28. Koch, S., Hendriksen, P.V., Mogensen, M., Liu, Y.L., Dekker, N., Rietveld, B., de Haart, B., and Tietz, F. (2006) Solid oxide fuel cell performance under severe operating conditions. *Fuel Cells*, **6**, 130–136.
29. Ivers-Tiffée, E., Weber, A., and Herbstritt, D. (2001) Materials and technologies for SOFC-components. *J. Eur. Ceram. Soc.*, **21**, 1805–1811.
30. Matsui, T., Kishida, R., Kim, J.Y., Muroyama, H., and Eguchi, K. (2010) Performance deterioration of Ni-YSZ anode induced by electrochemically generated steam in solid oxide fuel cells. *J. Electrochem. Soc.*, **157**, B776–B781.
31. Yamaji, K., Kishimoto, H., Xiong, Y., Horita, T., Sakai, N., and Yokokawa, H. (2004) Performance of anode-supported SOFCs fabricated with EPD techniques. *Solid State Ionics*, **175**, 165–169.
32. de Haart, L.G.J., Mayer, K., Stimming, U., and Vinke, I.C. (1998) Operation of anode-supported thin electrolyte film solid oxide fuel cells at 800 degrees C and below. *J. Power Sources*, **71**, 302–305.
33. Moon, J.W., Lee, H.L., Kim, J.D., Kim, G.D., Lee, D.A., and Lee, H.W. (1999) Preparation of ZrO_2-coated NiO powder using surface-induced coating. *Mater. Lett.*, **38**, 214–220.
34. Itoh, H., Yamamoto, T., Mori, M., Horita, T., Sakai, N., Yokokawa, H., and Dokiya, M. (1997) Configurational and electrical behavior of Ni-YSZ cermet with novel microstructure for solid oxide fuel cell anodes. *J. Electrochem. Soc.*, **144**, 641–646.
35. Itoh, H., Yamamoto, T., Mori, M., Watanabe, T., and Abe, T. (1996) Improved microstructure of Ni-YSZ cermet anode for SOFC with a long term stability. *Denki Kagaku*, **64**, 549–554.
36. Marinsek, M., Zupan, K., and Maèek, J. (2002) Ni-YSZ cermet anodes prepared by citrate/nitrate combustion synthesis. *J. Power Sources*, **106**, 178–188.
37. Esposito, V., de Florio, D.Z., Fonseca, F.C., Muccillo, E.N.S., Muccillo, R., and Traversa, E. (2005) Electrical properties of YSZ/NiO composites prepared by a liquid mixture technique. *J. Eur. Ceram. Soc.*, **25**, 2637–2641.
38. Sunagawa, Y., Yamamoto, K., and Muramatsu, A. (2006) Improvement in SOFC anode performance by finely-structured Ni/YSZ cermet prepared via heterocoagulation. *J. Phys. Chem. B*, **110**, 6224–6228.
39. Fukui, T., Ohara, S., Naito, M., and Nogi, K. (2002) Performance and stability of SOFC anode fabricated from NiO-YSZ composite particles. *J. Power Sources*, **110**, 91–95.
40. Jiang, S.P., Duan, Y.Y., and Love, J.G. (2002) Fabrication of high-performance $NiOY_2O_3$-ZrO_2 cermet anodes of solid oxide fuel cells by ion impregnation. *J. Electrochem. Soc.*, **149**, A1175–A1183.
41. Klemenso, T., Chung, C., Larsen, P.H., and Mogensen, M. (2005) The mechanism behind redox instability of anodes in high-temperature SOFCs. *J. Electrochem. Soc.*, **152**, A2186–A2192.
42. Busawon, A.N., Sarantaridis, D., and Atkinson, A. (2008) Ni infiltration as a possible solution to the redox problem of SOFC anodes. *Electrochem. Solid-State Lett.*, **11**, B186–B189.
43. Sarantaridis, D. and Atkinson, A. (2007) Redox cycling of Ni-based solid oxide fuel cell anodes: a review. *Fuel Cells*, **7**, 246–258.
44. Klemenso, T., Appel, C.C., and Mogensen, M. (2006) In situ observations of microstructural changes in SOFC anodes during redox cycling. *Electrochem. Solid-State Lett.*, **9**, A403–A407.
45. Malzbender, J., Wessel, E., and Steinbrech, R.W. (2005) Reduction and re-oxidation of anodes for solid oxide fuel cells. *Solid State Ionics*, **176**, 2201–2203.
46. Sarantaridis, D., Chater, R.J., and Atkinson, A. (2008) Changes in physical and mechanical properties of SOFCNi-YSZ composites caused by

redox cycling. *J. Electrochem. Soc.*, **155**, B467–B472.
47. Cassidy, M., Lindsay, G., and Kendall, K. (1996) The reduction of nickel-zirconia cermet anodes and the effects on supported thin electrolytes. *J. Power Sources*, **61**, 189–192.
48. Tucker, M.C., Lau, G.Y., Jacobson, C.P., DeJonghe, L.C., and Visco, S.J. (2008) Stability and robustness of metal-supported SOFCs. *J. Power Sources*, **175**, 447–451.
49. Vedasri, V., Young, J.L., and Birss, V.I. (2010) A possible solution to the mechanical degradation of Ni–yttria stabilized zirconia anode-supported solid oxide fuel cells due to redox cycling. *J. Power Sources.*, **195**, 5534–5542.
50. Waldbillig, D., Wood, A., and Ivey, D.G. (2005) Electrochemical and microstructural characterization of the redox tolerance of solid oxide fuel cell anodes. *J. Power Sources*, **145**, 206–215.
51. Kong, J., Sun, K., Zhou, D., Zhang, N., and Qiao, J. (2006) Electrochemical and microstructural characterization of cyclic redox behaviour of SOFC anodes. *Rare Met.*, **25**, 300–304.
52. Sumi, H., Kishida, R., Kim, J.Y., Muroyama, H., Matsui, T., and Eguchi, K. (2010) Correlation between microstructural and electrochemical characteristics during redox cycles for Ni-YSZ anode of SOFCs. *J. Electrochem. Soc.*, **157**, B1747–B1752.
53. Waldbillig, D., Wood, A., and Ivey, D.G. (2005) Thermal analysis of the cyclic reduction and oxidation behaviour of SOFC anodes. *Solid State Ionics*, **176**, 847–859.
54. Waldbillig, D., Wood, A., and Ivey, D.G. (2007) Enhancing the redox tolerance of anode-supported SOFC by microstructural modification. *J. Electrochem. Soc.*, **154**, B133–B138.
55. Hagen, A., Poulsen, H.F., Klemenso, T., Martins, R.V., Honkimaki, V., Buslaps, T., and Feidenshans'l, R. (2006) A depth-resolved in-situ study of the reduction and oxidation of Ni-based anodes in solid oxide fuel cells. *Fuel Cells*, **6**, 361–366.
56. Hanifi, A.R., Torabi, A., Zazulak, M., Etsell, T.H., Yamarte, L., Sarkar, P., and Tucker, M.C. (2011) Improved redox and thermal cycling resistant tubular ceramic fuel cells. *ECS Trans.*, **35**, 409–418.
57. Kim, S.-D., Moon, H., Hyun, S.-H., Moon, J., Kim, J., and Lee, H.-W. (2006) Performance and durability of Ni-coated YSZ anodes for intermediate temperature solid oxide fuel cells. *Solid State Ionics*, **177**, 931–938.
58. Tao, S.W. and Irvine, J.T.S. (2003) A redox-stable efficient anode for solid-oxide fuel cells. *Nat. Mater.*, **2**, 320–323.
59. Marina, O.A., Canfield, N.L., and Stevenson, J.W. (2002) Thermal, electrical, and electrocatalytical properties of lanthanum-doped strontium titanate. *Solid State Ionics*, **149**, 21–28.
60. Ong, K.P., Wu, P., Liu, L., and Jiang, S.P. (2007) Optimization of electrical conductivity of $LaCrO_3$ through doping: a combined study of molecular modeling and experiment. *Appl. Phys. Lett.*, **90**, 044109.
61. Tao, S.W. and Irvine, J.T.S. (2004) Synthesis and characterization of $(La_{0.75}Sr_{0.25})Cr_{0.5}Mn_{0.5}O_{3-\delta}$, a redox-stable, efficient perovskite anode for SOFCs. *J. Electrochem. Soc.*, **151**, A252–A259.
62. Moos, R. and Hardtl, K.H. (1996) Electronic transport properties of Sr_1-xLaxTiO_3 ceramics. *J. Appl. Phys.*, **80**, 393–400.
63. Jiang, S.P. and Chan, S.H. (2004) Development of Ni/Y_2O_3-ZrO_2 cermet anodes for solid oxide fuel cells. *Mater. Sci. Technol.*, **20**, 1109–1118.
64. Agnolucci, P. (2007) Hydrogen infrastructure for the transport sector. *Int. J. Hydrogen Energy*, **32**, 3526–3544.
65. Jiang, S.P. and Chan, S.H. (2004) A review of anode materials development in solid oxide fuel cells. *J. Mater. Sci.*, **39**, 4405–4439.
66. Murray, E.P., Tsai, T., and Barnett, S.A. (1999) A direct-methane fuel cell with a ceria-based anode. *Nature*, **400**, 649–651.
67. Park, S., Craciun, R., Vohs, J.M., and Gorte, R.J. (1999) Direct oxidation of

hydrocarbons in a solid oxide fuel cell I. Methane oxidation. *J. Electrochem. Soc.*, **146**, 3603–3605.

68. Takeguchi, T., Kani, Y., Yano, T., Kikuchi, R., Eguchi, K., Tsujimoto, K., Uchida, Y., Ueno, A., Omoshiki, K., and Aizawa, M. (2002) Study on steam reforming of CH_4 and C_2 hydrocarbons and carbon deposition on Ni-YSZ cermets. *J. Power Sources*, **112**, 588–595.
69. Gunji, A., Wen, C., Otomo, J., Kobayashi, T., Ukai, K., Mizutani, Y., and Takahashi, H. (2004) Carbon deposition behaviour on Ni-ScSZ anodes for internal reforming solid oxide fuel cells. *J. Power Sources*, **131**, 285–288.
70. He, H.P. and Hill, J.M. (2007) Carbon deposition on Ni/YSZ composites exposed to humidified methane. *Appl. Catal., A*, **317**, 284–292.
71. Iida, T., Kawano, M., Matsui, T., Kikuchi, R., and Eguchi, K. (2007) Internal reforming of SOFCs – carbon deposition on fuel electrode and subsequent deterioration of cell. *J. Electrochem. Soc.*, **154**, B234–B241.
72. Horita, T., Kishimoto, H., Yamaji, K., Sakai, N., Xiong, Y.P., Brito, M.E., Yokokawa, H., Rai, M., Amezawa, K., and Uchimoto, Y. (2006) Active parts for CH_4 decomposition and electrochemical oxidation at metal/oxide interfaces by isotope labeling-secondary ion mass spectrometry. *Solid State Ionics*, **177**, 3179–3185.
73. Chun, C.M., Mumford, J.D., and Ramanarayanan, T.A. (2000) Carbon-induced corrosion of nickel anode. *J. Electrochem. Soc.*, **147**, 3680–3686.
74. Macek, J., Novosel, B., and Marinsek, M. (2007) Ni-YSZ SOFC anodes–minimization of carbon deposition. *J. Eur. Ceram. Soc.*, **27**, 487–491.
75. Buergler, B.E., Grundy, A.N., and Gauckler, L.J. (2006) Thermodynamic equilibrium of single-chamber SOFC relevant methane-air mixtures. *J. Electrochem. Soc.*, **153**, A1378–A1385.
76. Joensen, F. and Rostrup-Nielsen, J.R. (2002) Conversion of hydrocarbons and alcohols for fuel cells. *J. Power Sources*, **105**, 195–201.
77. Sasaki, K., Kojo, H., Hori, Y., Kikuchi, R., and Eguchi, K. (2002) Direct-alcohol/hydrocarbon SOFCs: comparison of power generation characteristics for various fuels. *Electrochemistry*, **70**, 18–22.
78. Eguchi, K., Kojo, H., Takeguchi, T., Kikuchi, R., and Sasaki, K. (2002) Fuel flexibility in power generation by solid oxide fuel cells. *Solid State Ionics*, **152**, 411–416.
79. Timmermann, H., Sawady, W., Campbell, D., Weber, A., Reimert, R., and Ivers-Tiffee, E. (2007) Coke formation in hydrocarbons-containing fuel Gas and effects on SOFC degradation phenomena. *ECS Trans.*, **7**, 1429–1435.
80. Saunders, G.J., Preece, J., and Kendall, K. (2004) Formulating liquid hydrocarbon fuels for SOFCs. *J. Power Sources*, **131**, 23–26.
81. Atkinson, A., Barnett, S., Gorte, R.J., Irvine, J.T.S., McEvoy, A.J., Mogensen, M., Singhal, S.C., and Vohs, J. (2004) Advanced anodes for high-temperature fuel cells. *Nat. Mater.*, **3**, 17–27.
82. Lin, Y.B., Zhan, Z.L., Liu, J., and Barnett, S.A. (2005) Direct operation of solid oxide fuel cells with methane fuel. *Solid State Ionics*, **176**, 1827–1835.
83. Koh, J.-H., Yoo, Y.-S., Park, J.-W., and Lim, H.C. (2002) Carbon deposition and cell performance of Ni-YSZ anode support SOFC with methane fuel. *Solid State Ionics*, **149**, 157–166.
84. Trimm, D.L. (1997) Coke formation and minimisation during steam reforming reactions. *Catal. Today*, **37**, 233–238.
85. Kawano, M., Matsui, T., Kikuchi, R., Yoshida, H., Inagaki, T., and Eguchi, K. (2008) Steam reforming on Ni-samaria-doped ceria cermet anode for practical size solid oxide fuel cell at intermediate temperatures. *J. Power Sources*, **182**, 496–502.
86. Kawano, M., Matsui, T., Kikuchi, R., Yoshida, H., Inagaki, T., and Eguchi, K. (2007) Direct internal steam reforming at SOFC anodes composed of NiO-SDC composite particles. *J. Electrochem. Soc.*, **154**, B460–B465.
87. Finnerty, C.M., Coe, N.J., Cunningham, R.H., and Ormerod,

R.M. (1998) Carbon formation on and deactivation of nickel-based/zirconia anodes in solid oxide fuel cells running on methane. *Catal. Today*, **46**, 137–145.

88. Bae, G., Bae, J., Kim-Lohsoontorn, P., and Jeong, J. (2010) Performance of SOFC coupled with n-C_4H_{10} autothermal reformer: carbon deposition and development of anode structure. *Int. J. Hydrogen Energy*, **35**, 12346–12358.
89. Zhan, Z.L. and Barnett, S.A. (2005) An octane-fueled solid oxide fuel cell. *Science*, **308**, 844–847.
90. Kishimoto, H., Horita, T., Yamaji, K., Xiong, Y.P., Sakai, N., and Yokokawa, H. (2004) Attempt of utilizing liquid fuels with Ni-ScSZ anode in SOFCs. *Solid State Ionics*, **175**, 107–111.
91. Yamaji, K., Kishimoto, H., Yueping, X., Horita, T., Sakai, N., Brito, M.E., and Yokokawa, H. (2007) Stability of nickel-based cermet anode for carbon deposition during cell operation with slightly humidified gaseous hydrocarbon fuels. *ECS Trans.*, **7**, 1661–1668.
92. Gorte, R.J., Kim, H., and Vohs, J.M. (2002) Novel SOFC anodes for the direct electrochemical oxidation of hydrocarbon. *J. Power Sources*, **106**, 10–15.
93. Ahmed, K. and Foger, K. (2000) Kinetics of internal steam reforming of methane on Ni/YSZ-based anodes for solid oxide fuel cells. *Catal. Today*, **63**, 479–487.
94. Dicks, A.L. (1998) Advances in catalysts for internal reforming in high temperature fuel cells. *J. Power Sources*, **71**, 111–122.
95. Wang, J.B., Jang, J.-C., and Huang, T.-J. (2003) Study of Ni-samaria-doped ceria anode for direct oxidation of methane in solid oxide fuel cells. *J. Power Sources*, **122**, 122–131.
96. Mallon, C. and Kendall, K. (2005) Sensitivity of nickel cermet anodes to reduction conditions. *J. Power Sources*, **145**, 154–160.
97. Sumi, H., Ukai, K., Mizutani, Y., Mori, H., Wen, C.J., Takahashi, H., and Yamamoto, O. (2004) Performance of nickel-scandia-stabilized zirconia cermet anodes for SOFCs in 3% H_2O-CH_4. *Solid State Ionics*, **174**, 151–156.
98. Kishimoto, H., Xiong, Y.-P., Yamaji, K., Horita, T., Sakai, N., Brito, M.E., and Yokokawa, H. (2007) Direct feeding of liquid fuel in SOFC. *ECS Trans.*, **7**, 1669–1674.
99. Kishimoto, H., Yamaji, K., Horita, T., Xiong, Y.P., Sakai, N., Brito, M.E., and Yokokawa, H. (2006) Reaction process in the Ni-ScSZ anode for hydrocarbon fueled SOFCs. *J. Electrochem. Soc.*, **153**, A982–A988.
100. Matsui, T., Iida, T., Kikuchi, R., Kawano, M., Inagaki, T., and Eguchi, K. (2008) Carbon deposition over Ni-ScSZ anodes subjected to various heat-treatments for internal reforming of solid oxide fuel cells. *J. Electrochem. Soc.*, **155**, B1136–B1140.
101. Ke, K., Gunji, A., Mori, H., Tsuchida, S., Takahashi, H., Ukai, K., Mizutani, Y., Sumi, H., Yokoyama, M., and Waki, K. (2006) Effect of oxide on carbon deposition behavior of CH_4 fuel on Ni/ScSZ cermet anode in high temperature SOFCs. *Solid State Ionics*, **177**, 541–547.
102. Livermore, S.J.A., Cotton, J.W., and Ormerod, R.M. (2000) Fuel reforming and electrical performance studies in intermediate temperature ceria-gadolinia-based SOFCs. *J. Power Sources*, **86**, 411–416.
103. Kim, H., Lu, C., Worrell, W.L., Vohs, J.M., and Gorte, R.J. (2002) Cu-Ni cermet anodes for direct oxidation of methane in solid-oxide fuel cells. *J. Electrochem. Soc.*, **149**, A247–A250.
104. Cancellier, M., Sin, A., Morrone, M., Caracino, P., Sarkar, P., Yamarte, L., Lorne, J., Liu, M., Barker-Hemings, E., Caligiuri, A., and Cavallotti, C. (2007) SOFC anodes for direct oxidation of alcohols at intermediate temperatures. *ECS Trans.*, **7**, 1725–1732.
105. Xie, Z., Xia, C.R., Zhang, M.Y., Zhu, W., and Wang, H.T. (2006) Ni1-XCux alloy-based anodes for low-temperature solid oxide fuel cells with biomass-produced gas as fuel. *J. Power Sources*, **161**, 1056–1061.

106. Ai, N., Chen, K., Jiang, S.P., Lü, Z., and Su, W. (2011) Vacuum-assisted electroless copper plating on Ni/(Sm,Ce)O_2 anodes for intermediate temperature solid oxide fuel cells. *Int. J. Hydrogen Energy*, **36**, 7661–7669.

107. Triantafyllopoulos, N.C. and Neophytides, S.G. (2006) Dissociative adsorption of CH_4 on NiAu/YSZ: the nature of adsorbed carbonaceous species and the inhibition of graphitic C formation. *J. Catal.*, **239**, 187–199.

108. Tarancon, A., Morata, A., Peiro, F., and Dezanneau, G. (2011) A molecular dynamics study on the oxygen diffusion in doped fluorites: the effect of the dopant distribution. *Fuel Cells*, **11**, 26–37.

109. Boder, M. and Dittmeyer, R. (2006) Catalytic modification of conventional SOFC anodes with a view to reducing their activity for direct internal reforming of natural gas. *J. Power Sources*, **155**, 13–22.

110. Ishihara, T., Yan, J.W., Shinagawa, M., and Matsumoto, H. (2006) Ni-Fe bimetallic anode as an active anode for intermediate temperature SOFC using $LaGaO_3$ based electrolyte film. *Electrochim. Acta*, **52**, 1645–1650.

111. Zhong, H., Matsumoto, H., and Ishihara, T. (2009) Development of Ni-Fe based cermet anode for direct CH_4 fueled intermediate temperature SOFC using $LaGaO_3$ electrolyte. *Electrochemistry*, **77**, 155–157.

112. Borowiecki, T., Giecko, G., and Panczyk, M. (2002) Effects of small MoO_3 additions on the properties of nickel catalysts for the steam reforming of hydrocarbons: II. Ni—Mo/Al_2O_3 catalysts in reforming, hydrogenolysis and cracking of n-butane. *Appl. Catal. Gen.*, **230**, 85–97.

113. Triantafyllopoulos, N.C. and Neophytides, S.G. (2003) The nature and binding strength of carbon adspecies formed during the equilibrium dissociative adsorption of CH_4 on Ni-YSZ cermet catalysts. *J. Catal.*, **217**, 324–333.

114. Trimm, D.L. (1999) Catalysts for the control of coking during steam reforming. *Catal. Today*, **49**, 3–10.

115. Nikolla, E., Schwank, J.W., and Linic, S. (2008) Hydrocarbon steam reforming on Ni alloys at solid oxide fuel cell operating conditions. *Catal. Today*, **136**, 243–248.

116. Kan, H., Hyun, S.H., Shul, Y.G., and Lee, H. (2009) Improved solid oxide fuel cell anodes for the direct utilization of methane using Sn-doped Ni/YSZ catalysts. *Catal. Commun.*, **11**, 180–183.

117. Takeguchi, T., Kikuchi, R., Yano, T., Eguchi, K., and Murata, K. (2003) Effect of precious metal addition to Ni-YSZ cermet on reforming of CH_4 and electrochemical activity as SOFC anode. *Catal. Today*, **84**, 217–222.

118. Hibino, T., Hashimoto, A., Yano, M., Suzuki, M., and Sano, M. (2003) Ru-catalyzed anode materials for direct hydrocarbon SOFCs. *Electrochim. Acta*, **48**, 2531–2537.

119. Babaei, A., Jiang, S.P., and Li, J. (2009) Electrocatalytic promotion of palladium nanoparticles on hydrogen oxidation on Ni/GDC anodes of SOFCs via spillover. *J. Electrochem. Soc.*, **156**, B1022–B1029.

120. Nakagawa, N., Sagara, H., and Kato, K. (2001) Catalytic activity of Ni-YSZ-CeO_2 anode for the steam reforming of methane in a direct internal-reforming solid oxide fuel cell. *J. Power Sources*, **92**, 88–94.

121. Sadykov, V.A., Mezentseva, N.V., Bunina, R.V., Alikina, G.M., Lukashevich, A.I., Kharlamova, T.S., Rogov, V.A., Zaikovskii, V.I., Ishchenko, A.V., Krieger, T.A., Bobrenok, O.F., Smirnova, A., Irvine, J., and Vasylyev, O.D. (2008) Effect of complex oxide promoters and Pd on activity and stability of Ni/YSZ (ScSZ) cermets as anode materials for IT SOFC. *Catal. Today*, **131**, 226–237.

122. Yoon, S.P., Han, J., Nam, S.W., Lim, T.-H., and Hong, S.-A. (2004) Improvement of anode performance by surface modification for solid oxide fuel cell running on hydrocarbon fuel. *J. Power Sources*, **136**, 30–36.

123. Jiang, S.P., Wang, W., and Zhen, Y.D. (2005) Performance and electrode behaviour of nano-YSZ impregnated

nickel anodes used in solid oxide fuel cells. *J. Power Sources*, **147**, 1–7.
124. Jin, Y., Levy, C., Saito, H., Hasegawa, S., Yamahara, K., Hanamura, K., and Ihara, M. (2007) Electrochemical characteristics of anode with $SrZr_{0.95}Y_{0.05}O_{3-\alpha}$ for SOFC in dry methane fuel. *ECS Trans.*, **7**, 1753–1760.
125. Wang, W., Jiang, S.P., Tok, A.I.Y., and Luo, L. (2006) GDC-impregnated Ni anodes for direct utilization of methane in solid oxide fuel cells. *J. Power Sources*, **159**, 68–72.
126. Ding, D., Liu, Z.B., Li, L., and Xia, C.R. (2008) An octane-fueled low temperature solid oxide fuel cell with Ru-free anodes. *Electrochem. Commun.*, **10**, 1295–1298.
127. Zhan, Z.L., Lin, Y.B., Pillai, M., Kim, I., and Barnett, S.A. (2006) High-rate electrochemical partial oxidation of methane in solid oxide fuel cells. *J. Power Sources*, **161**, 460–465.
128. Ye, X.F., Wang, S.R., Wang, Z.R., Xiong, L., Sun, X.E., and Wen, T.L. (2008) Use of a catalyst layer for anode-supported SOFCs running on ethanol fuel. *J. Power Sources*, **177**, 419–425.
129. An, S., Lu, C., Worrell, W.L., Gorte, R.J., and Vohs, J.M. (2004) Characterization of Cu-CeO_2 direct hydrocarbon anodes in a solid oxide fuel cell with lanthanum gallate electrolyte. *Solid State Ionics*, **175**, 135–138.
130. McIntosh, S., Vohs, J.M., and Gorte, R.J. (2003) Effect of precious-metal dopants on SOFC anodes for direct utilization of hydrocarbons. *Electrochem. Solid-State Lett.*, **6**, A240–A243.
131. Gorte, R.J., Park, S., Vohs, J.M., and Wang, C.H. (2000) Anodes for direct oxidation of dry hydrocarbons in a solid-oxide fuel cell. *Adv. Mater.*, **12**, 1465–1469.
132. Park, S.D., Vohs, J.M., and Gorte, R.J. (2000) Direct oxidation of hydrocarbons in a solid-oxide fuel cell. *Nature*, **404**, 265–267.
133. McIntosh, S., Vohs, J.M., and Gorte, R.J. (2002) An examination of lanthanide additives on the performance of Cu-YSZ cermet anodes. *Electrochim. Acta*, **47**, 3815–3821.
134. Gross, M.D., Vohs, J.M., and Gorte, R.J. (2006) Enhanced thermal stability of Cu-based SOFC anodes by electrodeposition of Cr. *J. Electrochem. Soc.*, **153**, A1386–A1390.
135. Jung, S.W., Lu, C., He, H.P., Ahn, K.Y., Gorte, R.J., and Vohs, J.M. (2006) Influence of composition and Cu impregnation method on the performance of Cu/CeO_2/YSZ SOFC anodes. *J. Power Sources*, **154**, 42–50.
136. Goodenough, J.B. and Huang, Y.H. (2007) Alternative anode materials for solid oxide fuel cells. *J. Power Sources*, **173**, 1–10.
137. Fergus, J.W. (2006) Oxide anode materials for solid oxide fuel cells. *Solid State Ionics*, **177**, 1529–1541.
138. Steele, B.C.H. (1999) Fuel-cell technology – running on natural gas. *Nature*, **400**, 619–621.
139. Marina, O.A. and Mogensen, M. (1999) High-temperature conversion of methane on a composite gadolinia-doped ceria-gold electrode. *Appl. Catal. Gen.*, **189**, 117–126.
140. Tu, H., Apfel, H., and Stimming, U. (2006) Performance of alternative oxide anodes for the electrochemical oxidation of hydrogen and methane in solid oxide fuel cells. *Fuel Cells*, **6**, 303–306.
141. Uchida, H., Suzuki, S., and Watanabe, M. (2003) High performance electrode for medium-temperature solid oxide fuel cells – mixed conducting ceria-based anode with highly dispersed Ni electrocatalysts. *Electrochem. Solid-State Lett.*, **6**, A174–A177.
142. Putna, E.S., Stubenrauch, J., Vohs, J.M., and Gorte, R.J. (1995) Ceria-based anodes for the direct oxidation of methane in solid oxide fuel cells. *Langmuir*, **11**, 4832–4837.
143. Uchida, H., Suzuki, H., and Watanabe, M. (1998) High-performance electrode for medium-temperature solid oxide fuel cells – effects of composition and microstructures on performance of ceria-based anodes. *J. Electrochem. Soc.*, **145**, 615–620.
144. Uchida, H., Osuga, T., and Watanabe, M. (1999) High-performance electrode for medium-temperature solid oxide fuel cell control of microstructure of

ceria-based anodes with highly dispersed ruthenium electrocatalysts. *J. Electrochem. Soc.*, **146**, 1677–1682.

145. Kim, G., Vohs, J.M., and Gorte, R.J. (2008) Enhanced reducibility of ceria-YSZ composites in solid oxide electrodes. *J. Mater. Chem.*, **18**, 2386–2390.

146. Jiang, S.P., Chen, X.J., Chan, S.H., Kwok, J.T., and Khor, K.A. (2006) $(La_{0.75}Sr_{0.25})(Cr_{0.5}Mn_{0.5})_{O\text{-}3}$/YSZ composite anodes for methane oxidation reaction in solid oxide fuel cells. *Solid State Ionics*, **177**, 149–157.

147. Liu, J., Madsen, B.D., Ji, Z.Q., and Barnett, S.A. (2002) A fuel-flexible ceramic-based anode for solid oxide fuel cells. *Electrochem. Solid-State Lett.*, **5**, A122–A124.

148. Slater, P.R., Fagg, D.P., and Irvine, J.T.S. (1997) Synthesis and electrical characterisation of doped perovskite titanates as potential anode materials for solid oxide fuel cells. *J. Mater. Chem.*, **7**, 2495–2498.

149. Canales-Vázquez, J., Tao, S.W., and Irvine, J.T.S. (2003) Electrical properties in $La_2Sr_4Ti_6O_{19-\delta}$: a potential anode for high temperature fuel cells. *Solid State Ionics*, **159**, 159–165.

150. Hui, S.Q. and Petric, A. (2002) Electrical properties of yttrium-doped strontium titanate under reducing conditions. *J. Electrochem. Soc.*, **149**, J1–J10.

151. Ruiz-Morales, J.C., Canales-Vazquez, J., Savaniu, C., Marrero-Lopez, D., Zhou, W.Z., and Irvine, J.T.S. (2006) Disruption of extended defects in solid oxide fuel cell anodes for methane oxidation. *Nature*, **439**, 568–571.

152. Huang, Y.H., Dass, R.I., Denyszyn, J.C., and Goodenough, J.B. (2006) Synthesis and characterization of $Sr_2MgMoO_{6-\delta}$ – an anode material for the solid oxide fuel cell. *J. Electrochem. Soc.*, **153**, A1266–A1272.

153. Ji, Y., Huang, Y.H., Ying, J.R., and Goodenough, J.B. (2007) Electrochemical performance of La-doped $Sr_2MgMoO_{6-\delta}$ in natural gas. *Electrochem. Commun.*, **9**, 1881–1885.

154. Jiang, S.P., Liu, L., Ong, K.P., Wu, P., Li, J., and Pu, J. (2008) Electrical conductivity and performance of doped $LaCrO_3$ perovskite oxides for solid oxide fuel cells. *J. Power Sources*, **176**, 82–89.

155. Primdahl, S., Hansen, J.R., Grahl-Madsen, L., and Larsen, P.H. (2001) Sr-doped $LaCrO_3$ anode for solid oxide fuel cells. *J. Electrochem. Soc.*, **148**, A74–A81.

156. Canales-Vazquez, J., Tao, S.W., and Irvine, J.T.S. (2003) Electrical properties in $La_2Sr_4Ti_6O_{19-\delta}$: a potential anode for high temperature fuel cells. *Solid State Ionics*, **159**, 159–165.

157. Holtappels, P., Bradley, J., Irvine, J.T.S., Kaiser, A., and Mogensen, M. (2001) Electrochemical characterization of ceramic SOFC anodes. *J. Electrochem. Soc.*, **148**, A923–A929.

158. van den Bossche, M., Matthews, R., Lichtenberger, A., and McIntosh, S. (2010) Insights into the fuel oxidation mechanism of $La_{0.75}Sr_{0.25}Cr_{0.5}Mn_{0.5}O_{3-\delta}$ SOFC anodes. *J. Electrochem. Soc.*, **157**, B392–B399.

159. Jiang, S.P., Chen, X.J., Chan, S.H., and Kwok, J.T. (2006) GDC-impregnated, $(La_{0.75}Sr_{0.25})(Cr_{0.5}Mn_{0.5})O_3$ anodes for direct utilization of methane in solid oxide fuel cells. *J. Electrochem. Soc.*, **153**, A850–A856.

160. Jiang, S.P., Ye, Y.M., He, T.M., and Ho, S.B. (2008) Nanostructured palladium-$La_{0.75}Sr_{0.25}Cr_{0.5}Mn_{0.5}O_3/Y_2O_3\text{-}ZrO_2$ composite anodes for direct methane and ethanol solid oxide fuel cells. *J. Power Sources*, **185**, 179–182.

161. Ye, Y.M., He, T.M., Li, Y., Tang, E.H., Reitz, T.L., and Jiang, S.P. (2008) Pd-promoted $La_{0.75}Sr_{0.25}Cr_{0.5}Mn_{0.5}O_3$/YSZ composite anodes for direct utilization of methane in SOFCs. *J. Electrochem. Soc.*, **155**, B811–B818.

162. Trembly, J.P., Marquez, A.I., Ohrn, T.R., and Bayless, D.J. (2006) Effects of coal syngas and H_2S on the performance of solid oxide fuel cells: single-cell tests. *J. Power Sources*, **158**, 263–273.

163. Peterson, D.R. and Winnick, J. (1998) Utilization of hydrogen sulfide in an intermediate-temperature ceria-based

solid oxide fuel cell. *J. Electrochem. Soc.*, **145**, 1449–1454.

164. Gong, M.Y., Liu, X.B., Trembly, J., and Johnson, C. (2007) Sulfur-tolerant anode materials for solid oxide fuel cell application. *J. Power Sources*, **168**, 289–298.

165. Burke, A., Winnick, J., Xia, C.R., and Liu, M.L. (2002) Removal of hydrogen sulfide from a fuel gas stream by electrochemical membrane separation. *J. Electrochem. Soc.*, **149**, D160–D166.

166. Alexander, S.R. and Winnick, J. (1994) Electrochemical polishing of hydrogen-sulfide from coal synthesis gas. *J. Appl. Electrochem.*, **24**, 1092–1101.

167. Flytzani-Stephanopoulos, M., Sakbodin, M., and Wang, Z. (2006) Regenerative adsorption and removal of H_2S from hot fuel gas streams by rare earth oxides. *Science*, **312**, 1508–1510.

168. Trembly, J.P., Gemmen, R.S., and Bayless, D.J. (2007) The effect of coal syngas containing HCl on the performance of solid oxide fuel cells: investigations into the effect of operational temperature and HCl concentration. *J. Power Sources*, **169**, 347–354.

169. Matsuzaki, Y. and Yasuda, I. (2000) The poisoning effect of sulfur-containing impurity gas on a SOFC anode: part I. Dependence on temperature, time, and impurity concentration. *Solid State Ionics*, **132**, 261–269.

170. Lohsoontorn, R., Brett, D.J.L., and Brandon, N.P. (2008) The effect of fuel composition and temperature on the interaction of H_2S with nickel-ceria anodes for solid oxide fuel cells. *J. Power Sources*, **183**, 232–239.

171. Brightman, E., Ivey, D.G., Brett, D.J.L., and Brandon, N.P. (2011) The effect of current density on H_2S-poisoning of nickel-based solid oxide fuel cell anodes. *J. Power Sources*, **196**, 7182–7187.

172. Yoshizumi, T., Uryu, C., Oshima, T., Shiratoria, Y., Itoa, K., and Sasaki, K. (2011) Sulfur poisoning of SOFCs: dependence on operational parameters. *ECS Trans.*, **35**, 1717–1725.

173. Sasaki, K., Susuki, K., Iyoshi, A., Uchimura, M., Imamura, N., Kusaba, H., Teraoka, Y., Fuchino, H., Tsujimoto, K., Uchida, Y., and Jingo, N. (2006) H_2S poisoning of solid oxide fuel cells. *J. Electrochem. Soc.*, **153**, A2023–A2029.

174. Kuhn, J.N., Lakshminarayanan, N., and Ozkan, U.S. (2008) Effect of hydrogen sulfide on the catalytic activity of Ni-YSZ cermets. *J. Mol. Catal. A: Chem.*, **282**, 9–21.

175. Cheng, Z., Wang, J.H., Choi, Y.M., Yang, L., Lin, M.C., and Liu, M.L. (2011) From Ni-YSZ to sulfur-tolerant anode materials for SOFCs: electrochemical behavior, in situ characterization, modeling, and future perspectives. *Energy Environ. Sci.*, **4**, 4380–4409.

176. Rasmussen, J.F.B. and Hagen, A. (2009) The effect of H2S on the performance of Ni-YSZ anodes in solid oxide fuel cells. *J. Power Sources*, **191**, 534–541.

177. Zhang, L., Jiang, S.P., He, H.Q., Chen, X.B., Ma, J., and Song, X.C. (2010) A comparative study of H_2S poisoning on electrode behavior of Ni/YSZ and Ni/GDC anodes of solid oxide fuel cells. *Int. J. Hydrogen Energy*, **35**, 12359–12368.

178. Ishikura, A., Sakuno, S., Komiyama, N., Sasatsu, H., Masuyama, N., Itoh, H., and Yasumoto, K. (2007) *Solid Oxide Fuel Cells 10 (SOFC-X)*, Vol. **7** Parts 1 and 2, The Eletrochemical Society, Pennington, NJ, pp. 845–850.

179. Cheng, Z. and Liu, M.L. (2007) Characterization of sulfur poisoning of Ni-YSZ anodes for solid oxide fuel cells using in situ raman micro spectroscopy. *Solid State Ionics*, **178**, 925–935.

180. Wang, J.H. and Liu, M.L. (2007) Computational study of sulfur-nickel interactions: a new S-Ni phase diagram. *Electrochem. Commun.*, **9**, 2212–2217.

181. Cheng, Z., Abernathy, H., and Liu, M.L. (2007) Raman spectroscopy of nickel sulfide Ni_3S_2. *J. Phys. Chem. C*, **111**, 17997–18000.

182. Wang, J.H., Cheng, Z., Bredas, J.L., and Liu, M.L. (2007) Electronic and vibrational properties of nickel sulfides

from first principles. *J. Chem. Phys.*, **127**, 214705.

183. Choi, Y.M., Compson, C., Lin, M.C., and Liu, M.L. (2007) Ab initio analysis of sulfur tolerance of Ni, Cu, and Ni-Cu alloys for solid oxide fuel cells. *J. Alloys Compd.*, **427**, 25–29.
184. Shiratori, Y., Oshima, T., and Sasaki, K. (2008) Feasibility of direct-biogas SOFC. *Int. J. Hydrogen Energy*, **33**, 6316–6321.
185. Ouweltjes, J.P., Aravind, P.V., Woudstra, N., and Rietveld, G. (2006) Biosyngas utilization in solid oxide fuel cells with Ni/GDC anodes. *J. Fuel Cell Sci. Technol.*, **3**, 495–498.
186. Bartholomew, C.H. (2001) Mechanisms of catalyst deactivation. *Appl. Catal. Gen.*, **212**, 17–60.
187. Yang, L., Wang, S.Z., Blinn, K., Liu, M.F., Liu, Z., Cheng, Z., and Liu, M.L. (2009) Enhanced sulfur and coking tolerance of a mixed Ion conductor for SOFCs: $BaZr_{0.1}Ce_{0.7}Y_{0.2-x}Yb_xO_{3-\delta}$. *Science*, **326**, 126–129.
188. Kurokawa, H., Sholklapper, T.Z., Jacobson, C.P., De Jonghe, L.C., and Visco, S.J. (2007) Ceria nanocoating for sulfur tolerant Ni-based anodes of solid oxide fuel cells. *Electrochem. Solid-State Lett.*, **10**, B135–B138.
189. Choi, S.H., Wang, J.H., Cheng, Z., and Liu, M. (2008) Surface modification of Ni-YSZ using niobium oxide for sulfur-tolerant anodes in solid oxide fuel cells. *J. Electrochem. Soc.*, **155**, B449–B454.
190. Marina, O.A., Coyle, C.A., Engelhard, M.H., and Pederson, L.R. (2011) Mitigation of sulfur poisoning of Ni/zirconia SOFC anodes by antimony and Tin. *J. Electrochem. Soc.*, **158**, B424–B429.
191. Zheng, L.L., Wang, X., Zhang, L., Wang, J.Y., and Jiang, S.P. (2012) Effect of Pd-impregnation on performance, sulfur poisoning and tolerance of Ni/GDC anode of solid oxide fuel cells. *Int. J. Hydrogen Energy*, **37**, 10299–10310.
192. Kim, H., Vohs, J.M., and Gorte, R.J. (2001) Direct oxidation of sulfur-containing fuels in a solid oxide fuel cell. *Chem. Commun.*, 2334–2335.
193. He, H.P., Gorte, R.J., and Vohs, J.M. (2005) Highly sulfur tolerant Cu-ceria anodes for SOFCs. *Electrochem. Solid-State Lett.*, **8**, A279–A280.
194. Aguilar, L., Zha, S., Cheng, Z., Winnick, J., and Liu, M. (2004) A solid oxide fuel cell operating on hydrogen sulfide (H_2S) and sulfur-containing fuels. *J. Power Sources*, **135**, 17–24.
195. Cooper, M., Channa, K., De Silva, R., and Bayless, D.J. (2010) Comparison of LSV/YSZ and LSV/GDC SOFC anode performance in coal syngas containing H_2S. *J. Electrochem. Soc.*, **157**, B1713–B1718.
196. Huang, Y.H., Dass, R.I., Xing, Z.L., and Goodenough, J.B. (2006) Double perovskites as anode materials for solid-oxide fuel cells. *Science*, **312**, 254–257.
197. Zha, S.W., Cheng, Z., and Liu, M.L. (2005) A sulfur-tolerant anode material for SOFCs$Gd_2Ti_{1.4}Mo_{0.6}O_7$. *Electrochem. Solid-State Lett.*, **8**, A406–A408.
198. Mukundan, R., Brosha, E.L., and Garzon, F.H. (2004) Sulfur tolerant anodes for SOFCs. *Electrochem. Solid-State Lett.*, **7**, A5–A7.
199. Kurokawa, H., Yang, L., Jacobson, C.P., De Jonghe, L.C., and Visco, S.J. (2007) Y-doped $SrTiO_3$ based sulfur tolerant anode for solid oxide fuel cells. *J. Power Sources*, **164**, 510–518.
200. Pillai, M.R., Kim, I., Bierschenk, D.M., and Barnett, S.A. (2008) Fuel-flexible operation of a solid oxide fuel cell with $Sr_{0.8}La_{0.2}TiO_3$ support. *J. Power Sources*, **185**, 1086–1093.
201. Kivisaari, T., Björnbom, P., Sylwan, C., Jacquinot, B., Jansen, D., and de Groot, A. (2004) The feasibility of a coal gasifier combined with a high-temperature fuel cell. *Chem. Eng. J.*, **100**, 167–180.
202. Yi, Y., Rao, A.D., Brouwer, J., and Samuelsen, G.S. (2005) Fuel flexibility study of an integrated 25 kW SOFC reformer system. *J. Power Sources*, **144**, 67–76.
203. Verma, A., Rao, A.D., and Samuelsen, G.S. (2006) Sensitivity analysis of a vision 21 coal based zero emission power plant. *J. Power Sources*, **158**, 417–427.
204. Marina, O., Pederson, L.R., Edwards, D.J., Coyle, C.A., Templeton, J.,

Engelhard, M.H., and Zhu, Z. (2008) Effect of coal gas contaminants on solid oxide fuel cell operation. *ECS Trans.*, **11**, 63–70.

205. Cayan, F.N., Zhi, M.J., Pakalapati, S.R., Celik, I., Wu, N.Q., and Gemmen, R. (2008) Effects of coal syngas impurities on anodes of solid oxide fuel cells. *J. Power Sources*, **185**, 595–602.
206. Martinez, A., Gerdes, K., Gemmen, R., and Poston, J. (2010) Thermodynamic analysis of interactions between Ni-based solid oxide fuel cells (SOFC) anodes and trace species in a survey of coal syngas. *J. Power Sources*, **195**, 5206–5212.
207. Haga, K., Shiratori, Y., Nojiri, Y., Ito, K., and Sasaki, K. (2010) Phosphorus poisoning of Ni-cermet anodes in solid oxide fuel cells. *J. Electrochem. Soc.*, **157**, B1693–B1700.
208. Zhi, M.J., Chen, X.Q., Finklea, H., Celik, I., and Wu, N.Q.Q. (2008) Electrochemical and microstructural analysis of nickel-yttria-stabilized zirconia electrode operated in phosphorus-containing syngas. *J. Power Sources*, **183**, 485–490.
209. Gong, M.Y., Bierschenk, D., Haag, J., Poeppelmeier, K.R., Barnett, S.A., Xu, C.C., Zondlo, J.W., and Liu, X.B. (2010) Degradation of $LaSr_2Fe_2CrO_{9-\delta}$ solid oxide fuel cell anodes in phosphine-containing fuels. *J. Power Sources*, **195**, 4013–4021.
210. De Silva, K.C.R., Kaseman, B.J., and Bayless, D.J. (2011) Accelerated anode failure of a high temperature planar SOFC operated with reduced moisture and increased PH3 concentrations in coal syngas. *Int. J. Hydrogen Energy*, **36**, 9945–9955.
211. Marina, O.A., Coyle, C.A., Thomsen, E.C., Edwards, D.J., Coffey, G.W., and Pederson, L.R. (2010) Degradation mechanisms of SOFC anodes in coal gas containing phosphorus. *Solid State Ionics*, **181**, 430–440.
212. Liu, W.N., Sun, X., Pederson, L.R., Marina, O.A., and Khaleel, M.A. (2010) Effect of nickel-phosphorus interactions on structural integrity of anode-supported solid oxide fuel cells. *J. Power Sources*, **195**, 7140–7145.
213. Haga, K., Shiratori, Y., Ito, K., and Sasaki, K. (2008) Chlorine poisoning of SOFC Ni-cermet anodes. *J. Electrochem. Soc.*, **155**, B1233–B1239.
214. Xu, C., Gong, M., Zondlo, J.W., Liu, X., and Finklea, H.O. (2010) The effect of HCl in syngas on Ni-YSZ anode-supported solid oxide fuel cells. *J. Power Sources*, **195**, 2149–2158.
215. Marina, O.A., Pederson, L.R., Thomsen, E.C., Coyle, C.A., and Yoon, K.J. (2010) Reversible poisoning of nickel/zirconia solid oxide fuel cell anodes by hydrogen chloride in coal gas. *J. Power Sources*, **195**, 7033–7037.
216. Marina, O.A., Pederson, L.R., Coyle, C.A., Thomsen, E.C., and Edwards, D.J. (2011) Polarization-induced interfacial reactions between nickel and selenium in Ni/zirconia SOFC anodes and comparison with sulfur poisoning. *J. Electrochem. Soc.*, **158**, B36–B43.
217. Marina, O.A., Pederson, L.R., Coyle, C.A., Thomsen, E.C., Nachimuthu, P., and Edwards, D.J. (2011) Electrochemical, structural and surface characterization of nickel/zirconia solid oxide fuel cell anodes in coal gas containing antimony. *J. Power Sources*, **196**, 4911–4922.
218. Bao, J., Krishnan, G.N., Jayaweera, P., Perez-Mariano, J., and Sanjurjo, A. (2009) Effect of various coal contaminants on the performance of solid oxide fuel cells: Part I. Accelerated testing.. *J. Power Sources*, **193**, 607–616.
219. Bao, J.E., Krishnan, G.N., Jayaweera, P., and Sanjurjo, A. (2010) Effect of various coal gas contaminants on the performance of solid oxide fuel cells: Part III. Synergistic effects. *J. Power Sources*, **195**, 1316–1324.
220. Vels Jensen, K., Primdahl, S., Chorkendorff, I., and Mogensen, M. (2001) Microstructural and chemical changes at the Ni/YSZ interface. *Solid State Ionics*, **144**, 197–209.
221. Jensen, K.V., Wallenberg, R., Chorkendorff, I., and Mogensen, M. (2003) Effect of impurities on structural and electrochemical properties of the Ni-YSZ interface. *Solid State Ionics*, **160**, 27–37.

222. Schmidt, M.S., Hansen, K.V., Norrman, K., and Mogensen, M. (2008) Effects of trace elements at the Ni/ScYSZ interface in a model solid oxide fuel cell anode. *Solid State Ionics*, **179**, 1436–1441.

223. Liu, Y.L., Primdahl, S., and Mogensen, M. (2003) Effects of impurities on microstructure in Ni/YSZ-YSZ half-cells for SOFC. *Solid State Ionics*, **161**, 1–10.

224. Primdahl, S. and Mogensen, M. (2000) Durability and thermal cycling of Ni/YSZ cermet anodes for solid oxide fuel cells. *J. Appl. Electrochem.*, **30**, 247–257.

225. Liu, Y.L. and Jiao, C.G. (2005) Microstructure degradation of an anode/electrolyte interface in SOFC studied by transmission electron microscopy. *Solid State Ionics*, **176**, 435–442.

226. Gentile, P.S. and Sofie, S.W. (2011) Investigation of aluminosilicate as a solid oxide fuel cell refractory. *J. Power Sources*, **196**, 4545–4554.

227. Haga, K., Adachi, S., Shiratori, Y., Itoh, K., and Sasaki, K. (2008) Poisoning of SOFC anodes by various fuel impurities. *Solid State Ionics*, **179**, 1427–1431.

228. Brugnoni, C., Ducati, U., and Scagliotti, M. (1995) SOFC cathode/electrolyte interface. Part I: reactivity between $La_{0.85}Sr_{0.15}MnO_3$ and ZrO_2-Y_2O_3. *Solid State Ionics*, **76**, 177–182.

229. Willy Poulsen, F. and van der Puil, N. (1992) Phase relations and conductivity of Sr- and La-zirconates. *Solid State Ionics*, **53–56**, 777–783.

230. Lee, H.Y. and Oh, S.M. (1996) Origin of cathodic degradation and new phase formation at the $La_{0.9}Sr_{0.1}MnO_3$/YSZ interface. *Solid State Ionics*, **90**, 133–140.

231. Mitterdorfer, A. and Gauckler, L.J. (1998) $La_2Zr_2O_7$ formation and oxygen reduction kinetics of the $La_{0.85}Sr_{0.15}MnyO_3$, O_2(g)vertical bar YSZ system. *Solid State Ionics*, **111**, 185–218.

232. Jiang, S.P., Zhang, J.P., Ramprakash, Y., Milosevic, D., and Wilshier, K. (2000) An investigation of shelf-life of strontium doped $LaMnO_3$ materials. *J. Mater. Sci.*, **35**, 2735–2741.

233. Yokokawa, H., Sakai, N., Kawada, T., and Dokiya, M. (1991) Thermodynamic analysis of reaction profiles between Lamo3 (M=Ni,Co,Mn) and Zro_2. *J. Electrochem. Soc.*, **138**, 2719–2727.

234. Yang, C.C.T., Wei, W.C.J., and Roosen, A. (2003) Electrical conductivity and microstructures of $La_{0.65}Sr_{0.3}MnO_{3-8}$ mol% yttria-stabilized zirconia. *Mater. Chem. Phys.*, **81**, 134–142.

235. Haanappel, V.A.C., Mai, A., and Mertens, J. (2006) Electrode activation of anode-supported SOFCs with LSM-or LSCF-type cathodes. *Solid State Ionics*, **177**, 2033–2037.

236. Sahibzada, M., Benson, S.J., Rudkin, R.A., and Kilner, J.A. (1998) Pd-promoted $La_{0.6}Sr_{0.4}Co_{0.2}Fe_{0.8}O_3$ cathodes. *Solid State Ionics*, **113**, 285–290.

237. Murray, E.P., Sever, M.J., and Barnett, S.A. (2002) Electrochemical performance of (La,Sr)(Co,Fe)O_3-(Ce,Gd)O_3 composite cathodes. *Solid State Ionics*, **148**, 27–34.

238. Shao, Z.P. and Haile, S.M. (2004) A high-performance cathode for the next generation of solid-oxide fuel cells. *Nature*, **431**, 170–173.

239. Yokokawa, H. (2003) Understanding materials compatibility. *Ann. Rev. Mater. Res.*, **33**, 581–610.

240. Mai, A., Becker, M., Assenmacher, W., Tietz, F., Hathiramani, D., Ivers-Tiffee, E., Stover, D., and Mader, W. (2006) Time-dependent performance of mixed-conducting SOFC cathodes. *Solid State Ionics*, **177**, 1965–1968.

241. Zhu, Q.S., Jin, T.A., and Wang, Y. (2006) Thermal expansion behavior and chemical compatibility of $BaxSr_{1-x}Co_{1-y}FeyO_{3-}$delta with 8YSZ and 20GDC. *Solid State Ionics*, **177**, 1199–1204.

242. Yung, H., Jian, L., and Jiang, S.P. (2012) Polarization promoted chemical reaction between $Ba_{0.5}Sr_{0.5}C0_{0.8}Fe_{0.2}O_3$ Cathode and ceria based electrolytes of solid oxide fuel cells. *J. Electrochem. Soc.*, **159**, F794–F798.

243. Tietz, F., Haanappel, V.A.C., Mai, A., Mertens, J., and Stover, D. (2006)

Performance of LSCF cathodes in cell tests. *J. Power Sources*, **156**, 20–22.

244. Simner, S.P., Anderson, M.D., Engelhard, M.H., and Stevenson, J.W. (2006) Degradation mechanisms of La-Sr-Co-Fe-O_3SOFC cathodes. *Electrochem. Solid-State Lett.*, **9**, A478–A481.

245. Kostogloudis, G.C. and Ftikos, C. (1999) Properties of a-site-deficient $La_{0.6}Sr_{0.4}Co_{0.2}Fe_{0.8}O_{3-\delta}$-based perovskite oxides. *Solid State Ionics*, **126**, 143–151.

246. Wei, B., Lu, Z., Huang, X.Q., Miao, J.P., Sha, X.Q., Xin, X.S., and Su, W.H. (2006) Crystal structure, thermal expansion and electrical conductivity of perovskite oxides $Ba_xSr_{1-x}Co_{0.8}Fe_{0.2}O_{3-\delta}$ ($0.3 \leq x \leq 0.7$). *J. Eur. Ceram. Soc.*, **26**, 2827–2832.

247. Steele, B.C.H. (1995) Interfacial reactions associated with ceramic ion transport membranes. *Solid State Ionics*, **75**, 157–165.

248. Wu, Q.H., Liu, M.L., and Jaegermann, W. (2005) X-ray photoelectron spectroscopy of $La_{0.5}Sr_{0.5}MnO_3$. *Mater. Lett.*, **59**, 1980–1983.

249. Dulli, H., Dowben, P.A., Liou, S.H., and Plummer, E.W. (2000) Surface segregation and restructuring of colossal-magnetoresistant manganese perovskites $La_{0.65}Sr_{0.35}MnO_3$. *Phys. Rev. B*, **62**, 14629–14632.

250. de Jong, M.P., Dediu, V.A., Taliani, C., and Salaneck, W.R. (2003) Electronic structure of $La_{0.7}Sr_{0.3}MnO_3$ Thin films for hybrid organic/inorganic spintronics applications. *J. Appl. Phys.*, **94**, 7292–7296.

251. Katsiev, K., Yildiz, B., Balasubramaniam, K., and Salvador, P.A. (2009) Electron tunneling characteristics on $La_{0.7}Sr_{0.3}MnO_3$ Thin-film surfaces at high temperature. *Appl. Phys. Lett.*, **95**, 92106.

252. Vovk, G., Chen, X., and Mims, C.A. (2005) In situ XPS studies of perovskite oxide surfaces under electrochemical polarization. *J. Phys. Chem. B*, **109**, 2445–2454.

253. Norman, C. and Leach, C. (2011) In situ high temperature X-ray photoelectron spectroscopy study of barium strontium iron cobalt oxide. *J. Membr. Sci.*, **382**, 158–165.

254. Kim, Y.M., Chen, X.B., Jiang, S.P., and Bae, J. (2012) Effect of strontium content on chromium deposition and poisoning in $Ba_{1-x}Sr_xCo_{0.8}Fe_{0.2}O_{3-\delta}$ ($0.3 \leq x \leq 0.7$) cathodes of solid oxide fuel cells. *J. Electrochem. Soc.*, **159**, B185–B194.

255. Fang, S.M., Yoo, C.Y., and Bouwmeester, H.J.M. (2011) Performance and stability of niobium-substituted $Ba_{0.5}Sr_{0.5}Co_{0.8}Fe_{0.2}O_{3-\delta}$ membranes. *Solid State Ionics*, **195**, 1–6.

256. Jorgensen, M.J., Holtappels, P., and Appel, C.C. (2000) Durability test of SOFC cathodes. *J. Appl. Electrochem.*, **30**, 411–418.

257. Jiang, S.P. and Wang, W. (2005) Sintering and grain growth of (La,Sr)MnO_3 electrodes of solid oxide fuel cells under polarization. *Solid State Ionics*, **176**, 1185–1191.

258. Sholklapper, T.Z., Jacobson, C.P., Visco, S.J., and De Jonghe, L.C. (2008) Synthesis of dispersed and contiguous nanoparticles in solid oxide fuel cell electrodes. *Fuel Cells*, **8**, 303–312.

259. Jiang, S.P. (2006) A review of wet impregnation – an alternative method for the fabrication of high performance and nano-structured electrodes of solid oxide fuel cells. *Mater. Sci. Eng., A*, **418**, 199–210.

260. Huang, Y.Y., Ahn, K., Vohs, J.M., and Gorte, R.J. (2004) Characterization of Sr-doped $LaCoO_3$-YSZ composites prepared by impregnation methods. *J. Electrochem. Soc.*, **151**, A1592–A1597.

261. Peters, C., Weber, A., and Ivers-Tiffee, E. (2008) Nanoscaled $(La_{0.5}Sr_{0.5})CoO_{3-\delta}$ thin film cathodes for SOFC application at $500\,^{\circ}C < T < 700\,^{\circ}C$. *J. Electrochem. Soc.*, **155**, B730–B737.

262. Ai, N., Jiang, S.P., Lu, Z., Chen, K.F., and Su, W.H. (2010) Nanostructured (Ba,Sr)(Co,Fe)$O_{3-\delta}$ impregnated (La,Sr)MnO_3 cathode for intermediate-temperature solid oxide fuel cells. *J. Electrochem. Soc.*, **157**, B1033–B1039.

263. Kungas, R., Bidrawn, F., Vohs, J.M., and Gorte, R.J. (2010) Doped-ceria diffusion barriers prepared by infiltration for solid oxide fuel cells. *Electrochem. Solid-State Lett.*, **13**, B87–B90.

264. Shah, M., Hughes, G., Voorhees, P.W., and Barnett, S.A. (2011) Stability and performance of LSCF-infiltrated SOFC cathodes: effect of nano-particle coarsening. *ECS Trans.*, **35**, 2045–2053.

265. Yoon, S.P., Han, J., Nam, S.W., Lim, T.H., Oh, I.H., Hong, S.A., Yoo, Y.S., and Lim, H.C. (2002) Performance of anode-supported solid oxide fuel cell with $La_{0.85}Sr_{0.15}MnO_3$ Cathode modified by sol-gel coating technique. *J. Power Sources*, **106**, 160–166.

266. Wang, W.S., Gross, M.D., Vohs, J.M., and Gorte, R.J. (2007) The stability of LSF-YSZ electrodes prepared by infiltration. *J. Electrochem. Soc.*, **154**, B439–B445.

267. Chen, K.F., Tian, Y.T., Lu, Z., Ai, N., Huang, X.Q., and Su, W.H. (2009) Behavior of 3 mol% yttria-stabilized tetragonal zirconia polycrystal film prepared by slurry spin coating. *J. Power Sources*, **186**, 128–132.

268. Sholklapper, T.Z., Radmilovic, V., Jacobson, C.P., Visco, S.J., and De Jonghe, L.C. (2007) Synthesis and stability of a nanoparticle-infiltrated solid oxide fuel cell electrode. *Electrochem. Solid-State Lett.*, **10**, B74–B76.

269. Shah, M., Voorhees, P.W., and Barnett, S.A. (2011) Time-dependent performance changes in LSCF-infiltrated SOFC cathodes: the role of nanoparticle coarsening. *Solid State Ionics*, **187**, 64–67.

270. Sholklapper, T.Z., Radmilovic, V., Jacobson, C.P., Visco, S.J., and De Jonghe, L.C. (2008) Nanocomposite Ag-LSM solid oxide fuel cell electrodes. *J. Power Sources*, **175**, 206–210.

271. Liang, F.L., Chen, J., Cheng, J.L., Jiang, S.P., He, T.M., Pu, J., and Li, J. (2008) Novel nano-structured Pd plus yttrium doped ZrO_2 cathodes for intermediate temperature solid oxide fuel cells. *Electrochem. Commun.*, **10**, 42–46.

272. Brunetti, A., Barbieri, G., and Drioli, E. (2011) Integrated membrane system for pure hydrogen production: a Pd-Ag membrane reactor and a PEMFC. *Fuel Process. Technol.*, **92**, 166–174.

273. Liang, F.L., Chen, J., Jiang, S.P., Wang, F.Z., Chi, B., Pu, J., and Jian, L. (2009) Mn-stabilised microstructure and performance of Pd-impregnated YSZ cathode for intermediate temperature solid oxide fuel cells. *Fuel Cells*, **9**, 636–642.

274. Babaei, A., Zhang, L., Liu, E., and Jiang, S.P. (2011) Performance and stability of $La_{0.8}Sr_{0.2}MnO_3$ Cathode promoted with palladium based catalysts in solid oxide fuel cells. *J. Alloys Compd.*, **509**, 4781–4787.

275. Kim, J.-S., Wieder, N.L., Abraham, A.J., Cargnello, M., Fornasiero, P., Gorte, R.J., and Vohs, J.M. (2011) Highly active and thermally stable core-shell catalysts for solid oxide fuel cells. *J. Electrochem. Soc.*, **158**, B596–B600.

276. Lee, D., Jung, I., Lee, S.O., Hyun, S.H., Jang, J.H., and Moon, J. (2011) Durable high-performance $Sm_{0.5}Sr_{0.5}CoO_3$-$Sm_{0.2}Ce_{0.8}O_{1.9}$ core-shell type composite cathodes for low temperature solid oxide fuel cells. *Int. J. Hydrogen Energy*, **36**, 6875–6881.

277. Zhang, Y.X. and Xia, C.R. (2010) A durability model for solid oxide fuel cell electrodes in thermal cycle processes. *J. Power Sources*, **195**, 6611–6618.

278. Kjellqvist, L. and Selleby, M. (2010) Thermodynamic assessment of the Cr-Mn-O system. *J. Alloys Compd.*, **507**, 84–92.

279. Yang, Z.G. (2008) Recent advances in metallic interconnects for solid oxide fuel cells. *Int. Mater. Rev.*, **53**, 39–54.

280. Taniguchi, S., Kadowaki, M., Kawamura, H., Yasuo, T., Akiyama, Y., Miyake, Y., and Saitoh, T. (1995) Degradation phenomena in the cathode of a solid oxide fuel cell with an alloy separator. *J. Power Sources*, **55**, 73–79.

281. Quadakkers, W.J., Greiner, H., Hansel, M., Pattanaik, A., Khanna, A.S., and Mallener, W. (1996) Compatibility of perovskite contact layers between cathode and metallic interconnector plates of SOFCs. *Solid State Ionics*, **91**, 55–67.

282. Chen, X.B., Zhang, L., Liu, E., and Jiang, S.P. (2011) A fundamental study of chromium deposition and poisoning at $(La_{0.8}Sr_{0.2})_{0.95}(Mn_{1-x}Co_x)O_{3\pm\delta}$

($0.0 \leq x \leq 1.0$) cathodes of solid oxide fuel cells. *Int. J. Hydrogen Energy*, **36**, 805–821.

283. Horita, T., Xiong, Y.P., Yoshinaga, M., Kishimoto, H., Yamaji, K., Brito, M.E., and Yokokawa, H. (2009) Determination of chromium concentration in solid oxide fuel cell cathodes: (La,Sr)MnO_3 and (La,Sr)FeO_3. *Electrochem. Solid-State Lett.*, **12**, B146–B149.

284. Matsuzaki, Y. and Yasuda, I. (2001) Dependence of SOFC cathode degradation by chromium-containing alloy on compositions of electrodes and electrolytes. *J. Electrochem. Soc.*, **148**, A126–A131.

285. Jiang, S.-P., Zhang, J.-P., Apateanu, L., and Foger, K. (1999) Deposition of chromium species on Sr-doped $LaMnO_3$ cathodes in solid oxide fuel cells. *Electrochem. Commun.*, **1**, 394–397.

286. Jiang, S.P., Zhang, J.P., and Foger, K. (2001) Deposition of chromium species at Sr-doped $LaMnO_3$ electrodes in solid oxide fuel cells – III. Effect of air flow. *J. Electrochem. Soc.*, **148**, C447–C455.

287. Chen, X.B., Zhen, Y.D., Li, J., and Jiang, S.P. (2010) Chromium deposition and poisoning in dry and humidified air at $(La_{0.8}Sr_{0.2})_{0.9}MnO_{3+\delta}$ cathodes of solid oxide fuel cells. *Int. J. Hydrogen Energy*, **35**, 2477–2485.

288. Matsuzaki, Y. and Yasuda, I. (2000) Electrochemical properties of a SOFC cathode in contact with a chromium-containing alloy separator. *Solid State Ionics*, **132**, 271–278.

289. Jiang, S.P., Zhang, S., and Zhen, Y.D. (2006) Deposition of Cr species at (La,Sr)(Co,Fe)$O_{[3]}$ cathodes of solid oxide fuel cells. *J. Electrochem. Soc.*, **153**, A127–A134.

290. Konysheva, E., Penkalla, H., Wessel, E., Mertens, J., Seeling, U., Singheiser, L., and Hilpert, K. (2006) Chromium poisoning of perovskite cathodes by the ODS alloy $Cr_5Fe_1Y_{[2]}O_{[3]}$ and the high chromium ferritic steel Crofer22APU. *J. Electrochem. Soc.*, **153**, A765–A773.

291. Jiang, S.P., Zhang, J.P., and Foger, K. (2000) Deposition of chromium species at Sr-doped $LaMnO_3$ electrodes in solid oxide fuel cells – II. Effect on O-2 reduction reaction. *J. Electrochem. Soc.*, **147**, 3195–3205.

292. Jiang, S.P., Zhang, J.P., and Zheng, X.G. (2002) A comparative investigation of chromium deposition at air electrodes of solid oxide fuel cells. *J. Eur. Ceram. Soc.*, **22**, 361–373.

293. Jiang, S.P., Zhang, S., and Zhen, Y.D. (2005) Early interaction between Fe-Cr alloy metallic interconnect and Sr-doped $LaMnO_3$ cathodes of solid oxide fuel cells. *J. Mater. Res.*, **20**, 747–758.

294. Zhen, Y.D., Jiang, S.P., Zhang, S., and Tan, V. (2006) Interaction between metallic interconnect and constituent oxides of (La,Sr)MnO_3 electrodes of solid oxide fuel cells.. *J. Euro. Ceram. Soc.*, **26**, 3253–3264.

295. Chen, X.B., Zhang, L., Liu, E.J., and Jiang, S.P. (2011) A fundamental study of chromium deposition and poisoning at $(La_{0.8}Sr_{0.2})_{0.95}(Mn_{1-X}Co_x)O_{\pm\delta}$ ($0.11 \leq x \leq 1.0$) Cathodes of solid oxide fuel cells. *Int. J. Hydrogen Energy*, **36**, 805–821.

296. Kim, Y.M., Chen, X.B., Jiang, S.P., and Bae, J. (2011) Chromium deposition and poisoning at $Ba_{0.5}Sr_{0.5}Co_{0.8}Fe_{0.2}O_{3-\delta}$ cathode of solid oxide fuel cells. *Electrochem. Solid-State Lett.*, **14**, B41–B45.

297. Zhen, Y.D., Li, J., and Jiang, S.P. (2006) Oxygen reduction on strontium-doped $LaMnO_3$ cathodes in the absence and presence of an iron-chromium alloy interconnect. *J. Power Sources*, **162**, 1043–1052.

298. Matsuzaki, Y. and Yasuda, I. (2002) Electrochemical properties of reduced-temperature SOFCs with mixed ionic-electronic conductors in electrodes and/or interlayers. *Solid State Ionics*, **152-153**, 463–468.

299. Wachsman, E.D., Oh, D., Armstrong, E., Jung, D.W., and Kan, C. (2009) Mechanistic understanding of Cr poisoning on $La_{0.6}Sr_{0.4}Co_{0.2}Fe_{0.8}O_3$ (LSCF). *ECS Trans.*, **25**, 2871–2879.

300. Zhen, Y.D. and Jiang, S.P. (2006) Transition behavior for O_2 reduction reaction on (La,Sr)MnO_3/YSZ composite cathodes of solid oxide fuel cells. *J. Electrochem. Soc.*, **153**, A2245–A2254.

301. Jiang, S.P., Zhen, Y.D., and Zhang, S. (2006) Interaction between Fe-Cr metallic interconnect and (La, Sr) MnO_3/YSZ composite cathode of solid oxide fuel cells. *J. Electrochem. Soc.*, **153**, A1511–A1517.
302. Zhen, Y.D., Tok, A.I.Y., Jiang, S.P., and Boey, F.Y.C. (2007) La(Ni,Fe)O_3 as a cathode material with high tolerance to chromium poisoning for solid oxide fuel cells. *J. Power Sources*, **170**, 61–66.
303. Komatsu, T., Arai, H., Chiba, R., Nozawa, K., Arakawa, M., and Sato, K. (2006) Cr poisoning suppression in solid oxide fuel cells using LaNi(Fe)$O_{[3]}$ electrodes. *Electrochem. Solid-State Lett.*, **9**, A9–A12.
304. Komatsu, T., Yoshida, Y., Watanabe, K., Chiba, R., Taguchi, H., Orui, H., and Arai, H. (2010) Degradation behavior of anode-supported solid oxide fuel cell using LNF cathode as function of current load. *J. Power Sources*, **195**, 5601–5605.
305. Stodolny, M.K., Berkel, F.P.V., and Boukamp, B.A. (2009) La(Ni,Fe)O_3 stability in the presence of Cr species – solid-state reactivity study. *ECS Trans.*, **25**, 2915–2922.
306. Zhen, Y. and Jiang, S.P. (2008) Characterization and performance of (La,Ba)(Co,Fe)O_3 cathode for solid oxide fuel cells with iron-chromium metallic interconnect. *J. Power Sources*, **180**, 695–703.
307. Chen, X.B., Zhang, L., and Jiang, S.P. (2008) Chromium deposition and poisoning on $(La_{0.6}Sr_{0.4\text{-}X}Ba_x)(Co_{0.2}Fe_{0.8})O_3$ $(0 \leq x \leq 0.4)$ cathodes of solid oxide fuel cells. *J. Electrochem. Soc.*, **155**, B1093–B1101.
308. Yang, Z.G., Hardy, J.S., Walker, M.S., Xia, G.G., Simner, S.P., and Stevenson, J.W. (2004) Structure and conductivity of thermally grown scales on ferritic Fe-Cr-Mn steel for SOFC interconnect applications. *J. Electrochem. Soc.*, **151**, A1825–A1831.
309. Simner, S.P., Anderson, M.D., Xia, G.G., Yang, Z., Pederson, L.R., and Stevenson, J.W. (2005) SOFC performance with Fe-Cr-Mn alloy interconnect. *J. Electrochem. Soc.*, **152**, A740–A745.
310. Stanislowski, M., Wessel, E., Hilpert, K., Markus, T., and Singheiser, L. (2007) Chromium vaporization from high-temperature alloys I. Chromia-forming steels and the influence of outer oxide layers. *J. Electrochem. Soc.*, **154**, A295–A306.
311. Chen, X.B., Hua, B., Pu, J., Li, J., Zhang, L., and Jiang, S.P. (2009) Interaction between (La, Sr)MnO_3 cathode and Ni-Mo-Cr metallic interconnect with suppressed chromium vaporization for solid oxide fuel cells. *Int. J. Hydrogen Energy*, **34**, 5737–5748.
312. Fujita, K., Ogasawara, K., Matsuzaki, Y., and Sakurai, T. (2004) Prevention of SOFC cathode degradation in contact with Cr-containing alloy. *J. Power Sources*, **131**, 261–269.
313. Kurokawa, H., Jacobson, C.P., DeJonghe, L.C., and Visco, S.J. (2007) Chromium vaporization of bare and of coated iron-chromium alloys at 1073 K. *Solid State Ionics*, **178**, 287–296.
314. Shaigan, N., Ivey, D.G., and Chen, W. (2008) Co/$LaCrO_3$ composite coatings for AISI 430 stainless steel solid oxide fuel cell interconnects. *J. Power Sources*, **185**, 331–337.
315. Qu, W., Jian, L., Ivey, D.G., and Hill, J.M. (2006) Yttrium, cobalt and yttrium/cobalt oxide coatings on ferritic stainless steels for SOFC interconnects. *J. Power Sources*, **157**, 335–350.
316. Yang, Z.G., Xia, G.G., and Stevenson, J.W. (2005) $Mn_{1.5}Co_{1.5}O_4$ Spinel protection layers on ferritic stainless steels for SOFC interconnect applications. *Electrochem. Solid-State Lett.*, **8**, A168–A170.
317. Montero, X., Tietz, F., Sebold, D., Buchkremer, H.R., Ringuede, A., Cassir, M., Laresgoiti, A., and Villarreal, I. (2008) $MnCo_{1.9}Fe_{0.1}O_4$ spinel protection layer on commercial ferritic steels for interconnect applications in solid oxide fuel cells. *J. Power Sources*, **184**, 172–179.
318. Bi, Z.H., Zhu, J.H., and Batey, J.L. (2010) $CoFe_2O_4$ Spinel protection coating thermally converted from the electroplated Co-Fe alloy for solid oxide

fuel cell interconnect application. *J. Power Sources*, **195**, 3605–3611.

319. Hua, B., Zhang, W., Wu, J., Pu, J., Chi, B., and Jian, L. (2010) A promising $NiCo_2O_4$ protective coating for metallic interconnects of solid oxide fuel cells. *J. Power Sources*, **195**, 7375–7379.

320. Batfalsky, P., Haanappel, V.A.C., Malzbender, J., Menzler, N.H., Shemet, V., Vinke, I.C., and Steinbrech, R.W. (2006) Chemical interaction between glass-ceramic sealants and interconnect steels in SOFC stacks. *J. Power Sources*, **155**, 128–137.

321. Hauch, A., Ebbesen, S.D., Jensen, S.H., and Mogensen, M. (2008) Solid oxide electrolysis cells: microstructure and degradation of the Ni/yttria-stabilized zirconia electrode. *J. Electrochem. Soc.*, **155**, B1184–B1193.

322. Zhang, T., Fahrenholtz, W.G., Reis, S.T., and Brow, R.K. (2008) Borate volatility from SOFC sealing glasses. *J. Am. Ceram. Soc.*, **91**, 2564–2569.

323. Zhou, X.D., Templeton, J.W., Zhu, Z., Chou, Y.S., Maupin, G.D., Lu, Z., Brow, R.K., and Stevenson, J.W. (2010) Electrochemical performance and stability of the cathode for solid oxide fuel cells. III. Role of volatile boron species on LSM/YSZ and LSCF. *J. Electrochem. Soc.*, **157**, B1019–B1023.

324. Chen, K.F., Ai, N., Lievens, C., Love, J., and Jiang, S.P. (2012) Impact of volatile boron species on the microstructure and performance of nano-structured $(Gd,Ce)O_2$ infiltrated $(La,Sr)MnO_3$ cathodes of solid oxide fuel cells. *Electrochem. Commun.*, **23**, 129–132.

325. Bae, J.M. and Steele, B.C.H. (1998) Properties of $La_{0.6}Sr_{0.4}Co_{0.2}Fe_{0.8}O_{3-\delta}$ (LSCF) double layer cathodes on gadolinium-doped cerium oxide (CGO) electrolytes – I. Role of SiO_2. *Solid State Ionics*, **106**, 247–253.

326. Komatsu, T., Watanabe, K., Arakawa, M., and Arai, H. (2009) A long-term degradation study of power generation characteristics of anode-supported solid oxide fuel cells using $LaNi(Fe)O_3$ electrode. *J. Power Sources*, **193**, 585–588.

327. Liu, R.-R., Taniguchi, S., Shiratori, Y., Ito, K., and Sasaki, K. (2011) Influence of SO_2 on the long-term durability of SOFC cathodes. *ECS Trans.*, **35**, 2255–2260.

328. Schuler, A.J., Wuillemin, Z., Hessler-Wyser, A., and Herle, J.V. (2009) Sulfur as pollutant species on the cathode side of a SOFC system. *ECS Trans.*, **25**, 2845–2852.

329. Yamaji, K., Xiong, Y., Yoshinaga, M., Kishimoto, H., Brito, M., Horita, T., Yokokawa, H., Akikusa, J., and Kawano, M. (2009) Effect of SO_2 concentration on degradation of $Sm_{0.5}Sr_{0.5}CoO_3$ Cathode. *ECS Trans.*, **25**, 2853–2858.

330. Sakai, N., Yamaji, K., Horita, T., Xiong, Y.P., Kishimoto, H., and Yokokawa, H. (2003) Effect of water on oxygen transport properties on electrolyte surface in SOFCs. *J. Electrochem. Soc.*, **150**, A689–A694.

331. Liu, R.R., Kim, S.H., Shiratori, Y., Oshima, T., Ito, K., and Sasaki, K. (2009) The influence of water vapor and SO_2 on the durability of solid oxide fuel cell. *ECS Trans.*, **25**, 2859–2866.

332. Kim, S.H., Ohshima, T., Shiratori, Y., Itoh, K., and Sasaki, K. (2008) in *Life-Cycle Analysis for New Energy Conversion and Storage Systems* (eds V. Fthenakis, A. Dillon, and N. Savage), Materials Research Society, pp. 131–137.

333. Yan, A., Cheng, M., Dong, Y., Yang, W., Maragou, V., Song, S., and Tsiakaras, P. (2006) Investigation of a $Ba_{0.5}Sr_{0.5}Co_{0.8}Fe_{0.2}O_{3-\delta}$ based cathode IT-SOFC: I. The effect of CO_2 on the cell performance. *Appl. Catal., B*, **66**, 64–71.

334. Yan, A.Y., Yang, M., Hou, Z.F., Dong, Y.L., and Cheng, M.J. (2008) Investigation of $Ba_{1-x}Sr_xCo_{0.8}Fe_{0.2}O_{3-\delta}$ as cathodes for low-temperature solid oxide fuel cells both in the absence and presence of CO_2. *J. Power Sources*, **185**, 76–84.

335. Bucher, E., Egger, A., Caraman, G.B., and Sitte, W. (2008) Stability of the SOFC cathode material $(Ba,Sr)(Co,Fe)O_{3-\delta}$ in CO_2-containing atmospheres. *J. Electrochem. Soc.*, **155**, B1218–B1224.

336. Herring, J.S., O'Brien, J.E., Stoots, C.M., Hawkes, G.L., Hartvigsen, J.J., and Shahnam, M. (2007) Progress in high-temperature electrolysis for hydrogen production using planar SOFC technology. *Int. J. Hydrogen Energy*, **32**, 440–450.

337. Hauch, A., Ebbesen, S.D., Jensen, S.H., and Mogensen, M. (2008) Highly efficient high temperature electrolysis. *J. Mater. Chem.*, **18**, 2331–2340.
338. Kuhn, M., Napporn, T.W., Meunier, M., and Therriault, D. (2008) Experimental study of current collection in single-chamber micro solid oxide fuel cells with comblike electrodes. *J. Electrochem. Soc.*, **155**, B994–B1000.
339. Hauch, A., Jensen, S.H., Ramousse, S., and Mogensen, M. (2006) Performance and durability of solid oxide electrolysis cells. *J. Electrochem. Soc.*, **153**, A1741–A1747.
340. Marina, O.A., Pederson, L.R., Williams, M.C., Coffey, G.W., Meinhardt, K.D., Nguyen, C.D., and Thomsen, E.C. (2007) Electrode performance in reversible solid oxide fuel cells. *J. Electrochem. Soc.*, **154**, B452–B459.
341. Brisse, A., Schefold, J., and Zahid, M. (2008) High temperature water electrolysis in solid oxide cells. *Int. J. Hydrogen Energy*, **33**, 5375–5382.
342. Hauch, A., Jensen, S.H., Ebbesen, S.D., and Mogensen, M. (2009) Durability of solid oxide electrolysis cells for hydrogen production. *Risoe Rep.*, **1608**, 327–338.
343. Ebbesen, S.D., and Mogensen, M. (2009) Electrolysis of carbon dioxide in solid oxide electrolysis cells. *J. Power Sources.*, **193**, 349–358.
344. Ebbesen, S.D. and Mogensen, M. (2010) Exceptional durability of solid oxide cells. *Electrochem. Solid-State Lett.*, **13**, D106–D108.
345. Ebbesen, S.D., Graves, C., Hauch, A., Jensen, S.H., and Mogensen, M. (2010) Poisoning of solid oxide electrolysis cells by impurities. *J. Electrochem. Soc.*, **157**, B1419–B1429.
346. Ebbesen, S.D., Høgh, J., Nielsen, K.A., Nielsen, J.U., and Mogensen, M. (2011) Durable SOC stacks for production of hydrogen and synthesis gas by high temperature electrolysis. *Int. J. Hydrogen Energy*, **36**, 7363–7373.
347. Hauch, A., Jensen, S.H., Bilde-Sorensen, J.B., and Mogensen, M. (2007) Silica segregation in the Ni/YSZ electrode. *J. Electrochem. Soc.*, **154**, A619–A626.
348. Knibbe, R., Traulsen, M.L., Hauch, A., Ebbesen, S.D., and Mogensen, M. (2010) Solid oxide electrolysis cells: degradation at high current densities. *J. Electrochem. Soc.*, **157**, B1209–B1217.
349. Kim-Lohsoontorn, P., Kim, Y.M., Laosiripojana, N., and Bae, J. (2011) Gadolinium doped ceria-impregnated nickel-yttria stabilised zirconia cathode for solid oxide electrolysis cell. *Int. J. Hydrogen Energy*, **36**, 9420–9427.
350. Schiller, G., Ansar, A., Lang, M., and Patz, O. (2009) High temperature water electrolysis using metal supported solid oxide electrolyser cells (SOEC). *J. Appl. Electrochem.*, **39**, 293–301.
351. Yang, X. and Irvine, J.T.S. (2008) $(La_{0.75}Sr_{0.25})_{0.95}Mn_{0.5}Cr_{0.5}O_3$ As the cathode of solid oxide electrolysis cells for high temperature hydrogen production from steam. *J. Mater. Chem.*, **18**, 2349–2354.
352. Momma, A., Kato, T., Kaga, Y., and Nagata, S. (1997) Polarization behavior of high temperature solid oxide electrolysis cells (SOEC). *J. Ceram. Soc. Jpn.*, **105**, 369–373.
353. Mawdsley, J.R., Carter, J.D., Kropf, A.J., Yildiz, B., and Maroni, V.A. (2009) Post-test evaluation of oxygen electrodes from solid oxide electrolysis stacks. *Int. J. Hydrogen Energy*, **34**, 4198–4207.
354. Chen, K.F., Ai, N., and Jiang, S.P. (2010) Development of $(Gd,Ce)O_2$-impregnated $(La,Sr)MnO_3$ anodes of high temperature solid oxide electrolysis cells. *J. Electrochem. Soc.*, **157**, P89–P94.
355. Minh, N.Q. (2011) Development of reversible solid oxide fuel cells (RSOFCs)and stacks. *ECS Trans.*, **35**, 2897–2904.
356. Hino, R., Haga, K., Aita, H., and Sekita, K. (2004) R & D on hydrogen production by high-temperature electrolysis of steam. *Nucl. Eng. Des.*, **233**, 363–375.

357. Laguna-Bercero, M.A., Campana, R., Larrea, A., Kilner, J.A., and Orera, V.M. (2011) Electrolyte degradation in anode supported microtubular yttria stabilized zirconia-based solid oxide steam electrolysis cells at high voltages of operation. *J. Power Sources*, **196**, 8942–8947.
358. McIntosh, S., Adler, S.B., Vohs, J.M., and Gorte, R.J. (2004) Effect of polarization on and implications for characterization of LSM-YSZ composite cathodes. *Electrochem. Solid-State Lett.*, **7**, A111–A114.
359. McEvoy, A.J. (2000) Activation processes, electrocatalysis and operating protocols enhance SOFC performance. *Solid State Ionics*, **135**, 331–336.
360. Jiang, S.P. (2006) Activation, microstructure, and polarization of solid oxide fuel cell cathodes. *J. Solid State Electrochem.*, **11**, 93–102.
361. Jiang, S.P. (2003) Issues on development of (La,Sr)MnO_3 cathode for solid oxide fuel cells. *J. Power Sources*, **124**, 390–402.
362. Wang, W. and Jiang, S.P. (2004) Effect of polarization on the electrode behavior and microstructure of (La,Sr)MnO_3 electrodes of solid oxide fuel cells. *J. Solid State Electrochem.*, **8**, 914–922.
363. Wang, W.S., Huang, Y.Y., Jung, S.W., Vohs, J.M., and Gorte, R.J. (2006) A comparison of LSM, LSF, and LSCo for solid oxide electrolyzer anodes. *J. Electrochem. Soc.*, **153**, A2066–A2070.
364. Osada, N., Uchida, H., and Watanabe, M. (2006) Polarization behavior of SDC cathode with highly dispersed Ni catalysts for solid oxide electrolysis cells. *J. Electrochem. Soc.*, **153**, A816–A820.
365. Chen, K.F. and Jiang, S.P. (2011) Failure mechanism of (La,Sr)MnO_3 oxygen electrodes of solid oxide electrolysis cells. *Int. J. Hydrogen Energy*, **36**, 10541–10549.
366. Kim-Lohsoontorn, P., Brett, D.J.L., Laosiripojana, N., Kim, Y.M., and Bae, J.M. (2010) Performance of solid oxide electrolysis cells based on composite $La_{0.8}Sr_{0.2}MnO_{3-\delta}$ – yttria stabilized zirconia and $Ba_{0.5}Sr_{0.5}Co_{0.8}Fe_{0.2}O_{3-\delta}$ oxygen electrodes. *Int. J. Hydrogen Energy*, **35**, 3958–3966.
367. Graves, C., Ebbesen, S.D., and Mogensen, M. (2011) Co-electrolysis of CO_2 and H_2O in solid oxide cells: performance and durability. *Solid State Ionics*, **192**, 398–403.
368. Mizusaki, J., Saito, T., and Tagawa, H. (1996) A chemical diffusion-controlled electrode reaction at the compact $La_{1-x}Sr_xMnO_3$/stabilized zirconia interface in oxygen atmospheres. *J. Electrochem. Soc.*, **143**, 3065–3073.
369. Brichzin, V., Fleig, J., Habermeier, H.U., Cristiani, G., and Maier, J. (2002) The geometry dependence of the polarization resistance of Sr-doped $LaMnO_3$ microelectrodes on yttria-stabilized zirconia. *Solid State Ionics*, **152**, 499–507.
370. Virkar, A.V. (2010) Mechanism of oxygen electrode delamination in solid oxide electrolyzer cells. *Int. J. Hydrogen Energy*, **35**, 9527–9543.
371. Jiang, S.P. and Wang, W. (2005) Effect of polarization on the interface between (La,Sr) MnO_3 electrode and Y_2O_3-ZrO_2 electrolyte. *Electrochem. Solid-State Lett.*, **8**, A115–A118.
372. Chen, K.F., Ai, N., and Jiang, S.P. (2012) Performance and stability of (La,Sr)MnO_3-Y_2O_3-ZrO_2 composite oxygen electrodes under solid oxide electrolysis cell operation conditions. *Int. J. Hydrogen Energy*, **37**, 10517–10525.
373. Laguna-Bercero, M.A., Kilner, J.A., and Skinner, S.J. (2010) Performance and characterization of (La,Sr)MnO_3/YSZ and $La_{0.6}Sr_{0.4}Co_{0.2}Fe_{0.8}O_3$ electrodes for solid oxide electrolysis cells. *Chem. Mater.*, **22**, 1134–1141.
374. Guan, J., Minh, N., Ramamurthi, B., Ruud, J., Hong, J.K., Riley, P., and Weng, D.C. (2006) High Performance Flexible Reversible Solid Oxide Fuel Cell. Final Report for DOE Cooperative Agreement DE-FC36-04GO-14351, GE Global Research Center.
375. Sharma, V.I. and Yildiz, B. (2010) Degradation mechanism in $La_{0.8}Sr_{0.2}CoO_3$ As contact layer on the solid oxide electrolysis cell anode. *J. Electrochem. Soc.*, **157**, B441–B448.

376. Elangovan, S., Hartvigsen, J., Larsen, D., Bay, I., and Zhao, F. (2011) Materials for solid oxide electrolysis cells. *ECS Trans.*, **35**, 2875–2882.

377. Yang, C.H., Jin, C., Coffin, A., and Chen, F.L. (2010) Characterization of infiltrated $(La_{0.75}Sr_{0.25})_{0.95}MnO_3$ As oxygen electrode for solid oxide electrolysis cells. *Int. J. Hydrogen Energy*, **35**, 5187–5193.

378. Chen, K., Ai, N., and Jiang, S.P. (2012) Reasons for the high stability of nano-structured $(La,Sr)MnO_3$ infiltrated Y_2O_3-ZrO_2 composite oxygen electrodes of solid oxide electrolysis cells. *Electrochem. Commun.*, **19**, 119–122.

379. Liu, Q., Yang, C., Dong, X., and Chen, F. (2010) Perovskite $Sr_2Fe_{1.5}Mo_{0.5}O_{6-\delta}$ as electrode materials for symmetrical solid oxide electrolysis cells. *Int. J. Hydrogen Energy*, **35**, 10039–10044.

380. Cayan, F.N., Pakalapati, S.R., Celik, I., Xu, C., and Zondlo, J. (2012) A degradation model for solid oxide fuel cell anodes due to impurities in coal syngas: part I theory and validation. *Fuel Cells*, **12**, 464–473.

8
Materials and Processing for Metal-Supported Solid Oxide Fuel Cells

Rob Hui

8.1
Introduction

Solid oxide fuel cells (SOFCs) are promising devices for energy conversion due to their high efficiency and low emissions [1, 2]. SOFCs have the potential to convert a wide range of readily available fuels electrochemically to electricity, such as renewable biomass fuels, hydrocarbon fuels, and CO from coal gasification. However, the cost, reliability, and durability of the system have hindered the commercialization of SOFCs. On the way to address these challenges, different cell configurations have been developed, including the electrolyte-supported in the early stage, the most studied electrode-supported, and the most recent metal-supported solid oxide fuel cells (MS-SOFCs). MS-SOFCs offer many potential advantages over other types of SOFCs.

1) **Low cost**. The major material cost among the cell components is from the thick support materials of the cell. With the inexpensive ferric steels as support to replace the expensive ceramics, the materials cost for MS-SOFCs could be seven times lower than that of anode-supported ones [3]. Metallic substrates allow the use of conventional metal joining and forming techniques and could significantly reduce the manufacturing costs of SOFC stacks as well.
2) **High robustness**. Compared with the ceramic-supported SOFCs, MS-SOFCs are more robust against fast start-up, thermal cycling, and redox process.
   The structural instability resulting from start-up and thermal cycling is one of the main causes of cell breakage and stack failure [4]. Both ductility and thermal conductivity of metallic materials are beneficial to improve the structural instability of SOFCs. SOFCs can be made more robust by replacing the brittle ceramic support with a ductile metallic support. The high thermal conductivity is helpful in lowering the internal temperature gradients of the stack for improving thermal shock resistance. A quick start-up of MS-SOFCs has been demonstrated and compared with anode-supported SOFCs [5]. The quick start-up has led to cracked electrolyte in anode-supported cells, but not in the metal-supported ones. In addition, the mechanical strength of the stacks

Materials for High-Temperature Fuel Cells, First Edition. Edited by San Ping Jiang and Yushan Yan.

may be increased by metal joining techniques such as brazing, welding, and crimp sealing in MS-SOFCs rather than using fragile cement or glass seals as in the electrode-supported cells. A feasibility study of employing braze seal has been carried out [6]. No significant loss of open circuit voltage (OCV) was observed for a metal-supported cell with braze seal after thermal cycling over 30 times with a heating and cooling rate greater than 350 °C min^{-1}. When cermets such as Ni–YSZ (yttria-stabilized zirconia) are used as anode materials, the metallic component may experience redox cycles on a long-term operation. There are a couple of reasons that explain why redox happens: interrupted fuel supply due to the imperfect seals and oxygen activity over the equilibrium between Ni and NiO under great fuel utilization [7, 8]. The volume change during the redox process results in failure of structural integrity of the cells and the stacks. According to the mechanical modeling of damage due to redox stress, strain of the anode oxidation at MS-SOFCs can be tolerated without delamination [9], which shows that MS-SOFCs are more resistant to redox cycling than the electrode- or electrolyte-supported SOFCs.

3) **Reduced operating temperature**. The application of metallic substrates inevitably sets the SOFCs to operate at intermediate or low temperatures to avoid the significant oxidation of metallic substrates, which is beneficial to the overall materials reliability and long-term cost. However, high-performance electrode materials and high-conducting electrolyte materials are needed to operate MS-SOFCs at reduced temperatures.

MS-SOFCs are expected to find a broad application in residential, industrial, and military power supply. Compared with the ceramic-supported SOFCs, the mechanical robustness of MS-SOFCs makes them suitable for transport applications where severe vibration exists and quick start-up is required [10]. Although the concept of MS-SOFCs has been proposed in 1966 [11], and great efforts have been made during the past decades by many groups (such as the Aerospace Research Centre and Space Agency (DLR) [12–14], the Ceres Power and Imperial College [15–18], the Lawrence Berkeley National Laboratory (LBNL) [3, 5, 6, 19, 20], companies in Japan [21, 22], the National Research Council (NRC) of Canada [23–28], and the Risø National Laboratory [29]), there are still significant challenges for the development of MS-SOFC technologies [30]. Fabrication processes for the electrolyte layer and the available metallic substrate materials have been the major barriers among all the other issues such as interactions and bonding between different components in terms of cost, performance, and degradation. In this chapter, the challenges of materials and processing for MS-SOFCs are discussed and the R&D status of the MS-SOFCs is also reviewed for different groups worldwide.

8.2
Cell Architectures

Same as the typical configurations of electrode-supported SOFCs [1], the MS-SOFCs can be either tubular [5, 20] or planar [12, 21], as shown in Figure 8.1.

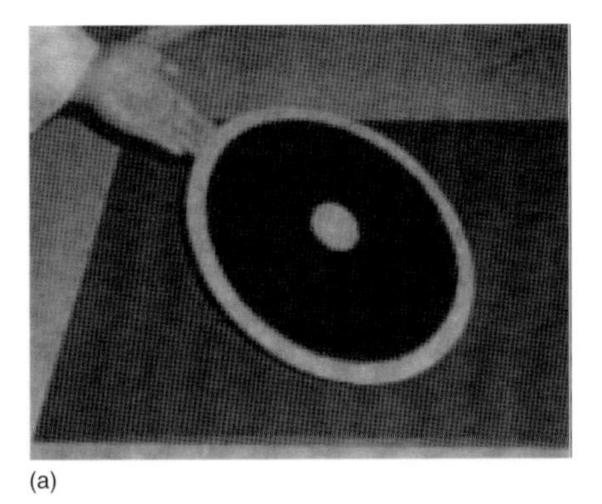

(a)

(b)

Figure 8.1 (a) Planar and (b) tubular (0.06 m^2 Ni-felt substrate) MS-SOFCs. (Source: Reproduced with permission from Refs. [20, 21].)

Although the performance is based on the same electrochemical processes, the operation characteristics and parameters vary considerably because of the different geometries [31]. Except gas seals by metal joining and robust structure, MS-SOFCs have the similar advantages and disadvantages for each design [32]. Relaxation times for different processes such as load change, start-up, and shutdown are shorter for the planar geometry due to its short current path. The tubular configuration is more advantageous concerning the level of temperature gradients within the cell due to its longer spatial extension. While planar SOFCs have showed improved power densities and more flexibility in design, the tubular ones have certain freedom in thermal expansion.

The metallic substrates can be porous or dense sheets with holes to allow gas penetration (Figure 8.2) or a combination of both for best mechanical strength and surface morphology. Some of the porous substrates have been widely used as filters and are commercially available in either tubular or planar shapes, which are mostly made by the powder metallurgy process. Perforated dense metal sheets can be made by chemical processes such as photoetching or physical methods such as e-beam and lazar microdrilling. Perforated dense sheets are mostly made as a request with limited length-to-diameter ratios of the sheet, such as 10 : 1. Compared with porous metals, perforated dense sheets may be more oxidation resistant, especially for a long-term operation.

The metallic substrates can be used on either the anode side or the cathode side, depending on the substrate properties, fabrication processes, electrode materials, and fuel compositions.

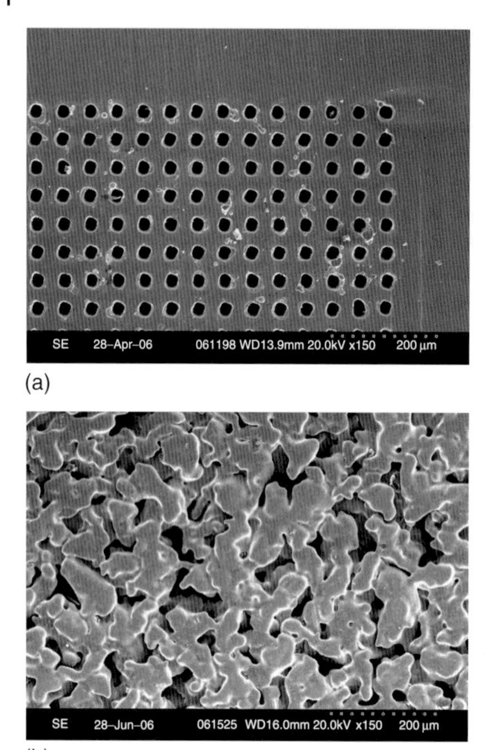

Figure 8.2 SEM images of two different substrates: (a) porous and (b) perforated.

1) **Substrates on the anode side**
   In this case, co-firing substrate, anode, and electrolyte at high temperature and low oxygen partial pressure are allowed. However, corrosion-resistant metals are required to operate in humid conditions. Co-firing at a high temperature may cause interdiffusion issues on the metal interface. This is discussed in the following section.
2) **Substrate on the cathode side**
   When metallic substrates are used on the cathode side, there is concern about the interaction of metallic materials with hydrocarbon fuels, and therefore, alternative anode materials can be used for hydrocarbon fuels. In contrast to the fuel conditions, corrosion-resistant metals are required in oxidizing conditions with less humidity. In addition, Cr poisoning may occur when Cr-based alloys are used and the cathode materials are not chromium tolerant. Coating the metal is a solution to prevent chromium migration [33, 34]. A promising Cr-resistant cathode material is $LaNi_{0.6}Fe_{0.4}O_3$ (LNF), developed at the NTT Corporation in Japan [35–37]. This material has desirable characteristics, including high electric conductivity, less coefficient of thermal expansion (CTE) mismatch with zirconia electrolyte, high catalytic activity, and high

durability against cathode poisoning by chromia from metallic components [38–40].

8.3 Substrate Materials and Challenges

8.3.1 Requirements for Substrates

The electrolyte and the electrode materials for conventional SOFCs are also suitable for MS-SOFCs. A series of review papers and books can be found for the components of SOFCs [30, 41–44], and the emphasis in this chapter is on the metallic materials suitable as substrates for MS-SOFCs because of the variety of selection and the challenges to satisfy the requirements. The requirements for the metallic substrate materials are similar to those of interconnect materials, such as electrical conductivity, thermal and chemical compatibility, mechanical strength, and cost concerns [30]. The differences between the metallic materials used as substrates and interconnects are the structure and application environments. To serve as substrate materials for SOFCs, the alloys are expected to have the following properties:

1) **Gas permeability**. In contrast to metallic interconnects that are gastight, the metals used as cell substrates should be gas permeable at either the anode side or the cathode side. Similar to the porosity requirement for a ceramic anode [30], the metallic substrates should have porosity around 30–35% to avoid polarization of mass transport.
2) **Oxidation resistance**. An interconnect is exposed to dual atmospheres for both anode and cathode sides, whereas substrates are only exposed to a single atmosphere, a reducing and more humid atmosphere on the anode side or an oxidizing and less humid environment on the cathode side.
3) **Electrically conductivity**. With the formation of protective scales on the surface of the metallic substrates, the area-specific resistance (ASR) is expected to be less than $0.1\ \Omega\ cm^2$ over tens of thousands of hours of operation to keep the power degradation rate less than 1% per 1000 h.
4) **Thermally compatibility**. The CTE of the substrate materials should match that of other component materials. The CTE for 8 mol% YSZ and 20 mol% Sm-doped ceria (SDC) is 10.5×10^{-6} and $12.5 \times 10^{-6}\ K^{-1}$, respectively [30, 45].
5) **Chemically compatibility**. The substrate materials have to be chemically compatible with other materials during contact under operating conditions.
6) **Mechanical strength**. The substrates have to be mechanically reliable and durable at operating conditions of SOFCs.
7) **Low cost**. The materials should be cost-effective.

8.3.2
Properties of Selected Alloys

8.3.2.1 Selected Alloys and Roles of Elements

High-temperature alloys have been considered as potential substrate materials for MS-SOFCs. These alloys resist oxidation by forming thin, adherent protective layers of SiO_2, Al_2O_3, or Cr_2O_3. Since Al_2O_3 scale is denser and less volatile than the Cr_2O_3 scale, oxidation rate of Al_2O_3-forming alloys are normally lower by one or two orders of magnitude than that of Cr_2O_3-forming alloys [46]. However, the electrical resistance of Al_2O_3 and SiO_2 are higher by a few orders of magnitude than that of Cr_2O_3 (1×10^2 Ω cm at 800 °C) [47–49]. The high resistance of the Al_2O_3 scales or the SiO_2 scales limits the Al_2O_3-forming or the SiO_2-forming alloys as substrates or interconnect for SOFCs. There are hundreds of compositions commercially available, and majority of the alloys studied as interconnect materials for SOFCs are Cr_2O_3-forming ones, which can be classified into Cr-, Fe-, Co-, or Ni-based alloys.

The oxidation rate of porous media as substrates could be much faster than the dense ones as interconnects under SOFC operating conditions [50–52]. However, the oxidation behavior of porous Cr_2O_3-forming alloys is much less studied in the literature. Table 8.1 summarizes nine alloys that have been well studied as potential interconnect materials and are expected to be the candidate compositions as substrate materials for SOFCs.

The alloy composition determines the growth rate, microstructure, and phase constitution of the oxide scales. With Cr, Fe, Co, or Ni as basic elements, the chemical and physical properties of the alloys can be modified by other elements [57–59]. The chromium content in the alloys plays the key role for oxidation resistance [60]. Minimum oxidation resistance is achieved when the chromium content is greater than 25% in the alloys at 1000 °C. The rare earth elements and Y as reactive elements can significantly improve oxidation resistance by enhancing preferential formation of chromia scales, reducing scale growth rate, improving scale adherence to the alloy substrate, and forming denser scales. The addition of their oxides into the matrix leads to superior creep properties of the alloys, known as the *oxide dispersion strengthened* (*ODS*) steels, such as Ducrolloy. Molybdenum, tungsten, and niobium are added to reduce the CTE of the alloy. Formation of volatile Cr(VI) species can be decreased by the addition of manganese and titanium to form spinel or rutile oxides to replace the chromia in the scale. The amount of aluminum and silicon has to be kept low to prevent the formation of their highly resistant oxides. A thin silica oxide subscale has been observed beneath the chromia scale, and the adherence of the oxide scale may be influenced when a continuous SiO_2 layer is formed [53]. The bond strength between the as-formed oxide scales and the underlying alloy substrate is critical for effective protection against oxidation and electrical conductivity of the alloys as substrate for SOFCs.

Some efforts have been made to develop alloys as interconnects for SOFCs [61–65], including Crofer 22 APU (Table 8.1). The composition of Crofer 22

Table 8.1 Nominal chemical composition (wt%) and CTE at 800 °C of selected alloys [53–56].

Alloy	Fe	Ni	Cr	Mn	Si	Al	Ti	Mo	La	Zr	Others	CTE (10^{-6} K^{-1})
Crofer 22 APU	Bal	—	22.78	0.4	0.02	0.006	0.7	—	0.086	—	S 0.02, P 0.05	11.0
F18TNb	Bal	—	19.4	0.12	0.46	0.02	0.12	1.7	—	—	Nb 0.17	11.0
ZMG232L	Bal	0.33	22.04	0.45	0.1	0.03	—	—	0.08	0.2	—	11.7
E-Brite	Bal	$\leq$0.5	26–27.5	$\leq$0.4	$\leq$0.4	—	—	0.75	—	—	Nb $\leq$ 0.2, S $\leq$ 0.02, P $\leq$ 0.02	11.9
AISI-SAE 430	Bal	—	16–18	$\leq$1	$\leq$1	—	—	—	—	—	S $\leq$ 0.03, P $\leq$ 0.04	11.4
IT-11	Bal	—	26.4	—	0.01	0.02	—	—	—	—	Y 0.08	11.9[a]
IT-14	Bal	—	26.3	—	0.02	0.02	—	—	—	—	Y 0.06	11.9[a]
Haynes 230	3	Bal	22–26	0.5–0.7	—	0.3	—	1–2	—	—	Co 5	17.1
Haynes 242	$\leq$2	Bal	8	$\leq$0.8	$\leq$0.8	$\leq$0.5	—	25	—	—	Co $\leq$ 2.5, C $\leq$ 0.03	11.1
Hastelloy X	19	Bal	24	1.0	—	—	—	5.3	—	—	Co 1.5	15.5
Ducrolloy	5	—	Bal	—	—	—	—	—	—	—	Y_2O_3 1.0	11.8

BAL, Balance.
[a] Estimated value.

APU has been selected to match the CTE of SOFC components without affecting oxidation resistance. An electrically conductive Cr–Mn oxide layer forms, while Cr_2O_3 is the only phase in the scale on E-Brite under SOFC operating conditions [46]. A low ASR for Crofer 22 APU has been reported [59, 66, 67]. However, scale growth dominated by an underlying Cr_2O_3 layer may increase electrical resistance for an extended operation [68, 69]. Ni-based alloys usually have high CTE. Efforts have been made to develop Ni-based alloys with low CTE and high oxidation resistance in air [64]. These alloys, such as J5 with a CTE of 12.6×10^{-6} K^{-1}, are promising candidates as substrates for MS-SOFCs.

8.3.2.2 **Oxidation in Oxidizing or Reducing Atmosphere**

The oxidation process on metal surfaces has been well established according to Wagner's theory of oxidation [70, 71]. The growth of oxide scales is dominated by diffusion process and normally follows the parabolic law:

$$x^2 = K_p t + C \tag{8.1}$$

where x can be expressed in terms of either the thickness of the oxide scale or the weight gain per area, K_p is the parabolic rate, t is the oxidation period, and C is a constant. The parabolic rate is used as the oxidation rate of alloys and is dependent on alloying composition, thermal history, and surface conditions. The parabolic rates of selected alloys oxidized in air have been plotted in Figure 8.3. The oxidation rates of Fe-based alloys are faster than those of Cr-based and Ni-based alloys with Cr content greater than 22%. The parabolic rate of Haynes 188, a Co-based alloy, was reported to be 0.712×10^{-13} and 0.327×10^{-13} $g^2\ cm^{-4}\ s^{-1}$ at 800 and 700 °C, respectively [46]. Therefore, the order of the resistance to oxidation in air for the selected alloys is Fe- > Co- > Ni- > Cr-based alloys. However, compared with Cr-based and Ni-based alloys, Fe-based alloys are attractive because of their low cost, high ductility, good workability, and excellent match of CTE with other fuel cell components.

Apart from the oxygen in the air at the cathode side, the gas environment for the metallic substrate may consist of H_2, H_2O, CO, CO_2, and CH_4 with a low oxygen partial pressure of 10^{-22} atm at the anode side and nitrogen at the cathode side. Carbon may exist when hydrocarbon fuel is used at the anode side as well. When CO or CH_4 exists, Cr-based alloys tend to form a sublayer containing chromium carbide [72, 73]. While most chromium carbides are electrically conducting [74], more investigation on the impact of the chromium carbides on the performance of alloys is required; of particular importance is the nitrogen and water vapor. The nitrogen may dissolve into the alloy or form nitride beneath the chromia scale during high-temperature exposure [75], which leads to environmentally induced embrittlement. However, the effect of nitrogen on the performance of the alloys has not been well studied for SOFC applications.

Cr_2O_3 is thermodynamically unstable and may evaporate according to the following reaction [76]:

$$2Cr_2O_3\,(s) + 2O_2\,(g) \rightarrow 4CrO_3\,(g) \tag{8.2}$$

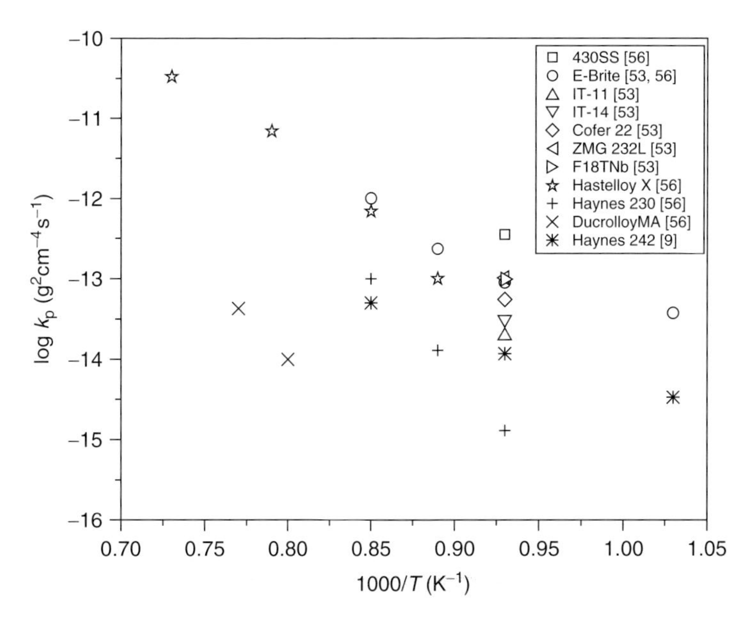

Figure 8.3 Oxidation rates for selected alloys in air or oxygen (numbers in square bracket indicate references).

The existence of water vapor enhances the volatilization of chromia because of the formation of volatile hydroxides and oxyhydroxides through the following primary reactions [76]:

$$2Cr_2O_3\,(s) + 3O_2\,(g) + 4H_2O\,(g) \rightarrow 4CrO_2(OH)_2\,(g) \quad (8.3)$$

$$Cr_2O_3\,(s) + O_2\,(g) + H_2O\,(g) \rightarrow 2CrO_2OH\,(g) \quad (8.4)$$

The calculated partial pressure of CrO_3 and $CrO_2(OH)_2$ is 1.0×10^{-4} and 2.2×10^{-2} Pa at 800 °C in air, respectively [46]. Increased water vapor pressure or temperature accelerates the volatilization process. With the formation of hydroxides and oxyhydroxides, the chromia scales become less protective on the metal surface [77]. Even worse, the volatile chromium species deposit and cover the active reaction sites in the porous electrodes and lead to rapid degradation in the electrochemical performance of SOFCs [30, 31]. The chromium is also known to react with some components in the electrodes to form low-conducting species such as $SrCrO_4$, which contribute to high polarization loss in SOFCs [78].

Compared to the studies on the oxidation of alloys in oxidizing atmospheres, the studies on fuel conditions are much less. In contrast to the oxidation rate in air, the parabolic rate constants in humid hydrogen [56, 79, 80] of selected alloys presented in Table 8.1 are shown in Figure 8.4. Comparing the results in Figure 8.3 and Figure 8.4, the order of the resistance to oxidation in hydrogen and water atmosphere is still as same as the order in air for the selected alloys. However, the parabolic rates in the presence of hydrogen and water are slightly higher than

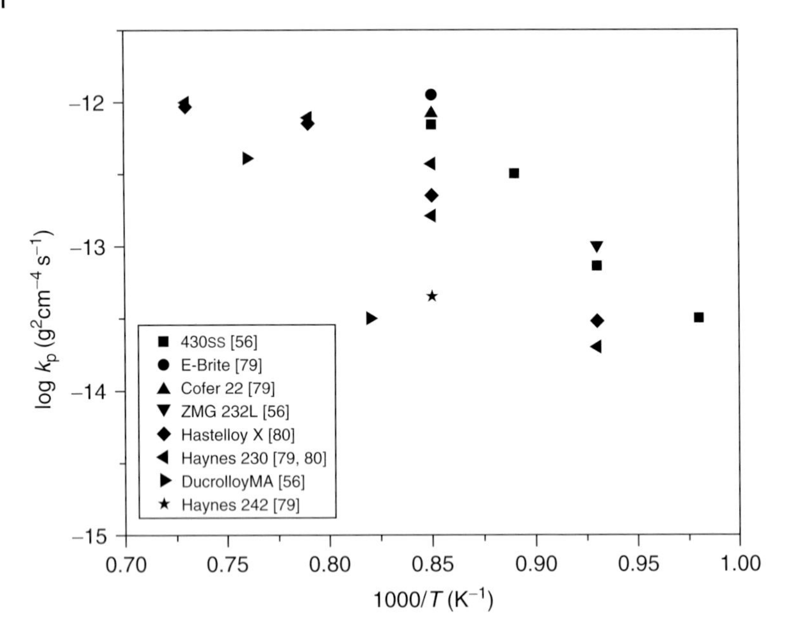

Figure 8.4 Oxidation rates for selected alloys in fuel conditions (numbers in square bracket indicate references).

the rate in the air. The increased oxidation rates in the humidified atmosphere are most likely due to the formation of less protective, less dense, and fast growing scales. This feature is especially harmful to the performance of SOFCs when the alloys with Cr former are employed as substrates at the anode side because the fuel gases invariably involve certain amounts of water vapor as a by-product of fuel cell reactions.

8.3.2.3 **Scale Conductivity**

The electrical conductivity of the metallic substrates is limited by the protective scales after oxidation, which can be measured by ASR. There are many factors that can contribute to ASR, such as composition, microstructure, adhesion, thickness, and the surrounding atmosphere of the scales. The amounts of aluminum and silicon have to be controlled in the alloy composition to avoid the formation of their highly insulating oxides [81, 82]. Compared with the ASR of ZMG232, the ASR of ZMG232L has been significantly improved by decreasing the content percentage of aluminum and silicon from 0.22 and 0.4 to 0.03 and 0.1, respectively, while increasing the lanthanum content from 0.04 to 0.08% [81]. Doping NiO, TiO_2, MgO, or other oxides can enhance the electrical conductivity of Cr_2O_3 scales [49, 83–91], as reported by Quadakkers *et al.* [58]. The chromia scales are more likely to spall when the scales grow up to 3–5 μm. Therefore, the time for spalling can be estimated when the thickness of the chromia scales reach 3 μm (Figure 8.5) [20]. Poor adhesion, cracks, thick coverage, and porous structure of the scale on the alloy surface all contribute to high electrical resistance and should be avoided. The

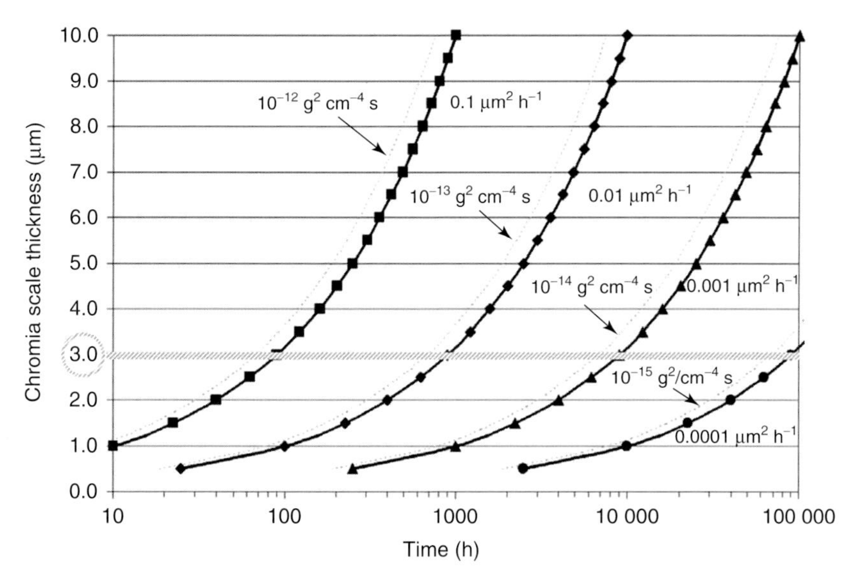

Figure 8.5 Chromia scale thickness as a function of time for various parabolic growth rate constants. (Source: Reproduced with permission from Ref. [20].)

bonding strength of the coating on metallic surfaces can be improved significantly, depending on the approach used for surface treatments [92]. While conventional sandblasting may not be applicable for porous metals, a heat treated reducing atmosphere could be the best choice for the adhesion of the coating. The ASRs of the selected alloys in oxidizing atmospheres [54] and reducing conditions (Ar + 4 vol% H_2 + 3 vol% H_2O, flow rate 45 ml min^{-1}) [79] are shown in Figure 8.6. The ASRs were measured after oxidation at 800 °C in air for 500 h or in reducing atmosphere at 900 °C for 1000 h. The lowest ASR among these alloys is of Haynes 242 at 800 °C: 3.55×10^{-3} Ω cm^2 in air for 500 h and 6.03×10^{-2} Ω cm^2 in hydrogen with 3% H_2O for 1000 h for. This value still may be promising when compared with the target value of 25×10^{-3} Ω cm^2 for interconnect materials [46, 58]. It should be noted that the values of ASR in air are lower than that in reducing atmosphere for selected alloys. One reason is that the alloys were treated in air at 800 °C for 500 h and in reducing atmosphere at 900 °C for 1000 h before ASR measurement. The other reason could be due to the higher oxidation rate in the reducing atmosphere with the presence of water, as discussed in the previous section. Figure 8.7 presents the ASR of different steels in air at 800 °C for 1000 h [53]. Although most commercially available alloys hardly meet the needs for SOFCs for long-term operation at temperatures higher than 700 °C [59, 62, 66, 93–99] and the degradation rates are as high as 2–25% per 1000 h, the selected alloys such as Crofer 22 APU, ZMG232L, E-Brite, IT-14, and Haynes 242 are potential candidate materials as metallic supports for MS-SOFCs.

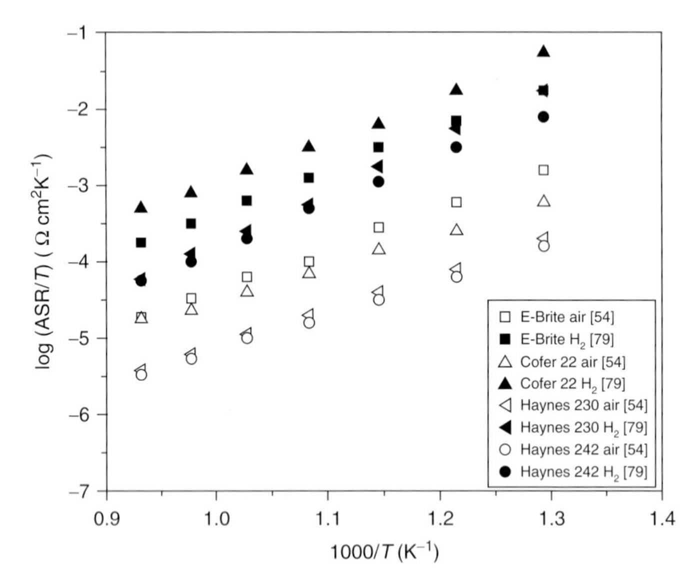

Figure 8.6 ASR as a function of temperature for selected alloys in air and reducing atmosphere after oxidation at 1073 K in air for 500 h or in reducing atmosphere at 1173 K for 1000 h (numbers in square bracket indicate references).

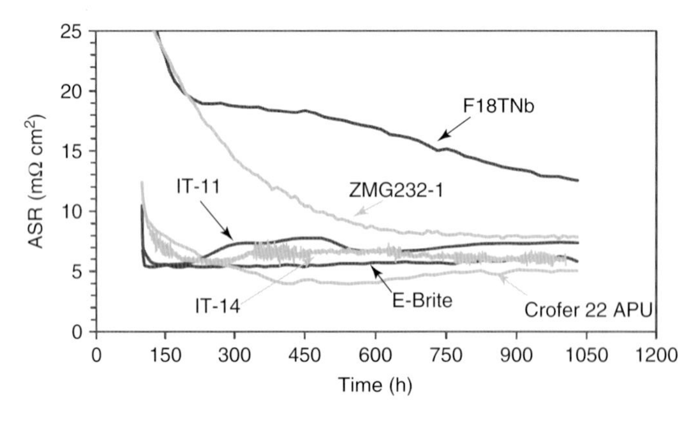

Figure 8.7 ASR measurement of different steels in air at 1073 K for 1000 h. (Source: Reproduced with permission from Ref. [53].)

8.3.2.4 Additional Improvement

It was a challenge to make the currently commercially available alloys to meet all the requirements as substrates for SOFCs, especially for porous alloys with accelerated oxidation rates and cost. In addition to the porous structure as substrates, it is desirable to develop new alloys and conductive ceramic coatings as performed for metallic interconnect materials in the future.

Although Fe-based alloys usually have more compatible CTE with other ceramic cell components, Ni-based or Cr-based alloys with chromia former are favored as substrates for SOFCs from oxidation resistance and ASR point of view. Development of Ni-based alloys with low CTE has been carried out and reported [64]. Based on the fact that the CTE of Ni alloys increases with increase in the content of Mn, Fe, Co, or Cr and decreases with decrease in the content of Mo, W, C, Al, and Ti [55, 100–104], new alloys have been produced with a CTE of $12.0–13.0 \times 10^{-6}\,^{\circ}C^{-1}$ from 23 to 800 °C. However, no ASR data have been reported and a protective coating may still be needed, because a cerium surface treatment has been performed and a ceria layer might form on the surface for the reduced oxidation rate.

Applications of conductive ceramic coatings on the alloys have been believed as the potential solution to decrease the growth rate of the scales, to prevent the evaporation of Cr species and reactions with electrode materials, and to provide low ASR for long-term operation [56, 105–111]. Significant efforts have been made to develop coating techniques and coating materials for SOFC applications [105, 111], including sol–gel, pulsed laser deposition (PLD), plasma spraying, screen printing, thermally growing, and electrodeposition and coating of materials such as $LaCrO_3$, $LaMnO_3$, CeO_2, and MCrAlYO (M=Mn or Co). Doped CeO_2 has been shown to be a promising coating material because of its high stability and higher electrical conductivity [112]. Topsoe Fuel Cells have demonstrated an operation of 13 000 h with minimal degradation for a stack operating with coated FeCr-based interconnects [113]. Thermally sprayed oxide coatings on Ducrolloy have led to a degradation rate of cell voltage less than 1% 1000 h^{-1} over 12 000 h [56, 114, 115]. A stable operation of MS-SOFCs has been achieved with a perovskite oxide as the diffusion barrier layer (DBL) for 2300 h [116]. A preferred situation would be an inexpensive ferritic stainless steel coated with conducting ceramics. It has been demonstrated that surface stability, ASR, and Cr volatility have been dramatically improved for coated 430SS and Crofer 22 APU [69]. A composite coating of 1 μm thick $Mn_{1.5}Co_{1.5}O_4$–0.3 μm thick $Cr_{6.5}Al_{30}Y_{0.15}O_{66.35}$ was applied and subsequently annealed in air on ferritic steels by the large area filtered arc deposition (FAD) and the hybrid filtered arc-assisted electron beam physical vapor deposition (FA-EBPVD) techniques, respectively. The coated 430SS has an ASR less than 8 mΩ cm^2 for 1000 h at 800 °C in air, which is about as same as the coated Crofer 22 APU, as shown in Figure 8.8. With only one layer of 1 μm thick $Mn_{1.5}Co_{1.5}O_4$, Cr evaporation has been reduced by over 30-fold compared with that in the uncoated 430SS. However, ASR of the 430SS was lower than that of the uncoated one with a single layer of protection, whereas the ASR of coated Crofer 22 APU was same no matter one or two segments.

8.4
Cell Fabrication and Challenges

One of the major challenges for the development of MS-SOFCs is the cell fabrication process without oxidizing the metallic substrate and with potentially low cost for commercialization. The microstructures of the cell components have to be controlled with either porous structures for electrode layers or dense structures

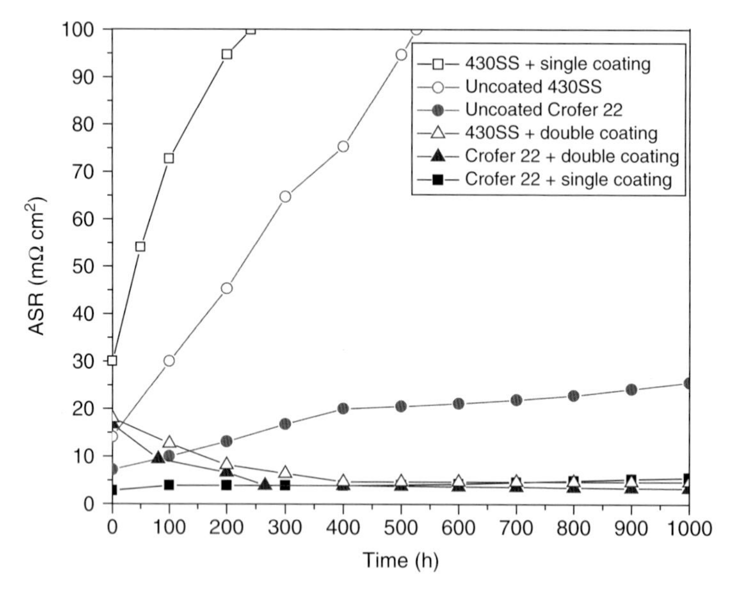

Figure 8.8 ASR for coated and uncoated 430SS and Crofer 22 APU in air at 1073 K. (Source: Reproduced with permission from Ref. [69].)

for the electrolyte layers. The fabrication of dense and gastight electrolyte layers is more technically difficult than that of porous layers. The MS-SOFCs can be potentially fabricated by a number of techniques, which fall into two groups: sintering processes and deposition processes according to the fabrication of dense electrolyte layers. These two major approaches have been adopted by different researchers: sintering processes by the LBNL in the United States and the Ceres Power in United Kingdom and deposition processes by the DLR in Germany and the NRC in Canada. The current status of the sintering and deposition processes, performance of the cells, and issues related to the fabrication and component materials are discussed in the following.

8.4.1
Sintering Approaches

To obtain a dense electrolyte and avoid oxidation of metallic substrates, the structure of the cells can only take the design of "substrate on anode side" and cells have to be sintered under reducing conditions at high temperatures above 1000 °C [3, 5, 6, 15–20, 29]. However, co-sintering can be carried out in air when oxides are used as the precursors to form the metallic substrate in a consequent reduction process [117–119]. The anode and the electrolyte are usually co-sintered with a green metal substrate, whereas the cathode and current collector at cathode side have to be applied and fired in oxidizing atmosphere to avoid decomposition. MS-SOFCs with the design of "substrate on cathode side" and tubular micro-MS-SOFCs have been proposed with sintering approaches [120]. The biggest advantage of sintering approaches is that the MS-SOFCs can be fabricated by the conventional powder processing techniques used for fabrication of ceramic-supported SOFCs, such as tape casting, screen printing, and extrusion. Co-sintering at high temperatures offers strong adhesion between the different functional layers. However, sintering processes may limit the applications of commercially available alloy sheets and lead to some potential issues including interfacial diffusion, chromia evaporation and poisoning on anodes, and nickel coarsening. Extra care may be required when sintering the green components with organic binders in reducing conditions because the formation of carbon may form low-melting carbon steel [3].

Both planar and tubular MS-SOFCs have been fabricated using the conventional sintering processes. For tubular cell Fe–Cr alloy/Ni–YSZ/YSZ/YSZ–LSM with infiltrated nickel and LSM at the LBNL, the maximum power density has reached 332 and 1300 mW cm^{-2} at 700 °C with air and oxygen as the oxidant, respectively [5, 121]. The polarization curves are presented in Figure 8.9 for the cell with oxygen as the oxidant. The cell impedance revealed a reduction in electrode polarization, especially the decrease of mass transport resistance when switching air to oxygen. Ikerlan in Spain has demonstrated a successful operation of tubular MS-SOFCs for 1000 h at 300 mA cm^{-2} at 800 °C, revealing that the failure of the cells was directly related to the composition of the supports and the sealing [122]. A combined deposition and sintering process has been used for the fabrication of planar cell alloy1.4509/Ni–$Ce_{0.9}Gd_{0.1}O_{2-x}$ (CGO)/CGO/$La_{0.6}Sr_{0.4}Co_{0.2}Fe_{0.8}O_{3-x}$ (LSCF) at Ceres Power [15–18]. The anode and electrolyte layers were deposited by wet spray/screen printing and electrophoretic deposition (EPD), respectively, followed by sintering under reducing atmosphere below 1000 °C. A peak power density of ~490 mW cm^{-2} at 570 °C was reported, using air as the oxidant and moist hydrogen as the fuel [123]. A stack with 40 cells has been operated at 200 mA cm^{-2} at 570 °C for more than 1000 h on simulated reformate gas and air [17, 18]. Modeling studies have predicted achievable high efficiencies for the cells and systems based on the SOFCs with CGO as the electrolyte [18]. However, the ceria reduction may lead to microcracks in the CGO electrolyte layer because of the volume expansion of Cr^{4+} to Cr^{3+}, which makes the cells structurally unstable for long-term operation.

Redox stability has been studied for the cell FeCr/Ni–YSZ/YSZ/LSM and compared with anode-supported cells [124]. By switching between air and moist hydrogen in the fuel stream, the anode-supported cell completely failed after two redox cycles, as shown in Figure 8.10. The stress from the volume change of anode during redox process may be restricted by the volume change of metallic substrates [9]. However, the electrolyte layer of the anode-supported cell cracked due to the tension from the large volume expansion associated with the oxidation of nickel [125, 126]. The application of infiltration of nickel catalyst into the preformed porous YSZ layer can avoid the impact of redox of nickel and also benefit redox stability [124]. In addition, the resistance against thermal shock has also been demonstrated for this cell, with heating and cooling rates greater than 350 °C min^{-1} [124]. The power density decreased from 370 to 325 mW cm^{-2} at 0.7 V after

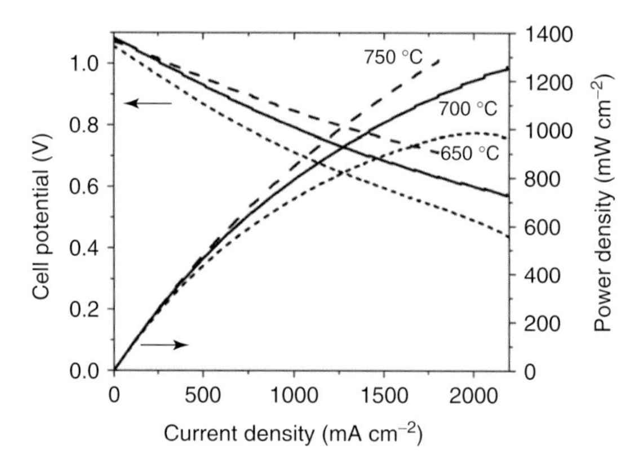

Figure 8.9 Polarization and power density of tubular cell fabricated by sintering processes with infiltrated nickel and LSM (moist hydrogen, Fe–Cr alloy/Ni–YSZ/YSZ/YSZ–LSM, oxygen). (Source: Reproduced with permission from Ref. [5].)

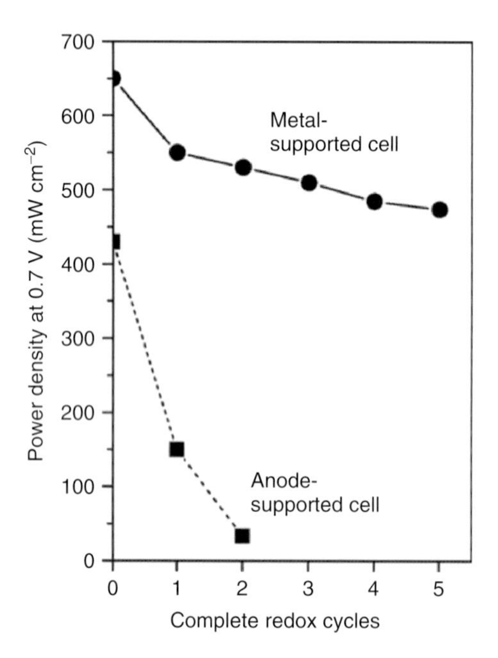

Figure 8.10 Redox stability of MS-SOFCs compared with anode-supported cells. (Source: Reproduced with permission from Ref. [124].)

five rapid thermal cycles. In contrast, an anode-supported cell exhibited structural failure after only one similar thermal cycle [6]. The abuse tolerance of MS-SOFCs allows for rapid start-up, simplified balance of plant, and tolerance to unintended fluctuations in operational parameters such as temperature, fuel flow rate, and fuel utilization.

The interfacial diffusion of elements in metallic substrates into the adjacent anode layer has been reported and causes fast degradation in cell performance [3, 19, 50]. Energy dispersive analysis by X-ray (EDAX) of cross sections of the cells revealed significant diffusion of Cr, Fe, and Ni into the anode layer when sintering Ni–YSZ on FeCr alloy in a reducing atmosphere at 1350 °C [3, 19]. Creating a DBL between the anode and metallic support may be a potential solution to stop interdiffusion. Then, there is need for electrically conducting and chemically compatible materials that may not be available. Nickel has been widely used as an electrocatalyst in the anode. Nickel coarsening is well known during sintering at high temperatures, which is detrimental to cell performance, including significantly enhanced anodic polarization losses due to disconnection of electrical contacts and loss of active surface areas.

Both interfacial diffusion and nickel coarsening can be potentially avoided by change of anode structure or materials. The infiltration technique has been employed to improve the electrode performance of ceramic-supported SOFCs [43, 127–130]. By creating a porous YSZ structure at the anode or cathode layers at high temperatures, the electrode materials can be infiltrated into the porous structure and fired at low temperatures [5, 43, 124, 131]. Figure 8.11 shows the scanning electron microscopic (SEM) images of the porous YSZ with infiltrated nickel nanoparticles. The stability of the cell performance can be significantly improved by presintering the infiltrated fine nickel particles at a temperature just above the operating temperature. Compared to the peak power density of 100 mW cm^{-2} at 800 °C of the conventional sintering processes, with infiltrated nickel and LSM, the peak power density was increased to 332 mW cm^{-2} at 700 °C [5]. By replacing air with oxygen, the maximum power density of the cell with infiltrated electrodes was significantly increased to 1300 mW cm^{-2} at 700 °C, as shown in Figure 8.9 [5]. Similarly, it has been demonstrated that cell Fe–Cr/Ni–CGO/ScYSZ/CGO–LSCF achieved a peak power density of 1200 mW cm^{-2} at 749 °C [29]. Another alternative approach to avoid interfacial diffusion and nickel coarsening is using oxides as starting materials for both metallic support and anode and co-sintering in air [117, 118]. However, these metallic supports may be limited to the alloying elements that can be easily reduced from oxides at relatively low temperatures. In addition, the redox stability and the mismatch in CTE between the metals could be potential issues. Starting from sintering a mixture of NiO–Fe_2O_3, a Ni–Fe-alloy-supported SOFC has been fabricated [117]. The peak power density decreased from 900 to 600 mW cm^{-2} after one thermal cycle at 800 °C. A Ni-supported half-cell produced by co-sintering NiO/NiO–YSZ/YSZ in air has also been reported. The peak power density of the cell with screen-printed $Sr_{0.5}Sm_{0.5}CoO_3$ (SSCo) cathode was 350 and 800 mW cm^{-2} at 700 and 800 °C, respectively [119]. However, no data were reported for thermal cycling and redox stability from this study. Interfacial diffusion and nickel coarsening can be avoided by using ceramic materials without nickel or with less nickel as anode [3]. The tolerance to sulfur poisoning would also be increased with ceramic anode materials that are desired to increase the fuel flexibility.

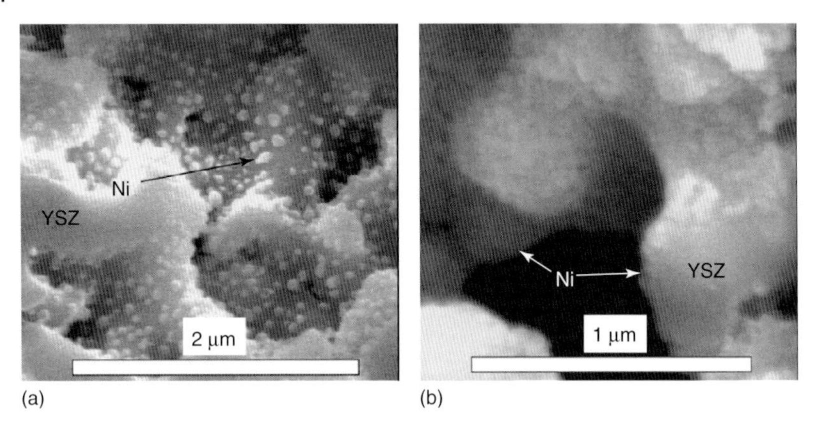

Figure 8.11 SEM images of porous YSZ structures infiltrated with Ni (a) once and (b) five times. (Source: Reproduced with permission from Ref. [5].)

8.4.2
Deposition Approaches

Instead of sintering the components at high temperatures, MS-SOFCs can be fabricated by deposition techniques without sintering processes [132]. Different deposition techniques have been employed for the fabrication of MS-SOFCs, such as thermal spray including plasma spray [23–27, 133, 134] and high-velocity oxy-fuel (HVOF) spray [23, 135], spray pyrolysis [136], PLD [24, 25, 117, 137], and EPD [15–18]. The porosities of the deposited layers depend on the deposition conditions, and the thickness of the layer is mainly a function of time. Most deposition techniques are easily scaled up. Additional features of different deposition approaches have been briefly summarized in Table 8.2. A consequent sintering process is required for the depositions with the EPD method and such cells have been discussed in the previous section. PLD has been employed to deposit $LaGaO_3$-based electrolyte on a NiO–Fe_2O_3 substrate that was sintered from powder of oxides in air at 1450 °C for 2 h [117]. The Ni–Fe alloy is formed in hydrogen by the reduction of presintered dense NiO–Fe_2O_3 substrate. The Ni–Fe alloy is used as both anode and support for the cell. With an electrolyte thickness around 5 μm, the maximum power density of the cell reached 1600 and 900 mW cm^{-2} at 700 and 600 °C, respectively, as shown in Figure 8.12. Thermal plasma spray and HVOF spray have been the most studied techniques for fabrication of MS-SOFCs during the past decades and are focused on in the following discussion.

Thermal plasma spray and HVOF spray are well-established and proven technologies for a variety of applications such as ceramic thermal barrier coatings on metals in aerospace industries [138, 139]. There are many advantages of using thermal spray techniques for MS-SOFCs [132]. Thermal spray potentially provides a much simplified and cost-effective choice for fabricating SOFC components and integrated cells [140, 141]. Thermal spray allows the entire multilayer SOFC to be processed in a consecutive spray process using only one piece of equipment and within

Table 8.2 Comparison of different deposition techniques used for SOFC fabrication.

Technique	Equipment cost	Technique maturity	Deposition rate	Deposition yield	Bonding strength	Post sintering
Thermal spray	Medium	High	High	Low	Medium	No
Pulsed laser deposition	Medium	High	Low	Medium	High	No
Electrophoretic deposition	Low	Medium	Medium	High	Low	Yes
Spray pyrolysis	Low	Low	Low	Medium	Medium	No

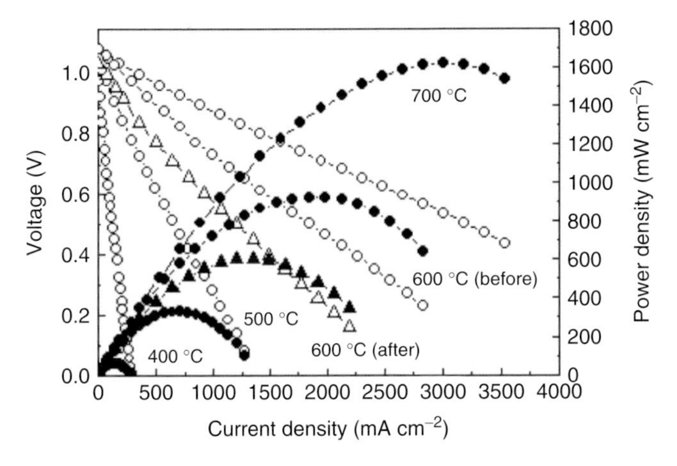

Figure 8.12 Polarization and power density planar cell fabricated by sintering and deposition processes (Ni–Fe/LSGM–SDC/$Sm_{0.5}Sr_{0.5}CoO_3$). (Source: Reproduced with permission from Ref. [117].)

a few minutes. The metallic substrates can be either planar or tubular shaped, with easy scale-up in terms of individual cell size and volume of production. Changing the spray conditions can easily modify the composition, porosity, and microstructure of the electrodes, such as dendrite structure [133]. The temperature of the metallic substrates can be controlled below 700 °C to avoid oxidation. The detrimental interfacial reactions and nickel coarsening in sintering processes can be avoided because there is no further sintering required for the deposited ceramic components.

The requirements for the application of thermally sprayed ceramic layers in SOFCs are much higher than those for the conventional applications. For example, the conventional thermally sprayed depositions generally have a porosity ranging from 5 to 15 vol%, which is either too porous for electrolyte layer or too dense for electrode structures for SOFCs. The conventional feedstock materials have an average particle size around 45 μm for high momentum, which makes it hard to get fully dense electrolyte layers with a thickness less than 20 μm. The lamellar microstructure formed by large splat stacking in the conventional plasma spray generates high residual defects and thermal stresses in SOFC components. In contrast to the conventional feedstock materials such as alumina, titania, and chromia, the cost of the materials for SOFCs is much higher and the materials' blowby has to be considered for reduction or recycling.

Significant efforts have been made to apply the thermal spray techniques for either materials production or component fabrication of SOFCs [132]. Most of the studies have focused on the deposition of porous electrodes and dense electrolytes through modifications of spray conditions and optimization of spray parameters. As in the cell fabrication by sintering processes, dense and thin electrolyte deposition has been the key for successful MS-SOFCs by deposition approaches. Suspensions or small particles have been employed to replace the conventional powder feedstock in thermal spray to obtain thin electrolyte deposition [142–148]. A suspension dispersed with desired materials is injected directly into the plasma flame. The plasma–liquid interaction atomizes the suspension into a fine mist and the suspension medium evaporates, thereby concentrating the solid content into microsized particles [149, 150]. The small particles are almost immediately accelerated to plasma gas velocity. At impact on the substrate, the particles form thin lamellae with rapid solidification rates. The splats produced by suspension plasma spray (SPS) are much smaller than those of conventional plasma spray [149, 150]. With a lower solid content of the suspensions, more electrolyte particles completely melt and a denser coating can be deposited. Thin electrolyte layers with 3–20 μm thickness and a more homogenous microstructure are achievable with SPS [143, 151, 152]. SPS processing can also be used for the preparation of electrode layers [153–155]. The smaller splats from SPS would benefit the occurrence of three-phase boundaries in the electrodes, which is desired for electrodes in SOFCs. Thin and dense depositions can also be achieved by directly spraying small particles less than 200 nm in diameter using a beveled injector [156, 157]. The particles are entrained into the outer shell of the plasma jet and are efficiently deposited onto a substrate. Deposition of YSZ has been achieved with a thickness of 5 μm and density greater than 98% using fine powders [158].

Compared to the plasma spray, the HVOF spray may offer much dense and crack/pinhole-free deposition. Suspensions as feedstock can also be combined with HVOF spray to reduce deposition thickness and improve density. Plasma spray offers a flame temperature above 8000 °C in a reducing atmosphere, which may lead to reduction of ceria-based electrolytes. The reduction of ceria from Ce^{4+} to Ce^{3+} is detrimental to both fuel utilization and structural stability of the cells, associated with increased electronic conduction and volume expansion. HVOF provides a much lower flame temperature (2600–3200 °C) but higher velocity of the sprayed particles [133]. The flame temperature in HVOF is lower than that in thermal plasma spray, but the temperature is still higher than the melting temperature of SDC (~2730 °C) or YSZ (~2700 °C). The high velocity of the flame would be helpful for the small particles in suspension to increase their momentum for denser coatings. The flame temperatures in HVOF can be adjusted according to the torch design, the fuel and fuel-to-oxygen ratio, the particle size and feed rate of feedstock, and the deposition distance. Tensile stresses that generally arise because of material shrinkage during solidification of the molten droplets often lead to transverse cracks in plasma-sprayed coatings [158]. The as-sprayed materials may also have a phase transformation from amorphous to crystalline post heat treatment or during application at elevated temperatures, which may lead to volume change and also cracks [158, 159]. Using HVOF, these stresses may be partially compensated by the plastic deformation of unmelted or partially molten and ductile ceramic particles impacting at high velocity [23, 160]. The top surface of the deposition is relatively smooth without the humps and defects often seen in SPS [23, 161]. The impacting colder particles may remove loosely attached material during deposition and result in low surface roughness.

Oberste Berghaus *et al.* [161, 162] and Killinger *et al.* [161, 162] have reported the feasibility of depositing ceramic oxides using HVOF spray and suspension feedstock with different fuels. The advantages of suspension feedstock over conventional powder feedstock for SOFC fabrication have also been demonstrated [160]. Using suspension feedstock, metal-supported half-cells have been produced by consequent depositions through air plasma spray (APS) for anode layer and HVOF spray for electrolyte layer on porous Hastelloy X [23]. The density of the electrolyte layer is greater than 98%, of the theoretical density estimated by SEM examination (Figure 8.13). The cathode materials were applied to the as-sprayed half-cells by screen printing and were heat treated at 800 °C for 2 h. Figure 8.14 summarized the electrochemical performance of the half-inch button cell Hastelloy X/NiO–SDC/SDC/SSCo with humidified hydrogen as the fuel and air as the oxidant at temperatures between 500 and 700 °C. The cell exhibits a maximum power density in excess of 0.50 W cm^{-2} at 600 °C and 0.92 W cm^{-2} at 700 °C when operated with humidified hydrogen fuel. The 5×5 cm^2 cells were also fabricated under the same conditions and they showed a maximum power density of 0.26 W cm^{-2} at 600 °C and 0.56 W cm^{-2} at 700 °C with dry hydrogen as the fuel. The impedance analysis showed that the ohmic resistance of the cells was the major polarization loss in all cells, whereas both ohmic and electrode polarizations were significantly increased when the operating temperature decreased from 700 to 500 °C. Table 8.3 summarizes the performance of the MS-SOFCs fabricated by sintering and deposition methods [5, 23]. The selected two cells are cell 1 that is fabricated by sintering processes (97% H_2 + 3% H_2O, Fe–Cr alloy/Ni–YSZ/YSZ/YSZ–LSM, O_2) and cell 2 that is fabricated by deposition processes (97% H_2 + 3% H_2O, Hastelloy X/NiO–SDC/SDC/SSCo, air). Even though the two cells are not completely comparable, there are still two things that may be concluded. First, either sintering approach or deposition approach provided reasonable interfacial and bulk resistance with compatible ohmic loss (R_e) for YSZ in cell 1 and SDC in cell 2. Second, although the electrode polarizations (R_p) of cell 2 are much lower than that of cell 1, the internal electronic conduction in SDC has led to decreased power density or fuel utilization.

Long-term operation and scale-up of MS-SOFCs have not been studied and reported very much. Intensive R&D to apply thermal spray for fabrication of SOFCs has been carried out since mid-1990s at the DLR [12–14, 163–174]. The

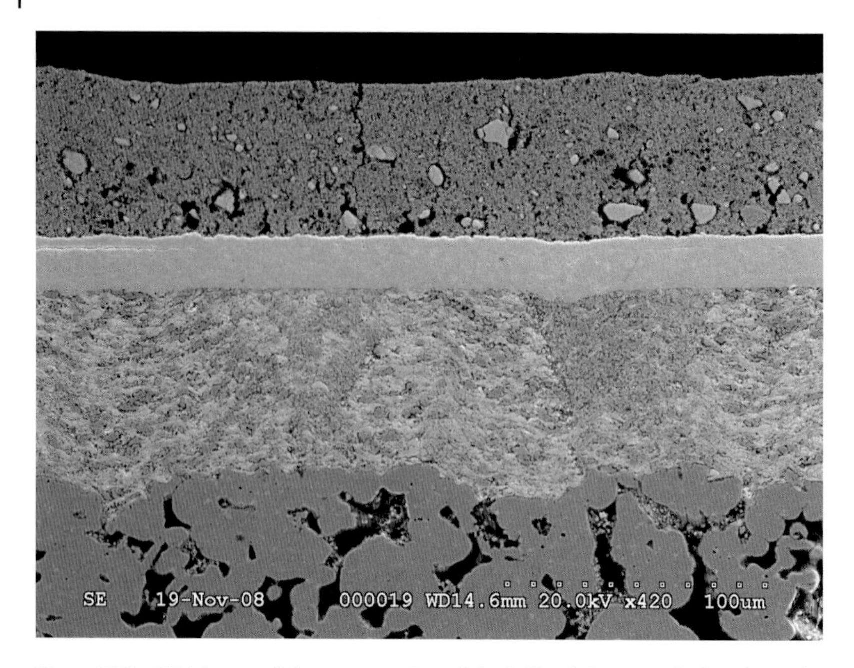

Figure 8.13 SEM image of the cross section of the half-inch button cell after electrochemical and thermal cycling tests. (Source: Reproduced with permission from Ref. [23].)

Table 8.3 Summary of MS-SOFCs by sintering (cell 1) and deposition (cell 2) methods [22, 129].

Temperature (°C)	Cell[a]	OCV (V)	R_e (Ω cm^2)	R_p (Ω cm^2)	Power density at 0.7 V (W cm^{-2})
750	Cell 1	1.08	0.118	0.139	0.98
	Cell 2	—	—	—	—
700	Cell 1	1.09	0.165	0.210	1.26
	Cell 2	0.81	0.101	0.014	0.56
650	Cell 1	1.06	0.257	0.155	0.70
	Cell 2	0.86	0.152	0.026	0.52
600	Cell 1	—	—	—	—
	Cell 2	0.90	0.262	0.070	0.37

[a]Cell 1: 97% H_2 + 3% H_2O, Fe–Cr alloy/Ni–YSZ/YSZ/YSZ–LSM, O_2; cell 2: 97% H_2 + 3% H_2O, Hastelloy X/NiO–SDC/SDC/SSCo, air.

DLR has evaluated the oxidation stability of chromium-, nickel-, and iron-based alloys and nickel felt under reducing atmospheres with water stream up to 50% at 900 °C. Among all the other candidate metals, nickel felt showed the best oxidation resistance as the potential support materials [12]. Based on nickel felt as substrates, the DLR has fabricated metal-supported cells with all the components by thermal

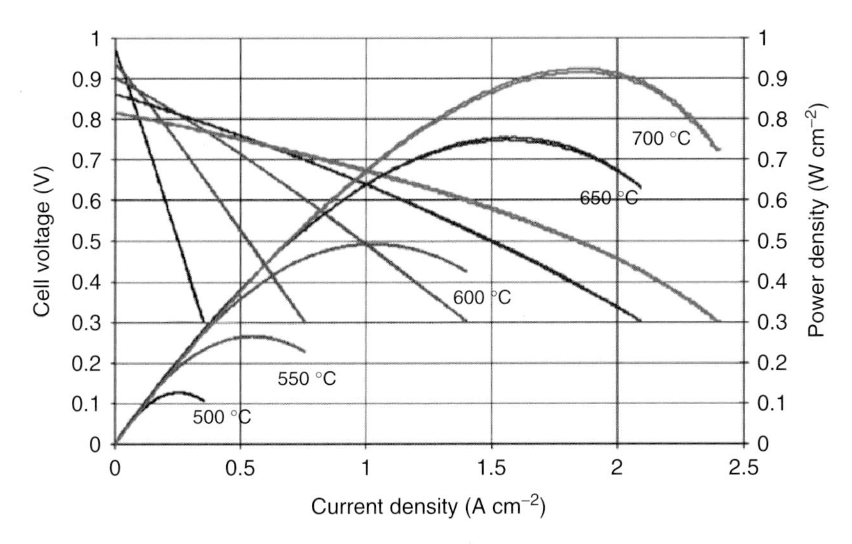

Figure 8.14 Electrochemical performance of the half-inch button cell made by thermal spray techniques (fuel: 97% H_2 with 3% H_2O, oxidant: air). (Source: Reproduced with permission from Ref. [23].)

spray techniques and scaled up the cell size to as large as 20×20 cm^2. The peak power density of the cell Ni-felt/Ni–ScSZ/ScSZ/ScSZ–LSM reached 300 mW cm^{-2} with hydrogen as the fuel and air as the oxidant at 750 °C [12]. In consortium with ElringKlinger AG, Plansee SE, and Sulzer Metco AG, the DLR is developing a lightweight 1 kW stack for auxiliary power unit (APU) [13]. Single cells of the structure Fe–Cr alloy/DBL/Ni–YSZ/8YSZ/DBL/LSCF have been produced and evaluated recently [13]. A new ODS ferritic steel (Fe–26Cr–Mn, Mo, Ti, Y_2O_3) has been developed as substrate in cooperation with Plansee SE (Reutte, Austria). The DLR claimed that the new alloy has shown considerably better high-temperature oxidation/corrosion behavior than Crofer 22 APU and the previous Plansee alloys [13]. The porous Fe–Cr metallic substrates were made by powder metallurgy. The materials $La_{0.6}Sr_{0.2}Ca_{0.2}CrO_3$ and $La_{0.6}Sr_{0.4}$ MnO_3 have been deposited by APS or physical vapor deposition as DBL at the anode and cathode sides, respectively. The DBLs prevented mutual diffusion of Cr, Fe, and Ni species between the substrate and the anode or between the electrolyte and the cathode, which assisted in enhancing the durability of the cells [173, 174] above 2000 h. The cells having DBL exhibited degradation rates less than 1% kh^{-1} after more than 2000 h of operation at 800 °C, as shown in Figure 8.15. The porous electrodes and the dense electrolyte layer were built by APS and reduced pressure spray with either vacuum plasma spray (VPS) or low-pressure plasma spray (LPPS), respectively. The cell performance has been improved with a new TripexPro spraying method and improved feedstock materials for electrodes. A 100 cm^2 single cell can be fabricated within a minute by the thermal spray method. The cell with 12.56 cm^2 at 800 °C has an OCV above 1.1 V and degradation in power density of less than 2.5% after 20 redox cycles. The power density has been increased from 170 to 619 mW

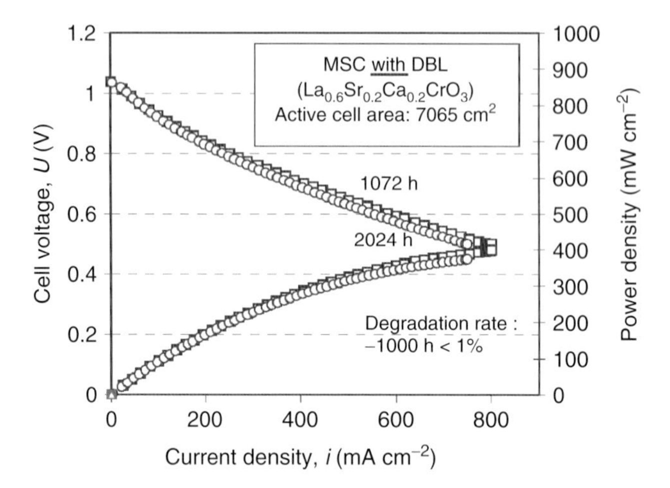

Figure 8.15 Long-term operations of metal-supported cell (MSC) at 1073 K under simulated reformate gas/air. (Source: Reproduced with permission from Ref. [13].)

cm^{-2} measured at 0.7 V for 12.56 cm^2 cells at 800 °C and upscaling to 100 cm^2 cells was successfully accomplished [13]. It has been reported that the ohmic loss and polarization loss for the 5 × 5 cm^2 cells was 0.221 and 0.014 Ω cm^2 at 700 °C, respectively. However, the ohmic loss and polarization loss for the 12.56 cm^2 cells measured by the DLR is 0.272 and 0.701 Ω cm^2 at 800 °C, respectively, which are relatively high and may limit the cell operation at low temperatures.

Figure 8.16 is a 3 kW class stack based on MS-SOFCs that was reported as early as in 2000 [21]. The anode and electrolyte layers were deposited on planar porous Ni-felt substrate by flame spraying (FS) and APS, respectively. The APS is also applied to form a (LaSr)MnO_3 protective coating on the metallic separators. The electrolyte layer has a bilayer structure of Y_2O_3-stabilized ZrO_2 (YSZ) and Y_2O_3-doped CeO_2 (YDC). The active electrode area of a single cell was 600 cm^2, as shown in Figure 8.1. The 3 kW class stack consisted of 30 cells with a seal-less structure and generated 3.3 kW at 970 °C at 300 mA cm^{-2} with fuel utilization of 50%. While the lower 15 cell block became unstable after 2100 h of operation, the upper 15 cell block was stably operated for 3200 h. The failure of the stack is most likely due to materials degradation at operating temperatures as high as 970 °C. There are many factors that could lead to performance degradation in MS-SOFCs, including increase in ohmic losses and electrode polarization losses such as substrate oxidation, delamination of the different layers, chemical reactions between electrode and metallic support, microstructure change of electrode, and chromium poisoning. Degradation studies at the Jülich Research Centre claimed that the high degradation rates were due to chromium diffusion from the metallic support into the anode layer [175]. However, the microstructure change or further sintering of cathode layers prepared at low temperatures may be the most pronounced cause for the rapid degradation at the beginning of cell evaluation [176]. The potential reaction between the electrode and the metallic support can be improved by using an electrically conducting

Figure 8.16 Image of a 3 kW class stack based on MS-SOFCs fabricated by plasma spray technique. (Source: Reproduced with permission from Ref. [21].)

and a chemically stable barrier layer. An MS-SOFC utilizing a barrier layer and plasma-sprayed YSZ electrolyte has achieved a power density of 400 mW cm^{-2} at 800 °C and was operated for 2300 h [116].

8.5 Summary

MS-SOFCs have many potential advantages compared with other types of SOFCs, such as low cost; high robustness against fast start-up, thermal cycling, and redox process; and improved overall materials reliability by reduced operating temperatures. These potentials have attracted increased worldwide interest during the past 20 years. MS-SOFCs have been successfully fabricated through either sintering processes or deposition approaches on many different types of metallic materials as substrates. The electrochemical performance of the cells has been reached to a level compatible with that of the conventional ceramic-supported SOFCs. A 3 kW class stack based on MS-SOFCs has also been demonstrated in 2000. However, MS-SOFCs are still in the early R&D stage, and there are currently no commercial products available. While there is still a need for high-performance electrolytes and electrode materials for any type of SOFCs, the following areas may be particularly challenging for the MS-SOFCs.

1) **Appropriate fabrication processes.** Both sintering and deposition approaches have demonstrated the feasibility to fabricate MS-SOFCs for high performance. However, more works are required for both methods in terms of repeatability of cell performance, optimization of process conditions, and detailed cost analysis. The most efficient fabrication processes may need to combine different techniques for appropriate components, such as thermal spray for porous

electrode deposition and PLD for dense and thin electrolyte deposition on commercial alloys.

2) **Commercially available porous metallic substrates, either tubular or planar ones**. The ferritic stainless steels with high Cr content are of particular interest as substrates for MS-SOFCs due to the thermal expansion match with that of the conventional SOFC components. However, nickel-based high-temperature alloys generally showed higher oxidation resistance and lower ASR than the ferritic stainless steels. Unfortunately, the CTE of commercially available Ni-based alloys limited their applications in SOFCs. By carefully selecting the alloy elements, Ni-based alloys with low thermal expansion could be designed [64, 177]. More work on the oxidation resistance in both oxidizing and reducing atmospheres and electrical properties may be needed for the new alloys and other potential candidate alloys such as Haynes HR-120, Incoloy 825, Incoloy 800, and Nicrofer 7520. In addition to the alloys with SiO_2, Al_2O_3, or Cr_2O_3 as the protective layer, alloys with TiO_2 former may benefit the ASR under reducing conditions because of the low electrical resistance of TiO_2. Obtaining a dense TiO_2 layer is key for acquiring high oxidation resistance [178].
3) **Ceramic coatings for the current commercially available alloys**. Although the commercially available ferritic stainless steels are most promising alloys as substrates for MS-SOFCs, ceramic coatings are required to avoid Cr evaporation and poisoning on the electrode materials and the continuous thermally grown oxides. Therefore, appropriate coating materials with high conductivity and chemical stability and coating techniques need to be further developed.
4) **Long-term stability of the cells and stacks**. Studies on the long-term stability and the degradation of MS-SOFCs are less reported due to the early stage of R&D. It has been demonstrated that either cells or stacks could operate steadily for a couple of thousand hours. Long-term stability and degradation mechanisms need to be addressed more in the future.

References

1. Singhal, S.C. and Kendall, K. (2004) *High-Temperature Solid Oxide Fuel Cells: Fundamentals, Design and Applications*, Elsevier, Oxford.
2. Williams, M.C., Strakey, J., and Surdoval, W. (2006) *J. Power Sources*, **159**, 1241–1247.
3. Visco, S.J., Jacobson, C.P., Villareal, I., Leming, A., Matus, Y., and Jonghe, L.C.D. (2003) *Proc. Electrochem. Soc.*, **07**, 1040–1050.
4. Singhal, S.C. (2002) *Solid State Ionics*, **152–153**, 405–410.
5. Tucker, M.C., Lau, G.Y., Jacobson, C.P., De Jonghe, L.C., and Visco, S.J. (2007) *J. Power Sources*, **171**, 477–482.
6. Tucker, M.C., Jacobson, C.P., De Jonghe, L.C., and Visco, S.J. (2006) *J. Power Sources*, **160**, 1049–1057.
7. Batawi, E., Doerk, T., Keller, M., Schuler, A., and Hager, C. (2000) in *Proceedings of the 4th European Solid Oxide Fuel Cell Forum* (ed. J.A. McEvoy), European Fuel Cell Forum, Lucerne, p. 735.
8. Robert, G., Kaiser, A., and Batawi, E. (2004) in *Proceedings of 6th European Solid Oxide Fuel Cell Forum* (ed. M. Mogensen), European Fuel Cell Forum, Lucerne, p. 193.
9. Sarantaridis, D. and Atkinson, A. (2007) *Fuel Cells*, **7**, 246–258.

10. Lamp, P., Tachtler, J., Finkenwirth, O., Mukerjee, S., and Shaffer, S. (2003) *Fuel Cells*, **3**, 146–152.
11. Williams, K.R. and Smith, J.G. (1966) Great Britain Patent GB 1,049,428.
12. Schiller, G., Franco, T., Henne, R., Lang, M., Ruckdäschel, R., Otschik, P., and Eichler, K. (2001) *Proceedings of the 7th International Symposium on Solid Oxide Fuel Cells (SOFC-VII)*, ECS Proceedings, Vol. 16, The Electrochemical Society, Pennington, NJ, pp. 885–894.
13. Szabo, P., Arnold, J., Franco, T., Gindrat, M., Refke, A., Zagst, A., and Ansar, A. (2009) *ECS Trans.*, **25**, 175–185.
14. Schiller, G., Henne, R., Lang, M., and Schaper, S. (1999) in *Proceedings of Solid Oxide Fuel Cells VI* (eds S.C. Singhal and M. Dokiya), The Electrochemical Society, Pennington, NJ, pp. 893–903.
15. Atkinson, S., Baron, N.P., Brandon, A., Esquirol, N., Kilner, J.A., Oishi, R., and Steele, B.C.H. (2003) in *Proceedings of 1st International Conference on Fuel Cell Science Engineering and Technology, Rochester, NY, April 21–23*, (eds R.K. Shah and S.G. Kandlikar), American Society of Mechanical Engineers, New York, ISBN: 0-7918-3668-1, 499–506.
16. Steele, B.C.H., Atkinson, A., Kilner, J.A., Brandon, N.P., and Rudkin, R.A. (2002) Great Britain Patents GB 2,368,450.
17. Attryde, P., Baker, A., Baron, S., Blake, A., Brandon, N.P., Corcoran, D., Cumming, D., Duckett, A., El-Koury, K., Haigh, D., Harrington, M., Kidd, C., Leah, R., Lewis, G., Matthews, C., Maynard, N., McColm, T., Selcuk, A., Schmidt, M., Trezona, R., and Verdugo, L. (2005) *Proc. Electrochem. Soc.*, **07**, 113–122.
18. Leah, R.T., Brandon, N.P., and Aguiar, P. (2005) *J. Power Sources*, **145**, 336–352.
19. Villareal, I., Jacobson, C.P., Leming, A., Matus, Y., Visco, S.J., and Jonghe, L.C.D. (2003) *Electrochem. Solid-State Lett.*, **6** (9), A178–A179.
20. Tucker, M., Sholklapper, T., Lau, G., Dejonghe, L., and Visco, S. (2009) *ECS Trans.*, **25** (2), 673–680.
21. Takenoiri, S., Kadokawa, N., and Koseki, K. (2000) *J. Therm. Spray Tech.*, **9**, 360–363.
22. Momma, A., Kaga, Y., Okuo, T., Fujii, K., Hohjyo, K., and Kanazawa, M. (1999) *Bull. Electrotech. Lab*, **63**, 1–11.
23. Hui, R., Berghaus, J.O., Decès-Petit, C., Qu, W., Yick, S., Legoux, J.G., and Moreau, C. (2009) *J. Power Sources*, **191**, 371–376.
24. Hui, S., Yang, D., Wang, Z., Yick, S., Decès-Petit, C., Qu, W., Tuck, A., Maric, R., and Ghosh, D. (2007) *J. Power Sources*, **167**, 336–339.
25. Huang, Q., Berghaus, J.O., Yang, D., Yick, S., Wang, Z., Wang, B., and Hui, R. (2008) *J. Power Sources*, **177**, 339–347.
26. Wang, Z., Berghaus, J.O., Yick, S., Decès-Petit, C., Qu, W., Hui, R., Maric, R., and Ghosh, D. (2008) *J. Power Sources*, **176**, 90–95.
27. Berghaus, J.O., Legoux, J.-G., Moreau, C., Hui, R., Decès-Petit, C., Qu, W., Yick, S., Wang, Z., Maric, R., and Ghosh, D. (2008) *J. Therm. Spray Tech.*, **17** (5), 700–706.
28. Oishi, N. and Yoo, Y. (2009) *ECS Trans.*, **25** (2), 739–744.
29. Blennow, P., Hjelm, J., Klemensø, T., Persson, Å., Brodersen, K., Srivastava, A., Frandsen, H., Lundberg, M., Ramousse, S., and Mogensen, M. (2009) *ECS Trans.*, **25** (2), 701–710.
30. Tucker, M.C. (2010) *J. Power Sources*, **195**, 4570–4582.
31. Miriam, K., Christoph, S., Azra, S., Bjørn, T., Tord, T., and Olav, B. (2005) *Proceedings of the International Symposium on Solid Oxide Fuel Cells 9*, The Electrochemical Society, Quebec City, Vol. 7 (2), pp. 659–669.
32. Gao, W. and Sammes, N.M. (1999) *An Introduction to Electronic and Ionic Materials*, World Scientific Publishing, Singapore, p. 340.
33. Mukerjee, S., Haltiner, K., Kerr, R., Chick, L., Sprenkle, V., Meinhardt, K., Lu, C., Kim, J.Y., and Weil, K. (2007) *Proceedings of Tenth International Symposium on Solid Oxide Fuel Cells, Nara,*

Japan, ECS Transactions, vol. 7 (1), The Electrochemical Society, Pennington, NJ, p. 59.

34. Steinberger-Wilckens, R., Blum, L., Buchkremer, H., Gross, S., de Haart, L., Hilpert, K., Nabielek, H., Quadakkers, W., Reisgen, U., Steinbrech, R.W., and Tietz, F. (2006) *Int. J. Appl. Ceram. Technol.*, **3**, 470–476.
35. Chiba, R., Yoshimura, F., and Sakurai, Y. (1999) *Solid State Ionics*, **124**, 281–288.
36. Chiba, R., Yoshimura, F., and Sakurai, Y. (2002) *Solid State Ionics*, **152–153**, 575–582.
37. Orui, H., Watanabe, K., Chiba, R., and Arakawa, M. (2004) *J. Electrochem. Soc.*, **151**, A1412–A1417.
38. Komatsu, T., Arai, H., Chiba, R., Nozawa, K., Arakawa, M., and Sato, K. (2006) *Electrochem. Solid-State Lett.*, **9**, A9–A12.
39. Komatsu, T., Arai, H., Chiba, R., Nozawa, K., Arakawa, M., and Sato, K. (2007) *J. Electrochem. Soc.*, **154**, B379–B382.
40. Komatsu, T., Chiba, R., Arai, H., and Sato, K. (2008) *J. Power Sources*, **176** (1), 132–137.
41. Fergus, J.W., Hui, R., Li, X., Wilkinson, D.P., and Zhang, J. (2009) *Solid Oxide Fuel Cells: Materials Properties and Performance*, CRC Press, New York.
42. Kharton, V.V., Marques, F.M.B., and Atkinson, A. (2004) *Solid State Ionics*, **174**, 135–149.
43. Jiang, S.P. (2006) *Mater. Sci. Eng. A*, **418**, 199–210.
44. Tietz, F.H., Buchkremer, P., and Stöver, D. (2002) *Solid State Ionics*, **152–153**, 373–381.
45. Sameshima, S., Kawaminami, M., and Hirata, Y. (2002) *J. Ceram. Soc. Jpn.*, **110** (1283), 597–600.
46. Yang, Z., Weil, K.S., Paxton, D.M., and Stevenson, J.W. (2003) *J. Electrochem. Soc.*, **150**, A1188–1201.
47. Kofstad, P. and Bredesen, R. (1992) *Solid State Ionics*, **52**, 69–75.
48. Srivastava, J.K., Prasad, M., and Wagner, J.B. Jr., (1985) *J. Electrochem. Soc.*, **132** (4), 955–963.
49. Nagai, H. and Ohbayashi, K. (1989) *J. Am. Ceram. Soc.*, **72**, 400.
50. Antepara, I., Villarreal, I., Rodriguez-Martinez, L.M., Lecanda, N., Castro, U., and Laresgoiti, A. (2005) *J. Power Sources*, **151**, 103–107.
51. Molin, S., Kusz, B., Gazda, M., and Jasinski, P. (2008) *J. Power Sources*, **181**, 31–37.
52. Bautista, A., Arahuetes, E., Velasco, F., Moral, C., and Calabres, R. (2008) *Oxid. Met.*, **70**, 267–286.
53. Montero, X., Tietz, F., Stover, D., Cassir, M., and Villarreal, I. (2009) *Corros. Sci.*, **51**, 110–118.
54. Geng, S.J., Zhu, J.H., and Lu, Z.G. (2006) *Solid State Ionics*, **177**, 559–568.
55. Wells, J.M., Hwang, S.K., and Hull, F.C. (1984) *Refractory Alloying Elements in Superalloys*, ASM International, Materials Park, OH, p. 175.
56. Kadowaki, T., Shiomitsu, T., Matsuda, E., Nakagawa, H., and Tsuneizumi, H. (1993) *Solid State Ionics*, **67**, 65–69.
57. Kofstad, P. (1988) *High Temperature Corrosion*, Elsevier Applied Science Publishers Ltd., London.
58. Quadakkers, W.J., Piron-Abellan, J., Shemet, V., and Singheiser, L. (2003) *High Temp.Mater.*, **20** (2), 115–127.
59. Quadakkers, W.J., Malkow, T., Pirón-Abbellán, J., Flesch, U., Shemet, V., and Singheiser, L. (2000) in *European Solid Oxide Fuel Cell Forum Proceedings, Lucerne, Switzerland*, Vol. 2 (ed. A.J. McEvoy), Elsevier, Amsterdam, pp. 827–836.
60. Khanna, A.S. (2005) in *Handbook of Environmental Degradation of Materials* (ed. M. Kutz), William Andrew Inc., New York, p. 124.
61. Crofer22APU (2004) High Temperature Alloy, MSDS No. 8005 June, ThyssenKrupp VDM.
62. Honegger, K., Plas, A., Diethelm, R., and Glatz, W. (2001) in *Solid Oxide Fuel Cells*, The Electrochemical Society Proceedings Series, Vol. PV 2001-lb (eds S.C. Singhal and M. Dokiya), The Electrochemical Society, Pennington, NJ, p. 803.
63. Horita, T., Xiong, Y., Yamaji, K., Sakai, N., and Yokokawa, H. (2003) *J. Electrochem. Soc.*, **150**, A243–248.

64. Jablonski, P.D. and Alman, D.E. (2007) *Int. J. Hydrogen Energy*, **32**, 3705–3712.
65. Church, B.C., Sanders, T.H., Speyer, R.F., and Cochran, J.K. (2007) *Mater. Sci. Eng. A*, **452–453**, 334–340.
66. Piron-Abellan, J., Shemet, V., Tietz, F., Singheiser, L., and Quadakkers, W.J. (2001) in *Solid Oxide Fuel Cells*, The Electrochemical Society Proceedings Series, Vol. PV 2001–16 (eds H. Yokokawa and S.C. Singhal), The Electrochemical Society, Pennington, NJ, p. 811.
67. Blum, L., de Haart, L.G.J., Vinke, I.C., Stolten, D., Buchkremer, H.-P., Tietz, F., Blap, G., Stover, D., Remmel, J., Cramer, A., and Sievering, R. (2002) in *Proceedings of the 5th European Solid Oxide Fuel Cell Forum* (ed. J. Huijsmans), European Fuel Cell Forum, Luzern, p. 784.
68. Yang, Z., Hardy, J.S., Walker, M.S., Xia, G., Simner, S.P., and Stevenson, J.W. (2004) *J. Electrochem. Soc.*, **151** (11), A1825–1831.
69. Gannon, P.E., Gorokhovsky, V.I., Deibert, M.C., Smith, R.J., Kayani, A., White, P.T., Sofie, S., Yang, Z., McCready, D., Visco, S., Jacobson, C., and Kurokawa, H. (2007) *Int. J. Hydrogen Energy*, **32** (16), 3672–3681.
70. Wagner, C. (1951) *Atom Movements*, American Society of Metals, Cleveland, OH.
71. Hauffe, K. (1965) *Oxidation of Metals*, Plenum Press, New York.
72. Quadakkers, W.J., Hänsel, M., and Rieck, T. (1998) *Mater. Corros.*, **49**, 252–257.
73. Thierfelder, W., Greiner, H., and Köck, W. (1997) in *Solid Oxide Fuel Cells (SOFC V)*, The Electrochemical Society Proceedings Series, Vol. PV 97–40 (eds U. Stimming, S.C. Singhal, H. Tagawa, and W. Lehnert), The Electrochemical Society, Pennington, NJ, p. 1306.
74. Hirota, K., Mitani, K., Yoshinaka, M., and Yamaguchi, O. (2005) *Mater. Sci. Eng. A*, **399**, 154–160.
75. Quadakkers, W.J., Greiner, H., and Köck, W. (1994) in *Proceedings of the First European Solid Oxide Fuel Cell Forum, European SOFC Forum*, Vol. 2 (ed. U. Bossel), European Fuel Cell Forum, Oberrohrdorf, p. 525.
76. Saunders, S.R.J., Monteiro, M., and Rizzo, F. (2008) *Prog. Mater. Sci.*, **53** (5), 775–837.
77. Hänsel, M., Quadakkers, W.J., and Young, D.J. (2003) *Oxid. Met.*, **59** (3/4), 285–301.
78. Min, K., Sun, C., Qu, W., Zhang, X., Decès-Petit, C., and Hui, R. (2009) *Int. J. Green Energy*, **6** (6), 627–637.
79. Liu, Y. (2008) *J. Power Sources*, **179**, 286–291.
80. England, D.M. and Virkar, A.V. (2001) *J. Electrochem. Soc.*, **148** (4), A330–A338.
81. Horita, T. (2008) *Int. J. Hydrogen Energy*, **33** (21), 6308–6315.
82. Horita, T., Yamaji, K., Xiong, Y., Kishimoto, H., Sakai, N., and Yokokawa, H. (2004) *J. Power Sources*, **131**, 293–298.
83. Holt, A. and Kofstadt, P. (1994) *Solid State Ionics*, **69**, 137–143.
84. Holt, A. and Kofstadt, P. (1999) *Solid State Ionics*, **117**, 21–25.
85. Holt, A. and Kofstadt, P. (1997) *Solid State Ionics*, **100**, 201–209.
86. Holt, A. and Kofstadt, P. (1994) *Solid State Ionics*, **69**, 127–136.
87. Liu, H., Stack, M., and Lyon, S. (1998) *Solid State Ionics*, **109**, 247–257.
88. Nagai, H. and Fujikawa, T. (1983) *Trans. JIM*, **24** (8), 581–588.
89. Nagai, H., Ishikawa, S., and Amano, N. (1985) *Trans. JIM*, **26** (10), 753–760.
90. Huang, K., Hou, P.Y., and Goodenough, J.B. (2001) *Mater. Res. Bull.*, **36**, 81–95.
91. Teller, O., Meulenberg, W.A., Tietz, F., Wessel, E., and Quadakkers, W.J. (2001) in *Solid Oxide Fuel Cells*, The Electrochemical Proceedings Series, Vol. PV 2001–16 (eds H. Yokokawa and S.C. Singhal), The Electrochemical Society, Pennington, NJ, p. 895.
92. Belogolovsky, I., Hou, P.Y., Jacobson, C.P., and Visco, S.J. (2008) *J. Power Sources*, **182**, 259–264.
93. Brylewski, T., Nanko, M., Maruyama, T., and Przybylski, K. (2001) *Solid State Ionics*, **143**, 131–150.
94. Piron-Abellan, J., Tietz, F., Shemet, V., Gil, A., Ladwein, T., Singheiser, L., and

Quadakkers, W.J. (2002) in *Proceedings of the 5th European Solid Oxide Fuel Cell Forum* (ed. J. Huijsmans), The European SOFC Forum, Lucerne, p. 248.

95. England, D.M. and Virkar, A.V. (1999) *J. Electrochem. Soc.*, **146**, 3196–3202.
96. Linderoth, S., Hendriksen, P.V., and Mogensen, M. (1996) *J. Mater. Sci.*, **31**, 5077–5082.
97. Badwal, S.P.S., Deller, R., Foger, K., Ramprakash, Y., and Zhang, J.P. (1997) *Solid State Ionics*, **99**, 297–310.
98. Badwal, S.P.S., Bolden, R., and Foger, K. (1998) in *Proceedings of the 3rd European Solid Oxide Fuel Cell Forum*, Vol. 1 (ed. P. Stevens), The European SOFC Forum, Oberrohrdorf, p. 105.
99. Buchkremer, H.P., Diekmann, U., de Haart, L.G.J., Kabs, H., Stover, D., and Vinke, I.C. (1998) in *Proceedings of the 3rd European Solid Oxide Fuel Cell Forum*, Vol. 1 (ed. P. Stevens), The European SOFC Forum, Lucerne, p. 143.
100. Harada, H., Yamagata, T., Yokokawa, T., Ohno, K., and Yamazaki, M. (2002) *Proceedings of the Seventh Leige Conference on Materials for Advanced Power Engineering, September 30–October 3, 2002*, Energy and Technology, Vol. 21, Forschungszentium Julich Gmbh Institut fur Werkstoffe und Verfahren der Energietechnik.
101. Muzyka, D.R., Whitney, C.R., and Schlosser, D.K. (1975) *J. Miner*, **11**, 11–15.
102. Hwang, S.K., Hull, F.C., and Wells, J.M. (1984) *Superalloys*, TMS, Warrendale, PA, p. 785.
103. Morrow, H., Sponseller, D.L., and Semchyhsen, M. (1975) *Metall. Trans. A*, **6A**, 477–485.
104. Sung, P.K. and Poirier, D.R. (1998) *Mater. Sci. Eng. A*, **245**, 135–141.
105. Shaigan, N., Qu, W., Ivey, D.G., and Chen, W. (2009) *J. Power Sources*, **195**, 1529–1542.
106. Gindorf, C., Singheiser, L., and Hilpert, K. (2000) *Fortschr.-Ber. VDI, Reihe*, **15** (2), 723–726.
107. Hilpert, K., Das, D., Miller, M., Peck, D.H., and Weib, R. (1996) *J. Electrochem. Soc.*, **143**, 3642–3647.
108. Quadakkers, W.J., Greiner, H., Hansel, M., Pattanaik, A., Khanna, A.S., and Mallener, W. (1996) *Solid State Ionics*, **91**, 55–67.
109. Batawi, E., Plas, A., Strab, W., Honegger, K., and Diethelm, R. (1999) in *Solid Oxide Fuel Cells*, The Electrochemical Society Proceedings Series, Vol. PV 99–19 (eds S.C. Singhal and D. Dokiya), The Electrochemical Society, Pennington, NJ, p. 767.
110. Linderoth, S. (1996) *Surf. Coat.Technol.*, **80**, 185–190.
111. Fergus, J.W. (2005) *Mater. Sci. Eng. A*, **397**, 271–283.
112. Brandner, M., Bram, M., Froitzheim, J., Buchkremer, H.P., and Stoever, D. (2008) *Solid State Ionics*, **179**, 1501–1504.
113. Christiansen, N. (2006) Presentation in Fuel Cell Seminar Conference, Honolulu, HI.
114. Diethelm, R., Schmidt, M., Honegger, K., and Batawi, E. (1999) in *Solid Oxide Fuel Cells VI*, The Electrochemical Society Proceedings Series, Vol. PV 99–19 (eds S.C. Singhal and M. Dokiya), The Electrochemical Society, Pennington, NJ, p. 60.
115. Frei, J., kruschwitz, R., and Voisard, C. (2005) in *Solid Oxide Fuel Cells VI*, The Electrochemical Society Proceedings Series, Vol. PV 2005–07 (eds J. Mizusaki and S.C. Singhal), The Electrochemical Society, Pennington, NJ, pp. 1781–1788.
116. Franco, T., Schibinger, K., Ilhan, Z., Schiller, G., and Venskutonis, A. (2007) *ECS Trans.*, **7** (1), 771–780.
117. Ishihara, T., Yan, J., Enoki, M., Okada, S., and Matsumoto, H. (2008) *J. Fuel Cell Sci. Tech.*, **5**, 031205.
118. Kong, Y., Hua, B., Pu, J., Chi, B., and Li, J. (2009) *Int. J. Hydrogen Energy*, **35** (10), 4592–4596.
119. Cho, H.J., Park, Y.M., and Choi, G.M. (2011) *Solid State Ionics* **192** (1), 519–522.
120. Rodriguez-Martinez, L.M., Rivas, M., Otaegi, L., Gomez, N., Alvarez, M.A., Sarasketa-Zabala, E., Manzanedo, J., Burgosc, N., Castroc, F., Laresgoiti, A., and Villarreal, I. (2011) *ECS Trans.*, **35** (1), 445–450.

121. Sakuno, S., Takahashi, S., and Sasatsu, H. (2009) *ECS Trans.*, **25** (2), 731–737.
122. Rodriguez-Martinez, L.M., Otaegi, L., Alvarez, M., Rivas, M., Gomez, N., Zabala, A., Arizmendiarrieta, N., Antepara, I., Urriolabeitia, A., Olave, M., Villarreal, I., and Laresgoiti, A. (2009) *ECS Trans.*, **25** (2), 745–752.
123. Brandon, N.P. (2007) Hydrogen and Fuel Cells 2007, International Conference and Trade Show, Vancouver, BC, Canada.
124. Tucker, M.C., Lau, G.Y., Jacobson, C.P., DeJonghe, L.C., and Visco, S.J. (2008) *J. Power Sources*, **175** (1), 447–451.
125. Cassidy, M., Lindsay, G., and Kendall, K. (1996) *J. Power Sources*, **61**, 189–192.
126. Malzbender, J., Wessel, E., and Steinbrech, R. (2005) *Solid State Ionics*, **176** (29/30), 2201–2203.
127. Sholklapper, T.Z., Kurokawa, H., Jacobson, C.P., Visco, S.J., and De Jonghe, L.C. (2007) *Nano Lett.*, **7** (7), 2136–2141.
128. Sholklapper, T.Z., Lu, C., Jacobson, C.P., Visco, S.J., and DeJonghe, L.C. (2006) *Electrochem. Solid-State Lett.*, **9** (8), A376–A378.
129. Sholklapper, T.Z., Radmilovic, V., Jacobson, C.P., Visco, S.J., and De Jonghe, L.C. (2007) *Electrochem. Solid-State Lett.*, **10** (4), B74–B76.
130. Jung, S., Lu, C., He, H., Ahn, K., Gorte, R.J., and Vohs, J.M. (2006) *J. Power Sources*, **154**, 42–50.
131. Matus, Y.B., Jonghe, L.C.D., Jacobson, C.P., and Visco, S.J. (2005) *Solid State Ionics*, **176**, 443–449.
132. Hui, R., Wang, Z., Kesler, O., Rose, L., Jankovic, J., Yick, S., Maric, R., and Ghosh, D. (2007) *J. Power Sources*, **170**, 308–323.
133. Müller, M., Bouyer, E., Bradke, M.V., Branston, D.W., Heimann, R.B., Henne, R., Lins, G., and Schiller, G. (2002) *Materialwiss. Werkstofftech.*, **33** (6), 322–330.
134. Hwang, C.S., Tsai, C.H., Yu, J.F., Chang, C.L., Lin, J.M., Shiu, Y.H., and Cheng, S.W. (2011) *J. Power Sources*, **196**, 1932–1939.
135. Gadow, R., Killinger, A., Candel Ruiz, A., Weckmann, H., Öllinger, A., and Patz, O. (2007) International Thermal Spray Conference and Exposition (ITSC'2007), Beijing, Paper 15443.
136. Xie, Y., Roberto, N., Ching-shiung, H., Zhang, X., and Cyrille, D. (2008) *J. Electrochem. Soc.*, **155**, B407–B410.
137. Coocia, L.G., Tyrrell, G.C., Kilner, J.A., Waller, D., Chater, R.J., and Boyd, I.W. (1996) *Appl. Surf. Sci.*, **96–98**, 795–801.
138. Kesler, O., Finot, M., Suresh, S., and Sampath, S. (1997) *Acta Mater.*, **45**, 3123–3134.
139. Kesler, O., Matejicek, J., Sampath, S., Suresh, S., Gnaeupel-Herold, T., Brand, P.C., and Prask, H.J. (1998) *Mater. Sci. Eng. A*, **257**, 215–224.
140. Itoh, H., Mori, M., Mori, N., and Abe, T. (1994) *J. Power Sources*, **49**, 315–332.
141. Gauckler, L.J. (2000) *Solid State Ionics*, **131**, 79–96.
142. Bonneau, M.E., Gitzhofer, F., and Boulos, M.I. (2000) International Thermal Spray Conference and Exposition (ITSC'2000), Montreal, Canada.
143. Fauchais, P., Rat, V., Delbos, C., Coudert, J.F., Chartier, T., and Bianchi, L. (2005) *IEEE Trans. Plasma Sci.*, **33**, 920–930.
144. Berghaus, J.O., Legoux, J.G., Moreau, C., Hui, R., and Ghosh, D. (2006) Thermec' 2006, International Conference Proceedings on Advanced Materials and Manufacturing, Vancouver, BC, Canada, p. 92.
145. Waldbillig, D. and Kesler, O. (2009) *Surf. Coat. Technol.*, **203**, 2098–2101.
146. Waldbillig, D. and Kesler, O. (2009) *J. Power Sources*, **191**, 320–329.
147. Karthikeyan, J., Berndt, C.C., Tikkanen, J., Reddy, S., and Herman, H. (1997) *Mater. Sci. Eng. A*, **238**, 275–286.
148. Mizoguchi, Y., Kagawa, M., Suzuki, M., Syono, Y., and Hirai, T. (1994) *Nanostruct. Mater.*, **4** (5), 591–596.
149. Blazdell, P. and Kuroda, S. (2000) *Surf. Coat. Technol.*, **123**, 239–246.
150. Fazilleau, J., Delbos, C., Violier, M., Coudert, J.-F., and Fauchais, P. (2003) in *Thermal Spray 2003: Advancing the Science and Applying the Technology* (eds B.R. Marple and C. Moreau), ASM International, Materials Park, OH, pp. 889–893.

151. Fauchais, P., Etchart-Salas, R., Delbos, C., Tognonvi, M., Rat, V., Coudert, J.F., and Chartier, T. (2007) *J. Phy. D: Appl. Phys.*, **40**, 2394–2406.
152. Delbos, C., Fazilleau, J., Coudert, J.F., Fauchais, P., and Bianchi, L. (2003) in *Thermal Spray 2003: Advancing the Science and Applying the Technology* (eds B.R. Marple and C. Moreau), ASM International, Materials Park, OH, pp. 661–669.
153. Monterrubio-Badillo, C., Ageorges, H., Chartier, T., Coudert, J.F., and Fauchais, P. (2006) *Surf. Coat. Technol.*, **200**, 3743–3756.
154. Wang, Y. and Coyle, T.W. (2007) *J. Therm. Spray Tech.*, **16** (5–6), 899–904.
155. Wang, Y. and Coyle, T.W. (2008) *J. Therm. Spray Tech.*, **17** (5–6), 692–699.
156. Mawdsley, J.R., Jennifer Su, Y., Faber, K.T., and Bernecki, T.F. (2001) *Mater. Sci. Eng. A*, **308**, 189–199.
157. Boss, D.E., Bernecki, T., Kaufman, D., and Barnett, S. (1997) Proceedings of the Fuel Cell'97 Review Meeting, Morgantown, West Virginia, August 26–28.
158. Hui Rob, S., Zhang, H., Dai, J., Ma, X., Xiao, T.D., and Reisner, D.E. (2003) Eighth International Symposium on Solid Oxide Fuel Cells (SOFC-VIII), Paris, France.
159. Hui Rob, S., Dai, J., Roth, J., and Xiao, D. (2003) MRS Fall Meeting & Exhibit, Boston, MA.
160. Gadow, R., Killinger, A., Candel Ruiz, A., Weckmann, H., Öllinger, A., and Patz, O. (2007) *Proceedings of 2007 International Thermal Spray Conference, Global Coatings Solutions*, ASM International, Beijing, pp. 1053–1058.
161. Oberste Berghaus, J., Legoux, J.G., Moreau, C., Tarasi, F., and Chraska, T. (2008) *J. Therm. Spray Tech.*, **17** (1), 91–104.
162. Killinger, A., Kuhn, M., and Gadow, R. (2006) *Surf. Coat.Technol.*, **201**, 1922–1929.
163. Lang, M., Bilgin, M., Henne, R., Schaper, S., and Schiller, G. (1998) Proceedings of the 3rd European Solid Oxide Fuel Cell Forum, Nantes, France, pp. 161–170.
164. Schiller, G., Henne, R.H., Lang, M., Ruckdäschel, R., and Schaper, S. (2000) *Fuel Cells Bull.*, **3**, 7–12.
165. Lang, M., Franco, T., Henne, R., Schaper, S., and Schiller, G. (2000) Proceedings of the 4th European Solid Oxide Fuel Cell Forum, Lucerne, Switzerland, pp. 231–240.
166. Lang, M., Henne, R., Schaper, S., and Schiller, G. (2001) *J. Therm. Spray Tech.*, **10**, 618–625.
167. Lang, M., Franco, T., Schiller, G., and Wagner, N. (2002) *J. Appl. Electrochem.*, **32**, 871–874.
168. Lang, M., Franco, T., Henne, R., Schiller, G., and Ziehm, S. (2003) Fuel Cell Seminar Conference, Florida, pp. 794–797.
169. Schiller, G., Franco, T., Henne, R., Lang, M., and Szabo, P. (2003) *Proc. Electrochem. Soc.*, 1051–1058.
170. Schiller, G., Henne, R., Lang, M., and Müller, M. (2004) *Fuel Cells*, **4**, 56–61.
171. Henne, R. (2007) *J. Therm. Spray Tech.*, **16** (3), 381–403.
172. Schiller, G., Franco, T., Lang, M., Metzger, P., and Störmer, A.O. (2005) *Proc. Electrochem. Soc.*, 66–75.
173. Franco, T., Hoshiar Din, Z., Szabo, P., Lang, M., and Schiller, G. (2007) *J. Fuel Cell Sci. Technol.*, **4**, 406–412.
174. Schiller, G., Ansar, A., Lang, M., and Patz, O. (2009) *J. Appl. Electrochem.*, **39**, 293–301.
175. Hathiramani, D., Vaßen, R., Mertens, J., Sebold, D., Haanappel, V.A.C., and Stöver, D. (2007) *Adv. Solid Oxide Fuel Cells II: Ceram. Eng. Sci. Proc.*, **27** (4), 55–65.
176. Zhang, X., Robertson, M., Deces-Petit, C., and Kesler, O. (2009) *ECS Trans.*, **25** (2), 701–710.
177. Ryotaro, M., Hisataka, K., Yoshikuni, K., Ryuichi, Y., Toshiharu, N., Susumu, I., and Michio, O. (2007) US Patent 7,160,400.
178. Stanislowski, M., Froitzheim, J., Niewolak, L., Quadakkers, W.J., Hilpert, K., Markus, T., and Singheiser, L. (2007) *J. Power Sources*, **164**, 578–589.

9
Molten Carbonate Fuel Cells

Stephen J. McPhail, Ping-Hsun Hsieh, and Jan Robert Selman

9.1
Introduction

With the rapid ascent of developing countries and the continued growth of the industrialized world, the pressure on the supply of energy is ever increasing. It is becoming critical to maximize efficiency and minimize environmental impact. Reuse and sustainability are crucial in establishing the future energy infrastructure, and new technologies and approaches are called for. Fuel cells have the potential to become key applications in order to rationalize the production and distribution of usable energy, but persistent issues tied to durability and production costs slow down the effective implementation of these electrochemical devices. A major drawback in the immediate use of fuel cells is the scarce availability of sufficiently pure hydrogen as a generally adopted energy carrier. In this respect, high-temperature fuel cells, such as the molten carbonate fuel cells (MCFCs), have a clear advantage over their low-temperature counterparts, since the tolerance toward fuel impurities is higher and it is possible to utilize hydrocarbon fuels as well as hydrogen. Nevertheless, reducing the carbon footprint of our society is a necessity, especially given the developments in climate change. This can be achieved by capturing and confining anthropogenic CO_2 emissions (an immediate measure) and by replacing fossil-based fuels with renewable or waste-derived fuels (a more sustainable solution). MCFCs are unique in being able to do both these things. Thanks to their operating principle, CO_2 can be extracted from a gas stream on the cathode side and hydrocarbon fuels such as biogas can be converted to electricity on the anode side.

In this chapter, after a brief overview of MCFC development history, we first look at the basic operating principles of MCFCs (Section 9.2), and then present a review of the state-of-the-art materials and components (Section 9.3). In Section 9.4, the major challenges and most pressing needs of the technology are discussed in relation to the most promising applications of the MCFCs. Nevertheless, MCFC systems and plants are edging their way into the highly competitive market of distributed heat and power generation, and the status of their deployment is reviewed in Section 9.5.

Materials for High-Temperature Fuel Cells, First Edition. Edited by San Ping Jiang and Yushan Yan.

9.1.1
Development History of the MCFCs

The basic principle of fuel cell operation was demonstrated in 1839 by Sir William Grove (Figure 9.1) by reversing the water electrolysis reaction by means of a prototype that operated with two separate platinum rods surrounded by closed tubes containing oxygen and hydrogen gas submerged in a dilute sulfuric acid electrolyte solution.

The limited usefulness of the invention at the time along with the technological eruption of the internal combustion engine in the late nineteenth century made fuel cells to remain in the realm of scientific curiosity. Despite this, important progress was made in this period toward the fundamental understanding of the electrochemical mechanisms in high-temperature fuel cells. In Europe, Nernst and Schottky were the scientific pioneers who focused on electrolytes of solid oxides. The first scientists in Japan to develop a fuel cell, in 1935, were Tamaru and Ochiai, using a charcoal anode in a eutectic of molten carbonates [2].

It was not until the late 1950s when the interest for fuel cells increased, thanks to the necessity to generate auxiliary power devices for space vehicles in the space race between the United States and the Soviet Union. Scientists G. H. J. Broers and J. A. A. Ketelaar focused their attention on the use of molten carbonate salts as electrolyte and were the first ones to demonstrate the functioning of a modern style MCFC at laboratory scale. In 1960, they reported to have run a cell for 6 months nonstop, and the test had to be interrupted due to continuous electrolyte leakage. In the mid-1960s Texas Instruments, associated with the US Army, developed a series of MCFC prototypes ranging from 100 W to 1 kW running on battlefield fuel, which was transformed to hydrogen via an external reformer; however, further development was letup. Figure 9.2 shows an MCFC made for the US Army in 1966.

The petrol crisis in the early 1970s and the consequent rise in energy costs was a turning point for the fuel cell industry, as more efficient methods of power generation started to gain the attention of the general public. By the end of

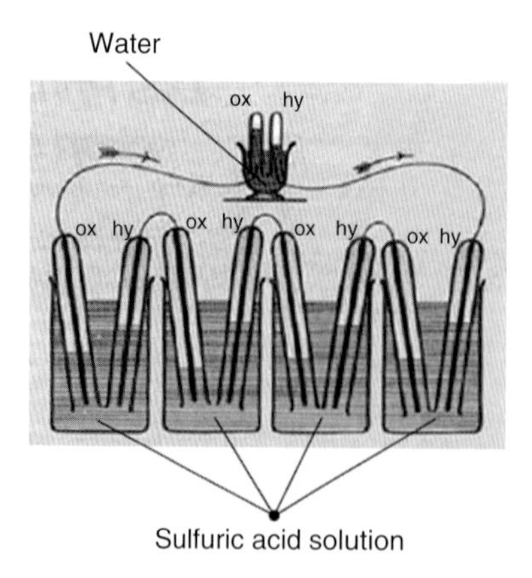

Figure 9.1 Illustration of Grove's fuel cell. (Source: After Ref. [1].)

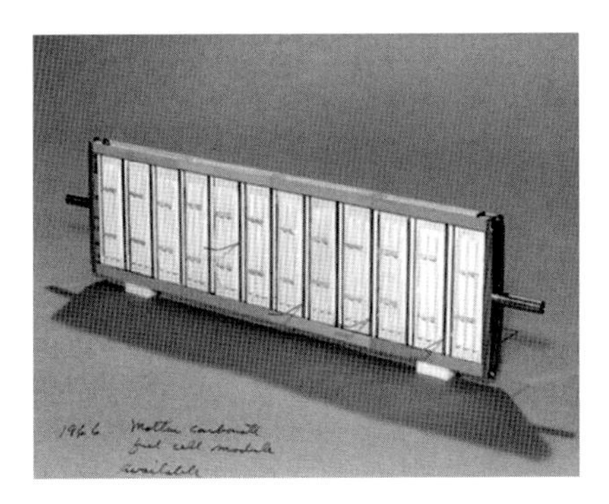

Figure 9.2 MCFC manufactured for the US Army in 1966. (Source: After Ref. [3].)

the decade and the following one, research and development of MCFC toward commercialization had a considerable growth in the United States, Japan, and Europe, concentrating the efforts on high-performance and high-pressure stack technologies as the basis for power generation and cogeneration. Several individual scale-up and proof-of-concept pilot systems were tested during the 1990s. In the first lustrum, the Ishikawajima-Heavy Industries in Japan successfully operated a 1 kW MCFC for 10 000 continuous hours, while FuelCell Energy (FCE) Inc. demonstrated in 1992 a 120 kW high-temperature carbonate fuel cell system, and in the period of 1996–1997, the company operated a 2 MW demonstration plant in Santa Clara (California) for 3000 h [3, 4].

In 2000, the Ishikawajima-Harima Heavy Industries started a mission to develop systems ready for commercialization. In 2002–2003, the demonstration phase started and four 300 kW MCFC systems were installed, operating at a pressure of 4 bar and a current density of 200 mA cm^{-2}. At the Aichi International Exposition (2005), two of these systems achieved a 51% gross efficiency operating on digester gas produced from waste collected within the exhibition area.

After development and field demonstration, in the past decade, several MCFC "HotModules" of 250 kW_{el} were installed by the German company MTU, incorporating the internal reforming stacks produced by FCE. The electrical efficiency achieved was approximately 50% and all test units achieved long operation times, one of more than 23 000 h and others running more than 18 000 h [5].

Ansaldo Fuel Cells, situated in Italy, developed from 1999 to 2008 the demonstration phase of the "2TW Series," a hybrid plant combining four MCFC stacks and a microturbine. These components are integrated in a hot pressurized vessel. The rated power is up to 500 kW, with an operating pressure of 3.5 bar and an integrated (external) reformer for natural gas; however, the use of alternative fuels such as diesel oil and simulated biogas was validated.

The year 2011 saw a stark contrast in development of the MCFCs as a commercial product. While MTU and Ansaldo Fuel Cells had to close down their activities due to strategic decisions by their respective holding groups (Tognum and Finmeccanica), Franco Cell emerged as a new original equipment manufacturing (OEM) player, presenting an ambitious project for the installation of 30 MW of MCFC power in Guadeloupe, while FCE generated a gross profit of $0.2 million for the third quarter of 2011 from MCFC-related products and services. This is an important milestone on the path to profitability as it is the first quarterly gross profit since FCE started to commercialize their MCFC technology. The product cost to revenue ratio dropped below unity because of the increased production volume, achieving a rate of 56 MW annually compared to the 22 MW of production for 2010 [6].

The $129 million order for 70 MW of fuel cell kits and other equipment and services received during the third quarter of 2011 illustrates the growing demand and sizeable market potential for clean baseload distributed generation in South Korea. Under this contract, FCE will export 2.8 MW of fuel cell kits monthly beginning in October 2011 through October 2013 to POSCO Power. South Korea is aggressively installing new and renewable power generation to reduce pollutants and carbon emissions while simultaneously developing a clean technology industry to create jobs. South Korea adopted a renewable portfolio standard that takes effect in 2012 and mandates approximately 6000 MW of new and renewable power through 2022, including fuel cells operating on either natural gas or renewable biogas.

Using fuel cell components supplied by FCE, POSCO Power assembled their first fuel cell stack during the third quarter of 2011 in their recently built fuel cell module assembly plant and installed the completed power plant at a customer site. The POSCO Power fuel cell module assembly and balance of plant (BOP) facilities are designed for 100 MW annual capacity, using fuel cell components purchased from FCE. Since 2007, POSCO Power has ordered 140 MW of fuel cell power plants, modules, and components [6].

For more details on MCFC deployment and on the variety of their applications, see Section 9.5.

9.2 Operating Principle

What distinguishes the MCFCs from other hydrogen-oxygen fuel cells is the employment of a molten salt electrolyte. The MCFC is operated at 580–700 °C to keep the salt in liquid state and exploit the high conductivity of the electrolyte in this temperature range. As has been underlined, high temperature offers distinct advantages: (electro)chemical reactions are more rapid resulting in faster reduction and oxidation kinetics, thereby eliminating the necessity for noble metal catalysts. In addition to cost reduction, this implies that carbon monoxide does not exhibit any poisoning effect on the fuel cell and, on the contrary, can be used as an additional fuel. The operating temperature is also sufficient to carry out hydrocarbon reforming, which can take place directly inside the cell. Thus, it is possible to feed the MCFCs with a vast spectrum of fuels, on condition of adequate design of the fuel supply system, both in terms of thermal management as composition and purity requirements.

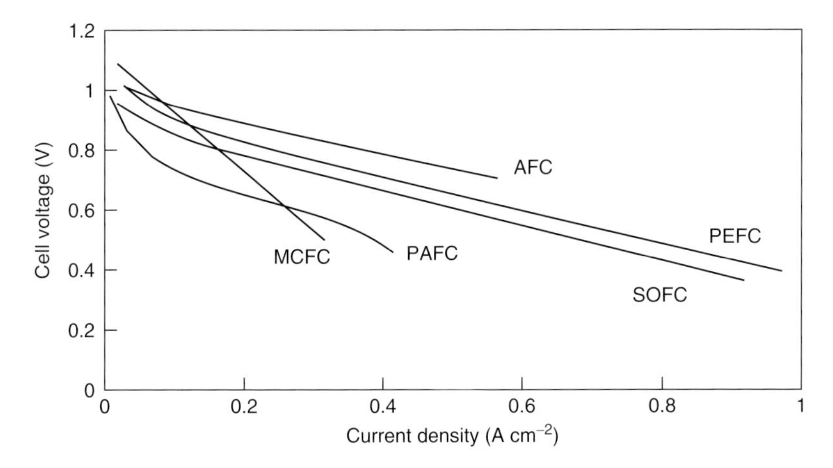

Figure 9.3 *V–I* characteristics of different fuel cell principles. (Source: After Ref. [7].)

MCFCs offer considerable opportunities especially for medium- to large-scale power generation because of their high electrical conversion efficiency (>45%), potential for cogeneration, quiet operation, and essentially clean products with low environmental impact. Compared to other fuel cell technologies, MCFCs have the steepest polarization curve (V–I characteristic). This means they are advantageous at low current density operation, resulting, however, in relatively low power densities (Figure 9.3).

The typical structure of an MCFC is schematically illustrated in Figure 9.4, where fuel (in particular, hydrogen, and also carbon monoxide in case of hydrocarbon fuels are used) enters at the anode side and air enters at the cathode side. The anode and cathode and the respective gas streams are separated by the electrolyte tile.

The overall reaction that takes place in the MCFC is

$$H_2 + CO_3^{2-} \rightarrow H_2O + CO_2 + 2e^- \tag{9.1}$$

which corresponds to the oxidation mechanism on the anode side. As mentioned, carbon monoxide may be employed as a fuel, in which case direct electrochemical oxidation of CO can take place,

$$CO + CO_3^{2-} \rightarrow 2CO_2 + 2e^- \tag{9.2}$$

or, because of the thermodynamic equilibrium of the species at the anode side, chemical transformation through the water gas shift reaction can take place

$$H_2O + CO \rightleftharpoons H_2 + CO_2 \tag{9.3}$$

where hydrogen subsequently reacts with the electrolyte according to mechanism (9.1). Owing to reaction kinetics, the latter mechanism of CO consumption is usually predominant.

The hydrogen oxidation reaction is exothermic, and therefore, thermal management of an MCFC is of great importance, especially in stack configuration. The

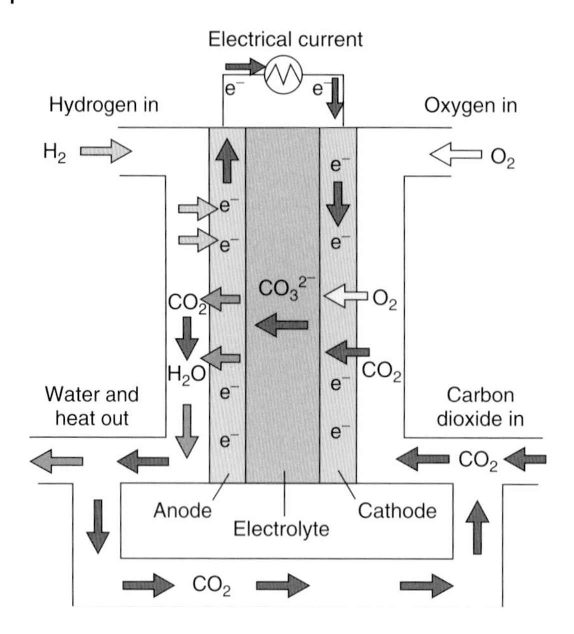

Figure 9.4 Schematic of the MCFC operating principle.

airflow on the cathode side is therefore mainly used as a coolant, so that several times the stoichiometric amount has to be utilized for temperature regulation. Internal reforming of hydrocarbon fuels (such as methane) reduces the need for external coolant since the water produced in reaction (9.1) can be used directly in the anode for steam reforming:

$$CH_4 + H_2O \rightleftharpoons 3H_2 + CO \tag{9.4}$$

This reaction is endothermic and acts as a sink for the heat generated by the oxidation reaction. The need for external cooling may thus be reduced by about 50%.

Ionic transfer inside the electrolyte takes place via CO_3^{2-} ions that migrate to the anode from the cathode, where they are created from air enriched with CO_2 through the reduction reaction:

$$CO_2 + \frac{1}{2}O_2 + 2e^- \rightarrow CO_3{}^{2-} \tag{9.5}$$

Since the CO_2 required for reaction (9.5) is the same that is formed as a consequence of reaction (9.1), anodic gas is generally recycled from the anode to the cathode, though, CO_2 from any source may be employed. This particular feature of MCFCs creates the opportunity for their application in active separation of CO_2 from the flue gas of conventional combined cycle power plants. This is discussed in more detail in Section 9.5.

9.3 State-of-the-Art Components

The state-of-the-art MCFCs use porous gas-diffusion electrodes partially filled with molten carbonate electrolyte to maximize the three-phase boundary, that is, solid electrode catalyst (and electronic conductor) in contact with reactant or product gas, and liquid electrolyte, for electrochemical reactions. Table 9.1 shows the materials and properties of the state-of-the-art components used in MCFCs.

The electrolyte in MCFCs is a eutectic melt of lithium carbonate with either potassium carbonate or sodium carbonate, which has a melting point around 500 °C. The electrolyte is usually impregnated into a porous solid matrix made of $LiAlO_2$, which is sandwiched between the electrodes in the cell. The chemistry and composition of molten carbonate mixtures have a strong effect on the performance and endurance of MCFCs. Molten carbonate is, in general, a very corrosive medium; therefore, considerable development efforts have been made over the past decades to produce stable cell components and fabrication techniques that ensure stability and performance.

In a reducing atmosphere, many metals are compatible with molten carbonate electrolytes, and several among them are suitable as electrocatalyst for hydrogen oxidation. Nickel in the form of sintered powders of a nickel-based alloy containing small amounts of Cr or Al is currently used as the anode material. In the oxidizing

Table 9.1 State-of-the-art MCFC cell components [8, 9].

Anode	Material	Ni–Cr/Ni–Al/Ni–Al–Cr
	Thickness (mm)	0.2–0.5
	Porosity (%)	45–70, Initial
	Pore size (μm)	3–6
	Surface area ($m^2\ g^{-1}$)	0.1–1
Cathode	Material	Lithiated NiO
	Thickness (mm)	0.5–1
	Porosity (%)	70–80, Initial
		60–65, After lithiation and oxidation
	Pore size (μm)	7–15
	Surface area ($m^2\ g^{-1}$)	0.15 (Ni pretest)
		0.5 (posttest)
Electrolyte support	Material	γ-$LiAlO_2$, α-$LiAlO_2$
	Thickness (mm)	0.5–1
	Surface area ($m^2\ g^{-1}$)	0.1–12
Electrolyte	Composition (mol%)	62 Li–38 K
		72 Li–28 K
		52 Li–48 Na
Current collector	Anode	Ni or Ni-plated steel, 1 mm thick
	Cathode	316SS, 1 mm thick

atmosphere of the cathode, however, only a few noble metals have adequate stability to serve as cathode materials. Therefore, the only practical choice for cathode materials is particular oxides that are sufficiently insoluble in the carbonate electrolyte. State-of-the art cathodes are made of lithiated NiO, formed from porous nickel by *in situ* oxidation and lithiation occurring during initial cell operation when nickel is in contact with lithium carbonate melt under an oxygen-containing atmosphere. However, the dissolution of NiO in molten carbonate electrolyte, even though it is only slightly soluble in carbonate, is generally recognized as one of the limiting factors of the cell lifetime. Alternative cathode materials are available but have not been generally adopted. This is further discussed below.

Each of the porous components in MCFCs is usually made by tape casting. This process is amenable to scale-up, and structures down to a few millimeters thickness can be produced with ease. Cells of up to 1 m^2 have been produced by several stack manufacturers (Figure 9.5).

The MCFC operating temperature allows the use of transition metals or metal alloys for the current collector and cell housing. A current collector, usually stainless

Figure 9.5 Example of state-of-the-art MCFC cells and stacks. (Source: Courtesy of FuelCell Energy.)

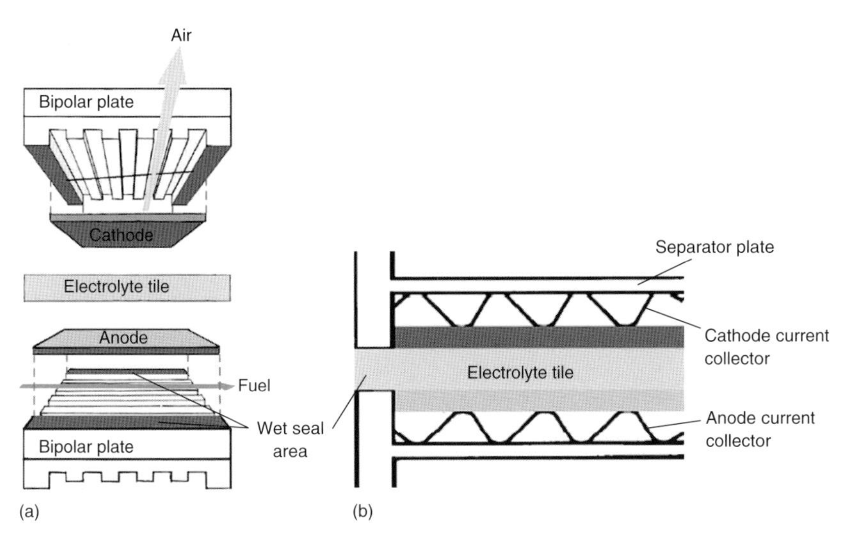

Figure 9.6 Schematic of (a) single cell assembly and (b) location of the wet seal. (Source: After Ref. [10].)

steel or nickel metal screen, is on one side in intimate contact with the back side of the electrodes while on the opposite side it is in uniform contact with the metal housing of the cell (made of stainless steel) or with the adjacent bipolar plate in the case of a cell within a stack. Inside the housing shell, manifolds and gas channels provide the gas flow to the cell. These gas ducts must be carefully designed to ensure a uniform distribution of the gas flow. The entire cell, after assembly of the components, is put under compression to minimize contact resistance between the active and structural components, which usually have a thin oxide surface layer. Also, to ensure gastightness of the cell with respect to the ambient atmosphere, a layer of liquid electrolyte forms a seal ("wet seal"; Figure 9.6) between the two half-shells of the housing (in a single cell) or between the edges of adjacent bipolar plates or end plates (in a cell stack).

9.3.1
Electrolytes

The electrolyte composition affects cell internal resistance, gas solubility, cathode kinetics, corrosion, and electrolyte loss rate; therefore, the choice of electrolyte is of crucial importance for cell performance and cell life. Since about 1980, the standard electrolyte composition has been Li_2CO_3–K_2CO_3 (62 : 38 mol%). However, in the past decade, Li_2CO_3–Na_2CO_3 (52 : 48 mol%) was increasingly considered a better choice, especially under pressurized operating conditions. The merits of Li–Na electrolyte include improved NiO cathode stability, reduced electrolyte evaporation, and decreased ohmic polarization. Especially the first two characteristics make Li–Na electrolyte a better choice for high-pressure operation than Li–K because of better performance over an extended time. However, cell performance with

Li–Na electrolyte shows greater temperature sensitivity than with Li–K, showing strong decline in performance as the operating temperature is decreased to 600 °C or below. This may be related to the wetting characteristics at the three-phase boundary inside the porous structure [11].

Material degradation of the components, especially the dissolution of NiO, is recognized as one of the limiting factors of lifetime. The solubility of NiO has been shown to depend strongly on the chemistry, in particular, the acid/base properties, of the molten carbonate, which are determined mainly by the cationic composition of the melt as well as by the gas atmosphere in contact with the melt. Considerable effort [12–19] has been devoted to the exploration of additives to the Li–K or Li–Na eutectics, such as alkaline earth carbonates or rare-earth oxides. These additives lower the electrolyte acidity and thereby decrease the driving force for cathode dissolution. Some also may help accelerate the kinetics of oxygen reduction reaction by favoring rapid reaction paths. Their effect on the temperature dependence of cell performance therefore is more complex.

The following examples are typical for the effect of additives at standard operating temperatures (650 °C) or slightly below. Tanimoto *et al.* [12–14] reported the effect of alkaline earth carbonate ($CaCO_3$, $SrCO_3$, and $BaCO_3$) on the stability of the nickel oxide cathode. The addition of alkaline earth carbonate effectively reduced the solubility of NiO in the electrolyte and improved the endurance of the system. In small size cell testing, the lifetime increased 15–20% by a 9 mol% addition of $CaCO_3$ or $BaCO_3$ to a Li–Na eutectic melt under pressurized conditions. Scaccia [15] reported that the addition of 3 mol% of $SrCO_3$–$BaCO_3$ mixture to Li–Na eutectic carbonate reduced the NiO solubility by half compared to the same electrolyte without additives. Mitsushima *et al.* [18] showed that an addition of 5% MgO to the carbonate melt reduces NiO solubility by 22%. Although it has a rather low solubility (0.5 wt%) in the melt, addition of MgO has a large effect, reducing the solubility of nickel oxide by more than half. A commercial-sized 10 kW scale MCFC stack with a Li–Na electrolyte has been tested successfully, establishing a record for longevity in pressurized operation [20]. More recently, rare-earth oxides, such as lanthanum oxide, are being investigated, which reduce Ni dissolution even further [21]. Ota and coworkers also explored the suitability of nickel ferrates as cathode materials, with good results.

It has long been known [22–24] that at sustained high current densities in a cell with eutectic Li–K electrolyte, electrolyte segregation occurs, that is, a difference in cationic composition of the electrolyte adjacent to the positive and negative electrodes develops. This is due to the difference in cation (Li^+, K^+) mobilities in the Li–K eutectic. It leads to nonuniformity of important physical properties, such as melting point, basicity, and wetting characteristics of the electrolyte at different locations in the MCFC. The enriched potassium near the cathode side causes increased cathode dissolution and a decline in performance. By adding alkaline earth elements in suitable proportions to the eutectic, one may, in addition to decreasing the cathode solubility, balance the Li^+ and K^+ mobilities so as to maintain nonsegregating behavior [25].

Of course, "alternative electrolytes" may be created by adding small amounts of alkaline earth carbonates or rare-earth oxides to any of the well-known eutectics such as Li–K as well as the ternary Li–Na–K eutectic.

Kaun *et al.* [25] reported that the melting point for the quaternary carbonate Ba–Ca–Li–Na (Ba–Ca = 3.5/3.5 mol% in Li–Na eutectic) is around 440 °C, which is significantly lower than the melting point (500 °C) for Li–Na eutectic carbonate. Among the molten carbonate mixtures, the Li–Na–K ternary eutectic system presents the advantage of having the lowest melting point (397 °C). Ternary carbonate electrolytes are currently under renewed investigation, especially under pressurized conditions [26].

Although these alternative electrolyte formulations promise benefits, a decrease in conductivity with alkaline earth addition has been observed [27]. Thus optimization of electrolyte composition remains a complicated task. The effect of additives on cell performance in long-term operation needs further systematic investigation. In practice, adoption of a new electrolyte composition necessitates a complete inspection for compatibility with structural as well as electroactive cell components and fine-tuning, if not complete overhaul, of the choice of materials.

However, improvement can be achieved by also following a system-based approach. For practical reasons (especially nonuniformity of the temperature distribution in a large-scale cell), the minimum operating temperature of an MCFC with a given cationic composition of the electrolyte is about 100 °C above the melting point of its electrolyte. By making possible operation at temperatures lower than 600–650 °C, "alternative electrolytes" can play an important role in extending the lifetime of an MCFC. Aside from lowering NiO solubility and minimizing electrolyte segregation, corrosion of structural materials can be significantly decreased and lifetime thereby extended. But also careful thermal management and control of the electrolyte distribution, in order to minimize temperature nonuniformity, can allow a reduction in operating temperature. This approach was adopted by FCE to reduce the problems associated with corrosion and electrolyte volatility and has led to significant improvement of stack lifetime and performance [28].

9.3.2 Electrolyte Support

In state-of-the-art MCFCs, an electrolyte "tile" or "matrix" of porous lithium aluminate (α- or γ-$LiAlO_2$) is used to support the molten electrolyte. The matrix requires very fine particle size, high porosity (50–70%) and a narrow, uniform pore size distribution (0.1–0.5 μm) to effectively immobilize the electrolyte. However, it has been known that the support material degrades; for example, particle growth and phase transformation occur over time in $LiAlO_2$ during cell operation, leading to detrimental changes in the pore structure. The stability is related to temperature, partial pressure of CO_2, melt composition, and the particle size of the starting powders [29, 30]. Generally, the particles grow faster at higher temperatures, in low CO_2 gas atmospheres, and in strongly basic melts. In the early 1980s, based on short-term testing, γ-$LiAlO_2$ was believed to be the most stable phase [31], and hence, γ-$LiAlO_2$ was chosen as the matrix support material in the past. Recently, there is renewed interest in $LiAlO_2$ stability because γ- to α-phase transformation was observed in long-term testing [32–34]. Consequently, there is an incentive for industry toward replacing γ-$LiAlO_2$ with α-$LiAlO_2$ for better long-term phase and particle size stabilities. For the same reason, alternative support materials other than $LiAlO_2$ receive new attention.

While a thin $LiAlO_2$ matrix is desirable for minimizing ohmic resistance, commercial requirements dictate that stacks have large-area cells to minimize the capital cost per kilowatt produced. This limits minimization of electrolyte thickness because the matrix must be strong enough to resist mechanical and thermal stresses during start-up and long-term operation. Besides, the matrix has to remain substantially crack free to maintain the gas sealing. There is, therefore, a double-edged benefit in optimizing electrolyte thickness and power density: higher power density would mean smaller area cells could be sufficient, so that the electrolyte tile could be thinner, thus in turn increasing power density through lower ohmic losses.

The use of reinforcements for the matrix construction can improve the thermomechanical capability [35, 36]. Also, various ways of preparing thin and robust $LiAlO_2$ matrix are being pursued. To date, the greatest success has been reported with tape casting, which is capable of scale-up and thin electrolyte structures with 0.25–0.5 mm thickness.

9.3.3 Anode Materials

Although the active material is nickel, state-of-the-art anodes are made of Ni–Cr or Ni–Al alloy and consist of a nickel particle sinter containing microdispersions of chromium or aluminum oxide. The porosity and pore size are ultimately determined by *in situ* pressure and stress conditions during cell or stack operation. The main problem with nickel as the anode material is that the pure metal does not sufficiently resist creeping and sintering under the compressive force required to minimize contact resistance between cell components. The microdispersed oxide particles formed in Ni–10 wt% Cr anodes and Ni–(5–10) wt% Al anodes have proved to maintain adequate stability of the electroactive microstructure in the anode. These days, they are used almost exclusively in commercially produced stacks.

The oxide dispersion stabilization (ODS) approach to the mechanical stability issue of nickel is not without problems. The addition of Cr can lead to the formation of dispersed coarse Cr_2O_3 particles during sintering, which may reduce creep resistance. Besides, Cr in the anode is easily lithiated by the electrolyte to produce $LiCrO_2$. This process consumes electrolyte and creates micropores that remain unstable and during long-term operation cause performance decline [37]. The Ni–Al alloy anode shows higher creep resistance than the Ni–Cr anode, with minimum electrolyte loss. The low creep rate with this alloy is attributed to the formation of $LiAlO_2$ particles that are finely dispersed in the network structure of Ni–Al, but remain electrochemically inert [38].

Creep deformation is related to dislocation movement, and therefore, creep resistance can be achieved in principle by imperfection strengthening techniques using solute atoms, precipitates, or oxide particles, which are capable of blocking the movement [39]. CeO_2, Ce, Dy, and Sn [40–43] have been studied as additives to Ni or Ni–Cr anode materials to inhibit corrosion, creep, and sintering of the anode electrode. In addition, composite anodes made of ceramic oxides show good performance, for example, using oxides of lanthanum (La_2O_3) and samarium (Sm_2O_3) with titanium powder (to provide electronic conductivity) [44]. These results suggest the possibility of using dry methane directly as MCFC fuel.

9.3.4 Cathode Materials

The currently used cathode material is lithiated NiO, formed from porous Ni by oxidation and lithiation of porous nickel in contact with lithium carbonate under an oxidizing atmosphere. The NiO cathode provides acceptable conductivity and structural strength; however, dissolution of the cathode occurs slowly (over thousands of hours). This is recognized as the primary life-limiting factor of MCFCs. The dissolved nickel precipitates at the anode, re-forms as dendrites across the electrolyte matrix, and eventually causes short-circuiting of the cell. This mechanism is accelerated if the cell is operated under pressure, thereby drastically shortening its lifetime.

Extensive efforts have been made to solve the NiO dissolution issue. These have led to significant advances in cathode longevity during recent years. These advances have not been due to alternative materials, although such materials have been identified and investigated. Metallic oxides, for example, $LiFeO_2$ or $LiCoO_2$, show very stable properties under the cathode environment and do not dissolve in the carbonate electrolyte. However, these materials have an inferior electronic conductivity and slower kinetics compared to NiO, so that they are an acceptable solution only under conditions of pressurized operation, where reaction kinetics are improved. The most effective research direction has turned out to be continued use of NiO under controlled basicity of the electrolyte adjacent to the cathode particles. Various NiO composite materials have been reported to reduce NiO dissolution. Co, CoO, Co_3O_4, and $LiCoO_2$ [45–48] have been chosen to modify the surface of NiO. Adding rare-earth elements [49–52] such as CeO_2, La_2O_3, Pr_2O_3, Nd_2O_3, Dy_2O_3, and $La_{0.8}Sr_{0.2}CoO_3$ to a NiO cathode retards the dissolution rate of NiO by an order of magnitude. ZnO and $MgFe_2O_4$ [53, 54] composited with NiO have been studied as alternative cathode materials. These composite cathode materials show promising stability under cathode environments, but long-term in-cell tests of these materials are required to confirm their practical suitability.

9.4
General Needs

Improvements in lifetime and power density, as well as cost reduction, especially in manufacturing, are identified as key priorities to accelerate the commercialization of state-of-the-art MCFCs [55]. Although a lifetime of about 30 000 h has been achieved with today's technology on several occasions [56], commercial requirements dictate at least 40 000 h of operation with a total voltage loss of less than 10% (2 mV per 1000 h). According to FCE, their technology should be able to reach a significantly higher lifetime (over 60 000 h without stack replacement) as part of its next innovation [57].

In the present state of MCFC technology, deficiencies are still evident, which must be remedied to make this progress possible [55, 58]. For example, the development of superior corrosion-resistant and high-performance materials to sustain higher power density than the current ones (typically 110–150 mA cm^{-2}) over a longer lifetime is considered achievable. Also possible is the reduction of system cost with a system design that maximizes the benefit of internal reforming.

On the cell level, systematic work is needed to identify innate limiting causes of performance decay. These can be summarized as

- cathode dissolution and microstructural instability of both electrodes;
- electrolyte loss;
- corrosion of structural cell components (metal parts);
- sensitivity to contaminants, in particular, sulfur compounds.

To understand and remedy these innate limitations, it is necessary to rely on fundamental R&D, from microscopic understanding to system thermal integration. This R&D should go hand in hand with continuous commercialization of the technology in order to maintain constant incremental improvement of the products that are already approaching the market.

9.4.1
NiO Dissolution from the Cathode

NiO dissolution from the cathode into the carbonate electrolyte occurs according to one of the following mechanisms:

$$NiO + CO_2 \rightarrow Ni^{2+} + CO_3{}^{2-} \tag{9.6}$$

or

$$NiO \rightarrow Ni^{2+} + O^{2-} \tag{9.7}$$

Therefore, the driving force for the dissolution depends on CO_2 partial pressure and the acid–base properties of the molten carbonate. The basicity of the carbonate melt is defined as pO^{2-} ($=-\log[O^{2-}]$). The dissolution rate increases with CO_2 partial pressures, whereas increased activity of O^{2-} in the melt retards dissolution. The dissolution of NiO has been shown to be the cause of low-ohmic shorts that

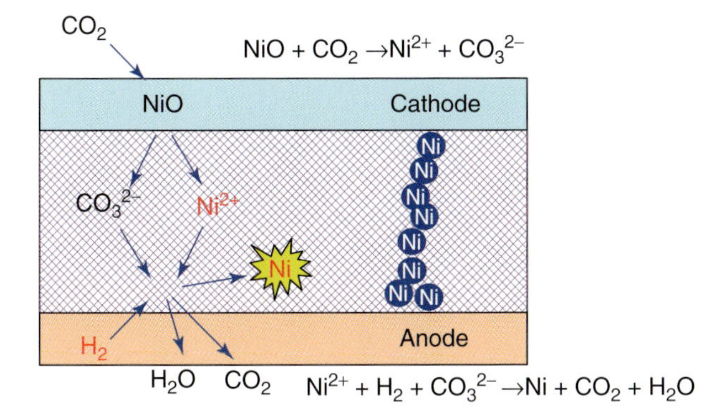

Figure 9.7 Schematic of Ni shorting in the electrolyte matrix.

develop gradually during long-term operation. Nickel dissolved from the cathode as Ni^{2+} ions diffuses into the matrix and is reduced by hydrogen diffusing into the matrix from the anode side:

$$Ni^{2+} + H_2 + CO_3{}^{2-} \rightarrow Ni + H_2O + CO_2 \tag{9.8}$$

Nickel is deposited as very fine metal particles on the reaction zone, which is the plane (parallel to cathode and anode) where the flux of Ni^{2+} ions and the flux of dissolved hydrogen meet. The deposited Ni particles do not stay but move across the matrix under the influence of the electric field (electrophoresis). Since the reduction reaction (9.8) continues, a string of Ni particles is thereby created that tends to stretch from cathode to anode. The particles may agglomerate, forming a metallic dendrite that grows out of the NiO substrate of the cathode toward the anode, or they may stay separate with conducting bridges between them. Eventually, shorting between anode and cathode occurs and causes a rapid decline in cell voltage and performance. The mechanism is illustrated in Figure 9.7. NiO dissolution is one of the two limiting factors of long-term operation. The other is electrolyte loss by corrosion and evaporation.

Nowadays, to decrease the dissolution rate of NiO, NiO–composite cathode materials are the most promising. This applies especially to coating or doping the NiO surface with lower solubility elements, possibly combined with basicity-enhancing additives in the carbonate electrolyte, as discussed in Sections 9.3.1 and 9.3.4.

9.4.2
Creeping in the Anode

Corrosion of the anode materials, that is, Ni–Cr or Ni–Al alloys, is less likely than that of cathode due to the more negative potential and the reducing gas environment. However, anode creep under the compressive force necessary for stack operation is one of the factors that limits the lifetime of an MCFC stack. It is well known [59] that a metal will deform plastically over a period of time if

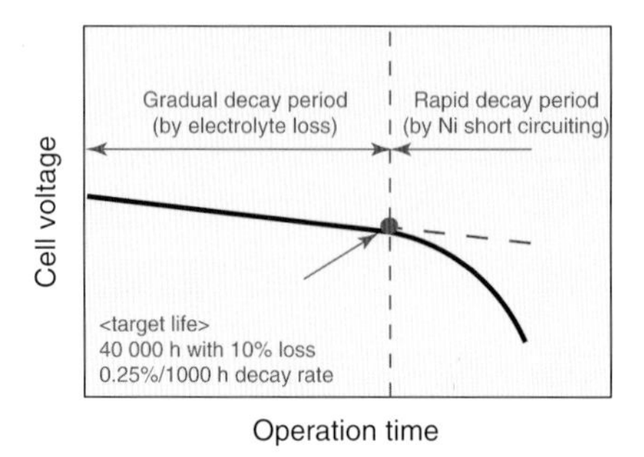

Figure 9.8 Schematic of MCFC degradation with time at a constant current level. (Source: After Ref. [60].)

the temperature is higher than half the melting point (in kelvin) of that metal, even when the applied stress is less than its yield strength. This time-dependent deformation is called *creep*. The high operating temperature of MCFCs causes Ni creep under a torque load during operation. Creep deformation takes place because of the movement of dislocations through a crystal lattice. Therefore, creep resistance can be enhanced by lattice imperfections such as solute atoms, precipitates, or oxide particles, which block the movement. Recently, numerous material strengthening methods such as dispersed-oxide-metal composites have been studied for increasing the stability of anode materials (discussed in Section 9.3.3).

9.4.3
Electrolyte Loss

A significant loss of carbonate electrolyte occurs at the start-up of an MCFC because of the lithiation of cell components. This is then followed by a continuous slow loss during its lifetime. The continuous loss is caused by a variety of factors, including direct vaporization, electrolyte creepage (movement out of the cell onto the exterior of hardware), and slow corrosion of the structural components, which are metals [60, 61]. Together these factors lead to increased ohmic resistance and electrode polarization, and eventually to gas crossover and gas leaking from cell to ambient atmosphere. These factors occur together, and electrolyte loss causes a continuous decrease in cell voltage during operation. From long-term experiments, two phases can be distinguished clearly, as shown schematically in Figure 9.8. The first stage shows constant-rate degradation because of the increasing internal resistance and polarization associated with slow electrolyte loss. The second phase shows a rapid decay ascribed to the Ni shorting between electrodes.

Excessive injection of electrolyte into a cell is not an effective solution for electrolyte loss since the electrolyte pooling in the cathode leads to increased mass transfer resistance and rapidly reduces cell performance. Reducing the electrolyte loss can be achieved by optimization of the electrolyte composition, for example, replacing the Li–K electrolyte by the less volatile Li–Na electrolyte or an alternative degradation-retardant electrolyte. The corrosion resistance of structural cell components (especially steels) varies depending on composition and can be fine-tuned to minimize overall corrosion rate. Also, highly corrosion-resistant alloys have been developed but are often expensive in application. This is further discussed in the following section.

9.4.4
Corrosion of Cell Hardware

The operating temperature of the MCFCs allows one to use commonly available metals for fabrication of cell and stack components. Conventional austenitic 316L and 310S stainless steels and various nickel alloys are the principal materials for structural hardware of MCFCs [8, 61]. Corrosion of hardware usually combined with electrolyte loss affects both performance and lifetime of MCFCs. Corrosion rate is dependent on various factors such as operating temperature, pressure, gas composition, moisture, contamination, and the surface area of the hardware parts directly exposed to molten carbonate as well as fuel and oxidant gas channels. The presence of H_2S and HCl in the fuel gas may cause a significant acceleration of corrosion. In the early stages of operation, corrosion is mostly due to the formation of $LiCrO_2$ or $LiFeO_2$ on steel parts and $LiAlO_2$ on Al-coated steel parts (usually in the wet seal area). This leads to a high initial rate of electrolyte loss (mostly Li_2CO_3).

The corrosion of the bipolar plate, which consists of a separator plate, current collectors, and the wet seal, is critical for performance and lifetime. Corrosion near the wet seal area may cause a slow electrolyte flow out of the cell, which is called *electrolyte creepage loss.*

Electrolyte loss rates due to corrosion are dependent on the material used and on the design (geometry) of the separator plate, as shown in Figure 9.9 [61]. By decreasing the contact area of the separator plate with the electrolyte,

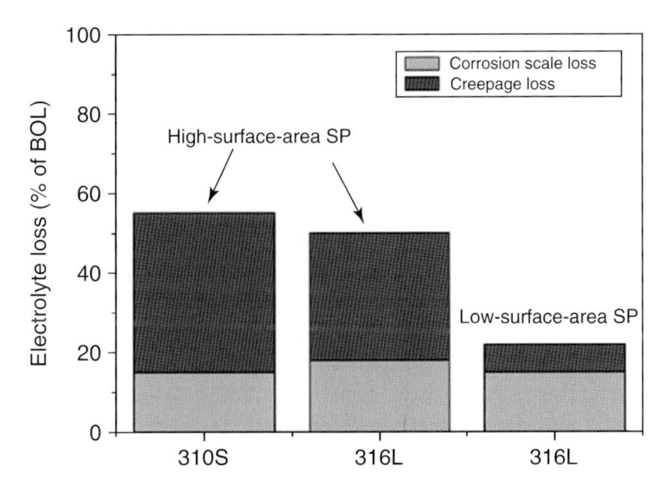

Figure 9.9 Electrolyte loss projections for 40 000 h operation (BOL, beginning of life). (Source: After Ref. [61].)

electrolyte loss can be significantly reduced. The stainless steel separator plate and the current collectors are usually coated or cladded on the anode side, which is more corrosion prone, with a thin layer of dense nickel. The wet seal area, where a thin layer of electrolyte is in direct contact with opposite surfaces of adjacent bipolar plates or end plates, is also very vulnerable to corrosion. It is best protected by applying a layer of alumina to electrically insulate the metallic parts from the electrolyte. Other corrosion protection materials, such as borosilicate glass [62] and co-deposited chromium and aluminum [63], have been investigated. In addition, decreasing the operating temperature (600 °C) to avoid severe corrosion and to increase the durability of components has been shown to be effective [60] in a bench-scale cell, yielding up to 66 000 h of operation. However, at such a low temperature, performance with state-of-the-art components and electrolyte is not adequate. Low-melting electrolyte may provide a solution for this issue.

9.4.5
Electrolyte Optimization

As briefly discussed earlier, the composition of the electrolyte affects the short-term performance of MCFCs as well as long-term endurance. During the development of the MCFC technology, the favored electrolyte composition has evolved from the early use of ternary eutectic (Li–Na–K carbonate) to Li–K eutectic carbonate, which has been the preferred choice for several decades, to Li–Na eutectic and to alternative electrolyte compositions containing additives. The performance (i.e., cell voltage) is related to the internal resistance of the cell and to the polarization of the electrodes. This is dependent on various factors, as discussed previously; most of which are directly related to the chemical, physical, and electrochemical properties of the electrolyte. Lower ohmic resistance can be achieved by using Li-rich electrolytes because of the relatively high conductivity of Li_2CO_3 compared to Na_2CO_3 and K_2CO_3. On the other hand, polarization (due to slow kinetics and diffusion) may not be so favorable in high-Li low-K electrolytes since the solubility and diffusivity of dissolved gaseous reactants, such as H_2, O_2, and CO_2, appear to be lower in Li_2CO_3-rich electrolyte. Moreover, the extent of migrational segregation of electrolyte due to differences in cationic mobility, as discussed earlier, varies strongly with cationic composition. Electrolyte segregation leads to nonuniformity of important physical properties, such as melting point, basicity, and wetting characteristics, of the electrolyte at different locations in the MCFC and may result in performance decline [25]. If we also take into consideration the corrosion effects on long-term electrolyte distribution among electrodes due to changes in structure and composition of electrodes and electrolyte, we must draw the conclusion that the overall process taking place in the cell assembly is very convoluted and a comprehensive approach is necessary for electrolyte optimization.

Many long-term performance studies with binary eutectic composition (62 Li–38 K and 52 Li–48 Na) are available. Performance with alternative electrolyte compositions has been investigated:

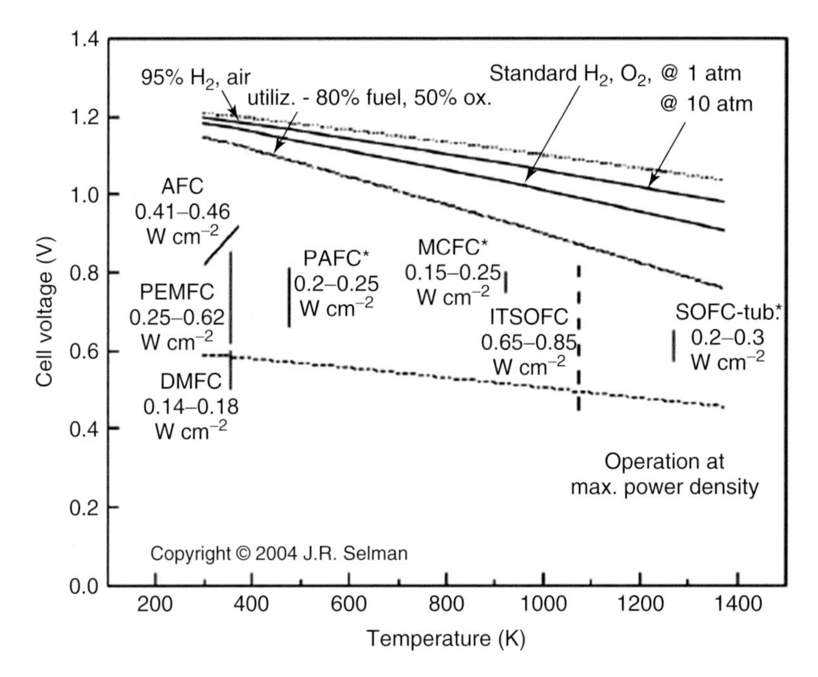

Figure 9.10 Typical performance ranges of different types of fuel cells. (Source: After Ref. [55].)

- lithium-rich (72 Li–28 K) binary carbonate;
- 52 Li–48 Na with 9 mol% addition of $CaCO_3$ and $BaCO_3$ [14];
- 52 Li–48 Na with 5 mol% MgO [18];
- 52 Li–48 Na or 62 Li–38 K with 0.5 mol% La_2O_3 or Gd_2O_3 [17].

9.4.6 Power Density

Figure 9.10 shows the range of cell voltage under load for the various types of fuel cell, as well as the dependence of ideal (thermodynamic) voltage on temperature, fuel, and oxidant utilization. The operating voltage for the high-temperature fuel cells (MCFC and SOFC) is closer to their ideal potential due to more rapid kinetics and smaller ohmic resistance compared to most low-temperature fuel cells.

Power density is important for practical applications because it has a major impact on capital cost. As shown in Figure 9.10, the power density of MCFCs, $\sim$0.15 W cm^{-2} operating at 110–160 mA cm^{-2}, is significantly below that of the competitors such as PEMFCs (proton exchange membrane fuel cells) (0.25–0.62 W cm^{-2}) and SOFCs (0.2–0.85 W cm^{-2}). From the viewpoint of cost, this is a major issue for MCFCs. The possibility of improving power density, for example, to almost 0.25 W cm^{-2} at 0.3 A cm^{-2}, has been demonstrated by operating under 0.7 MPa pressure [64]. Therefore, with state-of-the-art cell components and electrode microstructure, high power densities may be achieved in pressurized operation.

However, the necessary lifetime (>40 000 h) at a high current level, such as 0.3 A cm^{-2}, must still be demonstrated, especially the material stability under high pressure, as mentioned earlier.

Achieving high power density without pressurized operation may be possible also by increasing the active reaction area of the electrodes and using a high-performance electrolyte composition. The reactive area depends mainly on the pore structure of the electrode and the wetting characteristics of the electrode in the particular electrolyte employed. Fundamental understanding of the wetting characteristics and how the wetting depends on melt chemistry, gas environment, and polarization is essential for the optimal design of microstructure in porous electrodes. Wetting also plays a role in electrolyte distribution, and hence, it is important for long-term operation.

Electrode polarization, especially at the cathode, is a major loss factor in cell performance. According to the mathematical modeling results [65], power density could be increased by using catalyst particles of high internal area in the electrode. Obviously, state-of-the-art MCFC electrode materials (Ni-based alloys for the anode and lithiated NiO for the cathode), even if they initially have a high internal area, will probably not be capable of sustaining such structure over tens of thousands of hours. Therefore, the challenge of achieving high power density relies on finding stable high-internal-area or structure-strengthening materials. This requires, in the first place, a deeper understanding of the mechanisms of structure change. It ties the power density issue to the lifetime issue.

9.4.7
Tolerance to Contaminants

The electrocatalytic activity of nickel makes it possible to avoid the adoption of noble metals in high-temperature fuel cells such as MCFCs and SOFCs, but a severe drawback is encountered in the utilization of alternative fuels to hydrogen because of the affinity of nickel with contaminant compounds that poison its catalytic activity and degrade performance [66]. Especially in alternative and waste-derived fuels, the contaminants can be copious and disparate. Providing accurate tolerance limits for all the possible fuel impurities and their effects on MCFC materials can be quite difficult. However, an overview of contaminant effects can be attempted [67], as given in Table 9.2.

The tolerance levels are indicative (a margin of safety is included in the values in Table 9.2) and the extent of their harmful effect may depend on the partial pressure of other species in the gas (e.g., hydrogen, water), the current density at which the fuel cell is operated, the temperature, and the fuel utilization factor. Exposure time to the various impurities is also a determinant of the extent of damage caused and of the potential for its reversal. Elaborate investigations on endurance to contaminants are few since experimentation of these effects is necessarily destructive and takes long duration, but an accurate knowledge of the conditions that are deleterious is highly desired. Identification of the limits of safe operation would greatly enhance

Table 9.2 Contaminants and their tolerance limits for MCFCs [67].

Contaminant	Tolerance (ppm)	Effects
Sulfides, for example, H_2S, COS, and CS_2	0.5–1	Electrode deactivation, usurpation of electrolyte
Halides, for example, HCl and HF	0.1–1	Corrosion, usurpation of electrolyte
Siloxanes, for example, HDMS and D5	10–100	Silicate deposits
Particulates	10–100	Deposition, plugging
Tars	2000	Carbon deposition
Heavy metals, for example, As, Pb, Zn, Cd, and Hg	1–20	Deposition, usurpation of electrolyte

fuel cell durability and optimize the integration of the *fuel clean-up* stage in terms of requirements and cost.

One of the most pressing problems of MCFC poisoning is related to sulfur compounds, both in terms of their particularly deleterious impact as of their copious presence. Apart from naturally occurring sulfur species in alternative fuels, a sulfur-containing odorant is also often added in natural gas for leak detection. The effects of hydrogen sulfide (H_2S) on the nickel-based catalyst are dependent on many factors such as bulk concentration, concentration relative to hydrogen in the fuel, humidity, electrical load, and temperature. As temperature decreases, the propensity of Ni to react with sulfur tends to increase [68]. Although experiments have shown that the drastic effect of sulfur-containing compounds on electrode activity seems irreversible on prolonged exposure to concentrations of more than 10 ppm or even 5 ppm [9, 69], thermodynamic equilibrium calculations show that no permanent, bulk nickel sulfide phases should be formed until concentrations of over 100 ppm [68].

In MCFCs, hydrogen sulfide not only reacts with the anode material but also interacts with the electrolyte [70]. Reaction of hydrogen sulfide with nickel on the anode leads to blocking and deactivating the electrochemically active sites for hydrogen oxidation [71–73]. The affected sites give rise to morphological changes in the anode structure and can thereby cause further deterioration of cell performance through secondary effects such as impeded gas diffusion, volume change, or reduced wetting by the electrolyte. At the electrolyte, hydrogen sulfide can react *chemically* with carbonates to form either sulfide or sulfate ions [71], thereby using electrochemically active charge carriers, which would otherwise be available for hydrogen oxidation. This translates into reduced cell performance. However, hydrogen sulfide can also react *electrochemically* with carbonates [74, 75], releasing electrons, but yielding harmful ionized sulfate compounds. Due to the complexity of poisoning mechanisms and conditions, and the priority to reduce stack costs as much as possible, current MCFC developers prefer to rely on 100% effective gas clean-up and ensure that no sulfur at all reaches the electrodes.

9.5 Status of MCFC Systems Implementation

Competition in the field of power generation is highly challenging, with conventional combustion-based technologies benefiting from the many years of advantage in implementation and continuing to improve their performance to approach the efficiency limits posed by the Carnot cycle. The principle of conversion in an MCFC is fundamentally different and more efficient, but R&D on MCFCs is strongly needed, especially to make up the lag in terms of operational experience and manufacturing costs. However, there are many applications where MCFCs have already made economical sense, and several developers are working to corner these markets.

At present, MCFCs have been demonstrated at several sites, and in different sizes. Focus is mostly on the 200 kW to 1 MW range, while scale-up to multi-megawatt power plants are underway. High investment cost and reduced durability compared to conventional technologies are still two important issues to overcome, in order to ensure proper market penetration. R&D is therefore chiefly directed toward sturdiness of the technology and cost reduction of materials and fabrication processes, but operational experience is also helping to improve these aspects.

There are interesting applications where MCFCs have already made economical sense. The modular buildup helps exploit the high efficiency of MCFCs at a relatively small scale (down to 100 kW, below which the economy of the scale of the BOP starts to dominate [76]). Where requirements exist for power applications that are particularly stringent in terms of environmental legislation (e.g., Assembly Bill 32 in California [77]) or as a result of strong governmental policies (e.g., South Korea), the MCFCs are already competitive with conventional combined heat and power (CHP) technologies. This explains the strong growth of MCFC-based applications in the cited areas, exemplified by the alliance between FCE, based in Connecticut, USA, and the Korean POSCO Power: in South Korea, POSCO installed their first 250 kWe plant (with an FCE stack) in 2006; 7.8 MWe plant, in 2008; and 14.4 MWe plant, in 2009 [78].

Because of these characteristics, MCFCs find their prime application in the sizeable niche market of small-scale, ultraclean, high-quality, and autonomous CHP generation. Owing to the relatively slow start-up (>24 h) and issues related to electrolyte management and mechanical stresses, it is ideally suited for stationary, baseload generation.

The MCFC's silent operation makes it suitable for stand-alone CHP generation for advanced industries with significant electricity bills and/or demanding power supply requirements. Large-scale telecom utilities, for example, are suitable locations for installing MCFCs for self-provided premium power supply.

In the tracking of very high speed financial transactions, the downtime is so costly that a $10 000 per kilowatt fuel cell could easily be justified.

The size of auxiliary power supply units on board ferries and cruise ships can easily reach 0.5–1 MWe. Particularly for touristic stretches, there is a growing demand (or in certain locations, requirement) for silent and emissionless transport through water. The combination of these conditions led both the German developer MTU Onsite Energy and POSCO Power to take on pilot projects in this application: from September to December 2009, the Norwegian *Viking Lady* sailed with a 320 kW MCFC from the German company and in 2010, the Korean multinational has taken on a 5 year development project (at $30 million per year) for a similar type of system [79].

The most common fuel utilized in MCFCs is natural gas, which benefits from an expansive grid in most industrialized countries. "Getting the foot in the door" through the conventional gas grid helps acquire the operational experience and familiarity necessary to cut costs further and become competitive with more alternative applications. But MCFCs could operate on different fuels such as (i) biogas from anaerobic digestion of sewage sludge, organic waste, or dedicated biomass, (ii) landfill gas, and (iii) syngas from a thermal gasification or pyrolysis process using lignocellulosic biomass or waste material such as refuse-derived fuels (RDFs), industrial waste, or secondary process flows from refineries and chemical industries, where localized exploitation is feasible and readily useful.

Biogas plants represent a unique opportunity for fuel cell power plants. The methane produced from the anaerobic digester is used as the fuel to generate ultraclean electricity that can be used for the treatment plant while by-product heat from MCFCs can be used to heat the sludge to facilitate anaerobic digestion. This CHP application can thus result in up to 90% efficiency. Moreover, biogas (digester gas) is a renewable fuel eligible for incentive funding in many countries throughout the world. In many applications, digester gas production volume is variable. In such applications, the plant can be designed to operate with automatic blending with natural gas. FCE is a world leader in such applications, but biogas plants are operated also in Germany and by POSCO Power in South Korea. In Europe, the potential for use of biogas is enormous, given the increasing number of digester plants being built every year. In Germany alone, between 2000 and 2007, these digester plants increased in number from 1050 to 2800, especially in the size range 70–500 kW [79], which is the ideal size for current MCFC systems. An overview of plants running on waste-derived fuels and biogas installed by FCE is given in Table 9.3. A distinct feature of these data is that nearly all plants are situated in California, where energy and emissions policies have created suitable market conditions for widespread installation of such high-efficiency low-emission systems. Figure 9.11 represents one of their MW plants running on natural gas, installed in California, USA.

A technology such as that of the MCFC is eminently suitable for cutting-edge solutions to critical issues in our energy-hungry society. One of the simplest technologies is the utilization of heat from the stack for closed-circuit steam generation and desalination of seawater for the production of drinkable water. This merely requires the coupling of a desalination facility to the outlet of the MCFC power system (Figure 9.12a) [78].

Artificial CO_2 confinement (or carbon capture and sequestration, CCS) is probably going to be an important short-term solution applied by industrialized countries to temporarily contrast climate change through anthropogenic CO_2

Table 9.3 FCE installations running on waste-derived fuel and biofuel [80].

Location	Feedstock	Nominal power (kW)
Santa Barbara, CA, USA	Wastewater treatment – ADG	600
Sierra Nevada, CA, USA	Biogas (waste by-product of the brewing process)-ADG/NG fuel blending	1 000
Tulare, CA, USA	Wastewater treatment – ADG/NG fuel blending	900
Dublin San Ramon, Pleasanton, CA, USA	Wastewater treatment – ADG/NG fuel blending	600
Rialto, CA, USA	Wastewater treatment – ADG/NG fuel blending	900
Oxnard, CA, USA	Gills Onions' food waste processing facility-ADG/NG fuel blending	600
Riverside, CA, USA	Wastewater treatment – ADG/NG fuel blending	1 200
Moreno Valley, CA, USA	Wastewater treatment and waste treatment facility – ADG/NG fuel blending	750
Busan, South Korea	ADG/NG fuel blending	1 200
Total capacity		8 750
Backlog		
Orange County Sanitation District, Fountain Valley, CA, USA	Wastewater treatment and waste treatment facility – ADG/NG fuel blending	300
Tulare, CA, USA	Wastewater treatment – ADG/NG fuel blending	300
Olivera Egg Ranch, French Camp	Wastewater treatment – ADG/NG fuel blending	1 400
Eastern Municipal Water District, Perris Valley, CA, USA	Wastewater treatment – ADG/NG fuel blending	600
Total capacity + backlog		11 350

ADG, anaerobic digester gas; NG, natural gas.

emissions to the atmosphere. Here, MCFCs offer an interesting possibility since their intrinsic operating principle requires CO_2 to be transferred from the cathode to the anode. In CCS configuration, instead of recirculating the CO_2 produced at the anode to the cathode to close the ionic circuit, "fresh" CO_2 is supplied to the cathode through the flue gas of a combustion-based power plant (Figure 9.12b) [78]. The bulk of such flue gas consists of nitrogen from the combustion air saturated with water, but up to 15% CO_2 can be present, the separation of which from the bulk nitrogen is laborious. Using an MCFC, up to 90% CO_2 can be extracted from the nitrogen and combustion-product-containing stream in normal operation. The CO_2 is then transferred through the electrolyte to the anode (in the form of CO_3^{2-} ions), where it exits at a concentration of 30–40% and is mixed with essentially

Figure 9.11 Installation of a 1 MW MCFC power plant by FuelCell Energy. (Source: Courtesy of FuelCell Energy.)

water vapor. This makes the CO_2 sequestration process much easier and more efficient, and in the process, power is produced as well (increasing plant production up to 20%), by supplying the anode with an adequate amount of fuel, such as natural gas. Assuming equivalent conversion efficiencies between a combined cycle power plant and the MCFC, the overall electric efficiency of the power plant is thus not reduced, in contrast to passive solutions such as CO_2 separation using solvents such as ammines, where large quantities of additional refuse flows need to be processed and severe penalties result in net power production, causing up to 10% decrease in the plant's efficiency [81]. Paradoxically, this reduction in efficiency could even cause a net *increase* in CO_2 emissions since more fossil fuel has to be combusted for the same net power output.

Given the imminent, large-scale regulations on CCS in Europe and other parts of the world for the immediate checking of climate change induced by greenhouse gas emissions, this application is leveraging the further development of MCFC systems, as demonstrated by the great interest and preparation that is devoted by MCFC developers (FCE, Doosan Heavy Industries) to the possibilities of retrofitting existing power plants with MCFCs for active CCS with extra power production and minimal efficiency loss. The chief challenges posed by this application, compared to the general needs cited in Section 9.4, are the effects of contaminants in the power plant flue gas that flow through the cathode. Owing to the dissolution qualities of carbonate, contaminants can be transferred to the anode side during the ionic transport, where the virgin nickel is much more susceptible to poisoning. Also, the large flow at the cathode compared to the anode stream causes the effective concentration of pollutants migrating from the cathode to increase, thus aggravating their effects.

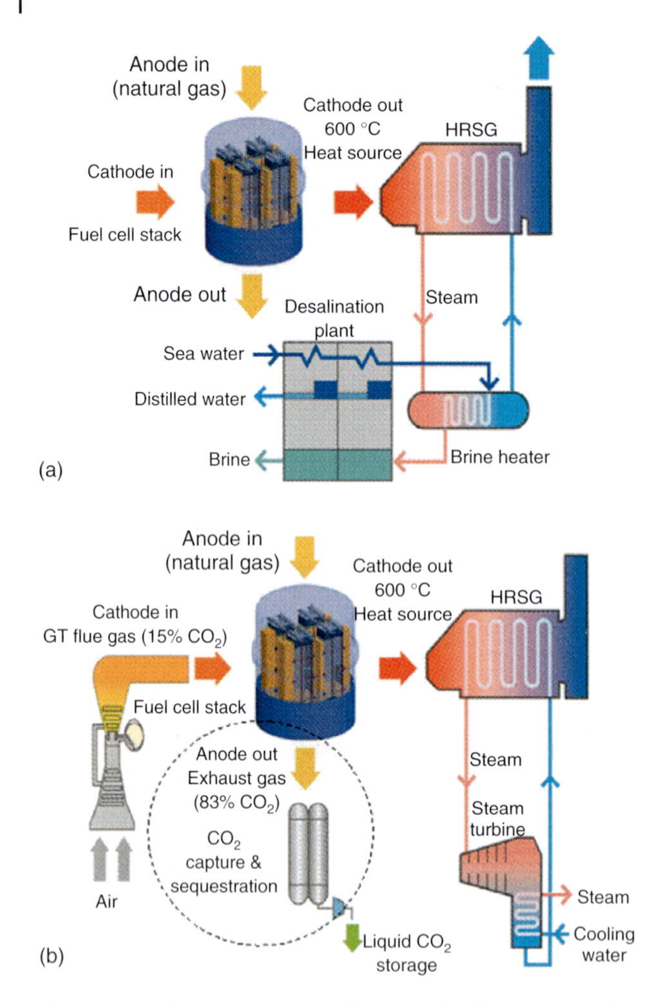

Figure 9.12 Alternative MCFC applications: desalination (a) and CO_2 separation (b). HRSG: heat recovering steam generator; GT: gas turbine (Source: After Ref. [78].)

MCFCs can also contribute to the advent of a hydrogen-based economy. The hydrocarbon fuels at the inlet are ultimately converted to hydrogen in the reforming processes integrated with the stack. But not all of this hydrogen is electrochemically oxidized in a single pass due to the low concentrations that would result near the anode outlet. Apart from diminishing local – and overall – performance, this could also lead to oxidation of the anode due to lack of fuel. Usually, the unspent hydrogen is, therefore, burned or recirculated to the inlet. It could be feasible, instead, given the price of hydrogen production, to separate the unspent hydrogen from the water and CO_2 present at the outlet and store it for separate use or resale. At a large scale, this form of energy storage could also be used to supplement the MCFC's baseload operation with peak power requests that could be provided by fast-reacting and cyclable low-temperature fuel cells such as the polymer electrolyte membrane or phosphoric acid fuel cells. In smaller stand-alone applications, battery aggregates will serve to supply power pulses in case of demanding variable-load profiles. The aim is thereby to achieve a fully independent and high-quality power supply system fit for the most demanding civilian as well as for military applications [78].

References

1. U.S. Department of Energy, Energy Efficiency & Renewable Energy *http://www1.eere.energy.gov* (accessed 1 September 2011).
2. Tamaru, S. and Ochiai, K. (1935) *Nippon Kagaku Kaishi*, **56**, 92.
3. Fuel Cells Molten Carbonate Fuel Cell History, *http://americanhistory.si.edu/fuelcells/mc/mcfc1.htm* (accessed 1 September 2011).
4. FuelCell Energy *http://www.fuelcellenergy.com/about-us.php* (accessed 1 September 2011).
5. G. Huppmann, *High Temperature Fuel Cell Trigeneration for Commercial and Municipal Buildings: The MTU Carbonate Fuel Cell HotModule®*, CIBSE CHP Group and LHP, 2005, London *http://www.cibse.org/pdfs/Gerhard%20Huppmann.pdf* (accessed 1 September 2011).
6. FuelCell Energy, Inc. (2011) Investor Relations Newsletter (Sept. 7).
7. Tomczyk, P. (2006) MCFC versus other fuel cells – characteristics, technologies and prospects. *J. Power Sources*, **160**, 858–862.
8. Selman, J.R. (1993) Molten carbonate fuel cells, in *Fuel Cell Systems* (eds L.J.M.J. Blomen and M.N. Mugerwa), Plenum Press, New York.
9. USA Department of Energy, Office of Fossil Energy, National Energy Technology Laboratory (2007) *Fuel Cell Handbook*, 7th edn.
10. Agüero, A., García de Blas, F., García, M., Muelas, R., and Román, A. (2001) Thermal spray coatings for molten carbonate fuel cells separator plates. *Surf. Coat. Technol.*, **146**, 578–585.
11. Hong, S.-G. and Selman, J.R. (2004) Thermal modeling and design considerations of lithium-ion batteries. *J. Electrochem. Soc.*, **151**, 77.
12. Tanimoto, K., Miyazaki, Y., Yanagida, M., Tanase, S., Kojima, T., Ohtori, N., Okuyama, H., and Kodama, T. (1991) Cell performance of molten-carbonate fuel cell with alkali and alkaline-earth carbonate mixtures. *Denki Kagaku*, **59**, 619.
13. Tanimoto, K., Miyazaki, Y., Yanagida, M., Tanase, S., Kojima, T., Ohtori, N., Okuyama, H., and Kodama, T. (1992) Cell performance of molten-carbonate fuel cell with alkali and alkaline-earth carbonate mixtures. *J. Power Sources*, **39**, 285.
14. Tanimoto, K., Kojima, T., Yanagida, M., Nomura, K., and Miyazaki, Y. (2004) Optimization of the electrolyte composition in a $(Li_{0.52}Na_{0.48})2-2xAExCO_3$ (AE = Ca and Ba) molten carbonate fuel cell. *J. Power Sources*, **131**, 256.
15. Scaccia, S. (2005) Investigation on NiO solubility in binary and ternary molten alkali metal carbonates containing additives. *J. Mol. Liq.*, **116**, 67.
16. Scaccia, S. and Frangini, S. (2006) Oxygen dissolution behaviour in (52/48) mol% Li_2CO_3/Na_2CO_3 electrolyte containing Ba and Ca additives. *J. Mol. Liq.*, **129**, 133.
17. Scaccia, S., Frangini, S., and Dellepiane, S. (2008) Enhanced O_2 solubility by RE_2O_3 (RE = La, Gd) additions in molten carbonate electrolytes for MCFC. *J. Mol. Liq.*, **138**, 107.
18. Mitsushima, S., Matsuzawa, K., Kamiya, N., and Ota, K. (2002) Improvement of MCFC cathode stability by additives. *Electrochim. Acta*, **47**, 3823.
19. Matsuzawa, K., Tatezawa, G., Matsuda, Y., Mitsushima, S.,

Kamiya, N., and Ota, K. (2005) Effect of rare earth oxides for improvement of MCFC. *J. Electrochem. Soc.*, **152**, A1116.

20. Yoshiba, F., Morita, H., Yoshikawa, M., Mugikura, Y., Izaki, Y., Watanabe, T., Komodab, M., Masudac, Y., and Zaima, N. (2004) Improvement of electricity generating performance and life expectancy of MCFC stack by applying Li/Na carbonate electrolyte: test results and analysis of 0.44 m2/10 kW- and 1.03 m2/10 kW-class stack. *J. Power Sources*, **128**, 152.
21. Ota, K.I., Matsuda, Y., Matsuzawa, K., Mitsushima, S., and Kamiya, N. (2006) Effect of rare earth oxides for improvement of MCFC. *J. Power Sources*, **160** (2), 811–815.
22. Brenscheidt, T. and Wendt, H. (1997) Segregation and transport of alkali and additive cations in binary carbonate melts in MCFC, in *Proceedings of the 4th International Symposium on Carbonate Fuel Cell Technology* (eds J.R. Selman, I. Uchida, H. Wendt, D.A. Shores, and T.F. Fuller), The Electrochemical Society, p. 353.
23. Okada, I. (1999) The chemla effect – from the separation of isotopes to the making of binary ionic liquids. *J. Mol. Liq.*, **83**, 5.
24. Yang, C., Takagi, R., Kawamura, K., and Okada, I. (1987) Internal cation mobilities in the molten binary system Li_2CO_3—K_2CO. *Electrochim. Acta*, **32**, 1607.
25. Kaun, T.D., Schoeler, A., Centeno, C.-J., and Krumpelt, M. (1999) Improved MCFC performance with Li/Na/Ba/Ca carbonate electrolyte, in *Carbonate Fuel Cell Technology*, Vol. 5 (eds J. Uchida, K. Hemmes, G. Lindbergh, D.A. Shores, and J.R. Selman), The Electrochemical Society, p. 219.
26. Mohamedi, M., Hisamitsu, Y., and Uchida, I. (2002) X-ray diffractometric study of in situ oxidation of Ni in Li/K and Li/Na carbonate eutectic. *J. Appl. Electrochem.*, **32**, 111.
27. Kojima, T., Yanagida, M., Tanase, S., Tanimoto, K., Tamiya, Y., Asai, T., and Miyazaki, Y. (1996) The electrical conductivity of molten Li_2CO_3-K_2CO_3 and Li_2CO_3-Na_2CO_3 containing alkaline earth (Ca, Sr and Ba) carbonates. *Denki Kagaku*, **64**, 471.
28. Farooque, M. (2007) FCE products status and durability experience, presentation at the International Workshop on Degradation Issues, Crete, Greece, September 19–21, 2007.
29. Takizawa, K. and Hagiwara, A. (2002) The transformation of $LiAlO_2$ crystal structure in molten Li/K carbonate. *J. Power Sources*, **109**, 127.
30. Tomimatsu, N., Ohzu, H., Akasaka, Y., and Nakagawa, K. (1997) Phase stability of $LiAlO_2$ in molten carbonate. *J. Electrochem. Soc.*, **144**, 4182.
31. Finn, P.A. (1980) The effects of different environments on the thermal stability of powdered samples of $LiAlO_2$. *J. Electrochem. Soc.*, **127**, 236.
32. Yuh, C.Y., Huang, C.M., and Farooque, M. (1997) Gas electrode reactions in molten carbonate media, in *Carbonate Fuel Cell Technology*, Vol. 4 (eds J.R. Selman, I. Uchida, H. Wendt, D.A. Shores, and T.F. Fuller), The Electrochemical Society.
33. Choi, H.J., Lee, J.J., Hyun, S.H., and Lim, H.C. (2010) Phase and microstructural stability of electrolyte matrix materials for molten carbonate fuel cells. *Fuel Cells*, **10**, 613.
34. Tanimoto, K., Yanagida, M., Kojima, T., Tamiya, Y., Matsumoto, H., and Miyazaki, Y. (1998) Long-term operation of small-sized single molten carbonate fuel cells. *J. Power Sources*, **72**, 77.
35. Lee, J.-J., Choi, H.-J., Hyun, S.-H., and Im, H.-C. (2008) Characteristics of aluminum-reinforced γ-$LiAlO_2$ matrices for molten carbonate fuel cells. *J. Power Sources*, **179**, 504.
36. Kim, S.-D., Hyun, S.-H., Lim, T.H., and Hong, S.A. (2004) Effective fabrication method of rod-shaped γ-$LiAlO_2$ particles for molten carbonate fuel cell matrices. *J. Power Sources*, **137**, 24.
37. Lee, D., Lee, I., and Chang, S. (2004) On the change of a Ni3Al phase in a Ni-12 wt.% Al MCFC anode during partial oxidation and reduction stages of sintering. *Electrochim. Acta*, **50** (2–3), 755–759.
38. Kim, G., Moon, Y., and Lee, D. (2002) Preparation of creep-resistant Ni–5 wt.%

Al anodes for molten carbonate fuel cells. *J. Power Sources*, **104**, 181–189.

39. Jung, D., Lee, I., Lim, H., and Lee, D. (2003) On the high creep resistant morphology and its formation mechanism in Ni–10 wt.% Cr anodes for molten carbonate fuel cells. *J. Mater. Chem.*, **13**, 1717.
40. Chauvaut, V., Albin, V., Schneider, H., Cassir, M., Ardéléan, H., and Galtayries, A. (2000) Study of cerium species in molten $Li_2CO_3 \pm Na_2CO_3$ in the conditions used in molten carbonate fuel cells. Part I: thermodynamic, chemical and surface properties. *J. Appl. Electrochem.*, **30**, 1405.
41. Zeng, C.L., Zhang, T., Guo, P.Y., and Wu, W.T. (2004) The corrosion behavior of two-phase Ni–Dy alloys in a eutectic (Li,K) $2CO_3$ mixture at 650 °C. *Corros. Sci.*, **46**, 2183.
42. Wee, J.-H. (2007) Effect of cerium addition to Ni–Cr anode electrode for molten carbonate fuel cells: surface fractal dimensions, wettability and cell performance. *Mater. Chem. Phys.*, **101**, 322.
43. Lee, D., Han, J.-H., Lim, H.-C., and Jang, S.-Y. (2010) A study on in situ sintering of Ni–5 wt% Al anode for molten carbonate fuel cell. *J. Electrochem. Soc.*, **157**, B1479.
44. Tagawa, T., Yanase, A., Goto, S., Yamaguchi, M., and Kondo, M. (2004) Ceramic anode catalyst for dry methane type molten carbonate fuel cell. *J. Power Sources*, **126**, 1.
45. Durairajan, A., Colon-Mercado, H., Haran, B., White, R., and Popov, B. (2002) Electrochemical characterization of cobalt-encapsulated nickel as cathodes for MCFC. *J. Power Sources*, **104**, 157.
46. Ganesan, P., Colon, H., Haran, B., White, R., and Popov, B.N. (2002) Study of cobalt-doped lithium–nickel oxides as cathodes for MCFC. *J. Power Sources*, **111**, 109.
47. Lee, H., Hong, M., Bae, S., Lee, H., Park, E., and Kim, K. (2003) A novel approach to preparing nano-size Co_3O_4-coated Ni powder by the pechini method for MCFC cathodes. *J. Mater. Chem.*, **13**, 2626.
48. Kuk, S.T., Song, Y.S., Suh, S., Kim, J.Y., and Kim, K. (2001) The formation of $LiCoO_2$ on a NiO cathode for a molten carbonate fuel cell using electroplating. *J. Mater. Chem.*, **11**, 630.
49. Ganesan, P., Colon, H., Haran, B., and Popov, B.N. (2003) Performance of $La_{0.8}Sr_{0.2}CoO_3$ Coated NiO as cathodes for molten carbonate fuel cells. *J. Power Sources*, **115**, 12.
50. Huang, B., Chen, G., Li, F., Yu, Q.-C., and Hua, K.-A. (2004) Study of NiO cathode modified by rare earth oxide additive for MCFC by electrochemical impedance spectroscopy. *Electrochim. Acta*, **49**, 5055.
51. Wee, J.-H. and Lee, K.-Y. (2006) Overview of the effects of rare-earth elements used as additive materials in molten carbonate fuel cell system. *J. Mater. Sci.*, **41**, 3585.
52. Liu, Z.P., Guo, P.Y., and Zeng, C.L. (2007) Effect of Dy on the corrosion of NiO/Ni in molten $(0.62Li,0.38K)2CO_3$. *J. Power Sources*, **166**, 348.
53. Huang, B., Li, F., Yu, Q.-C., Chen, G., Zhao, B.-Y., and Hu, K.-A. (2004) Study of NiO cathode modified by ZnO additive for MCFC. *J. Power Sources*, **128**, 135.
54. Okawa, H., Lee, J.H., Hotta, T., Ohara, S., Takahashi, S., Shibahashi, T., and Yamamasu, Y. (2004) Performance of $NiO/MgFe_2O_4$ composite cathode for a molten carbonate fuel cell. *J. Power Sources*, **131**, 251.
55. Selman, J.R. (2006) Molten-salt fuel cells – technical and economic challenges. *J. Power Sources*, **160** (2), 852–857.
56. Bischoff, M. (2006) Molten carbonate fuel cells: a high temperature fuel cell on the edge to commercialization. *J. Power Sources*, **160** (2), 842–845.
57. Hilmi, A. (2011) Emergence of the stationary DFC power plants. Presentation at the International Workshop on Molten Carbonates and Related Topics, Paris, France, March 21–22, 2011.
58. Dicks, A.L. (2004) Molten carbonate fuel cells. *Curr. Opin. Solid State Mater. Sci.*, **8** (5), 379–383.

59. Iacovangelo, C.D. (1986) Metal plated ceramic – a novel electrode material. *J. Electrochem. Soc.*, **133**, 2410.
60. Morita, H., Kawase, M., Mugikura, Y., and Asano, K. (2010) Degradation mechanism of molten carbonate fuel cell based on long-term performance: long-term operation by using bench-scale cell and post-test analysis of the cell. *J. Power Sources*, **195**, 6988.
61. Frangini, S. (2008) Corrosion of metallic stack components in molten carbonates: critical issues and recent findings. *J. Power Sources*, **182**, 462.
62. Pascual, M.J., Pascual, L., Valle, F.J., Durán, A., and Berjoan, R. (2003) Corrosion of borosilicate sealing glasses for molten carbonate fuel cells. *J. Am. Ceram. Soc.*, **86**, 1918.
63. Park, H., Lee, M., Yoon, J., Bae, I., and Kim, B. (2003) Corrosion resistance of austenitic stainless steel separator for molten carbonate fuel cell. *Met. Mater. Int.*, **9**, 311.
64. Mugikura, Y. (2003) Stack material and stack design, in *Handbook of Fuel Cells – Fundamentals, Technology and Application* (eds W. Vielstich, H.A. Gasteiger, and A. Lamm), John Wiley & Sons, Inc., pp. 908–919.
65. Hong, S.-G. and Selman, J.R. (2004) A stochastic structure model for liquid-electrolyte fuel cell electrodes, with special application to MCFCs. *J. Electrochem. Soc.*, **151**, 748.
66. Aarva, A., McPhail, S.J., and Moreno, A. (2009) From energy policies to active components in solid oxide fuel cells: state-of-the-art and the way ahead. *ECS Trans.*, **25** (2), 313–322.
67. Cigolotti, V., McPhail, S., and Moreno, A. (2009) Nonconventional fuels for high-temperature fuel cells: status and issues. *J. Fuel Cell Sci. Technol.*, **6** (2), 021311.
68. Lohsoontorn, P., Brett, D.J.L., and Brandon, N.P. (2008) Thermodynamic predictions of the impact of fuel composition on the propensity of sulfur to interact with Ni and ceria-based anodes for solid oxide fuel cells. *J. Power Sources*, **175** (1), 60–67.
69. Sasaki, K., Adachi, S., Haga, K., Uchikawa, M., Yamamoto, J., Iyoshi, A., Chou, J.T., Shiratori, Y., and Itoh, K. (2006) Fuel impurity tolerance of solid oxide fuel cells. Proceedings of the 7th European SOFC Forum, Lucerne, Switzerland, 2006.
70. Zaza, F., Paoletti, C., LoPresti, R., Simonetti, E., and Pasquali, M. (2008) Bioenergy from fuel cell: effects of hydrogen sulfide impurities on performance of MCFC fed with biogas. Proceedings of the Fundamentals and Developments of Fuel Cells Conference – FDFC2008, Nancy, France, December 10–12, 2008.
71. Weaver, D. and Winnick, J. (1989) Sulfation of the molten carbonate fuel cell anode. *J. Electrochem. Soc.*, **136** (6), 1679–1686.
72. Marianowski, L.G., Anderson, G.L., and Camara, E.H. (1991) Use of sulfur containing fuel in molten carbonate fuel cells. US Patent 5071718.
73. Dong, J., Cheng, Z., Zha, S., and Liu, M. (2006) Identification of nickel sulfides on Ni-YSZ cermet exposed to H_2 fuel containing H_2S using Raman spectroscopy. *J. Power Sources*, **156** (2), 461–465.
74. Townley, D., Winnick, J., and Huang, H.S. (1980) Mixed potential analysis of sulfation of molten carbonate fuel cells. *J. Electrochem. Soc.*, **127**, 1104–1106.
75. Zaza, F., Paoletti, C., LoPresti, R., Simonetti, E., and Pasquali, M. (2010) Studies on sulfur poisoning and development of advanced anodic materials for waste-to-energy fuel cells applications. *J. Power Sources*, **195** (13), 4043–4050.
76. Venkataraman, R., Farooque, M., and Ma, Z. (2007) Cost reduction through thermal management improvements in large scale carbonate fuel cells. *ECS Trans.*, **5** (1), 571–577.
77. California Environmental Protection Agency, Air Resources Board *http://www.arb.ca.gov/cc/ab32/ab32.htm* (accessed 26 June 2011).
78. Han, J. (2009) Status of MCFC development in Korea. Presentation at the IEA Advanced Fuel Cells Annex 23 Annual Meeting, Palm Springs, 2009.
79. Stegmann, H. (2008) Potentials of biological waste treatment technologies on energy production. Proceedings of the

2nd International Symposium on Energy from Biomass and Waste, Venice, Italy, November 17–20, 2008.

80. Moreno, A., McPhail, S., and Bove, R. (2008) International Status of Molten Carbonate Fuel Cell (MCFC) Technology. Report No. EUR 23363 EN, European Commission Publication.

81. Macchi, E. (2010) The potential long-term contribution of Fuel Cells to high-efficiency low carbon- emission power plants. Presentation at the International Workshop "Fuel Cells in the Carbon Cycle", Naples, Italy, July 12–13, 2010.

Index

Materials for High-Temperature Fuel Cells, First Edition. Edited by San Ping Jiang and Yushan Yan.